# Compressible Fluid Flow

# Compressible Fluid Flow

## Patrick H. Oosthuizen

*Department of Mechanical Engineering*
*Queen's University*
*Kingston, Ontario, Canada*

## William E. Carscallen

*Institute for Aerospace Research*
*National Research Council of Canada*
*Ottawa, Ontario, Canada*

**THE McGRAW-HILL COMPANIES, INC.**

New York   St. Louis   San Francisco   Auckland   Bogotá   Caracas   Lisbon
London   Madrid   Mexico City   Milan   Montreal   New Delhi
San Juan   Singapore   Sydney   Tokyo   Toronto

# McGraw-Hill

A Division of The **McGraw·Hill** Companies

## COMPRESSIBLE FLUID FLOW

1 2 3 4 5 6 7 8 9 0 DOC DOC 9 0 9 8 7 6

ISBN 0-07-048197-0

*This book was printed in Times Roman by Keyword Publishing Services.*
*The editors were Debra Riegert and John M. Morriss;*
*the production supervisor was Leroy A. Young.*
*Project supervision was done by Keyword Publishing Services.*
*R. R. Donnelley & Sons Company was printer and binder.*

Library of Congress Cataloging-in-Publication Data

Oosthuizen, P. H.
    Compressible fluid flow/Patrick H. Oosthuizen, William E. Carscallen.
        p.    cm.
    Includes index.
    ISBN 0-07-048197-0.—ISBN 0-07-048198-9
    1. Fluid dynamics.   2. Compressibility.   I. Carscallen, William E.   II. Title.
    QA911.057   1997
    532'.05—dc20                                    96-9908

**http://www.mhcollege.com**

*To my wife Jane and my late mother Eileen for all their loving encouragement and help.*

P.H.O.

*To my loving wife, Elizabeth and our four children, Peter, Laleah, Emma, and Mather.*

W.E.C.

# ABOUT THE AUTHORS

**PATRICK H. OOSTHUIZEN** is a Professor of Mechanical Engineering at Queen's University in Kingston, Ontario, Canada. He received B.Sc.(Eng.), M.Sc.(Eng.), and Ph.D. degrees in Mechanical Engineering from the University of Cape Town, South Africa, and an M.A.Sc. degree in Aerospace Engineering from the University of Toronto, Canada. He joined Queen's University after teaching for several years at the University of Cape Town. He does research in the areas of heat transfer and fluid mechanics and is the author of more than 350 technical papers. He has received a number of teaching and research paper awards. He has been involved with the organization of many national and international conferences and has edited a number of conference proceedings.

**WILLIAM E. CARSCALLEN** is a Senior Research Officer and Head of the Turbomachinery Aerodynamics Group within the Aerodynamics Laboratory of the Institute for Aerospace Research, National Research Council of Canada (NRC). He has an Honors Diploma from the von Karman Institute for Fluid Dynamics and received his Ph.D. degree from Queen's University in Kingston, Ontario, Canada. He is a recipient of an NRC President's Fund Award. Dr. Carscallen has taught for a number of years as a sessional lecturer at Carleton University in Ottawa, Ontario, Canada, and is the author of numerous publications in journals and conference proceedings.

# CONTENTS

# PREFACE

Compressible flow occurs in many devices encountered in mechanical and aerospace engineering practice and a knowledge of the effects of compressibility on a flow is therefore required by many professional mechanical and aerospace engineers. Most conventional course sequences in fluid mechanics and thermodynamics deal with some aspects of compressible fluid flow but the treatment is usually relatively superficial. For this reason, many mechanical and aerospace engineering schools offer a course dealing with compressible fluid flow at the senior undergraduate or at the graduate level. The purpose of such courses is to expand and extend the coverage given in previous fluid mechanics and thermodynamics courses. The present book is intended to provide the background material for such courses. The book also lays the foundation for more advanced courses on specialized aspects of the subject such as hypersonic flow, thus complementing the more advanced books in this area such as those by Anderson.

The widespread use of computer software for the analysis of engineering problems has, in many ways, increased the need to understand the assumptions and the theory on which such analyses are based. Such an understanding is required to interpret the computer results and to judge whether a particular piece of software will give results that are of adequate accuracy for the application being considered. Therefore, while computer methods are discussed in this book and some computer software is outlined, the major emphasis is on developing an understanding of the material and of the assumptions conventionally used in analyzing compressible fluid flows. Compared to available textbooks on the subject, then, the present book is, it is hoped, distinguished by its attempt to develop a thorough understanding of the theory and of the assumptions on which this theory is based, by its attempt to develop in the student a fascination with the phenomena involved in compressible flow, and by the breadth of its coverage.

## Contents

Our goal in writing this new text was to provide students with a clear explanation of the physical phenomena encountered in compressible flow, to develop in them an awareness of practical situations in which compressibility effects are likely to be important, to provide a thorough explanation of the assumptions used in the analysis of compressible flows, to provide a broad coverage of the subject, and to provide a firm foundation for the study of more advanced and specialized aspects of the subject. We have also tried to adopt an approach that will develop in the student a fascination with the phenomena involved in compressible flow.

The first seven chapters of the book deal with the fundamental aspects of the subject. They review some background material and discuss the analysis of isentropic flows, of normal and oblique shock waves, and of expansion waves. The next three chapters discuss the application of this material to the study of nozzle characteristics, of friction effects, and of heat exchange effects. Chapters dealing with the analysis of generalized one-dimensional flow, with simple numerical methods, and with two-dimensional flows are then given. The last three chapters in the book are interrelated and provide an introduction to hypersonic flow, to high temperature gas effects, and to low density flows. Some discussion of experimental methods is incorporated into the book, mainly to illustrate the theoretical material being discussed. However, because a number of shadowgraph and Schlieren photographs are given in the text, a separate discussion of these and other related methods of flow visualization is provided in an appendix.

The first ten chapters together with selected material from the remaining chapters provide the basis of the typical undergraduate course in compressible fluid flow. A graduate level course will typically cover more of the material on two-dimensional flows and high temperature gas effects.

The subject of compressible fluid flow involves so many interesting physical phenomena, involves so many intriguing mathematical complexities, and utilizes so many interesting and novel numerical techniques that the difficulty faced in preparing a book on the subject is not that of deciding what material to include but rather in deciding what material to omit. In preparing this book an attempt has been made to give a coverage that is broad enough to meet the needs of most instructors but at the same time to try to avoid losing the students' interest by going too deeply into the specialized areas of the subject.

## Acknowledgments

This book is based on courses on gas dynamics and compressible fluid flow taught by one of the authors (P.H.O.) at the University of Cape Town and at Queen's University and by the other author (W.E.C.) at Carleton University. The students in these courses, by their questions, comments, advice, and encouragement, have had a major influence on the way in which the material is presented in this book and their help is very gratefully acknowledged. The indirect influence of the authors' own professors on this book must also be gratefully acknowledged. In particular, R. Stegen at the University of Cape Town and I. I. Glass at the University of Toronto introduced one of the authors (P.H.O.) to the exciting nature of the subject, as did W. Gilbert to the other author (W.E.C.). The authors would also like to gratefully acknowledge the advice and help they received from colleagues at Queen's University, at the National Research Council of Canada, and at Carleton University. Jane Paul undertook the tedious job of typing and checking most of the text and her effort and encouragement is also gratefully acknowledged, as is the help given by Michael Bishop, Wayne Turnbull, and Laleah Carscallen.

The authors would also like to express their gratitude to the following reviewers of the text for their contributions to the development of this textbook: Kavesh Tagavi, University of Kentucky; C. H. Marston, Villanova University; Jay Khadadadi, Auburn University; John Lloyd, Michigan State University; and Jechel Jagoda, Georgia Institute of Technology. The authors would also like to thank John Lloyd and John D. Anderson for their advice and encouragement during the early stages of the development of this textbook.

## Solutions Manual

A manual that contains complete solutions to all of the problems in this textbook is available. In writing this manual, we have used the same problem-solving methodology as adopted in the worked examples in the book. The solution manual also provides summaries of the major equations developed in each chapter.

## Software

An interactive computer program, **COMPROP**, for the calculation of the properties of various compressible flows was developed by A. J. Ghajar of the School of Mechanical and Aerospace Engineering at Oklahoma State University, and L. M. Tam of the Mechanical Engineering Department of the University of Macau, to support this textbook. The program has modules for Isentropic Flow, Normal Shock Waves, Oblique Shock Waves, Fanno Flow, and Rayleigh Flow. The use of this software is described in Appendix A. The software is available free of charge to adopters of this book via the McGraw-Hill Web Page. The web address to access the program is: http://www.mhcollege.com/engineering/mecheng.html.

Patrick H. Oosthuizen
William E. Carscallen

# NOMENCLATURE

The following is a list of definitions of the main symbols used in this book:

| | |
|---|---|
| $A$ | area |
| $A^*$ | area at section where $M = 1$ |
| $A_e$ | exit plane area |
| $A_{\text{throat}}$ | throat area |
| $a$ | local speed of sound |
| $a^*$ | speed of sound at section where $M = 1$ |
| $a_0$ | stagnation speed of sound |
| $C$ | a constant |
| $C_D$ | drag coefficient |
| $C_L$ | lift coefficient |
| $C_p$ | pressure coefficient |
| $c$ | speed of light |
| $c_m$ | mean molecular speed |
| $c_p$ | specific heat at constant pressure |
| $c_v$ | specific heat at constant volume |
| $D$ | diameter |
| $D_H$ | hydraulic diameter of a duct |
| $e$ | internal energy |
| $F$ | force on system in direction indicated by subscript |
| $F_D$ | drag force |
| $F_\mu$ | friction force |
| $f$ | friction factor |
| $\bar{f}$ | mean friction factor over length of duct considered |
| $f_D$ | Darcy friction factor |
| $h$ | enthalpy |
| $h_f$ | enthalpy of formation |

| | |
|---|---|
| $I$ | impulse function |
| $Kn$ | Knudsen number |
| $K_P$ | equilibrium constant |
| $k$ | thermal conductivity |
| $L$ | velocity component parallel to oblique shock wave or size of system |
| $l^*$ | length of duct required to produce choking |
| $M$ | Mach number |
| $M_{\text{crit}}$ | critical Mach number |
| $M_{\text{des}}$ | design Mach number |
| $M_{\text{div}}$ | divergence Mach number |
| $M_e$ | exit plane Mach number |
| $M_N$ | component of Mach number normal to wave |
| $M_R$ | reflected shock Mach number |
| $M_S$ | shock Mach number |
| $m$ | molar mass |
| $\dot{m}$ | mass flow rate |
| $N$ | velocity component normal to oblique shock wave |
| $Nu$ | Nusselt number |
| $n$ | index of refraction |
| $P$ | perimeter |
| $p$ | pressure |
| $p^*$ | pressure at section where $M = 1$ |
| $p_0$ | stagnation pressure |
| $p_b$ | back pressure |
| $p_{b\text{crit}}$ | back pressure at which sonic flow first occurs |
| $p_e$ | exit plane pressure |
| $Pr$ | Prandtl number |
| $q$ | heat transfer rate per unit mass flow rate or per unit area |
| $Q$ | heat transfer rate |
| $R$ | gas constant for gas being considered |
| $\mathscr{R}$ | universal gas constant |

| | |
|---|---|
| $Re$ | Reynolds number |
| $r$ | recovery factor or relaxation parameter |
| $S$ | $V/c$ |
| $s$ | entropy |
| $s^*$ | entropy at section where $M = 1$ |
| $s_0$ | stagnation entropy |
| $T$ | temperature |
| $T_0$ | stagnation temperature |
| $T_\infty$ | freestream temperature |
| $T_{wa}$ | adiabatic wall temperature |
| $T^*$ | temperature at section where $M = 1$ |
| $t$ | time |
| $U_S$ | shock velocity |
| $U_{SR}$ | reflected shock velocity |
| $u$ | velocity component in $x$ direction |
| $V$ | velocity in one-dimensional flow |
| $V^*$ | velocity at section where $M = 1$ |
| $V_e$ | exit plane velocity |
| $\hat{V}$ | maximum velocity |
| $v$ | velocity component in $y$ direction |
| $w$ | rate at which work is done per unit mass flow rate |
| $W$ | rate at which work is done |
| $x$ | coordinate direction |
| $y$ | coordinate normal to $x$ |
| $Z$ | compressibility factor, $p/R\rho T$ |
| | |
| $\alpha$ | Mach angle |
| $\beta$ | shock angle |
| $\gamma$ | specific heat ratio |
| $\delta$ | turning angle |
| $\delta_{max}$ | maximum oblique shock wave turning angle |

$\theta$          Prandtl–Meyer angle

$\theta_{\mathrm{vib}}$        vibrational excitation temperature

$\lambda$          mean free path

$\mu$          coefficient of viscosity

$\nu$          kinematic viscosity or Prandtl–Meyer angle

$\rho$          density

$\rho^*$         density at section where $M = 1$

$\rho_0$         stagnation density

$\rho_e$         exit plane density

$\tau_w$         wall shear stress

$\psi$          dimensionless stream function

$\Phi$          potential function

$\Phi_p$         perturbation potential function

$\omega$          vorticity

# Compressible Fluid Flow

# CHAPTER 1

# Introduction

## 1.1
## COMPRESSIBILITY

The compressibility of a fluid is, basically, a measure of the change in density that will be produced in the fluid by a specified change in pressure. Gases are, in general, highly compressible whereas most liquids have a very low compressibility. Now, in a fluid flow, there are usually changes in pressure associated, for example, with changes in the velocity in the flow. These pressure changes will, in general, induce density changes which will have an influence on the flow, i.e., the compressibility of the fluid involved will have an influence on the flow. If these density changes are important, the temperature changes in the flow that arise due to the kinetic energy changes associated with the velocity changes also usually influence the flow, i.e., when compressibility is important, the temperature changes in the flow are usually important. Although the density changes in a flow field can be very important, there exist many situations of great practical importance in which the effects of these density and temperature changes are negligible. Classical incompressible fluid mechanics deals with such flows in which the pressure and kinetic energy changes are so small that the effects of the consequent density and temperature changes on the fluid flow are negligible, i.e., the flow can be assumed to be incompressible. There are, however, a number of flows that are of great practical importance in which this assumption is not adequate, the density and temperature changes being so large that they have a very significant influence on the flow. In such cases, it is necessary to study the thermodynamics of the flow simultaneously with its dynamics. The study of these flows in which the changes in density and temperature are important is basically

(a)

(b)

(c)

**FIGURE 1.1**
Aircraft designed to operate at different speeds: (a) de Havilland Dash 8 (*courtesy of Bombardier Regional Aircraft*); (b) Canadair Regional Jet (*courtesy of Bombardier Inc., Canadair*); (c) Aérospatiale/British Aerospace Concorde (*courtesy of Air France*).

Shock
wave

Shock
wave

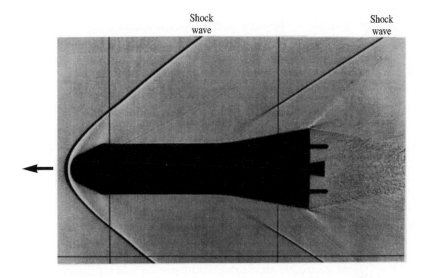

**FIGURE 1.2**
Photograph of supersonic flow over a body showing the presence of shock waves.
(Methods used to obtain such photographs are discussed in Appendix K.)

what is known as compressible fluid flow or gas dynamics, it usually only
being in gas flows that compressibility effects are important.

The fact that compressibility effects can have a large influence on a fluid
flow can be seen by considering the three aircraft shown in Fig. 1.1. The first
of the aircraft shown in Fig. 1.1 is designed for relatively low speed flight. It
has straight wings, it is propeller driven and the fuselage (the body of the
aircraft) has a "rounded" nose. The second aircraft is designed for higher
speeds. It has swept wings and tail surfaces and is powered by turbojet
engines. However, the fuselage still has a "rounded" nose and the intakes
to the engines also have rounded edges and are approximately at right angles
to the direction of flight. The third aircraft is designed for very high speed
flight. It has highly swept wings and a sharp nose, and the air intakes to the
engines have sharp edges and are of complex shape. These differences between
the aircraft are mainly due to the fact that compressibility effects become
increasingly important as the flight speed increases.

Although the most obvious applications of compressible fluid flow theory
are in the design of high speed aircraft, and this remains an important application of the subject, a knowledge of compressible fluid flow theory is required
in the design and operation of many devices commonly encountered in engineering practice. Among these applications are:

- Gas turbines: the flow in the blading and nozzles is compressible
- Steam turbines: here, too, the flow in the nozzles and blades must be treated
  as compressible

- Reciprocating engines: the flow of the gases through the valves and in the intake and exhaust systems must be treated as compressible
- Natural gas transmission lines: compressibility effects are important in calculating the flow through such pipelines
- Combustion chambers: the study of combustion, in many cases, requires a knowledge of compressible fluid flow

Compressibility effects are normally associated with gas flows in which, as discussed in the next chapter, the flow velocity is relatively high compared to the speed of sound in the gas and if the flow velocity exceeds the local speed of sound, i.e., if the flow is "supersonic," effects may arise which do not occur at all in "subsonic" flow, e.g., shock waves can exist in the flow as shown in Fig. 1.2. The nature of such shock waves will be discussed in later chapters.

**EXAMPLE 1.1**
Air flows down a variable area duct. Measurements indicate that the pressure is 80 kPa, the temperature is 5°C, and the velocity is 150 m/s at a certain section of the duct. Estimate, assuming incompressible flow, the velocity and pressure at a second section of the duct at which the duct area is half that of the section where the measurements were made. Comment on the validity of the incompressible flow assumption in this situation.

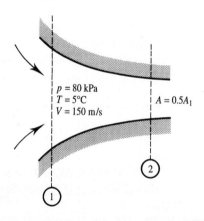

**FIGURE E1.1**

*Solution*
Because the flow is assumed to be incompressible, the continuity equation gives:

$$V_1 A_1 = V_2 A_2$$

Hence, using the supplied information:

$$150 \times A_1 = V_2(A_1/2)$$

Therefore:
$$V_2 = 300 \text{ m/s}$$

Assuming that the effects of friction on the flow are negligible, the pressure change can be found using Bernoulli's equation which gives:

$$p_1 + \rho \frac{V_1^2}{2} = p_2 + \rho \frac{V_2^2}{2}$$

This can be rearranged to give:

$$p_2 = \rho \left[ \frac{V_1^2}{2} - \frac{V_2^2}{2} \right] + p_1 \qquad \text{(a)}$$

The density, $\rho$, is evaluated using the initial conditions, i.e., using $p_1/\rho = RT_1$ which, since air flow is being considered, gives:

$$\rho = \frac{80 \times 10^3}{(287 \times 278)} = 1.003 \text{ kg/m}^3$$

Substituting this back into eq. (a) then gives:

$$p_2 = 1.003 \left[ \frac{150^2}{2} - \frac{300^2}{2} \right] + 80 \times 10^3$$

This gives:

$$p_2 = 4.62 \times 10^4 \text{ Pa} = 46.2 \text{ kPa}$$

In order to check the validity of the assumption that the flow is incompressible, it is noted that if the flow can be assumed to be incompressible, the temperature changes in the flow will normally be negligible so the temperature at the exit will also be approximately 5°C. The equation of state therefore gives at the exit:

$$\rho_2 = \frac{p_2}{RT_2} = \frac{46.2 \times 10^3}{(287) \times (278)} = 0.579 \text{ kg/m}^3$$

Since this indicates that the density changes by more than 40%, the incompressible flow assumption is not justified.

## 1.2
## FUNDAMENTAL ASSUMPTIONS

The concern in this book is essentially only with the flow of a gas. Now, in order to analyze an engineering problem, simplifying assumptions normally have to be made, i.e., a "model" of the situation being considered has to be introduced. The following assumptions will be adopted in the initial part of the present study of compressible fluid flow, some of these assumptions being relaxed in the later chapters.

1. The gas is continuous, i.e., the motion of individual molecules does not have to be considered, the gas being treated as a continuous medium. This assumption applies in flows in which the mean free path of the gas molecules is very small compared to all the important dimensions of the solid body through or over which the gas is flowing. This assumption will, of course, become invalid if the gas pressure and, hence, the density becomes very low as is the case with spacecraft operating at very high altitudes and in low density flows that can occur in high vacuum systems.

2. No chemical changes occur in the flow field. Chemical changes influence the flow because they result in a change in composition with resultant energy changes. One common chemical change is that which results from combustion in the flow field. Other chemical changes that can occur when there are large pressure and temperature changes in the flow field are dissociation and ionization of the gas molecules. These can occur, for example, in the flow near a satellite during re-entry to the earth's atmosphere.

3. The gas is perfect. This implies that:

   (a) The gas obeys the perfect gas law, i.e.

   $$\frac{p}{\rho} = RT = \frac{\mathscr{R}}{m} T \tag{1.1}$$

   In eq. (1.1), $\mathscr{R}$ is the universal gas constant, which has a value of 8314.3 J/kg mole K or 1545.3 ft-lbf/lbm mole °R, and $m$ is the molar mass. $R$ is the gas constant for a particular gas. For air, it is equal to $8314.3/28.966 = 287.04$ J/kg K or $1545.3/28.966 = 53.3$ ft-lbf/lbm °R. (A discussion of units will be given later in this chapter.)

   (b) The specific heats at constant pressure and constant volume, $c_p$ and $c_v$, are both constants, i.e., the gas is calorically perfect. The ratio of the two specific heats will be used extensively in the analyses presented in this book and is given by:

   $$\gamma = \frac{c_p}{c_v} \tag{1.2}$$

   The symbol $\gamma$ is used for the specific heat ratio instead of the frequently used symbol $k$ because the symbol $k$ is used for the thermal conductivity in this book. It should also be recalled that:

   $$R = c_p - c_v \tag{1.3}$$

   The assumption that the gas is calorically perfect may not apply if there are very large temperature changes in the flow or if the gas temperature is high. A discussion of this will be given in a later chapter. While a calorically perfect gas has specific heats that are constant, a thermally perfect gas has specific heats that depend only on temperature and which are thus not necessarily constant.

4. Gravitational effects on the flow field are negligible. This assumption is quite justified for gas flows.

5. Magnetic and electrical effects are negligible. These effects would normally only be important if the gas was electrically conducting, which is usually only true if the gas or a seeding material in the gas is ionized. In the so-called magnetohydrodynamic (MHD) generator, a hot gas which has been seeded with a substance that easily ionizes and which therefore makes the gas a conductor, is expanded through a nozzle to a high velocity. The high speed gas stream is then passed through a magnetic field which generates an

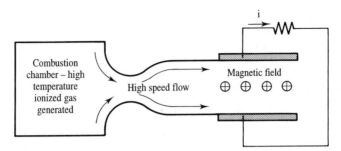

**FIGURE 1.3**
Magnetohydrodynamic generator.

emf. This induces an electrical current flow in a conductor connected across the gas flow normal to the magnetic field. The arrangement of such a device is shown very schematically in Fig. 1.3. Because there are no moving components in contact with the gas stream, this device can operate at very high gas temperatures. The magnetic and electrical effects that are important in this device are not considered in this book.

6. The effects of viscosity are negligible. This is never true close to a solid surface but in many cases the overall effects of viscosity remain small. The relaxation of this assumption and the effects of viscosity on compressible fluid flows will be discussed in later chapters.

When the above assumptions are adopted, the flow field is completely described by knowing the values of the following variables at all points in the flow field:

- Velocity vector, $V$
- Pressure, $p$
- Density, $\rho$
- Temperature, $T$

Therefore, in order to describe the flow field, four equations involving these four variables must be obtained. These equations are derived by applying the following principles:

- Conservation of mass (continuity equation)
- Conservation of momentum (Newton's Law)
- Conservation of energy (First Law of Thermodynamics)
- Equation of state

The present book describes the application of these principles to the prediction of various aspects of compressible flow.

**EXAMPLE 1.2**
The pressure and temperature in a gas in a large chamber are found to be 500 kPa and 60°C respectively. Find the density if the gas is (a) air and (b) hydrogen.

### Solution

The perfect gas law gives:

$$\rho = \frac{pm}{\mathscr{R}T}$$

Because $m$ is 28.97 for air and 2 for hydrogen this equation gives:

$$\rho_{air} = \frac{500 \times 1000 \times 28.97}{8314 \times 333} = 5.23 \text{ kg/m}^3$$

and

$$\rho_{hydrogen} = \frac{500 \times 1000 \times 2}{8314 \times 333} = 0.36 \text{ kg/m}^3$$

## 1.3
## UNITS

Although the SI system of units has virtually become the standard throughout the world, the "English" or "Imperial" system is still quite extensively used in industry and the engineer still needs to be familiar with both systems. Examples and problems based on both of these systems of units are, therefore, incorporated into this book.

Now, in modern systems of units, the units of mass, length, time and temperature are treated as fundamental units. The units of other quantities are then expressed in terms of these fundamental units. The relation between the derived units and the fundamental units is obtained by considering the "laws" governing the process and then equating the dimensions of the terms in the equation that follows from the "law," e.g., since:

$$\text{force} = \text{mass} \times \text{acceleration}$$

it follows that the dimensions of force are equal to the dimensions of mass multiplied by the dimensions of acceleration, i.e., equal to the dimensions of mass multiplied by the dimensions of velocity divided by the dimensions of time squared. From this it follows that in any system of units:

$$\text{units of force} = \text{units of mass} \times \text{units of length/units of time}^2$$

The fundamental units in the two systems used in this book are:

### SI Units

- Unit of mass = kilogram (kg)
- Unit of length = meter (m)
- Unit of time = second (s)
- Unit of temperature = Kelvin or degree Celsius (K or °C)

In this system, the unit of force is the newton, N, which is related to the fundamental units by:

$$1 \text{ N} = 1 \text{ kg m/s}^2$$

## English System

- Unit of mass = pound mass (lbm)
- Unit of length = foot (ft)
- Unit of time = second (sec)
- Unit of temperature = Rankine or degree Fahrenheit (R or °F)

In this system of units, therefore, the unit of force should be 1 lbm ft/sec$^2$. However, for historical reasons, it is usual in this system to take the unit of force as the pound force, lbf, which is related to the fundamental units by:

$$1 \text{ lbf} = 32.2 \text{ lbm ft/sec}^2$$

The force required to accelerate a 10 lbm mass at a rate of 10 ft/sec$^2$ is, therefore, 100 lbm ft/sec$^2$, i.e., 100/32.2 lbf. With the lbm as the unit of mass and the lbf as the unit of force, the mass of a body is equal in magnitude to the gravitational force acting on the body under standard sea-level gravitational conditions. This was a feature that was common to all earlier systems of units.

A list of derived units in both systems of units is given in Appendix J. This appendix also contains a list of conversion factors between the two systems of units. Whenever a pressure is given in this book it should be assumed to be an absolute pressure unless gauge pressure is clearly indicated.

**EXAMPLE 1.3**
A U-tube manometer containing water is used to measure the pressure difference between two points in a duct through which air is flowing. Under certain conditions, it is found that the difference in the heights of the water columns in the two legs of the manometer is 11 in. Find the pressure difference between the two points in the flow in psf, psi, and Pa.

**FIGURE E1.3**

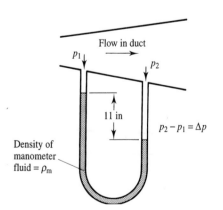

### Solution

It is shown in most books on fluid mechanics that, for a manometer, the pressure difference $\Delta p$ is related to the difference between the heights $\Delta h$ of the water columns in the two legs of the manometer by:

$$\Delta p = \rho_{man} g \Delta h$$

where $\rho_{man}$ is the density of the liquid in the U-tube manometer. In the situation being considered, the manometer contains water so $\rho_{man} = 62.4 \, \text{lbm/ft}^3$. Hence, since $g = 32.2 \, \text{ft/sec}^2$, it follows that:

$$\Delta p = 62.4 \times 32.2 \times (11/12)$$

the units being $(\text{lbm/ft}^3) \times (\text{ft/sec}^2) \times \text{ft}$, i.e., $(\text{lbm ft/sec}^2)/(\text{ft}^2)$. Hence, since $1 \, \text{lbf} = 32.2 \, \text{lbm ft/sec}^2$:

$$\Delta p = \frac{62.4 \times 32.2 \times (11/12)}{32.2} = 57.2 \, \text{lbf/ft}^2 \qquad (\text{i.e., psf})$$

Since $1 \, \text{lbf/in}^2 \, (\text{psi}) = 144 \, \text{psf}$ it follows that:

$$\Delta p = 57.2/144 = 0.397 \, \text{lbf/in}^2 \qquad (\text{i.e., psi})$$

Lastly, since the conversion tables give $1 \, \text{Pa} = 0.020886 \, \text{psf}$, it follows that:

$$\Delta p = 57.2/0.020886 = 2738.7 \, \text{Pa} = 2.739 \, \text{kPa}$$

## 1.4
## CONSERVATION LAWS

The analysis of compressible flows is, as mentioned previously, based on the application of the principles of conservation of mass, momentum and energy to the flow. These principles, in relatively general form, will be discussed in this section. These conservation principles or laws will here be applied to the flow through a control volume which is defined by an imaginary boundary drawn in the flow as shown in Fig. 1.4. For such a control volume, conservation of mass requires that:

$$\begin{array}{ccc} \text{Rate of increase} & & \\ \text{of mass of fluid} & = & \text{Rate mass enters} \quad - \quad \text{Rate mass leaves} \\ \text{in control volume} & & \text{control volume} \qquad \text{control volume} \end{array}$$

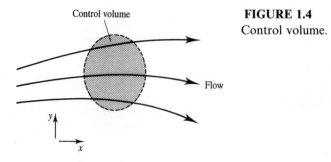

**FIGURE 1.4**
Control volume.

Next, consider conservation of momentum. Because momentum is a vector quantity, conservation of momentum must apply in any chosen direction. Hence, conservation of momentum requires that in any direction:

Net force on gas
in control volume    =    Rate of increase of momentum
in direction considered    in direction considered of
fluid in control volume

+    Rate momentum leaves
control volume in
direction considered    −    Rate momentum enters
control volume in
direction considered

Conservation of energy applied to the control volume requires that:

Rate of increase in internal
energy and kinetic energy    +    Rate enthalpy and
kinetic energy leave    −    Rate enthalpy and
kinetic energy enter
of gas in control volume    control volume    control volume

=    Rate heat is
transferred into    −    Rate work is
done by gas in
control volume    control volume

**EXAMPLE 1.4**
Air flows from a large chamber through a valve into an initially evacuated tank. The pressure and temperature in the large chamber are kept constant at 1000 kPa and 30°C respectively and the internal volume of the tank is 0.2 m$^3$. Find the time taken for the pressure in the initially evacuated tank to reach 160 kPa. Because of the pressures existing in the two vessels, the mass flow rate through the valve can be assumed to remain constant (see discussion of choking in Chapter 8) and equal to $0.9 \times 10^{-4}$ kg/s. Heat transfer from the tank can be neglected and the kinetic energy of the gas in the chamber and in the tank can also be neglected.

**FIGURE E1.4**

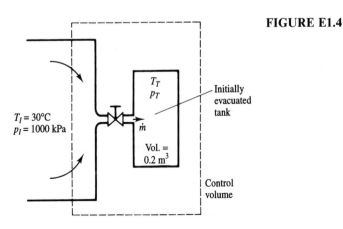

**Solution**

The situation being considered is shown in Fig. E1.4. Conservation of energy requires that, since there is no flow out of the tank and no heat loss from the tank:

rate of increase of internal energy in tank = rate of enthalpy flow into tank

i.e., since the enthalpy and internal energy per unit mass are equal to $c_p T$ and $c_v T$ respectively:

$$\frac{d}{dt}[M c_v T_T] = \dot{m} c_p T_I$$

where $M$ is the mass of air in the tank, $T_T$ is the temperature of the air in the tank, $\dot{m}$ is the mass flow rate into the tank and $T_I$ is the constant temperature of the flow into the tank. Because $\dot{m}$ and $T_I$ are constant, the above equation can be integrated to give:

$$M c_v T_T = \dot{m} c_p T_I t$$

But because $\dot{m}$ is constant and because the tank is initially evacuated:

$$M = \dot{m} t$$

Therefore:

$$\dot{m} t c_v T_T = \dot{m} c_p T_I t$$

i.e.,

$$T_T = c_p T_I / c_v = \gamma T_I$$

Here, $T_I$ is equal to 303 K so the above equation gives:

$$T_T = 1.4 \times 303 = 424.2 \text{ K}$$

Next, it is noted that the perfect gas law gives for the tank:

$$p_T = \rho_T R T_T = \frac{M R T_T}{V}$$

where $V$ is the volume of the tank. Hence, again using:

$$M = \dot{m} t$$

it follows that:

$$p_T = \frac{\dot{m} t R T_T}{V}$$

Therefore, the time taken to reach a pressure of 160 kPa is given by:

$$t = \frac{160000 \times 0.2}{0.00009 \times 287 \times 424.2} = 2921 \text{ s}$$

Hence the time taken for the pressure in the initially evacuated tank to reach 160 kPa is 0.81 h.

Attention will mainly be given to flows in which the coordinate system can be so chosen that the flow is steady, i.e., to flows in which a coordinate system can be so chosen that none of the flow properties are changing with time. In real situations, even when the flow is steady on the average, the flow is seldom truly steady. Instead, the flow variables fluctuate with time about the mean values either because the flow is turbulent or due to fluctuations in the system that produces the flow. It is assumed here that the mean values of the flow variables can be adequately described by equations that are based on the assumption of steady flow.

For steady flow, the above conservation equations give:

Conservation of mass:

$$\frac{\text{Rate mass enters}}{\text{control volume}} = \frac{\text{Rate mass leaves}}{\text{control volume}}$$

Conservation of momentum:

$$\begin{array}{c}\text{Net force on gas} \\ \text{in control volume} \\ \text{in direction considered}\end{array} = \begin{array}{c}\text{Rate momentum leaves} \\ \text{control volume in} \\ \text{direction considered}\end{array} - \begin{array}{c}\text{Rate momentum} \\ \text{enters control volume} \\ \text{in direction considered}\end{array}$$

Conservation of energy:

$$\begin{array}{c}\text{Rate enthalpy and} \\ \text{kinetic energy leave} \\ \text{control volume}\end{array} - \begin{array}{c}\text{Rate enthalpy and} \\ \text{kinetic energy enter} \\ \text{control volume}\end{array}$$

$$= \begin{array}{c}\text{Rate heat is} \\ \text{transferred into} \\ \text{control volume}\end{array} - \begin{array}{c}\text{Rate work is} \\ \text{done by gas in} \\ \text{control volume}\end{array}$$

In discussing the use of these conservation principles, attention will be directed to the flow through a system that, in general, has multiple inlets and outlets, an example of such a system being shown in Fig. 1.5. Here there are two inlets to the system, 1 and 2, and two outlets, 3 and 4.

A control volume, defined by an imaginary boundary drawn around the system, is introduced and the conservation principles are applied to the flow through this control volume.

As discussed above, the principle of conservation of mass requires that, since steady flow is being considered:

Rate mass leaves control volume = Rate mass enters control volume

Since the mass flow rate in a duct is given by:

$$\dot{m} = \rho V A$$

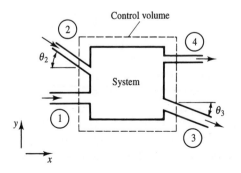

Control volume

System

**FIGURE 1.5**
Type of system considered. Control volume is also shown.

where $V$ is the mean velocity in the duct and $A$ is its cross-sectional area, conservation of mass requires for the system shown in Fig. 1.5:

$$\dot{m}_3 + \dot{m}_4 = \dot{m}_1 + \dot{m}_2$$

i.e.,
$$\rho_3 V_3 A_3 + \rho_4 V_4 A_4 = \rho_1 V_1 A_1 + \rho_2 V_2 A_2 \qquad (1.4)$$

Attention will next be directed to the conservation of momentum principle. As discussed above, because momentum is a vector quantity, conservation of momentum in a particular direction will be considered. In any direction it gives for steady flow:

| Net force on gas in control volume in direction considered | = | Rate momentum leaves control volume in direction considered | − | Rate momentum enters control volume in direction considered |

Consider the situation shown in Fig. 1.5. It will be assumed that the pressure around the surface of the control volume is the same everywhere except where the ducts carrying the gas into and out of the control volume cross this surface. Since the momentum flux in any direction is equal to the product of the mass flow rate and the velocity component in the direction considered, it follows that for the control volume shown in Fig. 1.5, conservation of momentum gives in the $x$-direction:

$$p_1 A_1 + p_2 A_2 \cos \theta_2 - p_3 A_3 \cos \theta_3 - p_4 A_4 + F_x$$
$$= \dot{m}_3 V_3 \cos \theta_3 + \dot{m}_4 V_4 - \dot{m}_1 V_1 - \dot{m}_2 V_2 \cos \theta_2 \quad (1.5)$$

while in the $y$-direction, conservation of momentum gives:

$$p_2 A_2 \sin \theta_2 - p_3 \sin \theta_3 - F_y = \dot{m}_3 V_3 \sin \theta_3 - \dot{m}_2 V_2 \cos \theta_2 \qquad (1.6)$$

In these equations, $F_x$ and $F_y$ are the $x$- and $y$-components of the force exerted by the system on the gas. This force is actually exerted on the gas at the surface of the system as shown in Fig. 1.6. Since, as also illustrated in Fig. 1.6, the gas will exert equal and opposite force components on the system, a force will have to be applied externally to the system and therefore to the control volume to keep it at rest.

Lastly, the application of the conservation of energy principle to the flow through the control volume will be considered. Conservation of energy applied to the control volume requires that:

| Rate enthalpy and kinetic energy leave control volume | − | Rate enthalpy and kinetic energy enter control volume |

| = | Rate heat is transferred into control volume | − | Rate work is done by gas in control volume |

This gives, since gravitational effects are being neglected:

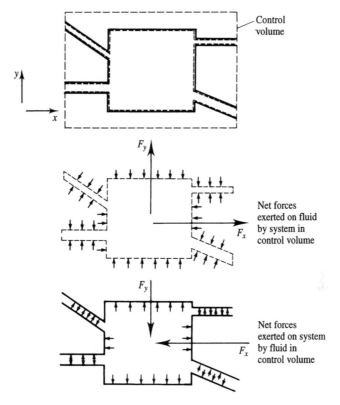

**FIGURE 1.6**
Force on fluid in control volume.

$$\dot{m}_3(h_3 + V_3^2/2) + \dot{m}_4(h_4 + V_4^2/2) - \dot{m}_1(h_1 + V_1^2/2) - \dot{m}_2(h_2 + V_2^2/2)$$

$$= Q - W \quad (1.7)$$

where $h$ is the enthalpy of the gas per unit mass, $Q$ is the rate of heat transfer to the system, and $W$ is the rate at which the system is doing work. The effect of work transfer will not be considered in the present book, i.e., $W$ will be taken as 0 and in most of the book it will be assumed that $h = c_p T$ to an adequate degree of accuracy. Equation (1.7) can then be written:

$$\dot{m}_3(c_p T_3 + V_3^2/2) + \dot{m}_4(c_p T_4 + V_4^2/2) - \dot{m}_1(c_p T_1 + V_1^2/2)$$

$$- \dot{m}_2(c_p T_2 + V_2^2/2) = Q \quad (1.8)$$

The above equations have been written for the particular situation shown in Fig. 1.5. The modification of these equations to deal with other flow situations is quite straightforward.

### EXAMPLE 1.5

Liquid oxygen and liquid hydrogen are both fed to the combustion chamber of a liquid fuelled rocket engine at a rate of 5 kg/s. The products of combustion from this chamber are exhausted through a convergent–divergent nozzle. The exit plane of this exhaust nozzle has a diameter of 0.3 m and the gases flowing through the nozzle are estimated to have a density of 0.1 kg/m$^3$ on the exit plane. Estimate the gas velocity on the nozzle exit plane assuming one-dimensional flow on the exit plane.

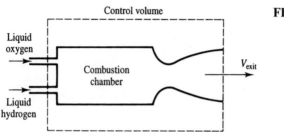

**FIGURE E1.5**

### Solution

Assuming that the flow is steady, conservation of mass requires that the rate mass leaves the system must be equal to the rate at which mass enters the system. The rate at which mass leaves the nozzle is, therefore, $5 + 5 = 10$ kg/s. But, since the flow is assumed to be one-dimensional:

$$\dot{m} = \rho V A$$

Hence, the velocity on the nozzle exit plane is given by:

$$V_{exit} = \frac{\dot{m}}{\rho A} = \frac{10}{0.1 \times \pi \times 0.3^2/4}$$

Hence:

$$V_{exit} = 1414.7 \text{ m/s}$$

### EXAMPLE 1.6

A solid fuelled rocket engine is fired on a test stand. The diameter of the exhaust nozzle on the discharge plane is 0.1 m. The exhaust plane velocity is estimated to be 650 m/s and the pressure on this exhaust plane is estimated to be 150 kPa. An estimation of the rate at which combustion occurs indicates that the gas flow rate through the exhaust nozzle is 40 kg/s. If the ambient pressure is 100 kPa, estimate the force exerted on the test stand by the rocket.

### Solution

The force $T$ shown in Fig. E1.6 is the force exerted on the test stand, i.e., the thrust. It is the force required to hold the engine in place. A force that is equal in magnitude to $T$ but opposite in direction to $T$ is exerted on the fluid in the engine.

Consider the application of the momentum equation to the control volume shown in Fig. E1.6. Since no momentum enters this control volume, it follows that:

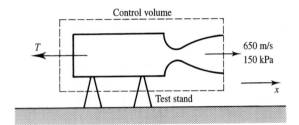

**FIGURE E1.6**

$$\frac{\text{Net force on control}}{\text{volume in } x\text{-direction}} = \frac{\text{Rate momentum}}{\text{leaves control volume}}$$

But the pressure is equal to ambient everywhere on the surface of the control volume except on the nozzle exit plane. Hence, if $T'$ is the force exerted on the fluid by the system, it follows that:

$$T' - (p_{\text{exit}} - p_{\text{ambient}})A_{\text{exit}} = \dot{m}V_{\text{exit}}$$

This gives:

$$T' - (150 \times 10^3 - 100 \times 10^3)\pi\frac{(0.1)^2}{4} = 40 \times 650$$

Hence:  $$T' = 26.4 \times 10^3 \text{ N} = 26.4 \text{ kN}$$

Therefore, the force exerted on the stand is:

$$T = -T' = -26.4 \text{ kN}$$

The negative sign indicates that it acts in the negative $x$-direction, i.e., that it acts in the direction shown in Fig. E1.6.

## 1.5
## CONCLUDING REMARKS

Compressible flows are flows in which the density changes induced by the pressure changes through the flow field have a significant influence on the flow. Compressibility effects are usually associated with the high speed flow of gases. Compressible gas flows are analyzed by applying the principles of conservation of mass, momentum, and energy together with the equation of state to deduce the variations of velocity, pressure, density, and temperature through the flow field.

## PROBLEMS

**1.1.** An air stream enters a variable area channel at a velocity of 30 m/s with a pressure of 120 kPa and a temperature of 10°C. At a certain point in the channel, the velocity is found to be 250 m/s. Using Bernoulli's equation (i.e., $p + \rho V^2/2 = \text{constant}$), which assumes incompressible flow, find the pressure at this point. In this calculation use the density evaluated at the inlet conditions. If the temperature of the air is assumed to remain constant, evaluate the air density at the point in the flow where the velocity is 250 m/s. Compare this density with the density at the inlet to the channel. On the basis of this comparison, do you think that the use of Bernoulli's equation is justified?

**1.2.** The gravitational acceleration on a large planet is 90 ft/sec². What is the gravitational force acting on a spacecraft with a mass of 8000 lbm on this planet?

**1.3.** The pressure and temperature at a certain point in an air flow are 130 kPa and 30°C respectively. Find the air density at this point in kg/m³ and lbm/ft³.

**1.4.** Two kilograms of air at an initial temperature and pressure of 30°C and 100 kPa undergoes an isentropic process, the final temperature attained being 850°C. Find the final pressure, the initial and final densities and the initial and final volumes.

**1.5.** Two jets of air, each having the same mass flow rate, are thoroughly mixed and then discharged into a large chamber. One jet has a temperature of 120°C and a velocity of 100 m/s while the other has a temperature of −50°C and a velocity of 300 m/s. Assuming that the process is steady and adiabatic, find the temperature of the air in the large chamber.

**1.6.** Two air streams are mixed in a chamber. One stream enters the chamber through a 5 cm diameter pipe at a velocity of 100 m/s with a pressure of 150 kPa and a temperature of 30°C. The other stream enters the chamber through a 1.5 cm diameter pipe at a velocity of 150 m/s with a pressure of 75 kPa and a temperature of 30°C. The air leaves the chamber through a 9 cm diameter pipe at a pressure of 90 kPa and a temperature of 30°C. Assuming that the flow is steady, find the velocity in the exit pipe.

**1.7.** The jet engine fitted to a small aircraft uses 35 kg/s of air when the aircraft is flying at a speed of 800 km/h. The jet efflux velocity is 590 m/s. If the pressure on the engine discharge plane is assumed to be equal to the ambient pressure and if effects of the mass of the fuel used are ignored, find the thrust developed by the engine.

**1.8.** The engine of a small jet aircraft develops a thrust of 18 kN when the aircraft is flying at a speed of 900 km/h at an altitude where the ambient pressure is 50 kPa. The air flow rate through the engine is 75 kg/s and the engine uses fuel at a rate of 3 kg/s. The pressure on the engine discharge plane is 55 kPa and the area of the engine exit is 0.2 m². Find the jet efflux velocity.

**1.9.** A small turbo-jet engine uses 50 kg/s of air and the air/fuel ratio is 90 : 1. The jet efflux velocity is 600 m/s. When the afterburner is used, the overall air/fuel ratio decreases to 50 : 1 and the jet efflux velocity increases to 730 m/s. Find the static thrust with and without the afterburner. The pressure on the engine discharge plane can be assumed to be equal to the ambient pressure in both cases.

**1.10.** A rocket used to study the atmosphere has a fuel consumption rate of 120 kg/s and a nozzle discharge velocity of 2300 m/s. The pressure on the nozzle discharge plane is 90 kPa. Find the thrust developed when the rocket is launched at sea level. The nozzle exit plane diameter is 0.3 m.

**1.11.** A solid fuelled rocket is fitted with a convergent-divergent nozzle with an exit plane diameter of 30 cm. The pressure and velocity on this nozzle exit plane are 75 kPa and 750 m/s respectively and the mass flow rate through the nozzle is 350 kg/s. Find the thrust developed by this engine when the ambient pressure is (a) 100 kPa and (b) 20 kPa.

**1.12.** In a hydrogen powered rocket, hydrogen enters a nozzle at a very low velocity with a temperature and pressure of 2000°C and 6.8 MPa respectively. The pressure on the exit plane of the nozzle is equal to the ambient pressure which is 10 kPa. If the required thrust is 10 MN, what hydrogen mass flow rate is required? The flow through the nozzle can be assumed to be isentropic and the specific heat ratio of the hydrogen can be assumed to be 1.4.

**1.13.** In a proposed jet propulsion system for an automobile, air is drawn in vertically through a large intake in the roof at a rate of 3 kg/s, the velocity through this intake being small. Ambient pressure and temperature are 100 kPa and 30°C respectively. This air is compressed and heated and then discharged horizontally out of a nozzle at the rear of the automobile at a velocity of 500 m/s and a pressure of 140 kPa. If the rate of heat addition to the air stream is 600 kW, find the nozzle discharge area and the thrust developed by the system.

**1.14.** Carbon dioxide flows through a constant area duct. At the inlet to the duct, the velocity is 120 m/s and the temperature and pressure are 200°C and 700 kPa respectively. Heat is added to the flow in the duct and at the exit of the duct the velocity is 240 m/s and the temperature is 450°C. Find the amount of heat being added to the carbon dioxide per unit mass of gas and the mass flow rate through the duct per unit cross-sectional area of the duct. Assume that for carbon dioxide, $\gamma = 1.3$.

**1.15.** Air enters a heat exchanger with a velocity of 120 m/s and a temperature and pressure of 225°C and 2.5 MPa. Heat is removed from the air in the heat exchanger and the air leaves with a velocity 30 m/s at a temperature and pressure of 80°C and 2.45 MPa. Find the heat removed per kilogram of air flowing through the heat exchanger and the density of the air at the inlet and the exit to the heat exchanger.

# CHAPTER 2

# The Equations of Steady One-Dimensional Compressible Fluid Flow

## 2.1
## INTRODUCTION

Many of the compressible flows that occur in engineering practice can be adequately modeled as a flow through a duct or streamtube whose cross-sectional area is changing relatively slowly in the flow direction. A duct is here taken to mean a solid walled channel, whereas a streamtube is defined by considering a closed curve drawn in a fluid flow. A series of streamlines will pass through this curve as shown in Fig. 2.1. Further downstream, these streamlines can be joined by another curve as shown in the figure. Since there is no flow normal to a streamline, in steady flow the rate at which fluid crosses the area defined by the first curve is equal to the rate at which fluid crosses the area defined by the second curve. The streamlines passing through the curves effectively, therefore, define the "walls" of a duct and this "duct" is called a streamtube. Of course, in the case of a duct with solid walls, streamlines lie along the walls and the duct is, effectively, also a streamtube. In

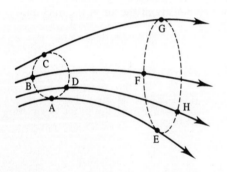

**FIGURE 2.1**
Definition of a streamtube. All streamlines that pass through curve ABCD also pass through curve EFGH, these streamlines then forming a streamtube.

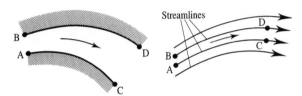

**FIGURE 2.2**
Solid walled channel and streamtube. ABCD is a solid walled channel. ABCD is a streamtube.

the case of both flow through a streamtube and flow through a solid-walled duct, there can be no flow through the "walls" of the duct, there being no flow through a solid wall and, by definition, no flow normal to a streamline. The two types of duct are shown in Fig. 2.2.

Examples of the type of flow being considered in this chapter are those through the blade passages in a turbine and the flow through a nozzle fitted to a rocket engine, these being shown in Fig. 2.3.

In many such practical situations, it is adequate to assume that the flow is steady and one-dimensional. As discussed in the previous chapter, steady flow implies that none of the properties of the flow are varying with time. In most real flows that are steady on the average, the instantaneous values of the flow properties, in fact, fluctuate about the mean values. However, an analysis of such flows based on the assumption of steady flow usually gives a good description of the mean values of the flow variables. One-dimensional flow is, strictly, flow in which the reference axes can be so chosen that the velocity vector has only one component over the portion of the flow field considered, i.e., if $u$, $v$, and $w$ are the $x$, $y$, and $z$ components of the velocity vector then, strictly, for the flow to be one-dimensional it is necessary that it be possible for the $x$ direction to be so chosen that the velocity components $v$ and $w$ are zero (see Fig. 2.4).

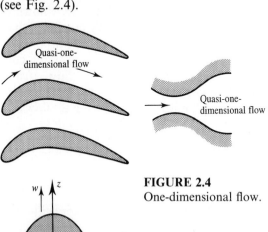

**FIGURE 2.3**
Typical duct flows.

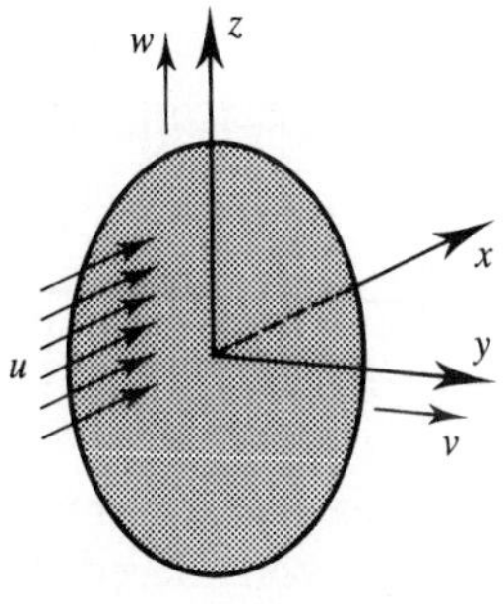

**FIGURE 2.4**
One-dimensional flow.

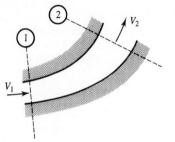

**FIGURE 2.5**
Definition of a velocity $V$.

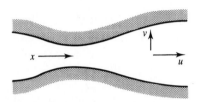

**FIGURE 2.6**
Flow situation that can be modeled as
one-dimensional.

In a one-dimensional flow the velocity at a section of the duct will here be given the symbol $V$, as indicated in Fig. 2.5.

Strictly speaking, the equations of one-dimensional flow are only applicable to flow in a straight pipe or stream tube of constant area. However, in many practical situations, the equations of one-dimensional flow can be applied with acceptable accuracy to flows with a variable area provided that the rate of change of area and the curvature are small enough for one component of the velocity vector to remain dominant over the other two components. For example, although the flow through a nozzle of the type shown in Fig. 2.6 is not strictly one-dimensional because $v$ remains very much less than $u$, the flow can be calculated with sufficient accuracy for most purposes by ignoring $v$ and assuming that the flow is one-dimensional, i.e., by only considering the variation of $u$ with $x$. Such flows in which the flow area is changing but in which the flow at any section can be treated as one-dimensional, are commonly referred to as "quasi-one-dimensional" flows.

## 2.2
## CONTROL VOLUME

The concept of a control volume is used in the derivation and application of many equations of compressible fluid flow. As discussed in the previous chapter, a control volume is an arbitrary imaginary volume fixed relative to the coordinate system being used (the coordinate system can be moving) and bounded by a control surface through which fluid may pass as shown in Fig. 2.7.

In applying the control volume concept, the effects of forces on the control surface and mass and energy transfers through this surface are considered. In

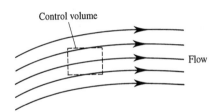

**FIGURE 2.7**
Control volume in a general two-dimensional flow.

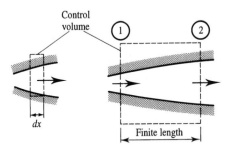

**FIGURE 2.8**
Types of control volume used in the analysis of one-dimensional duct flows.

general, it should be noted, it is possible for conditions in the control volume to be changing with time but for the reasons mentioned above attention will here be restricted to steady flow in which conditions inside and outside the control volume are constant in time in terms of the coordinate system being used.

In the case of one-dimensional duct flow that is here being considered, control volumes of the type shown in Fig. 2.8 are used. These control volumes either cover a differentially short length, $dx$, of the duct or a finite length of the duct as shown in Fig. 2.8.

In the case of the differentially short control volume, the changes in the flow variables through the control volume, such as those in velocity and pressure, i.e., $dV$ and $dp$, will also be small and in the analysis of the flow the products of these differentially small changes such as $dV \times d\rho$ will be neglected.

## 2.3
## CONTINUITY EQUATION

The continuity equation is obtained by applying the principle of conservation of mass to flow through a control volume. Consider the situation shown in Fig. 2.9. The changes through this control volume are indicated in Fig. 2.9, it being recalled that one-dimensional flow is being considered.

Since there is no mass transfer across the walls of the streamtube, the only mass transfer occurs through the ends of the control volume. If the possibility of a source of mass within the control volume is excluded, the principle of conservation of mass requires that the rate at which mass enters through the

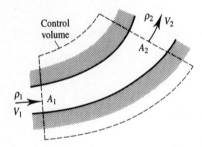

**FIGURE 2.9**
Control volume used in derivation of continuity equation.

left hand face of the control volume be equal to the rate at which mass leaves through the right hand face of the control volume, i.e., that:

$$\dot{m}_1 = \dot{m}_2 \qquad (2.1)$$

Since the rate at which mass crosses any section of the duct, i.e., $\dot{m}$, is equal to $\rho V A$ where $A$ is the cross-sectional area of the duct at the section considered, eq. (2.1) gives:

$$\rho_1 V_1 A_1 = \rho_2 V_2 A_2 \qquad (2.2)$$

For the differentially short control volume indicated in Fig. 2.10, this equation gives:

$$\rho V A = (\rho + d\rho)(V + dV)(A + dA)$$

i.e., neglecting higher order terms as discussed above:

$$VA\, d\rho + \rho A\, dV + \rho V\, dA = 0$$

Dividing this equation by $\rho V A$ then gives:

$$\frac{d\rho}{\rho} + \frac{dV}{V} + \frac{dA}{A} = 0 \qquad (2.3)$$

This equation relates the fractional changes in density, velocity, and area over a short length of the control volume. If the density can be assumed constant, this equation indicates that the fractional changes in velocity and area have opposite signs, i.e., if the area increases the velocity will decrease and vice versa. However, eq. (2.3) indicates that in compressible flow, where the

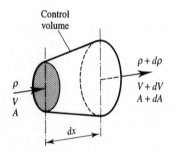

**FIGURE 2.10**
Differentially short control volume used in derivation of continuity equation.

fractional change in density is significant, no such simple relation between area and velocity changes exists.

## 2.4
## MOMENTUM EQUATION (EULER'S EQUATION)

Euler's equation is obtained by applying conservation of momentum to a control volume which again consists of a short length, $dx$, of a streamtube. Steady flow is again assumed. The forces acting on the control volume are shown in Fig. 2.11.

Because the flow is steady, conservation of momentum requires that for this control volume the net force in direction $x$ be equal to the rate at which momentum leaves the control volume in the $x$ direction minus the rate at which it enters in the $x$ direction since the flow is steady. Since, by the fundamental assumptions previously listed, gravitational forces are being neglected, the only forces acting on the control volume are the pressure forces and the frictional force exerted on the surface of the control volume. Thus, the net force on the control volume in the $x$ direction is:

$$pA - (p + dp)(A + dA) + \tfrac{1}{2}[p + (p + dp)][(A + dA) - A] - dF_\mu \qquad (2.4)$$

The third term in this equation represents the component of the force due to the pressure on the curved outer surface of the streamtube in the $x$ direction. It will be equal to the mean pressure on this curved surface multiplied by the projected area of this curved surface as illustrated in Fig. 2.12. Since $dx$ is small, the mean pressure on the curved surface can be taken as the average of the pressures acting on the two end surfaces, $0.5[p + (p + dp)]$, i.e., as $p + dp/2$ as indicated in Fig. 2.12. The term $dF_\mu$ is the frictional force.

Rearranging eq. (2.4) then gives the net force on the control volume in the $x$ direction as:

$$-A \, dp - dF_\mu \qquad (2.5)$$

In writing this equation, the higher order terms such as $dp \, dA$ have again been neglected since $dx$ is taken to be small.

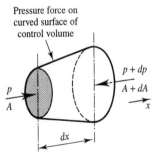

Pressure force on curved surface of control volume

**FIGURE 2.11**
Differentially short control volume used in derivation of momentum equation.

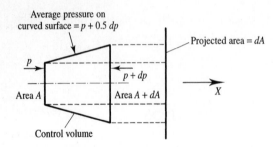

Average pressure on
curved surface $= p + 0.5\, dp$

Projected area $= dA$

$p$

$p + dp$

Area $A$

Area $A + dA$

$X$

Control volume

**FIGURE 2.12**
Pressure force on curved surface of control volume.

Since the rate at which momentum crosses any section of the duct is equal to $\dot{m}V$, the difference between the rate at which momentum leaves the control volume and the rate at which momentum enters the control volume is given by:

$$\rho V A[(V + dV) - V] = \rho V a\, dV \qquad (2.6)$$

since no momentum enters through the curved walls of the control volume.

Hence, since conservation of momentum requires that the net force on the control volume be equal to the rate at which momentum leaves the control volume minus the rate at which it enters the control volume, the above equations give:

$$-A\, dp - dF_\mu = \rho V A\, dV \qquad (2.7)$$

As discussed in the previous chapter, viscous friction effects will be neglected in the initial portion of this book, i.e., the term $dF_\mu$ in eq. (2.7) is assumed to be negligible. In this case, eq. (2.7) can be rearranged to give:

$$-\frac{dp}{\rho} = V\, dV \qquad (2.8)$$

This is Euler's equation for steady flow through a duct. Since $V$ is, by the choice of the $x$ direction, always positive, i.e., the $x$ direction is taken in the direction of the flow, this equation indicates that an increase in velocity is always associated with a decrease in pressure and vice versa. This is an obvious result because the decrease in pressure is required to generate the force needed to accelerate the flow, i.e., to increase the velocity, and vice versa.

If Euler's equation is integrated in the $x$ direction along the streamtube, it gives:

$$\frac{V^2}{2} + \int \frac{dp}{\rho} = \text{constant} \qquad (2.9)$$

In order to evaluate the integral, the variation of density with pressure has to be known. If the flow can be assumed to be incompressible, i.e., if the density can be assumed constant, this equation gives:

$$\frac{V^2}{2} + \frac{p}{\rho} = \text{constant} \qquad (2.10)$$

which is, of course, Bernoulli's equation. It should, therefore, be clearly under-stood that Bernoulli's equation only applies in incompressible flow.

## 2.5
## STEADY FLOW ENERGY EQUATION

This states that, for flow through the type of control volume considered above, if the fluid enters at section 1 with velocity $V_1$ and with enthalpy $h_1$ per unit mass, and leaves through section 2 with velocity $V_2$ and enthalpy $h_2$ then:

$$h_2 + \frac{V_2^2}{2} = h_1 + \frac{V_1^2}{2} + q - w \qquad (2.11)$$

where $q$ is the heat transferred into the control volume per unit mass of fluid flowing through it and $w$ is the work done by the fluid per unit mass in flowing through the control volume. In the present book attention will be restricted to flows in which no work is done so that $w$ is zero. Further, since only calorically perfect gases are being considered in this chapter:

$$h = c_p T \qquad (2.12)$$

Hence, the steady flow energy equation for the present purposes can be written as:

$$c_p T_2 + \frac{V_2^2}{2} = c_p T_1 + \frac{V_1^2}{2} + q \qquad (2.13)$$

Applying this equation to the flow through the differentially short control volume shown in Fig. 2.13 gives:

$$c_p T + \frac{V^2}{2} + dq = c_p(T + dT) + \frac{(V + dV)^2}{2} \qquad (2.14)$$

If higher order terms are again neglected, i.e., if $dV^2$ is neglected because the length of the control volume, $dx$, is very small, eq. (2.14) gives:

$$c_p \, dT + V \, dV = dq \qquad (2.15)$$

This equation indicates that in compressible flows, changes in velocity will, in general, induce changes in temperature and that heat addition can cause velocity changes as well as temperature changes.

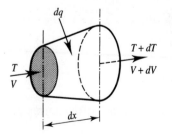

**FIGURE 2.13**
Differentially short control volume used in derivation of energy equation.

If the flow is adiabatic, i.e., if there is no heat transfer to or from the flow, equation (2.11) gives:

$$c_p T_2 + \frac{V_2^2}{2} = c_p T_1 + \frac{V_1^2}{2} \tag{2.16}$$

while equation (2.15) gives for adiabatic flow:

$$c_p \, dT + V \, dV = 0 \tag{2.17}$$

This equation shows that in adiabatic flow, an increase in velocity is always accompanied by a decrease in temperature.

**EXAMPLE 2.1**
Air flows down a variable area duct. Measurements indicate that the temperature is 5°C and the velocity is 150 m/s at a certain section of the duct. Measurements at a second section indicate that the temperature has decreased to −20°C. Assuming that the flow is adiabatic and one-dimensional, find the velocity at this second section.

*Solution*
The energy equation gives:

$$\frac{V_2^2}{2} = c_p T_1 + \frac{V_1^2}{2} - c_p T_2$$

Hence:     $$\frac{V_2^2}{2} = 1006 \times 278 + \frac{150^2}{2} - 1006 \times 253$$

$c_p$ having been taken as 1006 J/kg·°C. From the above equation it follows that:

$$V_2 = 269.8 \text{ m/s}$$

# 2.6
# EQUATION OF STATE

When applied between any two points in the flow, this equation gives:

$$\frac{p_1}{\rho_1 T_1} = \frac{p_2}{\rho_2 T_2} \tag{2.18}$$

When applied between the inlet and the exit of a differentially short control volume, this equation becomes:

$$\frac{p}{\rho T} = \frac{p + dp}{(\rho + d\rho)(T + dT)}$$

Since $dp/p$, $d\rho/\rho$, and $dT/T$ are small, this gives when higher order terms are neglected:

$$\frac{p}{\rho T} = \frac{p}{\rho T}\left(1 + \frac{dp}{p}\right)\left(1 - \frac{d\rho}{\rho}\right)\left(1 - \frac{dT}{T}\right)$$

i.e.
$$\frac{dp}{p} - \frac{d\rho}{\rho} - \frac{dT}{T} = 0 \qquad (2.19)$$

This equation shows how the changes in pressure, density, and temperature are interrelated in compressible flows.

### EXAMPLE 2.2
Consider adiabatic air flow through a duct. At a certain section of the duct, the flow area is 0.2 m², the pressure is 80 kPa, the temperature is 5°C and the velocity is 200 m/s. If, at this section, the duct area is changing at a rate of 0.3 m²/m (i.e., $dA/dx = 0.3$ m²/m) find $dp/dx$, $dV/dx$, and $d\rho/dx$ (a) assuming incompressible flow and (b) taking compressibility into account.

### Solution
The continuity equation gives:

$$\rho V \frac{dA}{dx} + \rho A \frac{dV}{dx} + VA\frac{d\rho}{dx} = 0$$

Hence, since using the information supplied:

$$\rho = \frac{p}{RT} = \frac{80 \times 10^3}{287 \times 278} = 1.003 \text{ kg/m}^3$$

it follows that

$$1.003 \times 200 \times 0.3 + 1.003 \times 0.2 \times \frac{dV}{dx} + 200 \times 0.2 \times \frac{d\rho}{dx} = 0 \qquad \text{(a)}$$

(a) Assuming incompressible flow, i.e., assuming that $d\rho/dx = 0$, the above equation gives:

$$\frac{dV}{dx} = -\frac{200 \times 0.3}{0.2} = -300 (\text{m/s})/\text{m}$$

Also, since in incompressible flow, the conservation of momentum equation gives:

$$-\frac{1}{\rho}\frac{dp}{dx} = V\frac{dV}{dx}$$

From this it follows that

$$\frac{dp}{dx} = 200 \times 300 \times 1.003$$

i.e.,
$$\frac{dp}{dx} = 6.02 \times 10^4 \text{ Pa/m} = 60.2 \text{ kPa/m}$$

(b) With compressibility effects accounted for, i.e., $dp/dx \neq 0$. The conservation of momentum equation still gives:

$$-\frac{1}{\rho}\frac{dp}{dx} = V\frac{dV}{dx}, \quad \text{i.e.,} \quad \frac{dp}{dx} = -\rho V\frac{dV}{dx}$$

Conservation of energy gives:

$$c_p \frac{dT}{dx} + V \frac{dV}{dx} = 0$$

From the perfect gas law, $p = \rho RT$, it follows that:

$$\frac{dp}{dx} = R\left\{ T \frac{d\rho}{dx} + \rho \frac{dT}{dx} \right\}$$

i.e., using the momentum equation result:

$$-\rho V \frac{dV}{dx} = R\left\{ T \frac{d\rho}{dx} + \rho \frac{dT}{dx} \right\}$$

But conservation of energy gives

$$\frac{dT}{dx} = -\frac{V}{c_p} \frac{dV}{dx}$$

so the above equation becomes:

$$-\rho V \frac{dV}{dx} = R\left\{ T \frac{d\rho}{dx} - \frac{\rho V}{c_p} \frac{dV}{dx} \right\}$$

which can be rearranged to give:

$$\frac{d\rho}{dx} = \left\{ \frac{V}{c_p} - \frac{V}{R} \right\} \frac{\rho}{T} \frac{dV}{dx}$$

Substituting the given values of the variables into this equation then gives:

$$\frac{d\rho}{dx} = \left\{ \frac{200}{1004} - \frac{200}{287} \right\} \frac{1.003}{278} \frac{dV}{dx} = -1.795 \times 10^{-3} \frac{dV}{dx}$$

Substituting this back into equation (a) then gives:

$$60.18 + 1.003 \times 0.2 \times \frac{dV}{dx} - 200 \times 0.2 \times (1.795 \times 10^{-3}) \frac{dV}{dx} = 0$$

Therefore:

$$\frac{dV}{dx} = -467.3 \frac{\text{m/s}}{\text{m}}$$

Substituting back then gives:

$$\frac{d\rho}{dx} = (-1.795 \times 10^{-3}) \div (-467.3) = 0.839 \frac{\text{kg/m}^3}{\text{m}}$$

and

$$\frac{dp}{dx} = -\rho V \frac{dV}{dx} = 9.37 \times 10^4 \text{ Pa/m} = 93.7 \text{ kPa/m}$$

The values obtained, taking compressibility into account, are therefore very different from those obtained when compressibility is ignored.

## 2.7
## ENTROPY CONSIDERATIONS

In studying compressible flows, another variable, the entropy, $s$, has to be introduced. The entropy basically places limitations on which flow processes are physically possible and which are physically excluded. The entropy change between any two points in the flow is given by:

$$s_2 - s_1 = c_p \ln \left[ \frac{T_2}{T_1} \right] - R \ln \left[ \frac{p_2}{p_1} \right] \tag{2.20}$$

Since $R = c_p - c_v$, this equation can be written:

$$\frac{s_2 - s_1}{c_p} = \ln \left[ \left( \frac{T_2}{T_1} \right) \left( \frac{p_2}{p_1} \right)^{-\frac{\gamma-1}{\gamma}} \right]$$

If there is no change in entropy, i.e., if the flow is isentropic, this equation requires that:

$$\frac{T_2}{T_1} = \left( \frac{p_2}{p_1} \right)^{\frac{\gamma-1}{\gamma}} \tag{2.21}$$

Hence, since the perfect gas law gives:

$$\frac{T_2}{T_1} = \frac{p_2 \, \rho_1}{p_1 \, \rho_2}$$

it follows that in isentropic flow:

$$\frac{p_2}{p_1} = \left( \frac{\rho_2}{\rho_1} \right)^{\gamma} \tag{2.22}$$

In isentropic flows, then, $p/\rho^\gamma$ is a constant.

If eq. (2.20) is applied between the inlet and the exit of a differentially short control volume, it gives:

$$(s + ds) - s = c_p \ln \left[ \frac{T + dT}{T} \right] - R \ln \left[ \frac{p + dp}{p} \right]$$

Since, if $\epsilon$ is a small quantity, $\ln (1 + \epsilon)$ is to first order equal to $\epsilon$, the above equation gives:

$$ds = c_p \frac{dT}{T} - R \frac{dp}{p} \tag{2.23}$$

which can be written as:

$$\frac{ds}{c_p} = \frac{dT}{T} - \left( \frac{\gamma - 1}{\gamma} \right) \frac{dp}{p} \tag{2.24}$$

Lastly, it is noted that in an isentropic flow, eq. (2.23) gives:

$$c_p \, dT = \frac{RT}{p} dp$$

i.e., using the perfect gas law:

$$c_p \, dT = \frac{dp}{\rho} \tag{2.25}$$

But the energy equation for isentropic flow, i.e., for flow with no heat transfer gives eq. (2.17):

$$c_p \, dT + V \, dV = 0$$

which using eq. (2.25) gives:

$$\frac{dp}{\rho} + V \, dV = 0$$

Comparing this to eq. (2.8) shows that this is identical to the result obtained using conservation of momentum considerations. In isentropic flow, then, it is not necessary to consider both conservation of energy and conservation of momentum since, when the "isentropic equation of state," i.e., eq. (2.22), is used they give the same result.

**EXAMPLE 2.3**

What is the form of the relation between the pressure and the velocity at two points in a flow if the flow is (a) isothermal and (b) isentropic?

*Solution*

The integrated Euler equation gives:

$$\frac{V^2}{2} + \int \frac{dp}{\rho} = \text{constant}$$

In isothermal flow, the perfect gas equation gives $p/\rho = RT$ a constant. Hence, in this case, the Euler equation gives:

$$\frac{V^2}{2} + RT \int \frac{dp}{p} = \text{constant}$$

i.e.,

$$\frac{V^2}{2} + RT \ln p = \text{constant}$$

Applying this equation between two points in a flow then gives

$$\frac{V_2^2}{2} - \frac{V_1^2}{2} = RT \ln \left( \frac{p_2}{p_1} \right)$$

Similarly, in isentropic flow, since $p \propto \rho^\gamma$, the integrated Euler equation gives:

$$\frac{V^2}{2} + \int \frac{dp}{p^{1/\gamma}} = \text{constant}$$

i.e.,

$$\frac{V^2}{2} + \frac{\gamma}{\gamma - 1} p^{\frac{\gamma-1}{\gamma}} = \text{constant}$$

Applying this equation between two points in a flow then gives

$$\frac{V_2^2}{2} - \frac{V_1^2}{2} = \frac{\gamma}{\gamma - 1} \left( p_2^{\frac{\gamma-1}{\gamma}} - p_1^{\frac{\gamma-1}{\gamma}} \right)$$

## 2.8
## USE OF THE ONE-DIMENSIONAL FLOW EQUATIONS

The most obvious application of the quasi-one-dimensional equations is to flow through a solid walled duct or a streamtube whose cross-sectional area is changing slowly with distance as shown in Fig. 2.14.

For the one-dimensional flow assumption to be valid, the rate of change of duct area with respect to the distance $x$ along the duct must remain small. However, in applying the one-dimensional flow equations to the flow through a duct, it should be noted that the flow does not have to be one-dimensional at all sections of the duct in order to use the one-dimensional flow equations. For

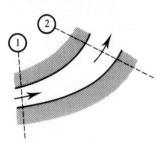

**FIGURE 2.14**
One-dimensional flow through a duct.

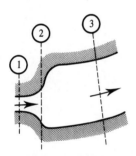

**FIGURE 2.15**
Duct in which one-dimensional flow assumptions are not valid throughout the flow.

example consider the situation shown in Fig. 2.15. The flow at section 2 in Fig. 2.15 cannot be assumed one-dimensional. However, the conditions at sections 1 and 3 can be related by the one-dimensional equation.

The one-dimensional equations also, as discussed above, apply to the flow through any streamtube. An example of a streamtube is shown in Fig. 2.16. This figure shows streamlines in a two-dimensional low speed flow over a "streamlined" cylinder. The body is long in the direction normal to the

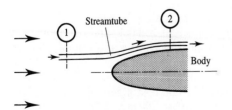

**FIGURE 2.16**
One-dimensional flow through a streamtube.

page and the flow pattern, therefore, essentially depends only on the $x$- and $y$-coordinates indicated. The flow along the streamtube shown in this figure will, however, be one-dimensional. As the fluid flows along this streamtube, its area changes and there are associated changes in the pressure, temperature, and density.

## 2.9
## CONCLUDING REMARKS

The equations discussed in the present chapter, although only strictly applicable to flows that are one-dimensional, still form the basis of the analysis to an acceptable degree of accuracy of many compressible fluid flows that occur in engineering practice. The equations clearly indicate how, in compressible flows, changes in temperature and density are interlinked with changes in the velocity field.

## PROBLEMS

**2.1.** Air enters a tank at a velocity of 100 m/s and leaves the tank at a velocity of 200 m/s. If the flow is adiabatic find the difference between the temperature of the air at exit and the temperature of the air at inlet.

**2.2.** Air at a temperature of 25°C is flowing at a velocity of 500 m/s. A shock wave (see later chapters) occurs in the flow reducing the velocity to 300 m/s. Assuming the flow through the shock wave to be adiabatic, find the temperature of the air behind the shock wave.

**2.3.** Air being released from a tire through the valve is found to have a temperature of 15°C. Assuming that the air in the tire is at the ambient temperature of 30°C, find the velocity of the air at the exit of the valve. The process can be assumed to be adiabatic.

**2.4.** A gas with a molecular weight of 4 and a specific heat ratio of 1.67 flows through a variable area duct. At some point in the flow the velocity is 180 m/s and the temperature is 10°C. At some other point in the flow the temperature is −10°C. Find the velocity at this point in the flow assuming that the flow is adiabatic.

**2.5.** At a section of a circular duct through which air is flowing the pressure is 150 kPa, the temperature is 35°C, the velocity is 250 m/s, and the diameter is 0.2 m. If, at this section, the duct diameter is increasing at a rate of 0.1 m/m, find $dp/dx$, $dV/dx$, and $d\rho/dx$.

**2.6.** Consider an isothermal air flow through a duct. At a certain section of the duct the velocity, temperature, and pressure are 200 m/s, 25°C, and 120 kPa, respectively. If the velocity is decreasing at this section at a rate of 30 percent per m, find $dp/dx$, $ds/dx$, and $d\rho/dx$.

**2.7.** Consider adiabatic air flow through a variable area duct. At a certain section of the duct the flow area is $0.1 \text{ m}^2$, the pressure is 120 kPa, the temperature is 15°C, and the duct area is changing at a rate of $0.1 \text{ m}^2/\text{m}$. Plot the variations of $dp/dx$, $dV/dx$, and $d\rho/dx$ with the velocity at the section for velocities between 50 m/s and 300 m/s.

# CHAPTER 3

# Some Fundamental Aspects of Compressible Flow

## 3.1
## INTRODUCTION

It was indicated in the previous chapters that compressibility effects become important in a gas flow when the velocity in the flow is high. An attempt will be made in the present chapter to show that it is not the value of the gas velocity itself but rather the ratio of the gas velocity to the speed of sound in the gas that determines when compressibility is important. This ratio is termed the Mach number, $M$, i.e.,

$$M = \frac{\text{gas velocity}}{\text{speed of sound}} = \frac{V}{a} \tag{3.1}$$

where $a$ is the speed of sound.

If $M < 1$ the flow is said to be subsonic, whereas if $M > 1$ the flow is said to be supersonic. If the Mach number is near 1 and there are regions of both subsonic and supersonic flow, the flow is said to be transonic. If the Mach number is very much greater than 1, the flow is said to be hypersonic. Hypersonic flow is normally associated with flows in which $M > 5$.

As will be shown later in this chapter, the speed of sound in a perfect gas is given by

$$a = \sqrt{\frac{\gamma p}{\rho}} = \sqrt{\gamma R T} \tag{3.2}$$

The speed of sound in a gas depends, therefore, only on the absolute temperature of the gas.

## 3.2
## ISENTROPIC FLOW IN A STREAMTUBE

In order to illustrate the importance of the Mach number in determining the conditions under which compressibility must be taken into account, isentropic flow, i.e., frictionless adiabatic flow, through a streamtube will be first considered. The changes in the flow variables over a short length, $dx$, of the streamtube shown in Fig. 3.1 are considered.

The Euler equation, eq. (2.8), derived in the previous chapter by applying the conservation of momentum principle and ignoring the effects of friction gives:

$$\frac{dp}{p} = -\frac{\rho V^2}{p}\frac{dV}{V} \tag{3.3}$$

Using the expression for the speed of sound, $a$, given in eq. (3.2) allows this equation to be written as:

$$\frac{dp}{p} = -\gamma\frac{V^2}{a^2}\frac{dV}{V} \tag{3.4}$$

but $M = V/a$ so the above equation can be written as:

$$\frac{dp}{p} = -\gamma M^2\frac{dV}{V} \tag{3.5}$$

This equation shows that the magnitude of the fractional pressure change, $dp/p$, induced by a given fractional velocity change, $dV/V$, depends on the square of the Mach number.

Next consider the energy equation. Since adiabatic flow is being considered, eq. (2.17) gives:

$$\frac{dT}{T} = -\frac{V^2}{c_p T}\frac{dV}{V} \tag{3.6}$$

which can be rearranged to give:

$$\frac{dT}{T} = -\frac{\gamma R}{c_p}M^2\frac{dV}{V} \tag{3.7}$$

However, since $R = c_p - c_v$, i.e., since $R = 1 - 1/\gamma$ it follows that:

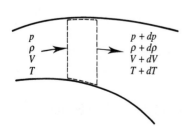

**FIGURE 3.1**
Portion of streamtube considered.

$$\frac{\gamma R}{c_p} = \gamma - 1 \tag{3.8}$$

Equation (3.7) can, therefore, be written as:

$$\frac{dT}{T} = -(\gamma - 1)M^2\frac{dV}{V} \tag{3.9}$$

This equation shows that the magnitude of the fractional temperature change, $dT/T$, induced by a given fractional velocity change, $dV/V$, also depends on the square of the Mach number.

Lastly, consider the equation of state. As shown in the previous chapter, this gives:

$$\frac{dp}{p} = \frac{d\rho}{\rho} + \frac{dT}{T} \tag{3.10}$$

Combining this equation with eqs. (3.5) and (3.9) then gives:

$$\frac{d\rho}{\rho} = -\gamma M^2\frac{dV}{V} + (\gamma - 1)M^2\frac{dV}{V} = -M^2\frac{dV}{V} \tag{3.11}$$

This equation indicates that:

$$\frac{d\rho/\rho}{dV/V} = -M^2$$

From this equation it will be seen that for a given fractional change in velocity, i.e., for a given $dV/V$, the corresponding induced fractional change in density will also depend on the square of the Mach number. For example, at a Mach number of 0.1 the fractional change in density will be 1 percent of the fractional change in velocity, at a Mach number of 0.33 it will be about 10 percent of this fractional change while at a Mach number of 0.4 it will be 16 percent of this fractional change. Therefore, at low Mach numbers, the density changes will be insignificant but as the Mach number increases, the density changes, i.e., compressibility effects, will become increasingly important. Hence, compressibility effects become important in high Mach number flows. The Mach number at which compressibility must start to be accounted for depends very much on the flow situation and the accuracy required in the solution. As a rough guide, it is sometimes assumed that if $M > 0.3$ then there is a possibility that compressibility effects should be considered.

It should also be noted that eq. (3.9) gives

$$\frac{dT/T}{dV/V} = -(\gamma - 1)M^2$$

This indicates that if the Mach number is high enough for density changes in the flow to be significant, the temperature changes in the flow will also be important.

It should be clear from the above results that the Mach number is the parameter that determines the importance of compressibility effects in a flow.

**EXAMPLE 3.1**
Consider the isentropic flow of air through a duct whose area is decreasing. Find the percentage changes in velocity, density, and pressure induced by a 1 percent reduction in area for Mach number between 0.1 and 0.95.

**Solution**
The continuity equation gives:

$$\frac{d\rho}{\rho} + \frac{dV}{V} + \frac{dA}{A} = 0$$

i.e., using the relation for $d\rho/\rho$ given above:

$$(1 - M^2)\frac{dV}{V} + \frac{dA}{A} = 0$$

therefore:

$$\frac{dV}{V} = -\frac{1}{(1 - M^2)}\frac{dA}{A}$$

Again using the expression for the density change given above gives:

$$\frac{d\rho}{\rho} = +\frac{M^2}{(1 - M^2)}\frac{dA}{A}$$

Then using the expression for the pressure change gives:

$$\frac{dp}{p} = +\frac{\gamma M^2}{(1 - M^2)}\frac{dA}{A}$$

In the present case $dA/A = -0.01$ so the above equations give:

$$\frac{dV}{V} = +\frac{0.01}{(1 - M^2)}$$

$$\frac{d\rho}{\rho} = -\frac{0.01 M^2}{(1 - M^2)}$$

$$\frac{dp}{p} = -\frac{0.01\gamma M^2}{(1 - M^2)}$$

Using these relations then gives the results shown in Table E3.1.

**TABLE E3.1**

| M | dV/V | dρ/ρ | dp/p |
|---|---|---|---|
| 0.1 | 0.010 10 | −0.000 10 | −0.000 14 |
| 0.2 | 0.010 42 | −0.000 42 | −0.000 58 |
| 0.3 | 0.010 99 | −0.000 99 | −0.001 39 |
| 0.4 | 0.011 91 | −0.001 91 | −0.002 67 |
| 0.5 | 0.013 33 | −0.003 33 | −0.004 67 |
| 0.6 | 0.015 63 | −0.005 63 | −0.007 88 |
| 0.7 | 0.019 61 | −0.009 61 | −0.013 45 |
| 0.8 | 0.027 78 | −0.017 78 | −0.024 89 |
| 0.9 | 0.052 63 | −0.042 63 | −0.059 68 |
| 0.95 | 0.102 56 | −0.092 56 | −0.129 59 |

The percentage change in density thus increases from 0.01 percent at a Mach number of 0.1 to 9 percent at a Mach number of 0.95. It will be seen that there is a singularity at $M = 1$. The consequences of this will be discussed later.

## 3.3
## SPEED OF SOUND

The importance of the Mach number in determining the characteristics of high speed gas flows was indicated in the preceding section. In order to calculate $M$, the local speed of sound in the gas has to be known. An expression for the speed of sound, which was used in the previous section, will therefore be derived in the present section.

The speed of sound is, of course, the speed at which very weak pressure waves are transmitted through the gas. Consider a plane infinitesimally weak pressure wave propagating through a gas. It could, for example, be thought of as a wave that is propagating down a duct after being generated by a small movement of a piston at one end of the duct as shown in Fig. 3.2.

More generally, a plane wave will be a small, effectively plane, portion of a spherical wave moving outwards through the gas from a point source of disturbance as indicated in Fig. 3.3.

Now from experience (this will actually be proved later in the present book), it is known that the wave remains steep, i.e., the longitudinal distance over which the changes produced by the wave occur remains small as indicated in Figs. 3.2 and 3.3.

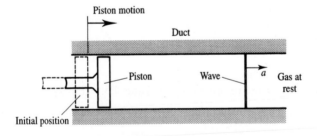

**FIGURE 3.2**
Generation of a weak pressure wave by the motion of a piston in a duct.

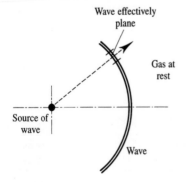

**FIGURE 3.3**
Portion of spherical pressure wave considered.

Let the pressure change across the wave be $dp$ and let the corresponding density and temperature changes be $d\rho$ and $dT$ respectively. The gas into which the wave is propagating is assumed to be at rest. The wave will then induce a gas velocity, $dV$, behind it as it moves through the gas. The changes across the wave are, therefore, as shown in Fig. 3.4.

In order to analyze the flow through the wave and thus to determine $a$, it is convenient to use a coordinate system that is attached to the wave, i.e., is moving with the wave. In this coordinate system, the wave will, of course, be at rest and the gas will effectively flow through it with the velocity, $a$, ahead of the wave and a velocity, $a - dV$, behind the wave. In this coordinate system, then, the changes through the wave are as shown in Fig. 3.5. The pressure, temperature, and density changes are, of course, independent of the coordinate system used.

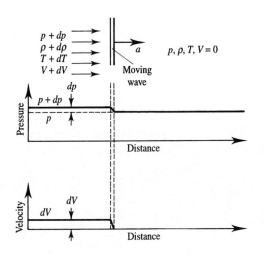

**FIGURE 3.4**
Changes through propagating weak pressure wave.

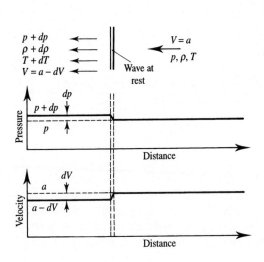

**FIGURE 3.5**
Changes relative to the weak pressure wave.

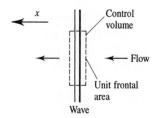

Control volume

Flow

Unit frontal area

Wave

**FIGURE 3.6**
Control volume used in analysis of flow through a weak pressure wave.

The continuity and momentum equations are applied to a control volume of unit area across the wave as indicated in Fig. 3.6. The continuity equation gives for this control volume:

$$\frac{\dot{m}}{A} = \rho a = (\rho + d\rho)(a - dV) \tag{3.12}$$

where $\dot{m}/A$ is the mass flow rate per unit area through the wave.

Since the case of a very weak wave is being considered, the second order term, i.e., $d\rho\, dV$ that arises in equation (3.12) can be neglected and this equation then gives:

$$d\rho = \frac{\rho}{a} dV \tag{3.13}$$

Conservation of momentum is next considered. The only forces acting on the control volume are the pressure forces. The momentum equation, therefore, gives since a control volume with unit area normal to the flow direction is being considered:

$$p - (p + dp) = \left(\frac{\dot{m}}{A}\right)[(a - dV) - a]$$

which leads to:

$$dp = \left(\frac{\dot{m}}{A}\right) dV \tag{3.14}$$

Hence, using eq. (3.12), conservation of momentum gives:

$$dp = \rho a\, dV \tag{3.15}$$

Dividing eq. (3.15) by eq. (3.13) then gives the speed of sound as:

$$\frac{dp}{d\rho} = a^2, \quad \text{i.e.,} \quad a = \sqrt{\frac{dp}{d\rho}} \tag{3.16}$$

In order to evaluate $a$ using the above equation, it is necessary to know the process that the gas undergoes in passing through the wave. Because a very weak wave is being considered, the temperature and velocity changes through the wave are very small and the gradients of temperature and velocity within the wave remain small. For this reason, heat transfer and viscous effects for flow through the wave are assumed to be negligible. Hence, in passing through the wave, the gas is assumed to undergo an isentropic process. The flow through the wave is, therefore, assumed to satisfy:

$$\frac{p}{\rho^\gamma} = C \qquad \text{(a constant)} \tag{3.17}$$

Differentiating this then gives:

$$\frac{dp}{d\rho} = \gamma C \rho^{\gamma-1} = \frac{\gamma p}{\rho} \tag{3.18}$$

Substituting this into eq. (3.16) then gives:

$$a = \sqrt{\frac{\gamma p}{\rho}} = \sqrt{\gamma R T} \tag{3.19}$$

Experimental measurements of the speed of sound in gases are in good agreement with the values given by this equation, thus confirming that the assumptions made in its derivation are justified.

For a given gas, therefore, the speed of sound depends only on the square root of the absolute temperature. Further, since the equation for the speed of sound can be written as:

$$a = \sqrt{\gamma \frac{\mathcal{R}}{m} T} \tag{3.20}$$

where $m$ is the molar mass of the gas, it follows that, since $\gamma$ does not vary greatly between gases, the speed of sound of a gas at a given temperature is approximately inversely proportional to its molar mass.

Some typical values for the speed of sound at 0°C are shown in Table 3.1.

**TABLE 3.1**

| Gas | Molar mass | $\gamma$ | Speed of sound at 0°C (m/s) |
|---|---|---|---|
| Air | 28.960 | 1.404 | 331 |
| Argon (Ar) | 39.940 | 1.667 | 308 |
| Carbon dioxide ($CO_2$) | 44.010 | 1.300 | 258 |
| Freon 12 ($CCl_2F_2$) | 120.900 | 1.139 | 146 |
| Helium (He) | 4.003 | 1.667 | 970 |
| Hydrogen ($H_2$) | 2.016 | 1.407 | 1270 |
| Xenon (Xe) | 131.300 | 1.667 | 170 |

**EXAMPLE 3.2**
An aircraft is capable of flying at a maximum Mach number of 0.91 at sea-level. Find the maximum velocity at which this aircraft can fly at sea-level if the air temperature is (a) 5°C and (b) 45°C.

***Solution***
Since

$$M_{max} = \frac{V_{max}}{a}$$

it follows that

$$V_{max}|_{\text{sea-level}} = M_{max} \times a_{\text{sea-level}} = 0.91\sqrt{\gamma R T_{\text{sea-level}}}$$

When $T_{\text{sea-level}} = 5°C = 278$ K the above equation gives:

$$V_{max}|_{\text{sea-level}} = 0.91\sqrt{1.4 \times 287 \times 278} = 304 \text{ m/s}$$

Similarly, when $T_{\text{sea-level}} = 45°C = 318$ K the above equation gives:

$$V_{max}|_{\text{sea-level}} = 0.91\sqrt{1.4 \times 287 \times 318} = 325 \text{ m/s}$$

In the days when world speed records for aircraft were established when flying at ground level, if there was a limiting Mach number on the aircraft it therefore paid to attempt to establish the record on a hot day.

### EXAMPLE 3.3
An aircraft is driven by propellers with a diameter of 4 m. At what engine speed will the tips of the propellers reach sonic velocity if the air temperature is 15°C?

### Solution
Since the tip moves through a distance of $\pi D$ in one revolution, the tip speed is given by:

$$V = n\pi D$$

where $n$ is the rotational speed in revolutions per second. Hence, if the Mach number at the tip is 1:

$$M = 1, \qquad V = a = \sqrt{\gamma R T}$$

Combining the above results then gives:

$$n = \frac{\sqrt{\gamma R T}}{\pi D} = \frac{\sqrt{1.4 \times (8314/29) \times 288}}{\pi \times 4}$$

$$= 27.06 \text{ revs/sec} = 1623 \text{ revs/min}$$

### EXAMPLE 3.4
In evaluating the performance of an aircraft, a "standard atmosphere" is usually introduced. The conditions in the "standard atmosphere" are meant to represent average conditions in the atmosphere. In the U.S. Standard Atmosphere, the temperature in the inner portion of the atmosphere is defined by the following two equations:

For altitudes, $H$, of from 0 m (sea-level) to 11 019 m:

$$T = 288.16 - 0.0065H$$

Above an altitude, $H$, of 11 019 m:

$$T = 216.66$$

The altitude, $H$, is measured in meters and the temperature, $T$, in K.

Plot a graph showing how the speed of sound varies with altitude in this atmosphere for altitudes from sea-level to 12 000 m.

### Solution
Using $a = \sqrt{\gamma R T}$ the following is obtained:

$$a = \sqrt{1.4 \times 287 \times (288.16 - 0.0065H)} \qquad \text{for } 0 \le H \le 11\,019 \text{ m}$$

$$a = \sqrt{1.4 \times 287 \times 216.66} \qquad \text{for } H > 11\,019 \text{ m}$$

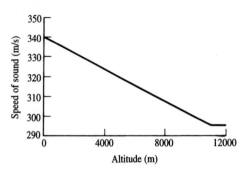

**FIGURE E3.4**
Variation of speed of sound with altitude.

This has been used to derive the variation of the speed of sound, $a$, with altitude, $H$. The result is shown in Fig. E3.4. It will be noted that

$$a|_{H=0} = 340 \text{ m/s}, \qquad a|_{H=11\,019} = 294.9 \text{ m/s}$$

**EXAMPLE 3.5**
Typical cruising speeds and altitudes for three commercial aircraft are:

- Dash 8: Cruising speed: 500 km/h at an altitude of 4570 m.
- Boeing 747: Cruising speed: 978 km/h at an altitude of 9150 m.
- Concorde: Cruising speed: 2340 km/h at an altitude of 16 600 m.

Find the Mach number of these three aircraft when flying at these cruise conditions. Use the properties of the standard atmosphere discussed in the previous problem.

**Solution**
Dash 8:

$$V = 500 \text{ km/h} = 138.9 \text{ m/s}$$

$$H = 4570 \text{ m}$$

$$T = 288.16 - 0.0065 \times 4570 = 258.4 \text{ K}$$

$$a = \sqrt{1.4 \times \frac{8314}{29} \times 258.4} = 322.0 \text{ m/s}$$

$$M = V/a = 0.431$$

Boeing 747:

$$V = 978 \text{ km/h} = 271.7 \text{ m/s}$$

$$H = 9150 \text{ m}$$

$$T = 288.16 - 0.0065 \times 9150 = 228.7 \text{ K}$$

$$a = \sqrt{1.4 \times \frac{8314}{29} \times 228.7} = 303.0 \text{ m/s}$$

$$M = V/a = 0.897$$

Concorde:

$$V = 2340 \text{ km/h} = 650.0 \text{ m/s}$$

$$H = 16\,600$$

$$T = 216.66 \text{ K}$$

$$a = \sqrt{1.4 \times \frac{8314}{29} \times 216.66} = 294.9 \text{ m/s}$$

$$M = V/a = 2.204$$

**EXAMPLE 3.6**

A weak pressure wave (a sound wave) across which the pressure rise is 0.05 kPa is traveling down a pipe into air at a temperature of 30°C and a pressure of 105 kPa. Estimate the velocity of the air behind the wave.

***Solution***

Consider the analysis of a sound wave presented above. As discussed before, conservation of mass gives for the flow through the control volume shown in Fig. 3.6, which has unit frontal area:

$$\dot{m} = \rho a = [(\rho + d\rho)(a - dV)]$$

while conservation of momentum gives:

$$p - (p + dp) = \dot{m}[(a - dV) - a]$$

Therefore, $dp = \dot{m}\, dV = \rho a\, dV$. The velocity behind the sound wave, $dV$, is therefore given by:

$$dV = \frac{1}{\rho a} dp$$

But, for the values given:

$$\rho = \frac{p}{RT} = \frac{105 \times 10^3}{287 \times 303} = 1.209 \text{ kg/m}^3$$

and:

$$a = \sqrt{\gamma RT} = \sqrt{1.4 \times 287 \times 303} = 348.7 \text{ m/s}$$

Therefore:

$$dV = \frac{1}{1.209 \times 348.7} \times 0.05 \times 10^3 = 0.119 \text{ m/s}$$

The velocity behind the wave is therefore 0.119 m/s.

## 3.4
## MACH WAVES

Consider a small solid body moving relative to a gas. In order for the gas to pass smoothly over the body, disturbances tend to be propagated ahead of the body to "warn" the gas of the approach of the body, i.e., because the pressure at the surface of the body is greater than that in the surrounding gas, pressure waves spread out from the body. Since these pressure waves are very weak except in the immediate vicinity of the body, they effectively move outwards at the speed of sound.

In order to illustrate the effect of the velocity of the body relative to the speed of sound on the flow field, consider the small body, i.e., essentially a point source of disturbance, to be moving at a uniform linear velocity, $u$, through the gas and let the speed of sound in the gas be $a$. Although the body is essentially emitting waves continuously, a series of waves emitted at time intervals $t$ will be considered. Since the body is moving through the gas, the origin of these waves will be continually changing. Waves generated at times 0, $t$, $2t$, and $3t$ will be considered. First, consider the case where the speed of the body is less than the speed of sound, i.e., the Mach number is less than 1. The body is at positions $a$, $b$, $c$, and $d$ at the four times considered. Since the waves spread radially outwards from their point of origin at the speed of sound, the wave pattern at time $3t$ will be as shown in Fig. 3.7.

Next consider the case where the small body is moving faster than the speed of sound, i.e., where the Mach number is greater than 1. In this case, the wave pattern at time $3t$ will be as shown in Fig. 3.8.

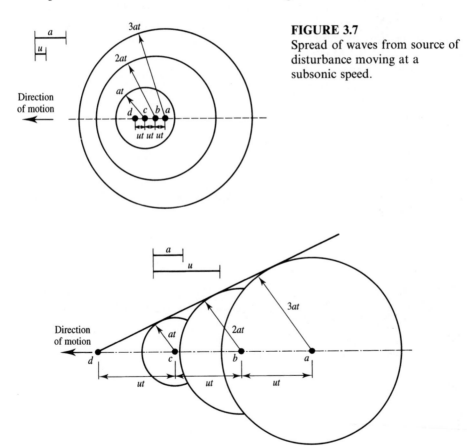

**FIGURE 3.7**
Spread of waves from source of disturbance moving at a subsonic speed.

**FIGURE 3.8**
Spread of waves from source of disturbance moving at a supersonic speed.

It will be seen that all the waves lie within the cone indicated, which has its vertex at the body at the time considered. Only the gas that lies within this cone is "aware" of the presence of the body. This cone has a vertex angle, $\alpha$, which will be seen to be given by:

$$\sin \alpha = \frac{a}{V} = \frac{1}{M} \tag{3.21}$$

The angle $\alpha$ is termed the Mach angle.

A comparison of the results shown in Figs. 3.7 and 3.8 shows the importance of the Mach number in determining the nature of the flow field.

If the body is at rest and the gas is moving over it at a supersonic velocity, all the disturbances generated by the body are swept downstream and lie within the Mach cone shown in Fig. 3.9. There will be essentially jumps in the values of the flow variables when the flow reaches the cone. The cone is therefore termed a conical Mach wave.

Similarly, in two-dimensional flow, all waves originating at a weak line source of disturbance will all lie behind a plane wave inclined at the Mach angle to the flow as shown in Fig. 3.10. This result is sometimes used in the measurement of the Mach number of a gas flow. A small irregularity is put in the surface or occurs in the surface and the Mach wave generated at this source of disturbance is made visible by a suitable optical technique. By measuring the angle made by the wave generated at the disturbance to the oncoming flow, the Mach number can be found using:

$$M = \frac{1}{\sin \alpha} \tag{3.22}$$

Mach waves originating from small changes in wall shape are shown in Fig. 3.11. In addition to the Mach waves, it will be seen from Fig. 3.11 that strong

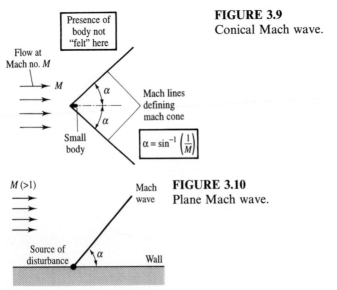

**FIGURE 3.9**
Conical Mach wave.

**FIGURE 3.10**
Plane Mach wave.

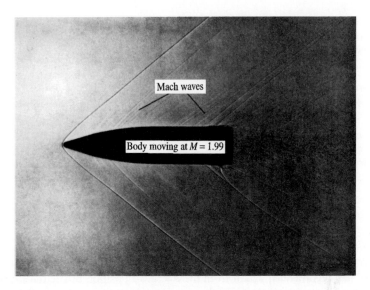

**FIGURE 3.11**
Mach waves originating at the surface of a body moving at supersonic velocity.

waves are generated near the leading edge of the body. The analysis of such waves will be discussed later.

**EXAMPLE 3.7**
Air at a temperature of −10°C flows through a supersonic wind tunnel. A shadowgraph (see Appendix K) photograph of the flow reveals weak waves originating at imperfections on the walls. These weak waves are at an angle of 40° to the flow. Find the Mach number and velocity in the wind tunnel.

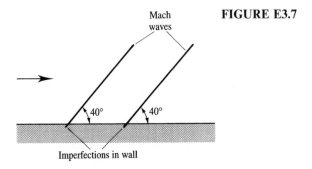

FIGURE E3.7

*Solution*

$$M = \frac{1}{\sin \alpha} = \frac{1}{\sin 40°} = 1.556$$

Using this then gives:

$$V = Ma = M\sqrt{\gamma RT} = 1.556 \times \sqrt{1.4 \times \frac{8314}{29} \times 263} = 505.5 \text{ m/s}$$

**EXAMPLE 3.8**
A gas has a molar mass of 44 and a specific heat ratio of 1.3. Find the speed of
sound in this gas if the gas temperature is $-30°C$. If this gas is flowing at a velocity
of 450 m/s, find the Mach number and the Mach angle.

*Solution*
The speed of sound is given by:

$$a = \sqrt{\gamma RT} = \sqrt{1.3 \times \frac{8314}{44} \times 243} = 244.3 \text{ m/s}$$

Hence the Mach number is given by:

$$M = V/a = 1.842$$

Therefore, the Mach angle is given by:

$$\alpha = \sin^{-1}(1/M) = 32.9°$$

**EXAMPLE 3.9**
An observer on the ground finds that an airplane flying horizontally at an altitude
of 5000 m has traveled 12 km from the overhead position before the sound of the
airplane is first heard. Estimate the speed at which the airplane is flying.

*Solution*
It is assumed that the net disturbance produced by the aircraft is weak, i.e., that,
as indicated by the wording of the question, basically what is being investigated is
how far the aircraft will have traveled from the overhead position when the sound
waves emitted by the aircraft are first heard by the observer. If the discussion of
Mach waves given above is considered, it will be seen that, as indicated in Fig.
E3.9, the aircraft will first be heard by the observer when the Mach wave emanating from the nose of the aircraft reaches the observer.

**FIGURE E3.9**

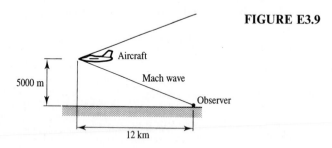

Now, since the temperature varies through the atmosphere, the speed of
sound varies as the sound waves pass down through the atmosphere which
means that the Mach waves from the aircraft are actually curved. This effect is,
however, small and will be neglected here, the sound speed at the average temperature between the ground and the aircraft being used to describe the Mach
wave.

Now as discussed in Example 3.3, for altitudes, $H$, of from 0 m (sea-
level) to 11 019 m the temperature in the atmosphere is given by $T =
288.16 - 0.0065H$ so, at the mean altitude of 2500 m, the temperature is
$288.16 - 0.0065 \times 2500 = 271.9$ K. Hence, the mean speed of sound is given by:

$$a = \sqrt{\gamma RT} = \sqrt{1.4 \times 287.04 \times 271.9} = 330.6 \text{ m/s}$$

From the above figure it will be seen that if $\alpha$ is the Mach angle based on the mean speed of sound then

$$\tan \alpha = 5000/12\,000 = 0.417$$

But since $\sin \alpha = 1/M$, it follows that $\tan \alpha = 1/\sqrt{M^2 - 1}$ so

$$M = \sqrt{(1/0.417)^2 + 1} = 2.6$$

Hence, it follows that:

$$\text{Velocity of aircraft} = 2.6 \times 330.6 = 859.6 \text{ m/s}$$

## 3.5
## CONCLUDING REMARKS

The discussion presented in this chapter, which was essentially only concerned with the flow of a gas, indicates that the Mach number $M$ is the parameter that determines the importance of compressibility effects on a flow. The speed of sound was shown to vary directly with the square root of the absolute temperature and inversely with the square root of the molar mass of the gas involved. The following terms were introduced:

- Incompressible flow: $M$ very much less than 1
- Subsonic flow: $M$ less than 1
- Transonic flow: $M$ approximately equal to 1
- Supersonic flow: $M$ greater than 1
- Hypersonic flow: $M$ very much greater than 1

It was shown that, in supersonic flow, disturbances are propagated along lines, termed Mach waves, and an expression for the angle such waves make to the flow was derived.

## PROBLEMS

**3.1.** The velocity of an air flow changes by 1 percent. Assuming that the flow is isentropic, plot the percentage changes in pressure, temperature, and density induced by this change in velocity with flow Mach number for Mach numbers between 0.2 and 2.

**3.2.** Calculate the speed of sound at 288 K in hydrogen, helium, and nitrogen. Under what conditions will the speed of sound in hydrogen be equal to that in helium?

**3.3.** Find the speed of sound in carbon dioxide at temperatures of 20°C and 600°C.

**3.4.** A very weak pressure wave, i.e., a sound wave, across which the pressure rise is 30 Pa moves through air which has a temperature of 30°C and a pressure of 101 kPa. Find the density change, the temperature change, and the velocity change across this wave.

**3.5.** An airplane is traveling at 1500 km/h at an altitude where the temperature is −60°C. What is the Mach at which the airplane is flying?

**3.6.** An airplane is flying at 2000 km/h at an altitude where the temperature is −50°C. Find the Mach number at which the airplane is flying.

**3.7.** An airplane can fly at a speed of 800 km/h at sea-level where the temperature is 15°C. If the airplane flies at the same Mach number at an altitude where the temperature is −44°C, find the speed at which the airplane is flying at this altitude.

**3.8.** The test section of a supersonic wind tunnel is square in cross-section with a side length of 1.22 m. The Mach number in the test section is 3.5, the temperature is −100°C, and the pressure is 20 kPa. Find the mass flow rate of air through the test section.

**3.9.** A certain aircraft flies at the same Mach number at all altitudes. If it flies at a speed that is 120 km/h slower at an altitude of 12 000 m than it does at sea-level, find the Mach number at which it flies. Assume standard atmospheric conditions.

**3.10.** Air at a temperature of 45°C flows in a supersonic wind-tunnel over a very narrow wedge. A shadowgraph photograph of the flow reveals weak waves emanating from the front of the wedge at an angle of 35° to the undisturbed flow. Find the Mach number and velocity in the flow approaching the wedge.

**3.11.** Air at a temperature of −10°C flows through a supersonic wind tunnel. A Schlieren photograph of the flow reveals weak waves originating at imperfections on the walls. These weak waves are at an angle of 35° to the flow. Find the air velocity in the wind tunnel.

**3.12.** A gas with a molar mass of 44 and a specific heat ratio 1.67 flows through a channel at supersonic speed. The temperature of the gas in the channel is 10°C. A photograph of the flow reveals weak waves originating at imperfections in the wall running across the flow at an angle to 45° to the flow direction. Find the Mach number and the velocity in the flow.

**3.13.** Air at 80°F is flowing at a Mach number of 1.9. Find the air velocity and the Mach angle.

**3.14.** Air at a temperature of 25°C is flowing with a velocity of 180 m/s. A projectile is fired into the air stream with a velocity of 800 m/s in the opposite direction to that of the air flow. Calculate the angle that the Mach waves from the projectile make to the direction of motion.

**3.15.** An observer at sea level does not hear an aircraft that is flying at an altitude of 7000 m until it is a distance of 13 km from the observer. Estimate the Mach number at which the aircraft is flying. In arriving at the answer, assume that the average temperature of the air between sea level and 7000 m is −10°C.

**3.16.** An aircraft is flying at an altitude of 6 km at a Mach number of 3. Find the distance behind the aircraft at which the disturbances created by the aircraft reach sea level.

**3.17.** An observer on the ground finds that an airplane flying horizontally at an altitude of 2500 m has traveled 6 km from the overhead position before the sound of the airplane is first heard. Assuming that, overall, the aircraft creates a small disturbance, estimate the speed at which the airplane is flying. The average air temperature between the ground and the altitude at which the airplane is flying is 10°C. Explain the assumptions you have made in arriving at the answer.

# CHAPTER 4

# One-Dimensional Isentropic Flow

## 4.1
## INTRODUCTION

Many flows that occur in engineering practice can be adequately modeled by assuming them to be steady, one-dimensional and isentropic. The equations that describe such flows will be discussed in this chapter. The applicability of the one-dimensional flow assumption was discussed in Chapter 2 and will not be discussed further here. An isentropic flow is, of course, an adiabatic flow (a flow in which there is no heat exchange) in which viscous losses are negligible, i.e., it is an adiabatic frictionless flow. Although no real flow is entirely isentropic, there are many flows of great practical importance in which the major portion of the flow can be assumed to be isentropic. For example, in internal duct flows there are many important cases where the effects of viscosity and heat transfer are restricted to thin layers adjacent to the walls, i.e., are only important in the wall boundary layers, and the rest of the flow can be assumed to be isentropic as indicated in Fig. 4.1.

Similarly in external flows, the effects of viscosity and heat transfer can be assumed to be restricted to the boundary layers, wakes, and shock waves and

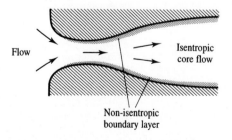

Flow

Isentropic core flow

Non-isentropic boundary layer

**FIGURE 4.1**
Region of duct flow that can be assumed to be isentropic.

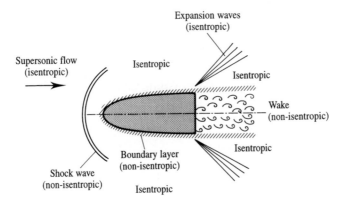

**FIGURE 4.2**
Region of external flow that can be assumed to be
isentropic.

the rest of the flow can be treated with adequate accuracy by assuming it to be
isentropic as indicated in Fig. 4.2.

Even when non-isentropic effects become important, it is often possible to
calculate the flow by assuming it to be isentropic and to then apply an
empirical correction factor to the solution so obtained to account for the
non-isentropic effects. This approach has been frequently adopted in the
past, for example, in the design of nozzles.

## 4.2
## GOVERNING EQUATIONS

By definition, the entropy remains constant in an isentropic flow. Using this
fact, it was shown in Chapter 2 that:

$$\frac{p}{\rho^\gamma} = c \quad \text{(a constant)} \tag{4.1}$$

If any two points, such as 1 and 2 shown in Fig. 4.3, in an isentropic flow

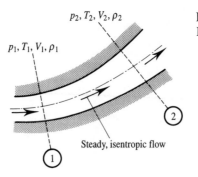

**FIGURE 4.3**
Flow situation considered.

are considered, it follows from eq. (4.1) that:

$$\frac{p_2}{p_1} = \left(\frac{\rho_2}{\rho_1}\right)^{\gamma}$$

(4.2)

Hence, since the general equation of state gives:

$$\frac{p_1}{\rho_1 T_1} = \frac{p_2}{\rho_2 T_2}, \quad \text{i.e.,} \quad \frac{T_2}{T_1} = \frac{p_2}{p_1} \frac{\rho_1}{\rho_2}$$

(4.3)

it follows that in isentropic flow:

$$\frac{T_2}{T_1} = \left(\frac{\rho_2}{\rho_1}\right)^{\gamma-1} = \left(\frac{p_2}{p_1}\right)^{\frac{\gamma-1}{\gamma}}$$

(4.4)

From this it then follows, recalling that $a = \sqrt{\gamma R T}$, that:

$$\frac{a_2}{a_1} = \left(\frac{T_2}{T_1}\right)^{\frac{1}{2}} = \left(\frac{\rho_2}{\rho_1}\right)^{\frac{\gamma-1}{2}} = \left(\frac{p_2}{p_1}\right)^{\frac{\gamma-1}{2\gamma}}$$

(4.5)

The steady flow adiabatic energy equation is next applied between the points 1 and 2. This gives:

$$c_p T_1 + \frac{V_1^2}{2} = c_p T_2 + \frac{V_2^2}{2}$$

i.e.,

$$\frac{T_2}{T_1} = \frac{1 + (V_1^2/2c_p T_1)}{1 + (V_2^2/2c_p T_2)}$$

But:

$$\frac{V^2}{2c_p T} = \left[\frac{V^2}{2\gamma RT}\right]\left[\frac{\gamma R}{c_p}\right] = \frac{\gamma - 1}{2} M^2$$

so it follows that:

$$\frac{T_2}{T_1} = \frac{1 + \frac{1}{2}(\gamma - 1)M_1^2}{1 + \frac{1}{2}(\gamma - 1)M_2^2}$$

(4.6)

This equation applies in adiabatic flow. If friction effects are also negligible, i.e., if the flow is isentropic, eq. (4.6) can be used in conjunction with the isentropic state relations given in eq. (4.5) to obtain:

$$\frac{p_2}{p_1} = \left[\frac{1 + \frac{1}{2}(\gamma - 1)M_1^2}{1 + \frac{1}{2}(\gamma - 1)M_2^2}\right]^{\frac{\gamma}{\gamma-1}}$$

(4.7)

and

$$\frac{\rho_2}{\rho_1} = \left[\frac{1 + \frac{1}{2}(\gamma - 1)M_1^2}{1 + \frac{1}{2}(\gamma - 1)M_2^2}\right]^{\frac{1}{\gamma-1}}$$

(4.8)

Lastly, it is recalled that the continuity equation gives:

$$\rho_1 V_1 A_1 = \rho_2 V_2 A_2$$

which can be rearranged to give:

$$\left(\frac{\rho_2}{\rho_1}\right)\left(\frac{V_2}{V_1}\right) = \frac{A_1}{A_2} \tag{4.9}$$

The above equations are, together, sufficient to determine all the characteristics of one-dimensional isentropic flow. It will be noted that the momentum equation was not used in the above analysis of isentropic flow. As discussed in the previous chapter, in isentropic flow, the momentum equation will always give the same result as the energy equation. This can be illustrated by using the integrated Euler equation (2.12), i.e.,

$$\frac{V_2^2}{2} - \frac{V_1^2}{2} + \int_1^2 \frac{dp}{\rho} = 0 \tag{4.10}$$

Since the relation between $\rho$ and $p$ is known in isentropic flow, the integral can be evaluated as follows:

$$\int_1^2 \frac{dp}{\rho} = \int_1^2 \frac{dp}{(p/p_1)^{\frac{1}{\gamma}}\rho_1} = \left(\frac{p_1}{\rho_1}\right)^{\frac{1}{\gamma}}\left(\frac{\gamma}{\gamma-1}\right)(p_2^{\left(1-\frac{1}{\gamma}\right)} - p_1^{\left(1-\frac{1}{\gamma}\right)})$$

$$= \left(\frac{\gamma}{\gamma-1}\right)\left(\frac{p_1}{\rho_1}\right)\left[\left(\frac{p_2}{p_1}\right)^{\frac{\gamma-1}{1}} - 1\right] \tag{4.11}$$

Substituting this result into eq. (4.10) then gives:

$$V_2^2 - V_1^2 + \left(\frac{2}{\gamma-1}\right)a_1^2\left[\left(\frac{p_2}{p_1}\right)^{\frac{\gamma-1}{\gamma}} - 1\right] = 0 \tag{4.12}$$

But it was shown before that the isentropic equation of state gives:

$$\left(\frac{a_2}{a_1}\right)^2 = \left(\frac{p_2}{p_1}\right)^{\frac{\gamma-1}{\gamma}}$$

Substituting this into eq. (4.12) then gives:

$$\left(\frac{p_2}{p_1}\right)^{\frac{\gamma-1}{\gamma}}\left[1 + \frac{\gamma-1}{2}M_2^2\right] = 1 + \left(\frac{\gamma-1}{2}\right)M_1^2 \tag{4.13}$$

which can be rearranged as:

$$\frac{p_2}{p_1} = \left[\frac{1 + \frac{1}{2}(\gamma-1)M_1^2}{1 + \frac{1}{2}(\gamma-1)M_2^2}\right]^{\frac{\gamma}{\gamma-1}} \tag{4.14}$$

This is the same as the result that was derived previously using the energy equation. Thus, in the analysis of isentropic flow either the momentum or the energy equation may be used. The energy equation is usually simpler to apply than the momentum equation.

**EXAMPLE 4.1**

A gas which has a molar mass of 39.9 and a specific heat ratio of 1.67 is discharged from a large chamber in which the pressure is 500 kPa and the temperature is 30°C through a nozzle. Assuming one-dimensional isentropic flow, find:

1. If the pressure at some section of the nozzle is 80 kPa, the Mach number, temperature, and velocity at this section.
2. If the nozzle has a circular cross-section and if its diameter is 12 mm at the section discussed in (a) above, the mass flow rate through the nozzle.

**Solution**

The flow situation is shown in Fig. E4.1. The following are given: $\gamma = 1.67$, $m = 39.9$, $p_0 = 500$ kPa, $T_0 = 303$ K. Steady, one-dimensional, isentropic flow is assumed.

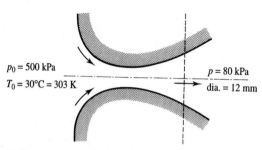

**FIGURE E4.1**

$p_0 = 500$ kPa
$T_0 = 30°C = 303$ K

$p = 80$ kPa
dia. = 12 mm

1. Isentropic relations give:

$$M = \sqrt{\left[\frac{2}{\gamma - 1}\right]\left\{\frac{p_0}{p}^{\frac{\gamma-1}{\gamma}} - 1\right\}}$$

*entire ratio should be raised to 1/γ*

Hence:

$$M = \sqrt{\frac{2}{0.67}\left\{\left[\frac{500}{80}\right]^{0.4} - 1\right\}} = 1.8$$

The following has also been established:

$$T = T_0\left\{1 + \frac{\gamma - 1}{2}M^2\right\}^{-1}$$

Hence:

$$T = 303 \text{ K}\left\{1 + \frac{0.67}{2}1 \times 8^2\right\}^{-1} = 145.3 \text{ K}$$

Using this value of the temperature then gives:

$$a = \sqrt{\frac{\gamma \mathcal{R} T}{m}} = \sqrt{1.67\frac{8314}{39.9}145.3} = 224.9 \text{ m/s}$$

Hence

$$V = aM = (1.8)(224.9 \text{ m/s}) = 404.7 \text{ m/s}$$

Therefore, the Mach number is 1.8, the temperature is 145.3 K ($= -127.7°C$) and the velocity is 404.7 m/s.

2. The mass flow rate is given by $\dot{m} = \rho V A$. But using the values of temperature and pressure established in (a), the density can be found as follows:

$$\rho = \frac{p}{RT} = \frac{80 \times 10^3}{(8314/39.9) \times 145.3} = 2.64 \text{ kg/m}^3$$

Hence:
$$\dot{m} = 2.64 \times 404.7 \frac{\pi \times (0.072)^2}{4} = 0.121 \text{ kg/s}$$

Therefore the mass flow rate through the nozzle is 0.121 kg/s.

### EXAMPLE 4.2

A gas with a molar mass of 4 and a specific heat ratio of 1.3 flows through a variable area duct. At some point in the flow, the velocity is 150 m/s, the pressure is 100 kPa and the temperature is 15°C. Find the Mach number at this point in the flow. At some other point in the flow the temperature is found to be −10°C. Find the Mach number, pressure, and velocity at this second point in the flow assuming the flow to be isentropic and one-dimensional.

### Solution

**FIGURE E4.2**

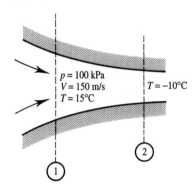

For the gas being considered:
$$R = \mathcal{R}/m = 8314/4 = 2078.5$$

Hence, the Mach number at section 1 is given by:
$$M_1 = \frac{V_1}{a_1} = \frac{150}{\sqrt{1.3 \times 2078.5 \times 288}} = \frac{150}{882.2} = 0.17$$

The speed of sound at section 2 will be given by:
$$a_2 = a_1(T_2/T_1)^{1/2} = 84.3 \text{ m/s} \quad 843 \text{ m/s}$$

Isentropic relations give:
$$\frac{T_2}{T_1} = \frac{1+(\gamma-1/2)M_1^2}{1+(\gamma-1/2)M_2^2}$$

which can be arranged to give:
$$M_2 = \sqrt{\frac{(1+(\gamma-1/2)M_1^2)T_1/T_2 - 1}{(\gamma-1)/2}} = 0.8157$$

But the speed of sound at section 2 is given by:

$$a_2 = 84.3 \text{ m/s} \quad 843 \text{ m/s}$$

so:
$$V_2 = M_2 a_2 = 687.6 \text{ m/s}$$

Isentropic relations also give:

$$p_2 = \left(\frac{a_2}{a_1}\right)^{2\gamma/\gamma-1} \times p_1 = \left(\frac{843}{882.15}\right)^{8.67} \times 100$$

$$= 0.6747 \times 100 = 67.5 \text{ kPa}$$

**EXAMPLE 4.3**
Air flows through a nozzle which has inlet areas of 10 cm². If the air has a velocity
of 80 m/s, a temperature of 28°C, and a pressure of 700 kPa at the inlet section
and a pressure of 250 kPa at the exit, find the mass flow rate through the nozzle
and, assuming one-dimensional isentropic flow, the velocity at the exit section of
the nozzle.

*Solution*

**FIGURE E4.3**

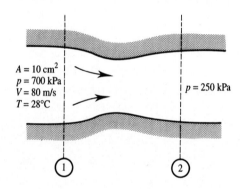

$A = 10 \text{ cm}^2$
$p = 700 \text{ kPa}$
$V = 80 \text{ m/s}$
$T = 28°C$

$p = 250 \text{ kPa}$

① ②

The mass flow rate is given by:

$$\dot{m} = \rho_1 V_1 a_1 = \frac{p_1}{RT_1} \times V_1 \times A_1 = \frac{700 \times 10^3}{287 \times 301} 80 \times 10^{-3}$$

Hence, the mass flow rate through the nozzle is given by:

$$\dot{m} = 0.648 \text{ kg/s}$$

Also:
$$M_1 = \frac{V_1}{a_1} = \frac{80}{\sqrt{\gamma RT}} = \frac{80}{\sqrt{1.4 \times 287 \times 301}}$$

$$= \frac{80}{347.77} = 0.23$$

Therefore, since isentropic relations give:

$$\frac{p_2}{p_1} = \left[\frac{1 + \dfrac{\gamma - 1}{2}M_1^2}{1 + \dfrac{\gamma - 1}{2}M_2^2}\right]^{\frac{\gamma}{\gamma-1}}$$

It follows that:

$$\frac{250}{700} = \left[ \frac{1 + \dfrac{1.4 - 1}{2} 0.23^2}{1 + \dfrac{1.4 - 1}{2} M_2^2} \right]^{\frac{1.4}{1.4-1}}$$

Solving this equation for $M_2$ then gives $M_2 = 1.335$. But since the flow is by assumption isentropic:

$$\frac{T_2}{T_1} = \left( \frac{p_2}{p_1} \right)^{\frac{\gamma-1}{\gamma}}$$

so $T_2 = 301 \times (250/700)^{1/3.5} = 224.3$ K. Hence:

$$V_2 = M_2 \times a_2 = 1.335 \times \sqrt{1.4 \times 287 \times 224.3 \text{ K}} = 400.8 \text{ m/s}$$

Hence, the exit velocity is 400.8 m/s.

## 4.3
## STAGNATION CONDITIONS

Stagnation conditions are those that would exist if the flow at any point in a fluid stream was isentropically brought to rest. (To define the stagnation temperature, it is actually only necessary to require that the flow be adiabatically brought to rest. To define the stagnation pressure and density, it is necessary, however, to require that the flow be brought to rest isentropically.)

If the entire flow is essentially isentropic and if the velocity is essentially zero at some point in the flow, then the stagnation conditions will be those existing at the zero velocity point as indicated in Fig. 4.4.

However, even when the flow is non-isentropic, the concept of the stagnation conditions is still useful, the stagnation conditions at a point then being the conditions that would exist if the local flow were brought to rest isentropically as indicated in Fig. 4.5.

If the equations derived in the previous section are applied between a point in the flow where the pressure, density, temperature, and Mach number

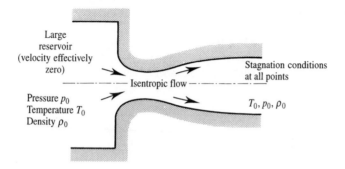

Large
reservoir
(velocity effectively
zero)

Stagnation conditions
at all points

— · — · Isentropic flow — · — · — · —

Pressure $p_0$
Temperature $T_0$
Density $\rho_0$

$T_0, p_0, \rho_0$

**FIGURE 4.4**
Stagnation conditions in an isentropic flow.

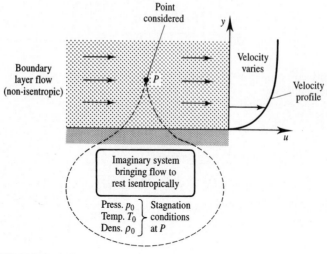

**FIGURE 4.5**
Stagnation conditions at a point in a non-isentropic flow.

are $p$, $\rho$, $T$, and $M$ respectively, then if the stagnation conditions are denoted by the subscript 0, the stagnation pressure, density and temperature will, since the Mach number is zero at the point where the stagnation conditions exist, be given by:

$$\frac{p_0}{p} = \left[ 1 + \frac{\gamma - 1}{2} M^2 \right]^{\frac{\gamma}{\gamma - 1}} \tag{4.15}$$

$$\frac{\rho_0}{\rho} = \left[ 1 + \frac{\gamma - 1}{2} M^2 \right]^{\frac{1}{\gamma - 1}} \tag{4.16}$$

$$\frac{T_0}{T} = \left[ 1 + \frac{\gamma - 1}{2} M^2 \right] \tag{4.17}$$

The variations of $p_0/p$ and $T_0/T$ with $M$ for the particular case of $\gamma = 1.4$ as given by eqs. (4.15) and (4.17) is shown in Fig. 4.6.

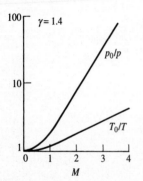

**FIGURE 4.6**
Variation of stagnation pressure and temperature ratios with Mach number.

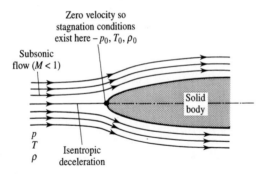

**FIGURE 4.7**
Stagnation conditions at leading edge of submerged body.

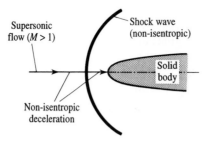

**FIGURE 4.8**
Supersonic flow near leading edge of submerged body.

The stagnation conditions will effectively exist at the leading edge of a bluff body in subsonic flow, the deceleration of the flow ahead of the body being essentially isentropic. The situation is shown in Fig. 4.7.

This is not true in supersonic flow because in this case shock waves form ahead of the body and produce part of the deceleration of the flow. This is shown in Fig. 4.8. As discussed in the next chapter, the flow through the shock wave is not isentropic which means that the stagnation conditions for the flow ahead of the shock wave are different from those downstream of the shock wave. This supersonic flow situation will be discussed in the next chapter.

### EXAMPLE 4.4

Air flows over a body. The air flow in the freestream ahead of the body has a Mach number of 0.85 and a static pressure of 80 kPa. Find the highest pressure acting on the surface of the body.

### Solution

**FIGURE E4.4**

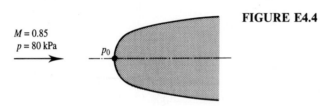

The highest pressure will be the stagnation pressure. But for $M = 0.85$ isentropic flow relations give:

$$\frac{p_0}{p} = \left[1 + \frac{\gamma - 1}{2} M^2\right]^{\frac{\gamma}{\gamma - 1}}$$

i.e.,

$$\frac{p_0}{p} = \left[1 + \frac{1.4 - 1}{2} 0.85^2\right]^{\frac{1.4}{1.4 - 1}} = 1.604$$

Therefore:     $p_0 = 80\ \text{kPa} \times 1.604 = 128.3\ \text{kPa}$

Therefore the highest pressure acting on the surface of the body is 128.3 kPa.

Returning to a consideration of subsonic stagnation point flow, it follows from the above discussion that a pitot tube placed in a subsonic compressible flow will register the stagnation pressure. This is illustrated in Fig. 4.9.

Since the disturbance produced by the tube is restricted to the stagnation point region, a pitot-static tube can be used to measure the Mach number in subsonic flow, the static hole measuring essentially the static pressure existing in the free stream. This is shown in Fig. 4.10.

Since

$$\frac{p_0}{p} = \left[1 + \frac{\gamma - 1}{2} M^2\right]^{\frac{\gamma}{\gamma - 1}}$$

it follows that the Mach number is given by:

$$M = \sqrt{\left(\frac{2}{\gamma - 1}\right)\left[\left(\frac{p_0}{p}\right)^{\frac{\gamma - 1}{\gamma}} - 1\right]} \qquad (4.18)$$

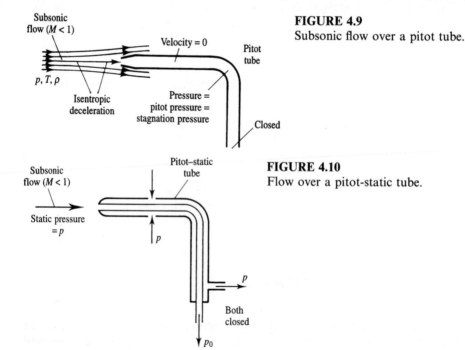

Subsonic flow $(M < 1)$

$p, T, \rho$

Isentropic deceleration

Velocity $= 0$

Pressure = pitot pressure = stagnation pressure

Pitot tube

Closed

**FIGURE 4.9**
Subsonic flow over a pitot tube.

Subsonic flow $(M < 1)$

Static pressure $= p$

Pitot–static tube

$p$

$p$

Both closed

$p_0$

**FIGURE 4.10**
Flow over a pitot-static tube.

**EXAMPLE 4.5**
A pitot-static tube is placed in a subsonic air flow. The static pressure and temperature in the flow are 80 kPa and 12°C respectively. The difference between the pitot and static pressures is measured using a manometer and found to be 200 mm of mercury. Find the air velocity and the Mach number.

**Solution**

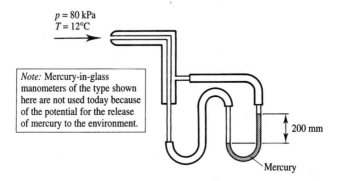

$p = 80$ kPa
$T = 12°C$

Note: Mercury-in-glass manometers of the type shown here are not used today because of the potential for the release of mercury to the environment.

200 mm

Mercury

**FIGURE E4.5**

The pressure difference is found from the manometer reading. This gives:

$$p_0 - p = \rho_m \times g \times \Delta H$$

where $\rho_m$ is the density of the liquid in the manometer which for mercury is 13 580 kg/m³. Therefore:

$$p_0 - p = 13\,580 \times 9.81 \times 0.2 = 26.64 \text{ kPa}$$

But:
$$\frac{p_0}{p} - 1 = \frac{p_0 - p}{p} = \frac{26.64}{80}$$

Hence:
$$\frac{p_0}{p} = 1.3331$$

Now, isentropic flow relations give, as shown above:

$$M = \sqrt{\left(\frac{2}{\gamma - 1}\right)\left[\left(\frac{p_0}{p}\right)^{\frac{\gamma - 1}{\gamma}} - 1\right]}$$

Hence:
$$M = \sqrt{\left(\frac{2}{0.4}\right)[(1.3331)^{0.4/1.4} - 1]} = 0.654$$

Hence the Mach number is 0.654. Since the temperature is given as 12°C = 285 K, the velocity can be found using:

$$V = Ma = M\sqrt{\gamma RT} = 0.654 \times \sqrt{1.4 \times 287 \times 285} = 221.3 \text{ m/s}$$

Therefore the velocity is 221.3 m/s.

In order to find the Mach number using eq. (4.18), $p_0$ and $p$ have to be separately measured and, if the velocity is required, the temperature will normally also have to be measured in order to find the speed of sound. In incompressible flow, of course, the Bernoulli equation gives:

$$V = \sqrt{2\frac{(p_0 - p)}{\rho}} \qquad (4.19)$$

which indicates that in order to determine the velocity in incompressible flow only the difference between $p_0$ and $p$ has to be measured and not their individual values.

It is convenient for many purposes to know the error that would be incurred by using the incompressible pitot tube equation in a compressible flow. The magnitude of this error will also illustrate the magnitude of compressibility effects in a subsonic flow. Now, by using the full compressible flow pitot tube equation it follows that:

$$p_0 - p = p\left[\frac{p_0}{p} - 1\right]$$

$$= (\tfrac{1}{2}\rho V^2)\left(\frac{2p}{\rho V^2}\right)\left\{\left[1 + \left(\frac{\gamma - 1}{2}\right)M^2\right]^{\frac{\gamma}{\gamma-1}} - 1\right\}$$

$$= \tfrac{1}{2}\rho V^2\left\{\left(\frac{2}{\gamma M^2}\right)\left[\left(1 + \frac{\gamma - 1}{2}M^2\right)^{\frac{\gamma}{\gamma-1}} - 1\right]\right\} \qquad (4.20)$$

which gives the actual velocity as:

$$V = \sqrt{\frac{2(p_0 - p)}{\rho}}\left\{\left(\frac{2}{\gamma M^2}\right)\left[\left(1 + \frac{\gamma - 1}{2}M^2\right)^{\frac{\gamma}{\gamma-1}} - 1\right]\right\}^{-\frac{1}{2}} \qquad (4.21)$$

On the other hand, if the incompressible flow pitot tube equation is used, the velocity would be given by:

$$V = \sqrt{\frac{2(p_0 - p)}{\rho}} \qquad (4.22)$$

Therefore, the error incurred in using the incompressible flow equation to find the velocity from the measured pressure difference, i.e.,

$$\epsilon = \left|\frac{(V_{\text{actual}} - V_{\text{incom}})}{V_{\text{actual}}}\right| \qquad (4.23)$$

is given by:

$$\epsilon = \left|1 - \left\{\left(\frac{2}{\gamma M^2}\right)\left[\left(1 + \frac{\gamma - 1}{2}M^2\right)^{\frac{\gamma}{\gamma-1}} - 1\right]\right\}^{\frac{1}{2}}\right| \qquad (4.24)$$

Thus, the error depends only on the Mach number and its variation is indicated in Fig. 4.11.

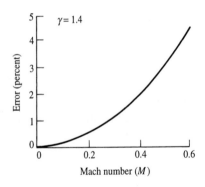

**FIGURE 4.11**
Variation of error incurred by use of incompressible pitot-static tube equation with Mach number.

From Fig. 4.11 it follows that the incompressible flow equation can be used to determine velocity with errors of less than 1 percent if the Mach number is less than roughly 0.3. However, the error rises to almost 5 percent when the Mach number is 0.6.

### EXAMPLE 4.6

A pitot-static tube is placed in a subsonic airflow. The static pressure and temperature in the flow are 96 kPa and 27°C respectively. The difference between the pitot and static pressures is measured and found to be 32 kPa. Find the air velocity (a) assuming an incompressible flow, (b) assuming compressible flow.

### Solution

The density in the flow is given by:

$$\rho = \frac{p}{RT} = \frac{96}{287 \times 300 \text{ K}} = 1.115 \text{ kg/m}^3$$

(a) If incompressible flow is assumed, the velocity is given by:

$$V = 2\sqrt{\frac{p_0 - p}{\rho}} = 2\sqrt{\frac{32 \times 10^3}{1.115}} = 239.6 \text{ m/s}$$

(b) When compressibility is accounted for, the velocity is found by noting that:

$$\frac{p_0 - p}{p} = \frac{p_0}{p} - 1 = \frac{32}{96}$$

Hence, $p_0/p = 1.3333$. But:

$$\frac{p_0}{p} = \left[1 + \frac{\gamma - 1}{2} M^2\right]^{\frac{\gamma}{\gamma - 1}}$$

So for $p_0/p = 1.3333$, this relation gives:

$$M^2 = \frac{2}{0.4} \times (1.333^{1/3.5} - 1)$$

which gives $M = 0.654$. The velocity is, therefore, given by:

$$V = Ma = 0.654\sqrt{1.4 \times 287 \times 300} = 225.7 \text{ m/s}$$

Hence, the actual velocity is 225.7 m/s whereas, when compressibility effects are neglected, the velocity is found to be 239.6 m/s.

## 4.4
## CRITICAL CONDITIONS

The critical conditions are those that would exist if the flow was isentropically accelerated or decelerated until the Mach number was unity, i.e., they are the conditions that would exist if the Mach number was isentropically changed from $M$ to 1. These critical conditions are usually denoted by an asterisk, i.e., they are denoted by the symbols $V^*$, $p^*$, $\rho^*$, $T^*$, and $A^*$. Using eqs. (4.6)–(4.8) and setting $M_2$ equal to 1 then gives the following relations for the critical conditions:

$$\frac{T^*}{T} = \left[ \frac{2}{\gamma + 1} + \frac{\gamma - 1}{\gamma + 1} M^2 \right] \tag{4.25}$$

$$\frac{a^*}{a} = \left[ \frac{2}{\gamma + 1} + \frac{\gamma - 1}{\gamma + 1} M^2 \right]^{\frac{1}{2}} \tag{4.26}$$

$$\frac{p^*}{p} = \left[ \frac{2}{\gamma + 1} + \frac{\gamma - 1}{\gamma + 1} M^2 \right]^{\frac{\gamma}{\gamma - 1}} \tag{4.27}$$

$$\frac{\rho^*}{\rho} = \left[ \frac{2}{\gamma + 1} + \frac{\gamma - 1}{\gamma + 1} M^2 \right]^{\frac{1}{\gamma - 1}} \tag{4.28}$$

The relation between the critical conditions and the stagnation conditions can be found by setting $M$ equal to zero in the above equations. This gives:

$$\frac{T^*}{T_0} = \frac{2}{\gamma + 1} \tag{4.29}$$

$$\frac{a^*}{a_0} = \sqrt{\frac{2}{\gamma + 1}} \tag{4.30}$$

$$\frac{p^*}{p_0} = \left( \frac{2}{\gamma + 1} \right)^{\frac{\gamma}{\gamma - 1}} \tag{4.31}$$

$$\frac{\rho^*}{\rho_0} = \left( \frac{2}{\gamma + 1} \right)^{\frac{1}{\gamma - 1}} \tag{4.32}$$

For the case of air flow, these equations give:

$$\frac{T^*}{T_0} = 0.833, \qquad \frac{p^*}{p_0} = 0.528, \qquad \frac{\rho^*}{\rho_0} = 0.634$$

**EXAMPLE 4.7**

A gas is contained in a large vessel at a pressure of 300 kPa and a temperature of 50°C. The gas is expanded from this vessel through a nozzle until the Mach number reaches a value of 1. Find the pressure, temperature and velocity at this point in the flow if the gas is (a) air and (b) helium.

**Solution**

It was shown above that:

$$\frac{p^*}{p_0} = \left(\frac{2}{\gamma+1}\right)^{\frac{\gamma}{\gamma-1}}$$

Hence, since $\gamma = 1.4$ for air and 1.667 for helium and since $p_0 = 300$ kPa, it follows that the pressure $p^*$ at the point where $M = 1$ is $300 \times (2/2.4)^{(1.4/0.4)} = 158.5$ kPa for air and $300 \times (2/2.667)^{(1.667/0.667)} = 146.1$ kPa for helium.

It was also shown above that:

$$\frac{T^*}{T_0} = \frac{2}{\gamma+1}$$

Hence since $T_0 = 323$ K, it follows that the temperature $T^*$ at the point where $M = 1$ is $323 \times (2/2.4) = 269.2$ K $= -3.8°C$ for air and $323 \times (2/2.667) = 242.2$ K $= -30.8°C$ for helium.

Lastly, since at the point considered $M = 1$,

$$V = Ma = a$$

Hence, using the temperature values already found,

$$V = \sqrt{1.4 \times (8314.3/28.97) \times 269.2} = 328.9 \text{ m/s}$$

for air and

$$V = \sqrt{1.667 \times (8314.3/4) \times 242.2} = 916.1 \text{ m/s}$$

for helium.

## 4.5
## MAXIMUM DISCHARGE VELOCITY

The "maximum discharge velocity" or "maximum escape velocity" is the velocity that would be generated if a gas was adiabatically expanded until its temperature had dropped to absolute zero. Using the adiabatic energy equation gives the maximum discharge velocity as:

$$\frac{\hat{V}^2}{2} = \frac{V^2}{2} + c_p T = c_p T_0 \tag{4.33}$$

This can be rearranged to give:

$$\hat{V} = \sqrt{(V^2 + 2c_p T)} = \sqrt{2c_p T_0}$$

$$= \sqrt{\left(V^2 + \frac{2a^2}{\gamma - 1}\right)} = \sqrt{\frac{2a_0^2}{\gamma - 1}} \tag{4.34}$$

There is, therefore, a definite maximum velocity that can be generated in a gas having a given stagnation temperature. However, since the temperature is zero when this maximum velocity is reached, the Mach number will be infinite since, under these conditions, the speed of sound is 0. It should be noted that the maximum discharge velocity given by the above equation could not be obtained in reality because at very low temperatures the assumptions used in deriving the above equations cease to apply and the gas would liquify. It should also be noted that the maximum discharge velocity has nothing to do with the existence of a maximum velocity at which a body can move relative to a gas, no such limit existing according to the laws of conventional mechanics.

### EXAMPLE 4.8
Consider the flow situation described in Example 4.7. What is the maximum velocity that could be generated by expanding the gas through a nozzle system?

### Solution
It was shown above that:

$$\hat{V} = \sqrt{\frac{2a_0^2}{\gamma - 1}}$$

But
$$T_0 = 273 + 50 = 323 \text{ K}$$

so
$$a_0 = \sqrt{1.4 \times (8314.3/28.97) \times 323} = 360.3 \text{ m/s}$$

for air and
$$a_0 = \sqrt{1.667 \times (8314.3/4) \times 323} = 1057.9 \text{ m/s}$$

for helium. Therefore

$$\hat{V} = \sqrt{2 \times 360.3^2/0.4} = 805.7 \text{ m/s}$$

for air and

$$\hat{V} = \sqrt{2 \times 1057.9^2/0.667} = 1831.9 \text{ m/s}$$

for helium.

### EXAMPLE 4.9
Consider the flow situation described in Example 4.1. What are the critical pressure, temperature, and speed of sound and what is the maximum velocity that could be generated by expanding the gas through the nozzle?

### Solution
Using the relations for the critical conditions given above, the following are obtained:

$$\dot{p} = p_0 \left[ \frac{2}{\gamma + 1} \right]^{\frac{\gamma}{\gamma - 1}} = 500 \text{ kPa} \left[ \frac{2}{2.67} \right]^{2.40} = 243.3 \text{ kPa}$$

$$T^* = T_0 \frac{2}{\gamma + 1} = 303 \times \frac{2}{2.67} = 227.0 \text{ K}$$

$$a^* = a_0 \sqrt{\frac{2}{\gamma + 1}} = \sqrt{\gamma RT} \sqrt{\frac{2}{\gamma + 1}} = 280.0 \text{ m/s}$$

Therefore, the critical pressure is 243.3 kPa and the critical temperature is 227 K ($= -46°C$). $M^*$ is of course equal to 1 by definition and $V^* = a^* = 280$ m/s.
Using the relation established above:

$$\hat{V} = \sqrt{2\frac{a_0^2}{\gamma - 1}} = \sqrt{2\frac{\gamma RT}{\gamma - 1}} = 560.9 \text{ m/s}$$

Therefore, the maximum possible velocity that could be generated is 560.9 m/s.

## 4.6
## ISENTROPIC RELATIONS IN TABULAR AND GRAPHICAL FORM

The equation derived above for one-dimensional, isentropic flow are relatively easily solved using a calculator or computer. Traditionally, however, isentropic flow calculations have been undertaken using sets of tables or graphs which give the variations of such quantities as $p_0/p$, and $T_0/T$ with $M$ in isentropic flow for a fixed value of the specific heat ratio $\gamma$. Such tables and graphs are available for various values of the specific heat ratio but care must be taken to ensure that a table for the correct value of $\gamma$ is used.

A typical set of tables would have the following headings:

| $M$ | $\dfrac{p_0}{p}$ | $\dfrac{T_0}{T}$ | $\dfrac{\rho_0}{\rho}$ | $\dfrac{a_0}{a}$ | $\dfrac{A}{A^*}$ |
|---|---|---|---|---|---|

The meaning of the entry $A/A^*$ will be discussed in a later chapter. Such a table can be conveniently used in the calculation of the properties of a one-dimensional isentropic flow. For example, if flow through a variable area channel is being considered and if the Mach number and pressure at one section are known, say $M_1$ and $p_1$ and if the pressure at some other section is known, say $p_2$, then to find the Mach number at the second section the value of $M_1$ is used with the table to find $p_1/p_0$. Then since in isentropic flow the stagnation pressure is constant:

$$\frac{p_2}{p_0} = \frac{p_2}{p_1} \times \frac{p_1}{p_0} \tag{4.35}$$

$p_0/p_2$ can be found. Then using the tables, the value of $M_2$ corresponding to this value of $p_0/p$ can be found.

A set of isentropic tables for air, i.e., $\gamma = 1.4$, is given in Appendix B.

As mentioned above, the widespread availability of programmable calculators and personal computers has led to a considerable reduction in the use of isentropic flow tables and charts for calculating the properties of isentropic flows. For example, the software provided to support this book allows the conditions in isentropic flow to be determined. The way in which a programmable calculator can be used to find conditions in isentropic flow is illustrated by the very simple BASIC program listed in Appendix L which allows any one of $M$, $p_0/p$, or $T_0/T$ to be entered and then finds the value of the other two quantities. This program is easily extended to apply to all the variables in standard isentropic tables. Where applicable, the worked examples in this book will be presented in such a way that they could have been solved using the equations directly or by using isentropic tables or by using the software.

**EXAMPLE 4.10**
Air flows from a large vessel in which the pressure is 300 kPa and the temperature is 40°C through a nozzle. If the pressure at some section of the discharge nozzle is measured as 200 kPa, find the temperature and velocity at this section. If the Mach number at some other section of the nozzle is 1.5, find the pressure, temperature, and velocity at this section. Assume that the flow is steady, isentropic, and one-dimensional.

**Solution**
In this case:

$$p_0 = 300 \text{ kPa}, \qquad T_0 = 313 \text{ K}$$

At section 2, $p_0/p = 300/200 = 1.5$ so isentropic relations or tables for $\gamma = 1.4$ or software gives $M = 0.78$ and $T_0/T = 1.12$ so:

$$T = 313/1.12 = 279 \text{ K} = +6°C$$

Hence, using $V = Ma$, it follows that:

$$V = 0.78 \times \sqrt{1.4 \times (8314.3/28.97) \times 279.2} = 336.9 \text{ m/s}$$

At section 3, $M = 1.5$ so isentropic relations or tables for $\gamma = 1.4$ or software give $T_0/T = 1.45$ and $p_0/p = 3.67$ so:

$$T = 313/1.45 = 216 \text{ K} = -57°C$$

$$p = 300/3.67 = 81.7 \text{ kPa}$$

$$V = 1.5 \times \sqrt{1.4 \times (8314.3/28.97) \times 216} = 441.9 \text{ m/s}$$

**EXAMPLE 4.11**
The velocity, pressure, and temperature at a certain point in a steady air flow, are 600 m/s, 70 kPa, and 5°C respectively. If the pressure at some other point in the flow is 30 kPa, find the Mach number, temperature, and velocity that exist at this second point. Assume that the flow is isentropic and one-dimensional.

**Solution**
In this case:

$$M_1 = 600/\sqrt{1.4 \times (8314.3/28.97) \times (273 + 5)} = 1.80$$

At section 1 using this value of $M$, isentropic relations or tables for $\gamma = 1.4$ or software give $T_0/T = 1.65$ and $p_0/p = 5.75$. But, because $p_0$ does not change in isentropic flow, it follows that:

$$\frac{p_2}{p_1} = \frac{p_0/p_1}{p_0/p_2}$$

Hence, since $p_2/p_1 = 30/70 = 0.4286$ it follows that:

$$\frac{p_0}{p_2} = \frac{5.75}{0.4286} = 13.42$$

For this value of $p_0/p$, isentropic relations or tables for $\gamma = 1.4$ or software give $M = 2.345$ and $T_0/T = 2.10$. Hence, because $T_0$ also does not change in isentropic flow, it follows that:

$$\frac{T_2}{T_1} = \frac{T_0/T_1}{T_0/T_2}$$

Hence, $T_2 = 278 \times 1.65/2.10 = 218.4 \, \text{K}$. From this and the fact that $M_2$ is 2.345, it follows that:

$$V_2 = 2.345 \times \sqrt{1.4 \times (8314.3/28.97) \times 218.4} = 694.7 \, \text{m/s}$$

## 4.7
## CONCLUDING REMARKS

Although no flow is, of course, truly isentropic the main characteristics of many practically significant flows can be predicted using the equations presented in this chapter. The concepts of stagnation point conditions and critical conditions have also been introduced in this chapter. The use of isentropic tables for the calculation of one-dimensional, steady, isentropic flows was also discussed. Such tables are sometimes convenient to use but it must always be remembered that a given table applies only to a specific value of $\gamma$.

## PROBLEMS

**4.1.** A gas with a molar mass of 4 and a specific heat ratio of 1.67 flows through a variable area duct. At some point in the flow the velocity is 200 m/s and the temperature is 10°C. Find the Mach number at this point in the flow. At some other point in the flow the temperature is −10°C. Find the velocity and Mach number at this point in the flow assuming that the flow is isentropic.

**4.2.** Air flows through a convergent–divergent duct with an inlet area of 5 cm² and an exit area of 3.8 cm². At the inlet section the air velocity is 100 m/s, the pressure is 680 kPa, and the temperature is 60°C. Find the mass flow rate through the nozzle and, assuming isentropic flow, the pressure and velocity at the exit section.

**4.3.** The exhaust gases from a rocket engine can be assumed to behave as a perfect gas with a specific heat ratio of 1.3 and a molecular weight of 32. The gas is expanded from the combustion chamber through the nozzle. At a point in the nozzle where the cross-sectional area is 0.2 m² the pressure, temperature, and Mach number are 1500 kPa, 800°C, and 0.2 respectively. At some other point in the nozzle, the pressure is found to be 80 kPa. Find the Mach number, temperature, and cross-sectional area at this point. Assume one-dimensional, isentropic flow.

**4.4.** The exhaust gases from a rocket engine have a molar mass of 14. They can be assumed to behave as a perfect gas with a specific heat ratio of 1.25. These gases are accelerated through a nozzle. At some point in the nozzle where the cross-sectional area of the nozzle is $0.7\,m^2$, the pressure is 1000 kPa, the temperature is 500°C and the velocity is 100 m/s. Find the mass flow rate through the nozzle and the stagnation pressure and temperature. Also find the highest velocity that could be generated by expanding this flow. If the pressure at some other point in the nozzle is 100 kPa, find the temperature and velocity at this point in the flow assuming the flow to be one-dimensional and isentropic.

**4.5.** A gas has a molar mass of 44 and a specific heat ratio of 1.3. At a certain point in the flow, the static pressure and temperature are 80 kPa and 15°C, respectively and the velocity is 100 m/s. The gas is then isentropically expanded until its velocity is 300 m/s. Find the pressure, temperature, and Mach number that exist in the resulting flow.

**4.6.** Carbon dioxide flows through a variable area duct. At a certain point in the duct the velocity is 200 m/s and the temperature is 60°C. At some other point in the duct, the temperature is 15°C. Find the Mach numbers and stagnation temperatures at the two points. Assume that the flow is adiabatic.

**4.7.** At a certain point in a gas flow, the velocity is 900 m/s, the pressure is 150 kPa, and the temperature is 60°C. Find the stagnation pressure and temperature if the gas is air and if it is carbon dioxide.

**4.8.** Helium, at a pressure of 120 kPa and a temperature of 20°C flows at a velocity of 800 m/s. Find the Mach number and the stagnation temperature and the stagnation pressure.

**4.9.** In an argon flow the temperature is 40°C and the pressure is half the stagnation pressure. Find the Mach number and the velocity in the flow.

**4.10.** If Concorde is flying at a Mach number of 2.2, at an altitude of 10 000 m in the standard atmosphere, find the stagnation pressure and temperature for the flow over the aircraft.

**4.11.** If a gas is flowing at 300 m/s and has a pressure and temperature of 90 kPa and 20°C, find the maximum possible velocity that could be generated by expansion of this gas if the gas is air and if it is helium.

**4.12.** A pitot-static tube is placed in a subsonic air flow. The static temperature and pressure in the air flow are 30°C and 101 kPa, respectively. The difference between the pitot and static pressures is measured using a manometer and is found to be 250 mm of mercury. Find the air velocity, assuming the flow to be incompressible and taking compressibility effects into account.

**4.13.** A pitot-static tube is placed in a subsonic air flow. The static pressure and temperature are 101 kPa and 30°C respectively. The difference between the pitot and static pressures is measured and found to be 37 kPa. Find the air velocity.

**4.14.** A pitot tube placed in an air stream indicates a pressure of 186 kPa. If the local temperature is 20°C and the local Mach number is 0.8, determine the static pressure.

**4.15.** A pitot tube indicates a pressure of 155 kPa when placed in an air stream in which the temperature is 15°C and the Mach number is 0.7. Find the static pressure in the flow. Also find the stagnation temperature in the flow.

**4.16.** A pitot tube is placed in an air stream in which the pressure is 60 kPa and the Mach number is 0.9. What will the pitot pressure be?

**4.17.** Consider one-dimensional isentropic flow through a duct. At a certain section of this duct, the velocity is 360 m/s, the temperature is 45°C, and the pressure is 120 kPa. Find the Mach number and the stagnation temperature and pressure at this point in the flow. If the temperature at some other point in the flow is 90°C, find the Mach number and pressure at this point in the flow.

**4.18.** A liquid fuelled rocket is fired on a test stand. The rocket nozzle has an exit diameter of 30 cm and the combustion gases leave the nozzle at a velocity of 3800 m/s and a pressure of 100 kPa, which is the same as the ambient pressure. The temperature of the gases in the combustion area is 2400°C. Find the temperature of the gases on the nozzle exit plane, the pressure in the combustion area, and the thrust developed. Assume that the gases have a specific heat ratio of 1.3 and a molar mass of 9. Assume that the flow in the nozzle is isentropic.

**4.19.** The pressure, temperature, and Mach number at the entrance to a duct through which air is flowing are 250 kPa, 26°C, and 1.4 respectively. At some other point in the duct, the Mach number is found to be 2.5. Assuming isentropic flow, find the temperature, velocity, and pressure at the second section. Also find the mass flow rate through the duct per square meter at the second section.

**4.20.** An aircraft is flying at a Mach number of 0.95 at an altitude where the pressure is 30 kPa and the temperature is −50°C. The diffuser at the intake to the engine decreases the Mach number to 0.3 at the inlet to the engine. Find the pressure and temperature at the inlet to the engine.

**4.21.** A conical diffuser has an inlet diameter of 15 cm. The pressure, temperature, and velocity at the inlet to the diffuser are 70 kPa, 60°C, and 180 m/s respectively. If the pressure at the diffuser exit is 78 kPa, find the exit diameter of the diffuser.

**4.22.** The control system for some smaller space vehicles uses nitrogen from a high-pressure bottle. When the vehicle has to be maneuvered, a valve is opened allowing nitrogen to flow out through a nozzle thus generating a thrust in the direction required to maneuver the vehicle. In a typical system, the pressure and temperature in the system ahead of the nozzle are about 1.6 MPa and 30°C respectively while the pressure in the jet on the nozzle exit plane is about 6 kPa. Assuming that the flow through the nozzle is isentropic and the gas velocity ahead of the nozzle is negligible, find the temperature and the velocity of the nitrogen on the nozzle exit plane. If the thrust required to manuever the vehicle is 1 kN, find the area of the nozzle exit plane and the required mass flow rate of nitrogen. It can be assumed that the vehicle is effectively operating in a vacuum.

**4.23.** Hydrogen enters a nozzle with a very low velocity and at a temperature and pressure of 3800 R and 1000 psia respectively. The pressure on the exit plane of the nozzle is 2 psia. Calculate the hydrogen flow rate per unit nozzle exit area. The flow through the nozzle can be assumed to be isentropic.

**4.24.** An aircraft flies at sea-level at a speed of 220 m/s. What is the highest pressure that can be acting on the surface of the aircraft.

**4.25.** Consider an air flow with a speed of 650 m/s, a pressure of 100 kPa, and a temperature of 20°C. What is the stagnation pressure and the stagnation temperature in the flow?

**4.26.** When an aircraft is flying at subsonic velocity, the pressure at its nose, i.e., at the stagnation point, is found to be 160 kPa. If the ambient pressure and temperature are 100 kPa and 25°C respectively, find the speed and the Mach number at which the aircraft is flying.

**4.27.** A body moves through air at a velocity of 200 m/s. The pressure and temperature in the air upstream of the body are 100 kPa and 30°C respectively. Find the pressure at a point on the body where the velocity of the air relative to the body is zero (1) accounting for compressibility and (2) assuming incompressible flow. Assume that the flow is isentropic.

**4.28.** Air enters a duct at a pressure of 30 psia, a temperature of 100°F, and a velocity of 580 ft/sec. At some other point in the duct the pressure is found to be 12 psia. Assuming that the flow is isentropic, find the temperature and Mach number at this point in the flow.

**4.29.** Consider a rocket engine that burns hydrogen and oxygen. The combustion chamber temperature and pressure are 3800 K and 1.5 MPa respectively, the velocity in the combustion chamber being very low. The pressure on the nozzle exit plane is 1.5 kPa. Assuming that the flow is isentropic, find the Mach number and the velocity on the exit plane. Assume that the products of combustion behave as a perfect gas with $\gamma = 1.22$ and $R = 519.6 \text{ J/kg} \cdot \text{K}$.

**4.30.** At a point in a supersonic air flow, the pressure and temperature are 5 kPa and −80°C. If the stagnation pressure at this point is 100 kPa, find the Mach number and the stagnation temperature.

# CHAPTER 5

---

# Normal Shock Waves

## 5.1
## SHOCK WAVES

It has been found experimentally that, under some circumstances, it is possible for an almost spontaneous change to occur in a flow, the velocity decreasing and the pressure increasing through this region of sharp change. The possibility that such a change can actually occur follows from the analysis given below. It has been found experimentally, and it also follows from the analysis given below, that such regions of sharp change can only occur if the initial flow is supersonic. The extremely thin region in which the transition from the supersonic velocity, relatively low pressure state to the state that involves a relatively low velocity and high pressure is termed a shock wave. The changes that occur through a normal shock wave, i.e., a shock wave which is straight with the flow at right angles to the wave, is shown in Fig. 5.1. A photograph of a normal shock wave is shown in Fig. 5.2.

A shock wave is extremely thin, usually only a few mean free paths thick. A shock wave is analogous in many ways to a "hydraulic jump" that occurs in free-surface liquid flows, a hydraulic jump being shown schematically in Fig. 5.3. A hydraulic jump occurs, for example, in the flow downstream of a weir.

A shock wave is, in general, curved. However, many shock waves that occur in practical situations are straight, being either at right angles (i.e., normal) to or at an angle to the upstream flow (see Fig. 5.4). A straight shock wave that is at right angles to the upstream flow is, as noted above, termed a normal shock while a straight shock wave that is at an angle to the upstream flow is termed an oblique shock wave.

In the case of a normal shock wave, the velocities both ahead (i.e., upstream) of the shock and after (i.e., downstream) of the shock are at

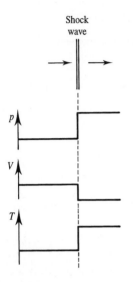

**FIGURE 5.1**
Changes through a normal shock wave.

**FIGURE 5.2**
Photograph of a normal shock wave.

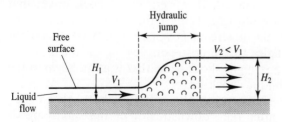

**FIGURE 5.3**
Hydraulic jump.

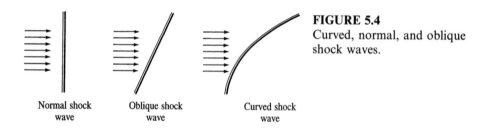

**FIGURE 5.4**
Curved, normal, and oblique shock waves.

right angles to the shock wave. In the case of an oblique shock wave there is a change in flow direction across the shock. This is illustrated in Fig. 5.5.

A complete shock wave may be effectively normal in part of the flow, curved in other parts of the flow, and effectively oblique in other parts of the flow, as shown in Fig. 5.6.

Because of its importance and because, as will be shown later, the oblique shock relations can be deduced from those for a normal shock wave, the normal shock wave will be first considered in the present chapter. Oblique shock waves will then be discussed in the next chapter. Curved shock waves are relatively difficult to analyse and they will not be discussed in detail in the present book.

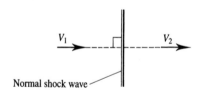

**FIGURE 5.5**
Velocity changes across normal and oblique shock waves.

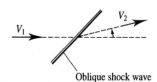

**FIGURE 5.6**
Shock wave with changing shape.

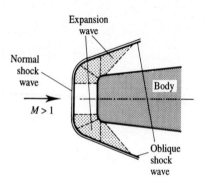

Normal shock waves occur, for example, in the intakes to the engines in some supersonic aircraft, in the exhaust system of reciprocating engines, in long distance gas pipelines, and in mine shafts as a result of the use of explosives.

When a normal shock wave occurs, for example, in a steady flow through duct, it is stationary with respect to the coordinate system which is fixed relative to the walls of the duct. Such a shock wave is called a stationary shock wave since it is not moving relative to the coordinate system used. On the other hand, when a sudden disturbance occurs in a flow, such as, for example, the sudden closing of a valve in a pipeline or an explosive release of energy at a point in a duct, a normal shock wave can be generated which is moving relative to the duct walls. This is illustrated in Fig. 5.7.

The analysis of stationary normal shock waves will first be considered and then the application of this analysis to moving normal shock waves will be discussed.

To illustrate how a shock wave can form, consider the generation of a sound wave as discussed in Chapter 3. There it was assumed that there was a long duct containing a gas at rest and that there was a piston at one end of this duct that was initially at rest. Then, at time 0, the piston was given a small velocity into the duct giving rise to a weak pressure pulse, i.e., a sound wave, that propagated down the duct into the gas (see Fig. 5.8).

If $dV$ was the velocity given to the piston, which is, of course, the same as the velocity of the gas behind the wave, then the increase in pressure and temperature behind the wave are equal to $\rho a\, dV$ and $(\gamma - 1)T\, dV/a$ respectively. Since $\rho$, $a$, and $T$ are all positive, this shows that the pressure and temperature both increase across the wave. It was also shown that the velocity at which the wave moves down the duct is equal to $\sqrt{\gamma RT}$, which is by definition the speed of sound. Therefore, since the temperature increases across the wave, the speed of sound behind the wave will be $a + da$, where $da$ is positive. Now consider what happens if some time after the piston is

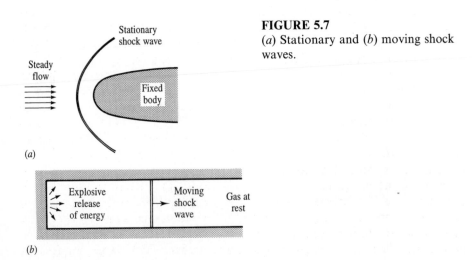

**FIGURE 5.7**
(a) Stationary and (b) moving shock waves.

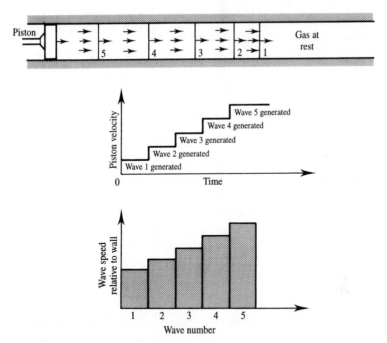

**FIGURE 5.8**
Generation of weak waves by piston movement.

given velocity $dV$ into the duct, its velocity is suddenly again increased to $2\,dV$. As a result of the second increase in piston speed, a second weak pressure wave will be generated that follows the first wave down the duct as shown in Fig. 5.8. This second wave will be moving relative to the gas ahead of it at the speed of sound in the gas through which it is propagating, but the gas ahead of the second wave has velocity $dV$. Hence, the second wave moves relative to the duct at a velocity of $a + da + dV$. The first wave is moving at a velocity of $a$ relative to the duct. Therefore, since both $da$ and $dV$ are positive, the second wave is moving faster than the first wave and, if the duct is long enough, the second wave will overtake the first wave. The second wave cannot pass through the first wave: Instead, the two waves merge into a single stronger wave. If, therefore, the piston is given a whole series of step increases in velocity, a series of weak pressure waves will be generated which will all eventually overtake each other and merge into a single strong wave if the duct is long enough. Since the "back of this wave" is always trying to move faster than the "front of this wave," the wave will remain thin. Because the change in pressure across the merged wave, i.e., $dp + dp + dp + \cdots$ will, in general, be large, the temperature gradients in the wave will not be small and the flow process, unlike that across a single weak wave, cannot be assumed to be isentropic. This thin merged single wave across which large changes in pressure, temperature, etc. occur and across which the flow is not isentropic is a shock wave.

## 5.2
## STATIONARY NORMAL SHOCK WAVES

Attention will first be given to the changes that occur through a stationary normal shock wave. In order to analyze the flow through a stationary normal shock wave, consider a control volume of the form indicated in Fig. 5.9.

This control volume has a cross-sectional area of $A$ normal to the flow direction. The shock wave relations are obtained by applying the laws of conservation of mass, momentum and energy to this control volume.

If the mass flow rate through the control volume is $\dot{m}$, conservation of mass gives:

$$\dot{m} = \rho_1 V_1 A = \rho_2 V_2 A$$

i.e.,
$$\rho_1 V_1 = \rho_2 V_2 \tag{5.1}$$

Since the only forces acting on the control volume in the flow direction are the pressure forces, conservation of momentum applied to the control volume gives:

$$p_1 A - p_2 A = \dot{m}(V_2 - V_1) \tag{5.2}$$

Combining this with eq. (5.1) then gives:

$$p_1 - p_2 = \rho_1 V_1(V_2 - V_1) \tag{5.3}$$

or
$$p_1 - p_2 = \rho_2 V_2(V_2 - V_1) \tag{5.4}$$

These two equations can be rearranged to give:

$$V_1 V_2 - V_1^2 = \frac{p_1 - p_2}{\rho_1} \tag{5.5}$$

or
$$V_2^2 - V_2 V_1 = \frac{p_1 - p_2}{\rho_2} \tag{5.6}$$

Adding these two equations together then gives:

$$V_2^2 - V_1^2 = (p_1 - p_2)\left(\frac{1}{\rho_1} + \frac{1}{\rho_2}\right) \tag{5.7}$$

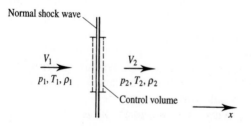

## FIGURE 5.9
Control volume used in analysis of a normal shock wave.

Lastly, consider the application of conservation of energy to the flow across the shock wave. Because one-dimensional flow is being considered there are no changes in the flow properties in any direction that is normal to that of the flow and, because the upstream and downstream surfaces of the control volume lie upstream and downstream of the shock wave, there are no temperature gradients normal to any surface of the control volume. The flow through the control volume is, therefore, adiabatic and the energy equation, therefore, gives:

$$\frac{V_1^2}{2} + c_p T_1 = \frac{V_2^2}{2} + c_p T_2 = c_p T_0 = \text{constant} \tag{5.8}$$

The stagnation temperature therefore does not change across the shock. Using, as before, $p/\rho = RT$ and $R = c_p - c_v$, this equation can be written as:

$$V_1^2 + \left(\frac{2\gamma}{\gamma - 1}\right)\frac{p_1}{\rho_1} = V_2^2 + \left(\frac{2\gamma}{\gamma - 1}\right)\frac{p_2}{\rho_2} \tag{5.9}$$

which can be rearranged to give:

$$V_2^2 - V_1^2 = \left(\frac{2\gamma}{\gamma - 1}\right)\left(\frac{p_1}{\rho_1} - \frac{p_2}{\rho_2}\right) \tag{5.10}$$

Using eqs. (5.10) and (5.7) together then gives:

$$\left(\frac{2\gamma}{\gamma - 1}\right)\left(\frac{p_2}{\rho_2} - \frac{p_1}{\rho_1}\right) = (p_2 - p_1)\left(\frac{1}{\rho_1} + \frac{1}{\rho_2}\right) \tag{5.11}$$

Now, a relationship between the density ratio, $\rho_2/\rho_1$, and the pressure ratio, $p_2/p_1$, is being sought. This can be obtained by multiplying eq. (5.11) by $\rho_2/\rho_1$ to give:

$$\left(\frac{2\gamma}{\gamma - 1}\right)\left(\frac{p_2}{p_1} - \frac{\rho_2}{\rho_1}\right) = \left(\frac{p_2}{p_1} - 1\right)\left(\frac{\rho_2}{\rho_1} + 1\right) \tag{5.12}$$

This can be rearranged to give:

$$\frac{p_2}{p_1} = \frac{\left[\left(\dfrac{\gamma + 1}{\gamma - 1}\right)\dfrac{\rho_2}{\rho_1} - 1\right]}{\left[\left(\dfrac{\gamma + 1}{\gamma - 1}\right) - \dfrac{\rho_2}{\rho_1}\right]} \tag{5.13}$$

Alternatively, it could have been arranged to give:

$$\frac{\rho_2}{\rho_1} = \frac{\left[\left(\dfrac{\gamma + 1}{\gamma - 1}\right)\dfrac{p_2}{p_1} + 1\right]}{\left[\left(\dfrac{\gamma + 1}{\gamma - 1}\right) + \dfrac{p_2}{p_1}\right]} \tag{5.14}$$

It is next noted that the continuity equation, i.e., eq. (5.1), gives:

$$\frac{V_1}{V_2} = \frac{\rho_2}{\rho_1} \tag{5.15}$$

Hence, using eq. (5.14) gives:

$$\frac{V_1}{V_2} = \frac{\left[\left(\dfrac{\gamma+1}{\gamma-1}\right)\dfrac{p_2}{p_1} + 1\right]}{\left[\left(\dfrac{\gamma+1}{\gamma-1}\right) + \dfrac{p_2}{p_1}\right]} \tag{5.16}$$

The temperature ratio across the shock wave is obtained by noting that the equation of state gives:

$$p_1 = \rho_1 R T_1, \qquad p_2 = \rho_2 R T_2 \tag{5.17}$$

which can be rearranged as:

$$\frac{T_2}{T_1} = \frac{p_2}{p_1}\frac{\rho_1}{\rho_2} \tag{5.18}$$

Using eq. (5.14), this equation gives:

$$\frac{T_2}{T_1} = \frac{\left[\left(\dfrac{\gamma+1}{\gamma-1}\right) + \dfrac{p_2}{p_1}\right]}{\left[\left(\dfrac{\gamma+1}{\gamma-1}\right) + \dfrac{p_1}{p_2}\right]} \tag{5.19}$$

Equations (5.14), (5.16), and (5.19) relate the changes in density, velocity, and temperature across a normal shock wave to the change in pressure across the shock wave. The pressure ratio, $p_2/p_1$, is often termed the strength of the shock wave. These equations, therefore, give $\rho_2/\rho_1$, $V_2/V_1$, and $T_2/T_1$ in terms of the shock strength. This set of equations is often termed the Rankine–Hugoniot normal shock wave relations. Now it will be noted that eq. (5.7) can be rearranged to give:

$$V_1^2\left[\left(\frac{V_2}{V_1}\right)^2 + 1\right] = \left(\frac{p_1}{\rho_1}\right)\left(1 - \frac{p_2}{p_1}\right)\left(1 + \frac{\rho_1}{\rho_2}\right)$$

Because $\rho_2/\rho_1$ and $V_2/V_1$ have been shown to be functions of $p_2/p_1$, it follows from this equation that for a particular value of $p_2/p_1$ there is an associated particular value of:

$$\frac{V_1^2}{(p_1/\rho_1)} = \frac{V_1^2}{a_1^2/\gamma} = \gamma M_1^2$$

i.e., a particular shock strength is associated with a particular upstream Mach number. This will be discussed further in the next section. Before doing this, the entropy changes across the shock will be discussed.

While the application of conservation of mass, momentum and energy principles shows that a shock wave can exist, it does not indicate whether the shock can be either compressive (i.e., $p_2/p_1 > 1$) or expansive (i.e.,

$p_2/p_1 < 1$). To examine this, the second law of thermodynamics must be used. Now the entropy change across the shock wave is given by:

$$s_2 - s_1 = c_p \ln \left( \frac{T_2}{T_1} \right) - R \ln \left( \frac{p_2}{p_1} \right)$$

$$= (R + c_v) \ln \left( \frac{p_2 \, \rho_1}{p_1 \, \rho_2} \right) - R \ln \left( \frac{p_2}{p_1} \right) \tag{5.20}$$

This equation can be rearranged to give:

$$\frac{s_2 - s_1}{R} = \left( 1 + \frac{1}{\gamma - 1} \right) \ln \left( \frac{p_2 \, \rho_1}{p_1 \, \rho_2} \right) - \ln \left( \frac{p_2}{p_1} \right)$$

$$= \ln \left[ \left( \frac{p_2}{p_1} \right)^{\frac{1}{\gamma - 1}} \left( \frac{\rho_2}{\rho_1} \right)^{\frac{-\gamma}{\gamma - 1}} \right] \tag{5.21}$$

Equation (5.14) can then be substituted into this equation to give the entropy increase as a function of the shock strength, $p_2/p_1$. This gives:

$$\frac{s_2 - s_1}{R} = \ln \left\{ \left( \frac{p_2}{p_1} \right)^{\frac{1}{\gamma - 1}} \left[ \frac{\left( \frac{\gamma + 1}{\gamma - 1} \right) \frac{p_2}{p_1} + 1}{\left( \frac{\gamma + 1}{\gamma - 1} \right) + \frac{p_2}{p_1}} \right]^{\frac{-\gamma}{\gamma - 1}} \right\}$$

i.e.: $$\frac{s_2 - s_1}{R} = \ln \left\{ \left( \frac{p_2}{p_1} \right)^{\frac{1}{\gamma - 1}} \left[ \frac{(\gamma + 1)\frac{p_2}{p_1} + (\gamma + 1)}{(\gamma + 1) + (\gamma - 1)\frac{p_2}{p_1}} \right]^{\frac{-\gamma}{\gamma - 1}} \right\} \tag{5.22}$$

Now, the second law of thermodynamics requires that the entropy must remain unchanged or must increase, i.e., it requires that

$$\frac{s_2 - s_1}{R} \geq 0 \tag{5.23}$$

The variation of $(s_2 - s_1)/R$ with $p_2/p_1$ for various values of $\gamma$ ($\gamma$ is always greater than 1) as given by eq. (5.22) is shown in Fig. 5.10.

It will be seen from the results given in Fig. 5.10 that for eq. (5.23) to be satisfied, it is necessary that

$$p_2/p_1 \geq 1 \tag{5.24}$$

It therefore follows that the shock wave must always be compressive, i.e., that $p_2/p_1$ must be greater than 1, i.e., the pressure must always increase across the shock wave. Using eqs. (5.14), (5.16), and (5.19) then indicates that the density always increases, the velocity always decreases, and the temperature always increases across a shock wave.

The entropy increase across the shock is, basically, the result of the fact that, because the shock wave is very thin, the gradients of velocity and

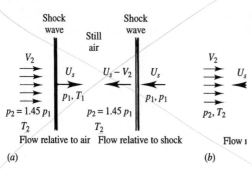

**FIGURE 5.10**
Effect of pressure ratio on entropy change across a normal shock wave.

temperature in the shock are very high. As a result, the effects of viscosity and heat conduction are important within the shock leading to the entropy increase across the shock wave.

Because the flow across a shock is adiabatic, the stagnation temperature does not change across a shock wave; see eq. (5.8). However, because of the entropy increase across a shock, the stagnation pressure always decreases across a shock wave. This is, perhaps, most easily shown by considering the flow situation shown in Fig. 5.11.

In the situation being considered, a gas flows from a large reservoir in which the velocity is effectively zero and is isentropically expanded until the Mach number is $M_1$. A normal shock wave then occurs. After the shock, the flow is isentropically decelerated until the velocity is again effectively zero in a second large reservoir. Since the flow is isentropic everywhere except across the shock wave, the pressure in the first reservoir, $p_{01}$, is the stagnation pressure everywhere in the flow ahead of the shock wave while the pressure in the second reservoir, $p_{02}$, is the stagnation pressure everywhere in the flow downstream of the shock wave. Now, eq. (5.20) applies between any two points in the flow. It can, therefore, be applied between a point in the first reservoir and a point in the second reservoir to give:

$$s_{02} - s_{01} = c_p \ln \left( \frac{T_{02}}{T_{01}} \right) - R \ln \left( \frac{p_{02}}{p_{01}} \right) \tag{5.25}$$

But the stagnation temperature does not change across the shock, so the first term on the right hand side of eq. (5.25) is zero, i.e., eq. (5.25) gives:

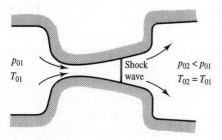

**FIGURE 5.11**
Flow situation used in analysis of stagnation pressure change across a normal shock wave.

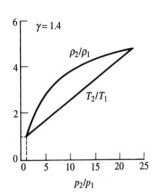

**FIGURE 5.12**
Variation of changes across a normal shock wave with the pressure ratio.

$$s_{02} - s_{01} = -R \ln \left( \frac{p_{02}}{p_{01}} \right) \tag{5.26}$$

However, since the flow is isentropic before and after the shock wave:

$$s_1 = s_{01} \quad \text{and} \quad s_2 = s_{02} \tag{5.27}$$

Therefore, eq. (5.26) gives:

$$\frac{s_2 - s_1}{R} = -\ln \left( \frac{p_{02}}{p_{01}} \right) \tag{5.28}$$

or:

$$\frac{p_{02}}{p_{01}} = \exp \left[ -(s_2 - s_1)/R \right] \tag{5.29}$$

Because the entropy must increase across the shock wave, this equation shows that the stagnation pressure must decrease across a shock.

The variations of the changes that occur across a normal shock wave with shock strength are illustrated in Fig. 5.12.

**EXAMPLE 5.1**
A normal shock occurs at a point in an air-flow where the pressure is 30 kPa and the temperature is $-30°C$. If the pressure ratio across this shock wave is 2.7, find the pressure and temperature downstream (i.e., after) this normal shock wave and the velocities both upstream of and downstream of the shock wave. Also find the change in the stagnation pressure across the shock.

**Solution**
For the shock wave being considered, $p_2/p_1 = 2.7$ and $\gamma = 1.4$. Hence since:

$$\frac{T_2}{T_1} = \frac{\left[ \left( \dfrac{\gamma + 1}{\gamma - 1} \right) + \dfrac{p_2}{p_1} \right]}{\left[ \left( \dfrac{\gamma + 1}{\gamma - 1} \right) + \dfrac{p_1}{p_2} \right]}$$

it follows that:

$$T_2/T_1 = (6 + 2.7)/(6 + 1/2.7) = 1.366$$

Also since:

$$\frac{p_2}{\rho_1} = \frac{\left[\left(\dfrac{\gamma+1}{\gamma-1}\right)\dfrac{p_2}{p_1}+1\right]}{\left[\left(\dfrac{\gamma+1}{\gamma-1}\right)+\dfrac{p_2}{p_1}\right]}$$

it follows that:

$$\rho_2/\rho_1 = (6 \times 2.7 + 1)/(6 + 2.7) = 1.977$$

Further since $V_1/V_2 = \rho_2/\rho_1$, it follows that:

$$V_1/V_2 = 1.977$$

From these results it follows that $p_2 = 2.7 \times 30 = 81\ \text{kPa}$ and $T_2 = (273 - 30) \times 1.366 = 331.9\ \text{K} = 58.9°\text{C}$.

One way to find the velocities is to recall that the energy equation gives:

$$V_2^2 - V_1^2 = \left(\frac{2\gamma}{\gamma-1}\right)\left(\frac{p_1}{\rho_1} - \frac{p_2}{\rho_2}\right)$$

But the perfect gas law gives $\rho_1 = p_1/RT_1 = 30\,000/(287.04 \times 243) = 0.43\ \text{kg/m}^3$. Therefore, since $V_1/V_2 = 1.977$, it follows that:

$$\left(\frac{1}{1.977^2} - 1\right)V_1^2 = 7\left(\frac{30\,000}{0.43} - \frac{81\,000}{1.977 \times 0.43}\right)$$

This equation gives $V_1 = 489.9\ \text{m/s}$ and so $V_2 = 489.9/1.977 = 247.8\ \text{m/s}$. Hence

$$M_1 = 489.9/\sqrt{1.4 \times 287.04 \times 243} = 1.568$$

and

$$M_2 = 247.8/\sqrt{1.4 \times 287.04 \times 331.9} = 0.679$$

Hence, since for a Mach number of 1.568, $p_0/p = 4.057$, whereas for a Mach number of 0.679, $p_0/p = 1.361$, these values being obtained either using the relationship given in Chapter 4 or using isentropic tables or using the software. The change in stagnation pressure across the shock wave is then given by:

$$\Delta p_0 = \left(\frac{p_{02}\,p_2}{p_2\,p_1} - \frac{p_{01}}{p_1}\right)p_1$$

$$= (1.361 \times 2.7 - 4.057) \times 30 = -11.47\ \text{kPa}$$

i.e., the stagnation pressure decreases by 11.47 kPa across the shock wave. Because the flow through the wave is adiabatic, there is, of course, no change in stagnation temperature through the wave.

## 5.3
## NORMAL SHOCK WAVE RELATIONS IN TERMS OF MACH NUMBER

Although the relations derived in the previous section for the changes across a normal shock in terms of the pressure ratio across the shock, i.e., in terms of the shock strength, are the most useful form of the normal shock wave relations for some purposes, it is often more convenient to have these relations in terms of the upstream Mach number, $M_1$. To obtain these forms of the normal shock wave relations, it is convenient to start again with a control volume across the shock wave such as that shown in Fig. 5.13, and to again apply conservation of mass, momentum, and energy to this control volume but in this case to rearrange the relations in terms of Mach number.

In expressing the conservation laws, no generality is lost by taking the area of the control volume parallel to the wave as unity. Conservation of mass then gives:

$$\rho_1 V_1 = \rho_2 V_2 \tag{5.30}$$

Dividing this equation by $a_1$ then gives:

$$\rho_1 \frac{V_1}{a_1} = \rho_2 \frac{V_2}{a_2} \frac{a_2}{a_1}$$

which can be rewritten in terms of Mach numbers as:

$$\frac{\rho_2}{\rho_1} = \frac{M_1}{M_2} \frac{a_1}{a_2} \tag{5.31}$$

Next consider conservation of momentum. This gives for the control volume shown in Fig. 5.13:

$$p_1 - p_2 = \rho_2 V_2^2 - \rho_1 V_1^2 \tag{5.32}$$

Hence, since:

$$a^2 = \frac{\gamma p}{\rho}, \quad \text{i.e.,} \quad p = \frac{a^2 \rho}{\gamma}$$

eq. (5.32) becomes:

$$\frac{a_1^2 \rho_1}{\gamma} + \rho_1 V_1^2 = \frac{a_2^2 \rho_2}{\gamma} + \rho_2 V_2^2$$

Dividing this through by $a_1^2/\gamma$ then gives:

$$\rho_1 + \gamma \rho_1 M_1^2 = \rho_2 \left(\frac{a_2}{a_1}\right)^2 + \gamma \rho_2 M_2^2 \left(\frac{a_2}{a_1}\right)^2$$

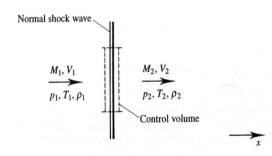

Normal shock wave

$M_1, V_1$

$p_1, T_1, \rho_1$

$M_2, V_2$

$p_2, T_2, \rho_2$

Control volume

$x$

**FIGURE 5.13**
Control volume used in deriving expressions for the changes across a normal shock wave in terms of the upstream Mach number.

which can be rearranged to give:

$$\frac{\rho_2}{\rho_1} = \left(\frac{1+\gamma M_1^2}{1+\gamma M_2^2}\right)\left(\frac{a_1}{a_2}\right)^2 \tag{5.33}$$

Lastly, consider the application of the conservation of energy principle to the control volume. This gives:

$$V_1^2 + \left(\frac{2}{\gamma - 1}\right)a_1^2 = V_2^2 + \left(\frac{2}{\gamma - 1}\right)a_2^2 \tag{5.34}$$

Dividing this equation by $2a_1^2/(\gamma - 1)$ then gives:

$$\left(\frac{\gamma - 1}{2}\right)M_1^2 + 1 = \left(\frac{\gamma - 1}{2}\right)M_2^2\left(\frac{a_2}{a_1}\right)^2 + \left(\frac{a_2}{a_1}\right)^2$$

which can be rearranged to give:

$$\left(\frac{a_2}{a_1}\right)^2 = \left[\frac{2+(\gamma - 1)M_1^2}{2+(\gamma - 1)M_2^2}\right] \tag{5.35}$$

The density ratio, $\rho_2/\rho_1$, is now eliminated between eqs (5.31) and (5.33) giving:

$$\left(\frac{a_2}{a_1}\right) = \left(\frac{1+\gamma M_1^2}{1+\gamma M_2^2}\right)\left(\frac{M_2}{M_1}\right) \tag{5.36}$$

The speed of sound ratio, $a_2/a_1$, is next eliminated between eqs (5.35) and (5.36) to give:

$$\frac{2+(\gamma - 1)M_1^2}{2+(\gamma - 1)M_2^2} = \left(\frac{1+\gamma M_1^2}{1+\gamma M_2^2}\right)^2\left(\frac{M_2}{M_1}\right)^2$$

This equation can be rearranged to give:

$$(\gamma - 1)(M_2^4 - M_1^4) - 2\gamma M_2^2 M_1^2(M_2^2 - M_1^2) + 2(M_2^2 - M_1^2) = 0 \tag{5.37}$$

But $(M_2^2 - M_1^2)$ cannot be zero as this would imply that there was no change in the Mach number across the shock wave. This term can therefore be cancelled out of eq. (5.37), giving:

$$M_2^2 = \frac{\left[M_1^2 + \left(\frac{2}{\gamma - 1}\right)\right]}{\left[\left(\frac{2\gamma}{\gamma - 1}\right)M_1^2 - 1\right]} = \left[\frac{(\gamma - 1)M_1^2 + 2}{2\gamma M_1^2 - (\gamma - 1)}\right] \tag{5.38}$$

This equation relates the downstream Mach number to the upstream Mach number. It can be used to derive expressions for the pressure ratio, the temperature ratio, and the density ratio in terms of the upstream Mach number. It is first noted that substituting eq. (5.38) into eq. (5.35) gives:

$$\left(\frac{a_2}{a_1}\right)^2 = \frac{T_2}{T_1} = \frac{[2 + (\gamma - 1)M_1^2]}{\left\{2 + (\gamma - 1)\left[\frac{(\gamma - 1)M_1^2 + 2}{2\gamma M_1^2 - (\gamma - 1)}\right]\right\}}$$

$$= \left\{\frac{[2\gamma M_1^2 - (\gamma - 1)][2 + (\gamma - 1)M_1^2]}{(\gamma + 1)^2 M_1^2}\right\} \tag{5.39}$$

In order to obtain an expression for the pressure ratio it is noted that the equation of state gives:

$$\frac{p_2}{p_1} = \left(\frac{\rho_2}{\rho_1}\right)\left(\frac{T_2}{T_1}\right) = \left(\frac{\rho_2}{\rho_1}\right)\left(\frac{a_2}{a_1}\right)^2 \tag{5.40}$$

Hence, using eq. (5.31), the following is obtained:

$$\frac{p_2}{p_1} = \frac{1 + \gamma M_1^2}{1 + \gamma M_2^2} = \frac{(1 + \gamma M_1^2)}{\left\{1 + \gamma\left[\frac{(\gamma - 1)M_1^2 + 2}{2\gamma M_1^2 - (\gamma - 1)}\right]\right\}} \tag{5.41}$$

The right hand side of this equation can be written as:

$$\frac{(1 + \gamma M_1^2)[2\gamma M_1^2 - (\gamma - 1)]}{[2\gamma M_1^2 - (\gamma - 1)] + \gamma[(\gamma - 1)M_1^2 + 2]}$$

i.e., as

$$\frac{(1 + \gamma M_1^2)[2\gamma M_1^2 - (\gamma - 1)]}{2\gamma M_1^2 - \gamma + 1 + \gamma^2 M_1^2 - \gamma M_1^2 + 2\gamma}$$

i.e., as

$$\frac{(1 + \gamma M_1^2)[2\gamma M_1^2 - (\gamma - 1)]}{\gamma M_1^2 + \gamma^2 M_1^2 + \gamma + 1}$$

i.e., as

$$\frac{(1 + \gamma M_1^2)[2\gamma M_1^2 - (\gamma - 1)]}{(\gamma + 1)(1 + \gamma M_1^2)}$$

Hence, eq. (5.41) gives the pressure ratio as:

$$\frac{p_2}{p_1} = \frac{2\gamma M_1^2 - (\gamma - 1)}{(\gamma + 1)} \tag{5.42}$$

The density ratio can now be directly obtained by again noting that the equation of state gives:

$$\frac{\rho_2}{\rho_1} = \left(\frac{p_2}{p_1}\right)\left(\frac{T_1}{T_2}\right)$$

Hence, using eqs. (5.42) and (5.39), the following is obtained:

$$\frac{\rho_2}{\rho_1} = \frac{(\gamma + 1)M_1^2}{2 + (\gamma - 1)M_1^2} \tag{5.43}$$

The stagnation pressure ratio across a normal shock wave is obtained by noting that:

$$\frac{p_0}{p} = \left(1 + \frac{(\gamma - 1)}{2} M^2\right)^{\gamma/(\gamma-1)} \tag{5.44}$$

Hence, since:

$$\frac{p_{02}}{p_{01}} = \frac{p_{02}/p_2}{p_{01}/p_1} \frac{p_2}{p_1}$$

using eqs. (5.42) and (5.44), the stagnation pressure change across a normal shock is given by:

$$\frac{p_{02}}{p_{01}} = \left\{ \frac{\left(1 + \frac{(\gamma - 1)}{2} M_2^2\right)}{\left(1 + \frac{(\gamma - 1)}{2} M_1^2\right)} \right\}^{\gamma/(\gamma-1)} \left\{ \frac{2\gamma M_1^2 - (\gamma - 1)}{(\gamma + 1)} \right\}$$

i.e., using eq. (5.38) to give $M_2$ and rearranging gives:

$$\frac{p_{02}}{p_{01}} = \left\{ \frac{(\gamma + 1)}{2} \frac{M_1^2}{\left(1 + \frac{(\gamma - 1)}{2} M_1^2\right)} \right\}^{\gamma/(\gamma-1)}$$

$$\left\{ \left(\frac{2\gamma}{\gamma + 1}\right) M_1^2 - \left(\frac{\gamma - 1}{\gamma + 1}\right) \right\}^{-1/(\gamma-1)} \tag{5.45}$$

The stagnation temperature does not, of course, as mentioned before change across the shock wave.

The above equations, some of which are summarized below, give the pressure ratio, the density ratio, the temperature ratio, the downstream Mach number, etc. in terms of the upstream Mach number for any gas, i.e., for any value of $\gamma$.

$$\frac{p_2}{p_1} = \frac{2\gamma M_1^2 - (\gamma - 1)}{(\gamma + 1)}, \quad \frac{\rho_2}{\rho_1} = \frac{(\gamma + 1)M_1^2}{2 + (\gamma - 1)M_1^2}$$

$$\frac{T_2}{T_1} = \left\{ \frac{[2\gamma M_1^2 - (\gamma - 1)][2 + (\gamma - 1)M_1^2]}{(\gamma + 1)^2 M_1^2} \right\}$$

$$M_2^2 = \frac{(\gamma - 1)M_1^2 + 2}{2\gamma M_1^2 - (\gamma - 1)}$$

The variations of pressure ratio, density ratio, temperature ratio, and downstream Mach number with upstream Mach number given by these equations are shown in Fig. 5.14 for the case of $\gamma = 1.4$.

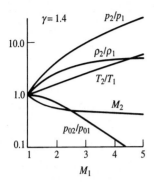

**FIGURE 5.14**
Variation of changes across normal shock with upstream Mach number.

## EXAMPLE 5.2

A gas which has a molar mass of 39.9 and a specific heat ratio of 1.67 is discharged through a nozzle. A normal shock wave occurs at a section of the flow at which the Mach number is 2.5, the pressure is 40 kPa and the temperature is $-20°C$. Find the Mach number, pressure, and temperature downstream (i.e., after) this normal shock wave.

**Solution**

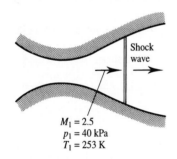

**FIGURE E5.2**

Shock
wave

$M_1 = 2.5$
$p_1 = 40$ kPa
$T_1 = 253$ K

Normal shock relations give:

$$M_2^2 = \frac{M_1^2 + \dfrac{2}{(\gamma - 1)}}{\dfrac{2\gamma}{(\gamma - 1)} M_1^2 - 1} = \frac{2.5^2 + \dfrac{2}{(1.67 - 1)}}{\dfrac{2 \times 1.67}{(1.67 - 1)} \times 2.5^2 - 1} = 0.306$$

Hence: $M_2 = 0.553$

Normal shock relations also give:

$$\frac{T_2}{T_1} = \frac{[2\gamma M_1^2 - (\gamma - 1)][2 + (\gamma - 1)M_1^2]}{(\gamma - 1)^2 \times M_1^2}$$

hence: $$\frac{T_2}{T_1} = \frac{[2 \times 1.67 \times 2.5^2 - (1.67 - 1)]}{(1.67 + 1)^2 \times 2.5^2} = 2.806$$

and so:
$$T_2 = 710 \text{ K} = 437°\text{C}$$

Also, using normal shock relations:

$$\frac{p_2}{p_1} = \frac{2\gamma M_1^2 - (\gamma - 1)}{(\gamma + 1)}$$

$$= \frac{2 \times 1.67 \times 2.5^2 - (1.67 - 1)}{1.67 + 1} = 7.5674$$

and so
$$p_2 = 303 \text{ kPa}$$

Therefore, the Mach number, the pressure and temperature behind the shock wave are 0.553, 303 kPa, and 437°C respectively.

### EXAMPLE 5.3

An aircraft flying at a Mach number of 1.2 at sea-level passes over a building. Estimate the highest force that could be exerted on a 1 m wide by 2 m high window in this building as a result of the aircraft flying over it.

### Solution

The highest possible pressure would be exerted if the aircraft essentially had a normal shock ahead of it which, as the aircraft flew over the building, sharply increased the pressure on the outside of the window to the value behind the shock wave and if the pressure inside the building essentially remained, for a short while, at the initial ambient pressure.

Assuming that $p_1 = 101.3$ kPa, the pressure behind the shock wave, which is assumed to be a normal shock wave, is given by:

$$\frac{p_2}{p_1} = \frac{2\gamma M_1^2 - (\gamma - 1)}{(\gamma + 1)}$$

i.e., $p_2/p_1 = 1.513$. Hence, the maximum force on the window is given by $(p_1 - p_2) \times$ Area, i.e., by $(1.513 \times 101.3 - 101.3) \times (1 \times 2) = 103.9$ kN. If it is recalled that the average weight of a person is approximately 0.67 kN, it will be realized that the force exerted on the window by the passage of the aircraft has the potential to break the window.

Consideration will lastly again be given to the change in entropy across a normal shock wave in terms of the upstream Mach number. Now it was shown above, see eq. (5.21), that the change in entropy is given by:

$$\frac{s_2 - s_1}{R} = \left(1 + \frac{1}{\gamma - 1}\right) \ln\left(\frac{p_2 \, \rho_1}{p_1 \, \rho_2}\right) - \ln\left(\frac{p_2}{p_1}\right)$$

$$= \ln\left[\left(\frac{p_2}{p_1}\right)^{\frac{1}{\gamma-1}} \left(\frac{p_2}{\rho_1}\right)^{\frac{-\gamma}{\gamma-1}}\right] \qquad (5.46)$$

The right hand side of this equation can be expressed in terms of the upstream Mach number by using the relationships derived above for the pressure and density ratios. Using these gives:

$$\frac{s_2 - s_1}{R} = \ln\left\{ \left[\frac{2\gamma}{\gamma+1}(M_1^2 - 1) + 1\right]^{\frac{1}{\gamma-1}} \left[\frac{(\gamma+1)M_1^2}{2+(\gamma-1)M_1^2}\right]^{\frac{-\gamma}{\gamma-1}} \right\} \tag{5.47}$$

The variation of $(s_2 - s_1)/R$ with $M_1$ as given by this equation for various values of $\gamma$ is shown in Fig. 5.15.

Now, the entropy must remain unchanged or must increase. It will be seen that this can only be the case if

$$M_1 \geq 1 \tag{5.48}$$

It therefore follows that the Mach number ahead of a shock wave must always be greater than 1 and that the shock wave must, therefore, as discussed above, always be compressive, i.e., $p_2/p_1$ must be greater than 1. It will also be noted that eq. (5.38) can be rearranged to give:

$$M_2^2 = \frac{M_1^2 + \left(\dfrac{2}{\gamma-1}\right)}{\left(\dfrac{2\gamma}{\gamma-1}\right)M_1^2 - 1} \tag{5.49}$$

Hence, since $\gamma$ will be between 1 and 2, and since $M_1$ is always greater than 1, it follows from this equation that $M_2$ will always be less than 1, i.e., the flow downstream of a normal shock wave will always be subsonic. These conclusions about a normal shock wave are summarized in Fig. 5.16.

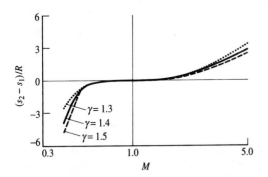

**FIGURE 5.15**
Variation of entropy change across normal shock with upstream Mach number.

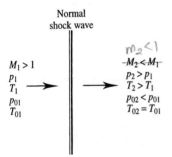

**FIGURE 5.16**
Required values of variables before and after a normal shock.

**Limiting Cases of Normal Shock Wave Relations**

It is instructive to consider the limiting case of a very strong normal shock, i.e., a normal shock wave for which $M_1$ is very large. Now, if $M_1$ is very large, the equations given above for a normal shock indicate that:

For $M_1 \gg 1$:

$$\frac{p_2}{p_1} = \frac{2\gamma M_1^2}{(\gamma + 1)}, \qquad \frac{\rho_2}{\rho_1} = \frac{(\gamma + 1)}{(\gamma - 1)}$$

$$\frac{T_2}{T_1} = \left\{ \frac{2\gamma(\gamma - 1)M_1^2}{(\gamma + 1)^2} \right\}$$

$$M_2^2 = \frac{(\gamma - 1)}{2\gamma}$$

Thus, if $M_1$ tends to infinity, $p_2/p_1$ and $T_2/T_1$ tend to infinity but $\rho_2/\rho_1$ tends to $(\gamma + 1)/(\gamma - 1)$ and $M_2$ tends to $\sqrt{(\gamma - 1)/2\gamma}$. In actual fact, the assumptions on which the above analysis of the changes across a normal shock wave are based, i.e., that the gas remains thermally and calorically perfect, will cease to be valid when the shock is very strong because very high temperatures will then usually exist behind the shock.

The above discussion concerned the flow across a very strong shock wave. Another limiting case is that of a very weak normal shock wave. Now the discussion of the entropy change across a normal shock indicates that in this weak shock case the flow is isentropic, i.e., that the relations for the pressure and density ratios derived for isentropic flow in Chapter 4 apply across such weak shocks. Now, the continuity equation applies across the shock whether or not the flow is isentropic. Therefore, even in the case of the weak shock limit, eq. (5.31) applies, i.e., the following applies:

$$\frac{\rho_2 a_2}{\rho_1 a_1} = \frac{M_1}{M_2} \tag{5.50}$$

But in isentropic flow:

$$\frac{\rho_2}{\rho_1} = \left( \frac{T_2}{T_1} \right)^{1/\gamma - 1}$$

Therefore, noting that:

$$\frac{a_2}{a_1} = \left( \frac{T_2}{T_1} \right)^{1/2}$$

eq. (5.50) gives:

$$\frac{M_2}{M_1} = \left( \frac{T_1}{T_2} \right)^{(\gamma+1)/2(\gamma-1)} \tag{5.51}$$

But, in adiabatic flow and therefore in isentropic flow the energy equation gives:

$$\frac{T_1}{T_2} = \frac{\left(1 + \frac{(\gamma - 1)}{2} M_2^2\right)}{\left(1 + \frac{(\gamma - 1)}{2} M_1^2\right)} \tag{5.52}$$

Substituting this into eq. (5.51) then gives:

$$\frac{M_2}{M_1} = \left\{ \frac{\left(1 + \frac{(\gamma - 1)}{2} M_2^2\right)}{\left(1 + \frac{(\gamma - 1)}{2} M_1^2\right)} \right\}^{(\gamma+1)/2(\gamma-1)} \tag{5.53}$$

This equation gives the value of the downstream Mach number, $M_2$, corresponding to any specified value of the upstream Mach number, $M_1$, for a very weak shock wave. Once $M_2$ is found, the changes in pressure, density, and temperature across the weak shock can be found by using the isentropic relations in conjunction with eq. (5.52). This procedure gives, for example,

$$\frac{p_2}{p_1} = \left\{ \frac{\left(1 + \frac{(\gamma - 1)}{2} M_1^2\right)}{\left(1 + \frac{(\gamma - 1)}{2} M_2^2\right)} \right\}^{\gamma/2(\gamma-1)} \tag{5.54}$$

To illustrate the relation between the strong shock, the weak shock, and the actual normal shock relations, the variations of $M_2$ with $M_1$ given by these relations is shown in Fig. 5.17 for the case of $\gamma = 1.4$.

It will be seen from the results given in Fig. 5.17 that the weak shock relations apply if $M_1 <$ about 1.1 while the strong shock relations only apply if $M_1$ is very large.

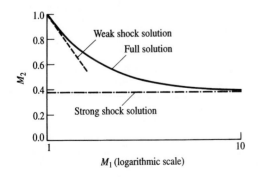

**FIGURE 5.17**
Relation between downstream Mach number values given by limiting normal shock wave solutions and the actual values given by the full normal shock relations ($\gamma = 1.4$).

## Normal Shock Wave Tables

A number of sets of tables and graphs are available which list the ratios of the various flow variables such as pressure, temperature, and density across a normal shock wave and the downstream Mach number as a function of the upstream Mach number for various gases, i.e., for various values of $\gamma$. Typical headings in such a set of normal shock tables are:

$$M_1 \qquad M_2 \qquad \frac{p_2}{p_1} \qquad \frac{T_2}{T_1} \qquad \frac{\rho_2}{\rho_1} \qquad \frac{a_2}{a_1} \qquad \frac{p_{20}}{p_{10}} \qquad \frac{p_{20}}{p_1}$$

The values in these tables and graphs are, of course, derived using the equations given in the previous section. Normal shock tables of this type are given in Appendix C for the case of $\gamma = 1.4$. As with isentropic flow, instead of using tables, it is often more convenient to use software, such as that available to support this book, to find the changes across a shock wave. Alternatively, most calculators can be programmed to give results for a normal shock wave. To illustrate how this can be done, a very simple BASIC program that allows $M_1$ or $p_2/p_1$ to be entered and which finds the values of some other changes across a normal shock wave is listed in Appendix L.

**EXAMPLE 5.4**
Air is expanded from a larger reservoir in which the pressure and temperature are 500 kPa and 35°C through a variable area duct. A normal shock occurs at a point in the duct where the Mach number is 2.5. Find the pressure and temperature in the flow just downstream of the shock wave. Downstream of the shock wave, the flow is brought to rest in another large reservoir. Find the pressure and temperature in this reservoir. Assume that the flow is one-dimensional and isentropic everywhere except through the shock wave.

*Solution*

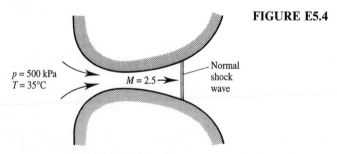

FIGURE E5.4

The flow upstream of the shock wave can be assumed to be isentropic so isentropic flow relations or tables or the software give for the flow upstream of the shock:

$$\frac{p_{01}}{p_1} = 17.085, \qquad \frac{T_{01}}{T_1} = 2.25$$

Using the specified stagnation point conditions, the following are then obtained:

$$p_1 = 500/17.085 = 29.3 \text{ kPa}, \qquad T_1 = 308/2.25 = 136.9 \text{ K}$$

Next consider the changes across the shock wave. For air flow at a Mach number of 2.5, normal shock relations or tables or the software gives:

$$M_2 = 0.5130, \qquad \frac{p_2}{p_1} = 7.215, \qquad \frac{T_2}{T_1} = 2.1375$$

From the above results, it follows that:

$$p_2 = 7.215 \times 29.3 = 208.8 \text{ kPa}$$

and: $$T_2 = 2.1375 \times 136.9 = 292.6 \text{ K} = 19.6°\text{C}$$

Therefore, the pressure and temperature immediately downstream of the shock wave are 208.8 kPa and 19.6°C respectively.

Lastly, consider the flow downstream of the shock. This flow is also assumed to be isentropic so isentropic flow relations or tables or the software give for the flow downstream of the shock:

$$\frac{p_{02}}{p_2} = 1.194, \qquad \frac{T_{02}}{T_2} = 1.052$$

Therefore the downstream stagnation conditions are as follows:

$$p_{02} = 208.8 \times 1.194 = 249.4 \text{ kPa}$$

$$T_{02} = 292.6 \times 1.052 = 308 \text{ K} = 35°\text{C}$$

Therefore, the pressure and temperature in the downstream reservoir are 249.4 kPa and 35°C respectively.

It will be noted that, because the entire flow is assumed to be adiabatic there is no change in the stagnation temperature. There is, however, as a result of the presence of the shock, an almost 50% loss of stagnation pressure.

## 5.4
## THE PITOT TUBE IN SUPERSONIC FLOW

Consider flow near the front of a blunt body placed in a supersonic flow as shown in Fig. 5.18. Because the flow is supersonic, a shock wave forms ahead of the body as shown in Fig. 5.18.

This shock wave is curved in general but ahead of the very front of the body, the shock is effectively normal to the flow. Hence, the conditions across the shock, i.e., between points 1 and 2 in Fig. 5.18, are related by the normal shock relations. Further, since the flow downstream of a normal shock wave is always subsonic, the deceleration from point 2 in Fig. 5.18 to point 3 in this figure where the velocity is effectively zero can, as discussed in the previous chapter, be assumed to be an isentropic process. Using this model of the flow, the pressure at the stagnation point can be calculated for any specified upstream conditions.

### EXAMPLE 5.5

Air flows over a blunt nosed body. The air flow in the freestream ahead of the body has a Mach number of 1.5 and a static pressure of 40 kPa. Find the pressure acting on the front of this body. Sketch the flow pattern near the nose.

### Solution

Here $M_1 = 1.5$ and $p_1 = 40$ kPa. But, normal shock relations or tables or software give:

$$\frac{p_{02}}{p_1} = 3.413$$

Hence:
$$p_{02} = 3.413 \times 40 = 136.5 \text{ kPa}$$

The flow pattern is as shown schematically in Fig. E5.5.

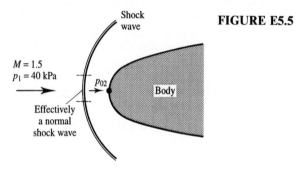

**FIGURE E5.5**

When a pitot tube is placed in a supersonic flow, a type of flow similar to that indicated in Fig. 5.18 occurs, i.e., the flow over a pitot tube in supersonic flow resembles that shown in Fig. 5.19.

Since there will be a change in stagnation pressure across the shock wave, it is not possible to use the subsonic pitot tube equation in supersonic flow. However, as noted above, over the small area of the flow covered by the pressure tap in the nose of the pitot tube the shock wave is effectively normal and the flow behind this portion of the shock wave is, therefore,

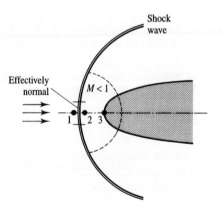

**FIGURE 5.18**
Supersonic flow over a blunt nosed body.

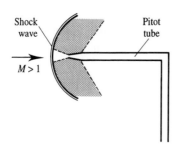

**FIGURE 5.19**

Supersonic flow near the nose of a pitot tube.

subsonic and the deceleration isentropic, these assumptions being illustrated in Fig. 5.20.

The flow can, therefore, be analyzed as follows:

1. The pressure ratio across the shock wave, $p_2/p_1$, can be found using normal shock wave relations.
2. The pressure at the stagnation point can be found by assuming that the isentropic relations apply between the flow behind the shock and the stagnation point.

Hence, since

$$\frac{p_{02}}{p_1} = \frac{p_{02}}{p_2}\frac{p_2}{p_1} \tag{5.55}$$

where the subscripts 1 and 2 denote the conditions upstream and downstream of the shock wave respectively, using the relations previously given, this equation becomes:

$$\frac{p_{02}}{p_1} = \left[1 + \left(\frac{\gamma - 1}{2}\right)M_2^2\right]^{\frac{\gamma}{\gamma-1}}\left[\frac{2\gamma M_1^2 - (\gamma - 1)}{(\gamma + 1)}\right] \tag{5.56}$$

Hence, using the expression for the downstream Mach number, it follows that:

$$\frac{p_{02}}{p_1} = \left\{1 + \left(\frac{\gamma - 1}{2}\left[\frac{(\gamma - 1)M_1^2 + 2}{2\gamma M_1^2 - (\gamma - 1)}\right]\right)\right\}\left[\frac{2\gamma M_1^2 - (\gamma - 1)}{(\gamma + 1)}\right]$$

$$= \frac{[(\gamma + 1)M_1^2/2]^{\frac{\gamma}{(\gamma-1)}}}{\left[\left(\frac{2\gamma M_1^2}{\gamma + 1}\right) - \left(\frac{\gamma - 1}{\gamma + 1}\right)\right]^{\frac{1}{(\gamma-1)}}} \tag{5.57}$$

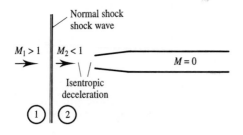

**FIGURE 5.20**

Assumed flow near the nose of a pitot tube.

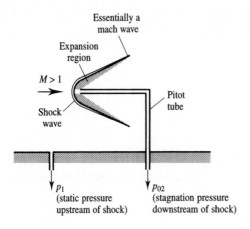

**FIGURE 5.21**
Use of a wall static pressure tap with a pitot tube in supersonic flow.

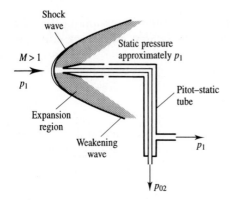

**FIGURE 5.22**
Pitot-static tube in supersonic flow.

This equation is known as the Rayleigh supersonic pitot tube equation. If $p_{02}$ and $p_1$ are measured, this equation allows $M_1$ to be found. The value of $p_{02}/p_1$ is usually listed in shock tables or given by software such as that included in this book. This fact was utilized in solving Example 5.5.

It should be noted that the static pressure ahead of the shock wave, i.e., $p_1$, must be measured. If the flow is very nearly parallel to a plane wall there will be essentially no static pressure changes normal to the flow direction and $p_1$ can then be found using a static hole in the wall as indicated in Fig. 5.21.

However, it has also been found that a pitot-static tube can also be used in supersonic flow since the shock wave interacts with the expansion waves (see later) decaying rapidly to a Mach wave and the pressure downstream of the vicinity of the nose of the pitot tube is thus essentially equal to $p_1$ again as indicated in Fig. 5.22.

**EXAMPLE 5.6**
A pitot-static tube is placed in a supersonic air flow. The static pressure and temperature in the flow are 45 kPa and −20°C respectively. The difference

between the pitot and static pressures is measured and found to be 350 kPa. Find the Mach number and the air velocity.

***Solution***
Using the supplied information gives:

$$p_{02}/p_1 = (350 + 45)/45 = 8.778$$

But for $p_{02}/p_1 = 8.778$, normal shock relations or tables or software give:

$$M_1 = 2.536$$

Therefore:     $$V_1 = M_1 \times a_1 = 2.536 \times \sqrt{1.4 \times 287 \times 253} = 809 \text{ m/s}$$

Hence the Mach number and velocity in the flow are 2.536 and 809 m/s respectively.

## 5.5
## MOVING NORMAL SHOCK WAVES

In the above discussion of normal shock waves, the coordinate system was so chosen that the shock wave was at rest. In many cases, however, it is necessary to derive results for the case where the shock wave is moving relative to the coordinate system. Consider the case where the gas ahead of the shock wave is stationary with respect to the coordinate system chosen and where the normal shock wave is moving into this stationary gas inducing a velocity in the direction of shock motion as indicated in Fig. 5.23. Such moving shock waves occur, for example, in the inlet and exhaust systems of internal combustion engines, in air compressors, as the result of explosions, and in pipelines following the opening or closing of a valve. The required results can be obtained from those that were derived above for a stationary normal shock wave by noting that the velocities relative to a coordinate system fixed to the shock wave are as indicated in Fig. 5.24. Hence, it follows that:

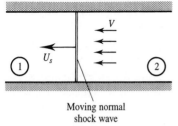

**FIGURE 5.23**
Moving normal shock wave.

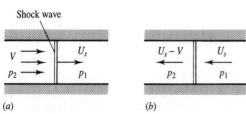

**FIGURE 5.24**
Relation between flow relative to walls and to moving normal shock wave. (*a*) Relative to walls; (*b*) relative to shock wave.

$$V_1 = U_s$$

$$V_2 = U_s - V \qquad (5.58)$$

Since the direction of the flow is obvious, only the magnitudes of the velocities will be considered here. The Mach numbers upstream and downstream of the shock wave relative to the shock wave are given by:

$$M_1 = \frac{U_s}{a_1} = M_s \qquad (5.59)$$

and
$$M_2 = \frac{U_s}{a_2} - \frac{V}{a_2} = \frac{U_s a_1}{a_1 a_2} - \frac{V}{a_2} = M_s \frac{a_1}{a_2} - M'_2 \qquad (5.60)$$

where
$$M_s = \frac{U_s}{a_1}$$

and
$$M'_2 = \frac{V}{a_2} \qquad (5.61)$$

$M_s$ is the "shock Mach number." Substituting the value of $M_1$ for the moving shock into the equations previously given for a stationary normal shock wave then gives:

$$\frac{p_2}{p_1} = \frac{2\gamma M_s^2 - (\gamma - 1)}{(\gamma + 1)} \qquad (5.62)$$

$$\frac{\rho_2}{\rho_1} = \frac{(\gamma + 1)M_s^2}{2 + (\gamma - 1)M_s^2} \qquad (5.63)$$

$$\frac{T_2}{T_1} = \left(\frac{a_2}{a_1}\right)^2 = \frac{[2 + (\gamma - 1)M_s^2][2\gamma M_s^2 - (\gamma - 1)]}{(\gamma + 1)^2 M_s^2} \qquad (5.64)$$

The gas velocity behind the shock wave can be obtained by substituting eq. (5.60) into eq. (5.38) and then using eq. (5.64). This leads to:

$$M'_2 = \frac{2(M_s^2 - 1)}{[2\gamma M_s^2 - (\gamma - 1)]^{0.5}[2 + (\gamma - 1)M_s^2]^{0.5}} \qquad (5.65)$$

It is perhaps worth noting that for an infinitely strong moving normal shock wave, i.e., for $M_s \to \infty$, the above equation shows that there is a limiting value for the Mach number downstream of the shock wave, $M'_2$, which is given by:

$$M'_2 \to \sqrt{\frac{2}{\gamma(\gamma - 1)}} \qquad (5.66)$$

Thus, for example, for air, a moving normal shock wave, no matter how strong, cannot generate a flow that has a Mach number that is greater than 1.89.

The actual velocity behind a moving normal shock wave, $V$, is given by:

$$V = M'_2 a_2 = M'_2 \left( \frac{a_2}{a_1} \right) a_1$$

$$= \frac{2(M_s^2 - 1)a_1[2 + (\gamma - 1)M_s^2]^{0.5}[2\gamma M_s^2 - (\gamma - 1)]^{0.5}}{[2\gamma M_s^2 - (\gamma - 1)]^{0.5}[2 + (\gamma - 1)M_s^2]^{0.5}(\gamma + 1)M_s}$$

$$= \frac{2(M_s^2 - 1)}{(\gamma + 1)M_s} a_1 \tag{5.67}$$

Normal shock wave tables and software can be used to evaluate the properties of a moving normal shock wave. To do this, $M_1$ is set equal to $M_s$ and the tables or software are then used to find the pressure, density, and temperature ratios directly. Further, since:

$$M'_2 = M_s \frac{a_1}{a_2} - M_2$$

and since $M_2$ is given by the normal shock tables or software, $M'_2$ can be found. Also since:

$$\frac{V}{a_1} = M'_2 \frac{a_2}{a_1} = M_s - M_2 \left( \frac{a_2}{a_1} \right)$$

$V$ can also be deduced using normal shock tables or software.

### EXAMPLE 5.7

A shock wave across which the pressure ratio is 1.25 is moving into still air at a pressure of 100 kPa and a temperature of 15°C. Find the velocity, pressure, and temperature of the air behind the shock wave.

### Solution

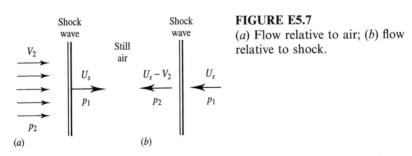

FIGURE E5.7
(a) Flow relative to air; (b) flow relative to shock.

Here:

$$p_1 = 100 \text{ kPa}, \qquad T_1 = 15°C, \qquad \frac{p_2}{p_1} = 1.25$$

For the given $p_2/p_1 = 1.25$, $M_1$, $M_2$, $T_2/T_1$ are obtained from the normal shock relations or tables or software. This gives:

$$M_1 = 1.102, \qquad M_2 = 0.9103, \qquad T_2/T_1 = 1.0662$$

But, considering the flow relative to the wave:

$$M_1 = \frac{U_s}{a_1}, \qquad M_2 = \frac{U_s - V}{a_2}$$

Therefore, the velocity downstream of the wave is given by:

$$V = U_s - M_2 \times a_2 = M_1 \times a_1 - M_2 \times a_2$$

$$= 1.102 \times \sqrt{1.4 \times 287 \times 288} - 0.9103$$

$$\times \sqrt{1.4 \times 287 \times 1.0662 \times 288}$$

$$= 374.87 - 319.71 = 55.2 \, \text{m/s}$$

The pressure downstream of the shock is given by:

$$p_2 = 1.25 \times 100 = 125 \, \text{kPa}$$

while the temperature downstream of the shock is given by:

$$T_2 = 1.0662 \times 288 = 307 \, \text{K}$$

Therefore, the velocity, pressure, and temperature behind the shock wave are 55.2 m/s, 125 kPa, and 307 K (24°C) respectively.

### EXAMPLE 5.8

A normal shock wave, across which the pressure ratio is 1.17, moves down a duct into still air at a pressure of 105 kPa and a temperature of 30°C. Find the pressure, temperature, and velocity of the air behind the shock wave. This shock wave passes over a small circular cylinder as shown in Fig. E5.8a. Assuming that the shock is unaffected by the small cylinder, find the pressure acting at the stagnation point on the cylinder after the shock has passed over it.

### Solution

The flow relative to the shock wave as shown in Fig. E5.8b is considered. Since the pressure ratio across the shock wave is 1.4, normal shock relations or tables or software give:

$$M_1 = 1.07, \qquad M_2 = 0.936, \qquad \frac{T_2}{T_1} = 1.046$$

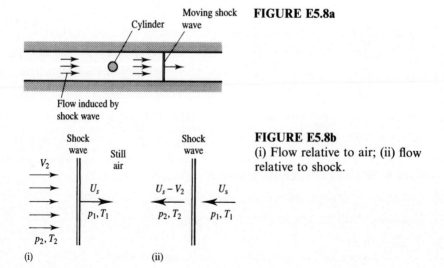

Moving shock wave    **FIGURE E5.8a**

Cylinder

Flow induced by shock wave

Shock wave   Still air    Shock wave   **FIGURE E5.8b**
(i) Flow relative to air; (ii) flow relative to shock.

$V_2$    $U_s$    $U_s - V_2$    $U_s$

$p_1, T_1$    $p_2, T_2$    $p_1, T_1$

$p_2, T_2$

(i)      (ii)

Using these pressure and temperature ratio values then gives:

$$p_2 = 1.17 \times 105 = 122.9 \text{ kPa}, \qquad T_2 = 1.046 \times 303 = 316.9 \text{ K}$$

It is also noted that the speed of sound in the air ahead of the wave is given by:

$$a_1 = \sqrt{\gamma R T_1} = \sqrt{1.4 \times 287 \times 303} = 348.9 \text{ m/s}$$

The speed of sound behind the wave is given by:

$$a_2 = \left[\frac{T_2}{T_1}\right]^{0.5} a_1 = 1.046^{0.5} \times 348.9 = 356.8 \text{ m/s}$$

Now
$$M_2 = \frac{U_s - V_2}{a_2}, \quad \text{i.e.,} \quad V_2 = U_s - M_2 a_2$$

But $M_s = M_1$ so the above equation gives:

$$V_2 = 1.07 \times 348.9 - 0.936 \times 356.8 = 39.38 \text{ m/s}$$

The cylinder is, therefore, exposed to a flow with a velocity of 39.37 m/s at a temperature of 316.9 K and a pressure of 122.9 kPa. The Mach number in this flow is equal to $39.37/356.8 = 0.11$. Now, isentropic relations or tables or software give, for a Mach number of 0.11, $p_0/p = 1.0085$. Therefore, the pressure at the stagnation point on the cylinder is $1.0085 \times 122.85 = 123.9$ kPa.

### EXAMPLE 5.9
A shock wave across which the pressure ratio is 1.15 moves down a duct into still air at a pressure of 50 kPa and a temperature of 30°C. Find the temperature and velocity of the air behind the shock wave. If instead of being at rest, the air ahead of the shock wave is moving towards the wave at a velocity of 100 m/s, what is the velocity of the air behind the shock wave?

### Solution
For the normal shock wave moving into still air which is shown in Fig. E5.9, the following are given:

$$p_2/p_1 = 1.15, \qquad p_1 = 50 \text{ kPa}, \qquad T_1 = 30°C$$

Considering the flow relative to the shock, for $p_2/p_1 = 1.15$, normal shock relations or tables or software give for $\gamma = 1.4$:

$$M_1 = 1.062, \qquad M_2 = 0.943, \qquad T_2/T_2 = 1.041$$

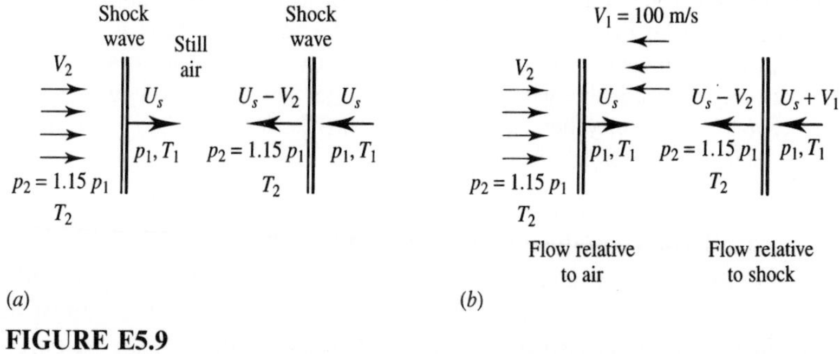

### FIGURE E5.9
(a) No flow ahead of shock; (b) with flow ahead of shock.

Therefore, using the known initial conditions:

$$T_2 = (273 + 30) \times 1.041 = 315.4 \text{ K } (42.4°C)$$

$$p_2 = 1.15 \times 50 = 57.5 \text{ kPa}$$

Also, since:

$$M_1 = U_s/a_1 \quad \text{and} \quad M_2 = (U_s - V)/a_2$$

it follows that:

$$V = M_1 a_1 - M_2 a_2 = 1.062 \times \sqrt{1.4 \times 287 \times 303}$$

$$- 0.943 \times \sqrt{1.4 \times 287 \times 315.4} = 35.1 \text{ m/s}$$

Hence, when the shock is moving into still air, the temperature and velocity behind the shock are 42.4°C and 35.1 m/s respectively.

Next consider the case where the air ahead of the shock is not at rest, this situation also being shown in Fig. E5.9. Because the pressure ratio across the shock wave is still 1.15, the following still apply:

$$M_1 = 1.062, \qquad M_2 = 0.943, \qquad T_2/T_2 = 1.041$$

So, again:

$$T_2 = (273 + 30) \times 1.041 = 315.4 \text{ K } (42.4°C)$$

But, since the flow relative to the wave is being considered it follows that:

$$M_1 = \frac{(U_s + V_1)}{a_1} \quad \text{and} \quad M_2 = \frac{(U_s - V_2)}{a_2}$$

Using these relations then gives:

$$V_2 = U_s - M_2 \times a_2 = (M_1 \times a_1 - V_1) - M_2 a_2$$

$$= (1.062 \times \sqrt{1.4 \times 287 \times 303} - 100)$$

$$- 0.943 \times \sqrt{1.4 \times 287 \times 315.4} = -64.9 \text{ m/s}$$

Therefore, the velocity behind the shock is −64.9 m/s. The negative sign indicates that the velocity behind the wave is in the opposite direction to that in which the wave is moving. Of course, the result for the second part of the question could have been directly deduced from that for the first part of the question by noting that, since the pressure ratio across the shock is the same in both cases, if the flow changes relative to the upstream flow is considered, the same situation as dealt with in the first part of the question is obtained. Therefore, the velocity behind the wave relative to the upstream flow will be 35.1 m/s. Therefore, the velocity behind the wave relative to the walls of the duct will be 35.1 − 100 = −64.9 m/s, as obtained before.

Brief consideration will now be given to the "reflection" of a moving shock wave off the closed end of a duct. Consider a moving normal shock wave propagating into a gas at rest in a duct. The shock, as discussed above, induces a flow behind it in the direction of shock motion. If the end of the duct is closed, however, there can be no flow out of the duct, i.e., the velocity of the gas in contact with the closed end must always be zero. Therefore, a normal

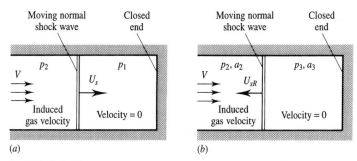

**FIGURE 5.25**
Reflection of a moving normal shock wave from the closed end of a duct: (a) before reflection; (b) after reflection.

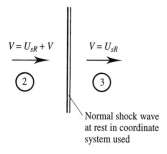

**FIGURE 5.26**
Gas velocities relative to reflected moving normal shock wave.

shock wave must be "reflected" off the closed end, the strength of this "reflected" shock wave being just sufficient to reduce the velocity to zero. This is illustrated in Fig. 5.25.

Consider a set of coordinates attached to the reflected normal shock wave. The gas velocities relative to this reflected shock wave are, therefore, as shown in Fig. 5.26.

Hence, since:

$$M_{R1} = \frac{(U_{sR} + V)}{a_2} = M_{sR} + M_2'$$

and

$$M_{R2} = \frac{U_{sR}}{a_3} = \left(\frac{U_{sR}}{a_2}\right)\left(\frac{a_2}{a_3}\right)$$

$$= M_{sR}\left(\frac{a_2}{a_3}\right)$$

These equations can be used in conjunction with the normal shock relations previously given or shock tables or the software provided to find the properties of the reflected shock. The procedure is illustrated in the following example.

**EXAMPLE 5.10**
A normal shock wave, across which the pressure ratio is 1.45, moves down a duct into still air at a pressure of 100 kPa and a temperature of 20°C. Find the pressure,

temperature, and velocity of the air behind the shock wave. If the end of the duct is closed, find the pressure acting on the end of the duct after the shock is reflected from it.

**Solution**

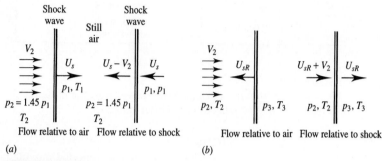

(a)                                                                (b)

**FIGURE E5.10**
(a) Initial shock wave; (b) reflected shock wave.

First consider the shock wave before the reflection. For this wave:

$$p_2/p_1 = 1.45, \qquad p_1 = 100 \text{ kPa}, \qquad T_1 = 20°C = 293 \text{ K}$$

Consider the flow relative to the wave as shown. For $p_2/p_1 = 1.45$, normal shock relations or tables or software give:

$$M_1 = 1.1772, \qquad M_2 = 0.8567, \qquad T_2/T_1 = 1.1137$$

Therefore:
$$p_2 = 1.45 \times 100 = 145 \text{ kPa}$$
$$T_2 = 1.1137 \times 293 = 326.3 \text{ K} \ (= 53.3°C)$$

Also, since:
$$M_1 = U_s/a_1 \quad \text{and} \quad M_2 = (U_s - V_2)/a_2$$

It follows that:
$$V_2 = M_1 a_1 - M_2 a_2$$

Hence:
$$V_2 = 1.1772 \times \sqrt{1.4 \times 287 \times 293} - 0.8567$$
$$\times \sqrt{1.4 \times 287 \times 326.3} = 93.8 \text{ m/s}$$

Therefore, the pressure, temperature, and velocity behind the initial shock wave are 145 kPa, 53.3°C, and 93.8 m/s.

Next consider the wave that is "reflected" off the closed end. The strength of this wave must be such that it brings the flow to rest. Hence, the Mach numbers of the air flow upstream and downstream of the reflected wave relative to this wave are:

$$M_{\text{up}} = \frac{V_2 + U_{sR}}{a_2} \quad \text{and} \quad M_{\text{down}} = \frac{U_{sR}}{a_3}$$

where $U_{sR}$ is the velocity of the reflected wave and $a_3$ is the speed of sound in the flow downstream of the reflected wave.

But:
$$a_2 = \sqrt{\gamma R T_2} = \sqrt{1.4 \times 287 \times 326.3} = 362.1 \text{ m/s}$$

**TABLE E5.10**

| $M_{up}$ (guessed) | $T_3/T_2$ (shock rels.) | $M_{down}$ (shock rels.) | $(M_{up} - 0.259)\sqrt{T_2/T_3}$ (calculated) |
|---|---|---|---|
| 1.000 | 1.000 | 1.000 | 0.741 |
| 1.100 | 1.065 | 0.912 | 0.802 |
| 1.200 | 1.128 | 0.842 | 0.886 |
| 1.500 | 1.320 | 0.701 | 1.080 |
| 1.250 | 1.159 | 0.813 | 0.920 |

Hence:
$$M_{up} = \frac{93.8}{362.1} + \frac{U_{sR}}{a_2} = 0.259 + \frac{U_{sR}}{a_2}$$

and
$$M_{down} = \frac{U_{sR}}{a_3} = \frac{U_{sR}}{a_2} \times \frac{a_2}{a_3} = \frac{U_{sR}}{a_2} \sqrt{\frac{T_2}{T_3}}$$

i.e.,
$$M_{down} = (M_{up} - 0.259)\sqrt{\frac{T_2}{T_3}}$$

Now the values of $M_{up}$ and $M_{down}$ are related by the normal shock equations, these equations also relating $T_2/T_3$ to $M_{up}$. These equations together, therefore, allow $M_{up}$ to be found. While elegant methods of finding the solution are available, a simple way of finding the solution is to guess a series of values of $M_{up}$ and then for each of these values to find $M_{down}$ and $T_2/T_3$ from shock relations or tables or software and then to derive the value of $M_{down} = (M_{up} - 0.259)\sqrt{T_2/T_3}$. The correct value of $M_{up}$ is that which has the value of $M_{down}$ as given directly by the normal shock relations equal to that given by the above equation. This correct value can be deduced from the results for various $M_{up}$. Results for various values of $M_{up}$ are shown in Table E5.10. From these results, it can be deduced that when $M_{up} = 1.17$ the values of $M_{down}$ given by the shock relations and by the above equation are the same (approximately 0.863). Now for an upstream Mach number of 1.17, normal shock relations or tables or software give $p_3/p_2 = 1.4304$. Hence, the pressure behind the reflected shock, i.e., the pressure acting on the closed end is given by:

$$p_3 = 1.4304 \times 145 = 207 \text{ kPa}$$

High pressures can therefore be generated when a moving shock wave is reflected off a closed end of a duct.

Lastly, consider what happens if a gas is flowing out of a duct at a steady rate when the end of the duct is suddenly closed. Since the velocity of the gas in contact with the closed end must again be zero, a shock wave is generated that moves into the moving gas bringing it to rest, i.e., the strength of the shock wave must be such that the velocity is reduced to zero behind it. This is illustrated in Fig. 5.27.

A shock wave generated in this way can be analyzed using the same procedure as used to analyze a normal shock wave reflected from a closed end of a duct. This is illustrated by the following example.

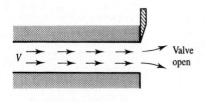

**FIGURE 5.27**
Moving normal shock wave generated by closure of valve.

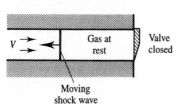

### EXAMPLE 5.11

Air is flowing out of a duct at a velocity of 250 m/s with a temperature of 0°C and a pressure of 70 kPa. A valve at the end of the duct is suddenly closed. Find the pressure acting on the valve immediately after the valve closure.

### *Solution*

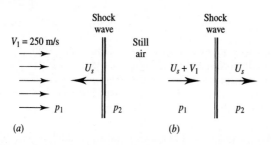

**FIGURE E5.11**
(*a*) Flow relative to air;
(*b*) flow relative to shock.

Consider the flow relative to the shock wave. Because the wave must bring the air to rest, it will be seen that if $U_s$ is the velocity of the shock wave,

$$M_1 = \frac{U_s + V}{a_1}$$

and

$$M_2 = \frac{U_s}{a_2} = \frac{U_s}{a_1} \times \frac{a_1}{a_2}$$

Substituting this into the equation for $M_1$ gives:

$$M_1 = M_2 \frac{a_2}{a_1} + \frac{V}{a_1}$$

But the speed of sound ahead of the wave is:

$$a_1 = \sqrt{\gamma RT} = \sqrt{1.4 \times 287 \times 273} = 331.2 \text{ m/s}$$

and $V = 250$ m/s so the above equation gives:

**TABLE E5.11**

| $M_1$ | $M_2$ | $a_2/a_1$ | RHS |
|-------|-------|-----------|------|
| 1.0 | 1.0 | 1.0 | 1.755 |
| 1.4 | 0.740 | 1.120 | 1.584 |
| 1.5 | 0.701 | 1.149 | 1.56 |
| 1.6 | 0.668 | 1.178 | 1.54 |

$$M_1 = M_2 \frac{a_2}{a_1} + \frac{250}{331.2} = M_2 \frac{a_2}{a_1} + 0.7548$$

Because the normal shock relations determine $M_2$ and $a_2/a_1$ as functions of $M_1$, the above equation together with the normal shock relations can be used to determine $M_1$. Although there are more elegant methods of obtaining the solution, the brute force approach is to choose a series of values of $M_1$ and then use normal shock relations to find the right hand side, i.e., to find the value of:

$$M_2 \frac{a_2}{a_1} + 0.7548$$

for each of these values of $M_1$. The value of $M_1$ that makes the right hand side equal to $M_1$ can then be deduced from these results. A set of results is shown in Table E5.11, the values of $M_2$ and $a_2/a_1$ being deduced for $\gamma = 1.4$ from the normal shock relations or from normal shock tables or using the software. "RHS" is the right hand side of the above equation.

From these results it will be seen that $M_1 = 1.55$. Normal shock relations or tables or software then gives $p_2/p_1 = 2.636$, so $p_2 = 2.636 \times 70 = 184.5 \, \text{kPa}$.

## 5.6
## CONCLUDING REMARKS

A normal shock wave is an extremely thin region at right angles to the flow across which large changes in the flow variables can occur. Although the flow within the shock wave is complex, it was shown that expressions for the overall changes across the shock can be relatively easily derived. It was shown that entropy considerations indicate that only compressive shock waves, i.e., shock waves across which the pressure increases, can occur and that the flow ahead of the shock must be supersonic. It was also shown that the flow downstream of the shock is always subsonic. The analysis of normal shock waves that are moving through a gas was also discussed.

## PROBLEMS

**5.1.** A normal shock wave occurs in an air flow at a point where the velocity is 680 m/s, the static pressure is 80 kPa, and the static temperature is 60°C. Find the velocity, static pressure, and static temperature downstream of the shock.

Also find the stagnation temperature and stagnation pressure upstream and downstream of the shock.

**5.2.** The pressure ratio across a normal shock wave that occurs in air is 1.25. Ahead of the shock wave, the pressure is 100 kPa and the temperature is 15°C. Find the velocity, pressure, and temperature of the air behind the shock wave.

**5.3.** A perfect gas flows through a stationary normal shock. The gas velocity decreases from 480 m/s to 160 m/s through the shock. If the pressure and the density upstream of the shock are 62 kPa and 1.5 kg/m$^3$, find the pressure and density downstream of the shock and the specific heat ratio of the gas.

**5.4.** A normal shock wave occurs in air at a point where the velocity is 600 m/s and the stagnation temperature and pressure are 200°C and 600 kPa respectively. Find the Mach numbers, pressures, and temperatures upstream and downstream of the shock wave.

**5.5.** Show that the downstream Mach number of a normal shock approaches a minimum value as the upstream Mach number increases towards infinity. What is this minimum Mach number for a gas with a specific heat ratio of 1.67?

**5.6.** Air is expanded isentropically from a reservoir in which the pressure is 1000 kPa to a pressure of 150 kPa. A normal shock occurs at this point in flow. Find the pressure downstream of the shock wave.

**5.7.** Air is expanded isentropically from a reservoir in which the pressure and temperature are 150 psia and 60°F to a static pressure of 20 psia. A normal shock occurs at this point in the flow. Find the static pressure, static temperature, and the air velocity behind the shock.

**5.8.** The exhaust gases from a rocket engine have a molar mass of 14. They can be assumed to behave as a perfect gas with a specific heat ratio of 1.25. These gases are accelerated through a convergent–divergent nozzle. A normal shock wave occurs in the nozzle at a point in the flow where the Mach number is 2. Find the pressure, temperature, density, and stagnation pressure ratio across this shock wave.

**5.9.** Air is expanded isentropically from a reservoir in which the pressure is 1000 kPa and the temperature is 30°C until the pressure has dropped to 25 kPa. A normal shock wave occurs at this point. Find the static pressure, the static temperature, the air velocity, and the stagnation pressure after the shock wave.

**5.10.** A gas with a molar mass of 4 and a specific heat ratio of 1.67 is expanded from a large reservoir in which the pressure and temperature are 600 kPa and 35°C respectively through a nozzle system until the Mach number is 1.5. A normal shock wave then occurs in the flow. Find the pressure and velocity behind the shock wave.

**5.11.** Air is expanded from a large chamber through a variable area duct. The pressure and temperature in the large chamber are 115 psia and 100°F respectively. At some point in the flow where the Mach number is 2.5 a normal shock wave occurs. Find the pressure, temperature, stagnation pressure, and velocity behind the shock wave.

**5.12.** A normal shock wave occurs in an air flow at a point where the velocity is 750 m/s, the pressure is 50 kPa, and the temperature is 10°C. Find the velocity, pressure, and static temperature downstream of the shock wave.

**5.13.** Air is isentropically expanded from a large chamber in which the pressure is 10 000 kPa and the temperature is 50°C until the Mach number reaches a value of 2. A normal shock wave then occurs in the flow. Following the shock wave, the air is isentropically decelerated until the velocity is again essentially zero. Find the pressure and temperature that then exist.

**5.14.** Air is expanded from a large reservoir in which the pressure and temperature are 500 kPa and 35°C through a variable area duct. A normal shock occurs at a point in the duct where the Mach number is 2.5. Find the pressure and temperature in the flow just downstream of the shock wave. Downstream of the shock wave, the flow is brought to rest in another large reservoir. Find the pressure and temperature in this reservoir. Assume that the flow is one-dimensional and isentropic everywhere except through the shock wave.

**5.15.** Air is expanded from a reservoir in which the pressure and temperature are maintained at 1000 kPa and 30°C. At a point in the flow at which the static pressure is 150 kPa a normal shock wave occurs. Find the static pressure, the static temperature, and the air velocity behind the shock wave. Assume the flow to be isentropic everywhere except through the shock wave.

**5.16.** Air is expanded through a convergent–divergent nozzle from a large chamber in which the pressure and temperature are 200 kPa and 310 K respectively. A normal shock wave occurs at a point in the nozzle where the Mach number is 2.5. The air is then brought to rest in a second large chamber. Find the pressure and temperature in this second chamber. Clearly state the assumptions you have made in arriving at the solution.

**5.17.** Air at a temperature of 10°C and a pressure of 50 kPa flows over a blunt nosed body at a velocity of 500 m/s. Estimate the pressure acting on the front of the body.

**5.18.** A pitot-static tube is placed in a supersonic air flow at a Mach number of 2.0. The static pressure and static temperature in the flow are 101 kPa and 30°C respectively. Estimate the difference between the pitot and static pressures.

**5.19.** A pitot-static tube is placed in a supersonic flow in which the static pressure and temperature are 60 kPa and −20°C respectively. The difference between the pitot and static pressures is measured and found to be 449 kPa. Find the Mach number and velocity in the flow. Discuss the assumptions used in deriving the answers.

**5.20.** A pitot-static tube is placed in an air flow in which the Mach number is 1.7. The static pressure in the flow is 55 kPa and the static temperature is −5°C. What will be the measured difference between the pitot and the static pressures?

**5.21.** A pitot-static tube is placed in a supersonic flow in which the static temperature is 0°C. Measurements indicate that the static pressure is 80 kPa and that the ratio of the pitot to the static pressure is 4.1. Find the Mach number and the velocity in the flow.

**5.22.** A pitot tube is placed in a stream of carbon dioxide in which the pressure is 60 kPa and the Mach number is 3.0. What will the pitot pressure be?

**5.23.** A thermocouple placed in the mouth of a pitot tube can be used to measure the stagnation temperature of a flow. Such an arrangement placed in an air flow gives the stagnation pressure as 180 kPa, the static pressure as 55 kPa, and the stagnation temperature as 95°C. Estimate the velocity of the stream assuming that the flow is supersonic.

**5.24.** A shock wave propagates down a constant area duct into stagnant air at a pressure of 101.3 kPa and a temperature of 25°C. If the pressure ratio across the shock wave is 3, find the shock speed and the velocity of the air downstream of the shock.

**5.25.** A normal shock wave propagates down a constant-area tube containing stagnant air at a temperature of 300 K. Find the velocity of the shock wave if the air behind the wave is accelerated to Mach number of 1.2.

**5.26.** A shock wave is moving down a constant area duct containing air. The air ahead of the shock wave is at rest and at a pressure and temperature of 100 kPa and 20°C respectively. If the pressure ratio across the shock wave is 2.5, find the velocity, pressure, and the temperature in the air behind the shock wave.

**5.27.** A normal shock wave propagates at a speed of 2600 m/s down a pipe that is filled with hydrogen. The hydrogen is at rest and at a pressure and temperature of 101.3 kPa and 25°C respectively upstream of the wave. Assuming hydrogen to behave as a perfect gas with constant specific heats, find the temperature, pressure, and velocity downstream of the wave.

**5.28.** A normal shock wave, across which the pressure ratio is 1.2, moves down a duct into still air at a pressure of 100 kPa and a temperature of 20°C. Find the pressure, temperature, and velocity of the air behind the shock wave. This shock wave passes over a small circular cylinder. Assuming that the shock is unaffected by the small cylinder, find the pressure acting at the stagnation point on the cylinder after the shock has passed over it.

**5.29.** As a result of a rapid chemical reaction a normal shock wave is generated which propagates down a duct in which there is air at a pressure of 100 kPa and a temperature of 30°C. The pressure behind this shock wave is 130 kPa. Half a second after the generation of this shock wave, a second normal shock wave is generated by another chemical reaction. This second shock wave follows the first one down the duct, the pressure behind this second wave being 190 kPa. Find the velocity of the air and the temperature behind the second shock wave. Also find the distance between the two waves at a time of 0.7 s after the generation of the first shock wave.

**5.30.** A normal shock wave across which the pressure ratio is 1.25 is propagating down a duct containing still air at a pressure of 120 kPa and a temperature of 35°C. This shock wave is reflected off the closed end of the duct. Find the pressure and temperature behind the reflected shock wave.

**5.31.** A normal shock wave is propagating down a duct in which $p = 110$ kPa and $T = 30$°C. The pressure ratio across the shock is 1.8. Find the velocity of the shock wave and the air velocity behind the shock. This moving shock strikes a closed end to the duct. Find the pressure on the closed end after the shock reflection.

**5.32.** A normal shock wave is moving down a duct into still air in which the pressure is 100 kPa and the temperature is 20°C. The pressure ratio across the shock is 1.8. Find the velocity of the shock and the velocity of the air behind the shock. If this shock strikes the closed end of the duct, find the pressure on this closed end after the shock reflection.

**5.33.** A shock across which the pressure ratio is 1.18 moves down a duct into still air at a pressure of 100 kPa and a temperature of 30°C. Find the temperature and velocity of the air behind the shock wave. If instead of being at rest, the air ahead of the shock wave is moving towards the wave at a velocity of 75 m/s, what would be the velocity of the air behind the shock wave?

**5.34.** Air at a pressure of 105 kPa and a temperature of 25°C is flowing out of a duct at a velocity of 250 m/s. A valve at the end of the duct is suddenly closed. Find the pressure acting on the valve.

**5.35.** Air is flowing out of a duct at a velocity of 250 m/s. The static temperature and pressure in the flow are 0°C and 70 kPa. A valve at the end of the duct is suddenly closed. Estimate the pressure acting on this valve immediately after it is closed.

**5.36.** Air is flowing down a duct at a velocity of 200 m/s. The pressure and the temperature in the flow are 85 kPa and 10°C respectively. If a valve at the end of the duct is suddenly closed, find pressure acting on the valve immediately after closure.

**5.37.** A piston in a pipe containing stagnant air at a pressure of 101 kPa and a temperature of 25°C is suddenly given a velocity of 100 m/s into the pipe causing a normal shock wave to propagate through the air down the pipe. Find the velocity of the shock wave and the pressure acting on the piston.

**5.38.** Air flows at a velocity of 90 m/s down a 20 cm diameter pipe. The air is at a pressure of 120 kPa and a temperature of 30°C. A valve at the end of the pipe is suddenly closed. This valve is held in place by eight mild steel bolts each with a diameter of 12 mm. Will the bolts hold the valve without yielding? Assume that the pipe is discharging the air to the atmosphere and that the ambient pressure is 101 kPa. It will be necessary to look up the yield strength of the steel in order to answer this question.

# CHAPTER 6

# Oblique Shock Waves

## 6.1
## INTRODUCTION

Attention is now turned from normal shock waves, which are straight and in which the flow before and after the wave is normal to the shock, to oblique shock waves. Such shock waves are, by definition, also straight but they are at an angle to the upstream flow and, in general, they produce a change in flow direction as indicated in Fig. 6.1.

The oblique shock relations can be deduced from the normal shock relations by noting that the oblique shock can produce no momentum change parallel to the plane in which it lies.

To show this, consider the control volume shown in Fig. 6.2. Because there are no changes in the flow variables in the direction parallel to the wave there is no net force on the control volume parallel to the wave and there is, consequently, no momentum change parallel to the wave. Because there is no momentum change parallel to the shock, $L_1$ must equal $L_2$. Hence, if the coordinate system moving parallel to the wave front at a velocity $L = L_1 = L_2$ is considered, the flow in this coordinate system through the wave is as shown in Fig. 6.3. In this coordinate system, the oblique shock

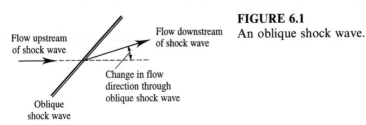

Flow upstream of shock wave

Flow downstream of shock wave

Change in flow direction through oblique shock wave

Oblique shock wave

**FIGURE 6.1**
An oblique shock wave.

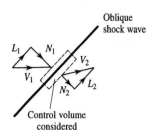

Oblique shock wave

**FIGURE 6.2**
Control volume considered. $L$, velocity component parallel to wave; $N$, velocity component normal to wave.

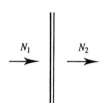

Control volume considered

**FIGURE 6.3**
Flow normal to an oblique shock wave.

$N_1$    $N_2$

has been reduced to a normal shock and the normal shock relations must, therefore, apply to the velocity components $N_1$ and $N_2$. Further, since the scalar flow properties $p$, $\rho$, and $T$ are unaffected by the coordinate system used, the Rankine–Hugonoit relations must apply without any modification to oblique shocks. Thus, all the properties of oblique shocks can be obtained by modification and manipulation of the normal shock relations provided that angle of the shock relative to the upstream flow is known. However, it is more instructive and, in some respects, simpler to deduce these oblique shock relations from the fundamental conservation of mass, momentum, and energy laws, using the normal shock relations when a parity is formally established.

## 6.2
## OBLIQUE SHOCK WAVE RELATIONS

Consider again flow through a control volume that spans the shock wave and which, without any loss of generality, can be assumed to have unit area parallel to the oblique shock wave. This control volume is shown in Fig. 6.4. As shown in this figure, $\beta$ is defined as the shock wave angle and $\delta$ is the change in flow direction induced by the shock wave.

The conservation of mass, momentum, and energy are now applied to the control volume shown. Since there is no change in velocity parallel to the wave, the $L$ velocity components can carry no net mass into the control volume. Conservation of mass therefore gives:

$$\rho_1 N_1 = \rho_2 N_2 \tag{6.1}$$

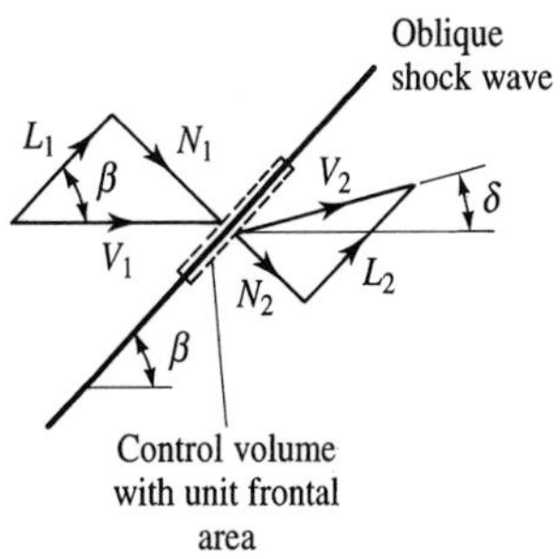

**FIGURE 6.4**
Control volume used in the analysis of an oblique shock wave.

The conservation of momentum equation is applied in the direction normal to the shock, giving:

$$p_1 - p_2 = \rho_2 N_2^2 - \rho_1 N_1^2 \tag{6.2}$$

Because there are no gradients of temperature upstream and downstream of the shock wave, the flow through the control volume must, as in the case of the normal shock, be adiabatic. The energy equation, therefore, gives:

$$\frac{2\gamma}{\gamma - 1}\frac{p_1}{\rho_1} + V_1^2 = \frac{2\gamma}{\gamma - 1}\frac{p_2}{\rho_1} + V_2^2 \tag{6.3}$$

But:
$$V^2 = L^2 + N^2$$

so, because $L_1 = L_2$, eq. (6.3) can be rearranged as:

$$\left(\frac{2\gamma}{\gamma - 1}\right)\left[\frac{p_2}{\rho_2} - \frac{p_1}{\rho_1}\right] = L_1^2 + N_1^2 - L_2^2 - N_2^2$$
$$= N_1^2 - N_2^2 \tag{6.4}$$

If eqs. (6.1), (6.2), and (6.4) are compared with the equations derived for normal shock waves it will be seen that they are identical in all respects except that $N_1$ and $N_2$ replace $V_1$ and $V_2$ respectively. It follows, therefore, that they can also be rearranged to give in the same way as for normal shocks:

$$\frac{p_2}{p_1} = \frac{\left[\left(\dfrac{\gamma + 1}{\gamma - 1}\right)\dfrac{\rho_2}{\rho_1} - 1\right]}{\left[\left(\dfrac{\gamma + 1}{\gamma - 1}\right) - \dfrac{\rho_2}{\rho_1}\right]}$$

$$\frac{\rho_2}{\rho_1} = \frac{\left[\left(\dfrac{\gamma + 1}{\gamma - 1}\right)\dfrac{p_2}{p_1} + 1\right]}{\left[\left(\dfrac{\gamma + 1}{\gamma - 1}\right) + \dfrac{p_2}{p_1}\right]}$$

$$\frac{N_1}{N_2} = \frac{\left[\left(\dfrac{\gamma+1}{\gamma-1}\right)\dfrac{p_2}{p_1} + 1\right]}{\left[\left(\dfrac{\gamma+1}{\gamma-1}\right) + \dfrac{p_2}{p_1}\right]}$$

$$\frac{T_2}{T_1} = \frac{\left[\left(\dfrac{\gamma+1}{\gamma-1}\right)\dfrac{p_2}{p_1}\right]}{\left[\left(\dfrac{\gamma+1}{\gamma-1}\right) + \dfrac{p_1}{p_2}\right]} \tag{6.5}$$

As with normal shock waves, it is very frequently convenient to know the relations between the changes across the shock wave and the upstream Mach number, $M_1$. Now since:

$$N_1 = V_1 \sin \beta$$

$$N_2 = V_2 \sin (\beta - \delta) \tag{6.6}$$

eqs. (6.1), (6.2), and (6.4) can be written as:

$$\rho_1 V_1 \sin \beta = \rho_2 V_2 \sin (\beta - \delta) \tag{6.7}$$

$$p_1 - p_2 = \rho_2 (V_1 \sin \beta)^2 - \rho_1 [V_2 \sin (\beta - \delta)]^2 \tag{6.8}$$

$$\frac{2\gamma}{(\gamma-1)}\left[\frac{p_2}{\rho_2} - \frac{p_1}{\rho_1}\right] = (V_1 \sin \beta)^2 - [V_2 \sin (\beta - \delta)]^2 \tag{6.9}$$

These are, of course, again identical to those used to study normal shocks, except that $V_1 \sin \beta$ occurs in place of $V_1$ and $V_2 \sin (\beta - \delta)$ occurs in place of $V_2$. Hence, if in the normal shock relations $M_1$ is replaced by $M_1 \sin \beta$ and $M_2$ by $M_2 \sin (\beta - \delta)$ the following relations for oblique shocks are obtained using equations given in Chapter 5:

$$\frac{p_2}{p_1} = \frac{2\gamma M_1^2 \sin^2 \beta - (\gamma-1)}{\gamma+1} \tag{6.10}$$

$$\frac{\rho_2}{\rho_1} = \frac{(\gamma+1)M_1^2 \sin^2 \beta}{2 + (\gamma-1)M_1^2 \sin^2 \beta} \tag{6.11}$$

$$\frac{T_2}{T_1} = \frac{[2 + (\gamma-1)M_1^2 \sin^2 \beta][2\gamma M_1^2 \sin^2 \beta - (\gamma-1)]}{(\gamma+1)^2 M_1^2 \sin^2 \beta} \tag{6.12}$$

$$M_2^2 \sin^2 (\beta - \delta) = \frac{M_1^2 \sin^2 \beta + 2/(\gamma-1)}{2\gamma M_1^2 \sin^2 \beta/(\gamma-1) - 1} \tag{6.13}$$

It should be noted that since it was proved using entropy considerations that for normal shocks $M_1$ had to be greater than 1, i.e., the flow ahead of the shock had to be supersonic it, therefore, follows from the above discussion that for oblique shocks it is necessary that:

$$M_1 \sin \beta \geq 1 \tag{6.14}$$

The minimum value that $\sin \beta$ can have is, therefore, $1/M_1$, i.e., the minimum shock angle is the Mach angle. When the shock has this angle, eq. (6.10) shows that $(p_2/p_1)$ is equal to 1, i.e., the shock wave is a Mach wave.

The maximum value that $\beta$ can have is, of course, $90°$, the wave then being a normal shock wave. Hence, the limits on $\beta$ are:

$$\sin^{-1}\left(\frac{1}{M_1}\right) \leq \beta \leq 90° \tag{6.15}$$

It should further be noted that since it was proved that the flow behind a normal shock wave is subsonic, i.e., $M_2$ was less than 1, it follows that for an oblique shock wave:

$$M_2 \sin (\beta - \delta) \leq 1 \tag{6.16}$$

Hence, for an oblique shock wave, $M_2$ can be greater than or less than 1.

In order to utilize the above relationships to find the properties of oblique shocks the relation between $\delta$, $\beta$, and $M_1$ has to be known. Now, it will be seen from Fig. 6.4 that:

$$\tan \beta = \frac{N_1}{L_1}, \qquad \tan (\beta - \delta) = \frac{N_2}{L_2} \tag{6.17}$$

But it was previously noted that:

$$L_1 = L_2$$

and from the continuity equation it follows that:

$$\frac{N_1}{N_2} = \frac{\rho_1}{\rho_2}$$

Equation (6.17), therefore, can be used to give:

$$\frac{\tan (\beta - \delta)}{\tan \beta} = \frac{\rho_1}{\rho_2} = \frac{2 + (\gamma - 1)M_1^2 \sin^2 \beta}{(\gamma + 1)M_1^2 \sin^2 \beta}$$

$$= X, \quad \text{say} \tag{6.18}$$

eq. (6.11) having been used.

But, since:

$$\tan (\beta - \delta) = \frac{\tan \beta - \tan \delta}{1 + \tan \beta \tan \delta}$$

it follows that:

$$\frac{\tan (\beta - \delta)}{\tan \beta} = \frac{1 - \tan \delta/\tan \beta}{1 + \tan \beta \tan \delta}$$

Substituting this into eq. (6.18) then gives:

$$1 - \frac{\tan \delta}{\tan \beta} = X + X \tan \beta \tan \delta$$

which becomes, on rearrangement:

$$\tan \delta = \frac{\tan \beta (1 - X)}{X \tan^2 \beta + 1}$$

Substituting for $X$ from eq. (6.18) then gives:

$$\tan \delta = \frac{(\gamma + 1)M_1^2 \sin^2 \beta - 2 - (\gamma - 1)M_1^2 \sin^2 \beta}{2 \tan^2 \beta + (\gamma - 1)M_1^2 \sin^2 \beta \tan^2 \beta + (\gamma + 1)M_1^2 \sin^2 \beta} \tan \beta$$

$$= \frac{2 \cot \beta (M_1^2 \sin^2 \beta - 1)}{2 + M_1^2 (\gamma + \cos 2\beta)} \tag{6.19}$$

This equation gives the variation of $\beta$ with $M_1$ and $\delta$. It will be noted that the turning angle, $\delta$, is zero when $\cot \beta = 0$ and also when $M_1 \sin \beta$ is equal to 1, i.e.,

$$\delta = 0 \qquad \text{when } \beta = 90°$$

$$\text{and } \beta = \sin^{-1}(1/M_1)$$

these two limits being, of course, a normal shock and an infinitely weak Mach wave. Thus as discussed before an oblique shock lies between a normal shock and a Mach wave as indicated in Fig. 6.5. In both of these two limiting cases, there is no turning of the flow. Between these two limits $\delta$ reaches a maximum.

The relation between $\delta$, $M_1$, and $\beta$ as given by eq. (6.18) is usually presented graphically and resembles that shown in Fig. 6.6. A larger and more accurate graph is given in Appendix G. The software that supports this book also allows the relationship to be found.

For flow over a body that has a surface set at an angle to the oncoming flow, this graph, or eq. (6.19) from which it is derived, or the software provided, allows the shock angle to be found for any value of $M_1$. This is illustrated in Fig. 6.7.

The normal shock limit and the Mach wave limit on the oblique shock at a given value of $M_1$ are given by the intercepts of the curves with the vertical axis at $\delta = 0$. This is illustrated in Fig. 6.8.

It will also be seen from the diagram given in Fig. 6.6 that, as already mentioned, there is a maximum angle through which a gas can be turned at a given $M_1$. The value of this maximum turning angle for a given $M_1$ can be obtained by differentiating eq. (6.19) with respect to $\delta$ for a fixed $M_1$ and setting $d\delta/d\beta$ equal to zero. This leads to the following expression for the

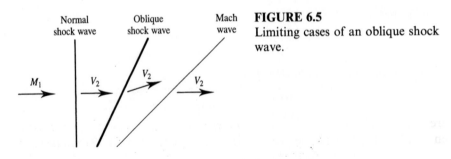

Normal shock wave    Oblique shock wave    Mach wave

$M_1$    $V_2$    $V_2$    $V_2$

**FIGURE 6.5**
Limiting cases of an oblique shock wave.

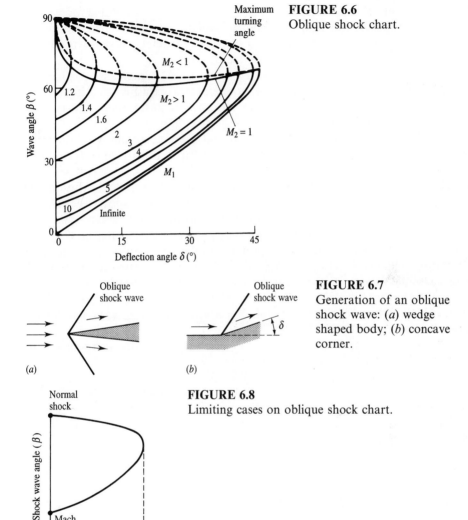

**FIGURE 6.6**
Oblique shock chart.

**FIGURE 6.7**
Generation of an oblique shock wave: (*a*) wedge shaped body; (*b*) concave corner.

**FIGURE 6.8**
Limiting cases on oblique shock chart.

maximum turning angle:

$$\sin^2 \beta_{\max} = \frac{\gamma+1}{4\gamma} - \frac{1}{\gamma M_1^2}$$

$$\left[ 1 - \sqrt{(\gamma+1)\left(1 + \frac{\gamma-1}{2}M_1^2 + \frac{\gamma+1}{16}M_1^4\right)} \right]$$

where $\beta_{\max}$ is the shock angle that exists when $\delta$ has its maximum value for a given $M_1$. Once $\beta_{\max}$ has been found using this equation, eq. (6.19) can be used to find the value of $\delta_{max}$. The variation of $\delta_{max}$ with $M_1$ so obtained for $\gamma = 1.4$ is shown in Fig. 6.9. For flow over bodies involving greater angles than this, a detached shock occurs as illustrated in Fig. 6.10. A detached shock is curved, in general, and not amenable to analytical treatment.

It should also be noted that as $M_1$ increases, $\delta_{max}$ increases so that if a body involving a given turning angle, accelerates from a low to a high Mach number, the shock can be detached at the low Mach numbers and become attached at the higher Mach numbers.

It will further be noted from Fig. 6.6 that if $\delta$ is less than $\delta_{max}$, there are two possible solutions, i.e. two possible values for $\beta$, for a given $M_1$ and $\delta$ as indicated in Fig. 6.11.

The solution giving the larger $\beta$ is termed the 'strong shock' solution and is indicated by the dotted line in Fig. 6.6. The figure also shows the $M_2 = 1$ locus and shows that $M_2$ is always less than 1 in the strong shock case.

Experimentally, it is found that for a given $M_1$ and $\delta$ in external flows the shock angle, $\beta$, is usually that corresponding to the 'weak' or non strong shock solutions. Under some circumstances, the conditions downstream of the shock may cause the strong shock solution to exist in part of the flow. In the event of no other information being available, the non-strong shock solution should be used. The software that supports this book gives both the weak and strong oblique shock solutions.

A curve that determines when the Mach number downstream of the

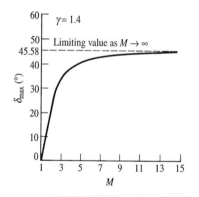

**FIGURE 6.9**
Variation of maximum turning angle with $M_1$ for $\gamma = 1.4$.

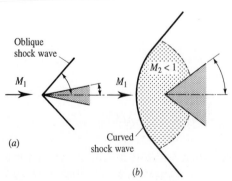

**FIGURE 6.10**
Detached oblique shock wave.
(a) $\delta < \delta_{max}$; (b) $\delta > \delta_{max}$.

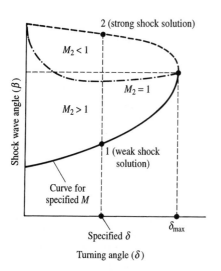

**FIGURE 6.11**
Strong and weak oblique shock wave solutions.

shock, i.e., $M_2$, is equal to 1 is also given in Fig. 6.6. The Mach number behind the shock in the 'strong' shock case will be seen to be always less than 1.

The oblique shock waves results are often presented in the form of a series of diagrams. However, all calculations can be carried out using a single graph giving the relation between $\delta$, $\beta$, and $M_1$ such as that shown in Fig. 6.6, together with normal shock tables. This procedure is, for a given $M_1$ and $\delta$, to use the diagram to give $\beta$ and then to calculate $M_{N1} = M_1 \sin \beta$ and to use the normal shock relations or the normal shock tables to find $p_2/p_1$, $\rho_1/\rho_1$, $T_2/T_1$ and $M_{N2}$ corresponding to this value of Mach number. $M_2$ is then obtained by setting it equal to $M_{N2}/\sin(\beta - \delta)$. Alternatively, the software provided can be used to carry out this procedure.

**EXAMPLE 6.1**
Air flowing with a Mach number of 2 with a pressure of 80 kPa and a temperature of 30°C passes over a component of an aircraft that can be modeled as a wedge with an included angle of 8° that is aligned with the flow, i.e., the flow is turned through an angle of 4°, leading to the generation of an oblique shock wave. Find the pressure acting on the surface of the wedge.

**Solution**

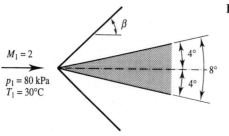

**FIGURE E6.1a**

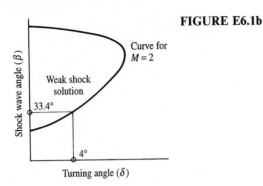

**FIGURE E6.1b**

Here the flow is turned through an angle $\delta = 4°$ and the Mach number upstream of the shock wave is $M_1 = 2.0$. Hence, using the oblique shock chart or the software gives $\beta = 33.4°$ (see Fig. E6.1b). Using this value then gives:

$$M_{N1} = M_1 \sin \beta = 2 \times \sin 33.4° = 1.10$$

This value is directly given by the software.

Normal shock relations or tables or the software give for an upstream Mach number of 1.10 ($M_{N1}$):

$$\frac{p_2}{p_1} = 1.245$$

Therefore:  $\qquad\qquad p_2 = 1.245 \times 80 = 99.6 \text{ kPa}$

i.e., the pressure downstream of the shock wave is 99.6 kPa. This will be the pressure acting on the surface of the wedge because the flow is uniform downstream of the shock wave.

## 6.3
## REFLECTION OF OBLIQUE SHOCK WAVES

Here, the "reflection" of an oblique shock wave from a plane wall will first be considered. An oblique shock is assumed to be generated from a body that turns the flow through an angle $\delta$ as shown in Fig. 6.12. The entire flow on passing through this wave is then turned "downwards" through an angle $\delta$. However, the flow adjacent to the lower flat wall must be parallel to the wall. This is only possible if a "reflected" wave is generated as shown in Fig. 6.12 that turns the flow back "up" through $\delta$. The changes through the initial and reflected waves are shown in Fig. 6.12, **a** being the initial wave and **b** the reflected wave.

Since the flow downstream of the "reflected" wave must again be parallel to the wall, both waves must produce the same change in flow direction. Thus, in order to determine the properties of this reflected wave, the following procedure is used:

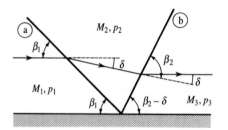

**FIGURE 6.12**
Reflected oblique shock wave.

1. For the given $M_1$ and $\delta$, determine $M_2$ and $p_2/p_1$.
2. For this value of $M_2$ and since the turning angle of the second wave is also $\delta$, determine $M_3$ and $p_3/p_2$.
3. The overall pressure ratio, $p_3/p_1$, is then found from:

$$\frac{p_3}{p_1} = \frac{p_3}{p_2}\frac{p_2}{p_1}$$

4. The angle that the reflected wave makes with the wall is $\beta_2 + \delta$ and since $\beta_2$ was found in step 2, this angle can be determined.

**EXAMPLE 6.2**
Air flowing with a Mach number of 2.5 with a pressure of 60 kPa and a temperature of −20°C passes over a wedge which turns the flow through an angle of 4° leading to the generation of an oblique shock wave. This oblique shock wave impinges on a flat wall, which is parallel to the flow upstream of the wedge, and is "reflected" from it. Find the pressure and velocity behind the reflected shock wave.

**Solution**
The situation under consideration is shown in Fig. E6.2a.
   Upstream of the initial wave the following conditions exist:

$$p_1 = 60\,\text{kPa}, \qquad T_1 = 253\,\text{K}, \qquad M_1 = 2.5$$

The conditions downstream of the initial wave, i.e., in region 2, are first obtained. Now for:

$$M_1 = 2.5, \qquad \delta = 4°$$

**FIGURE E6.2a**

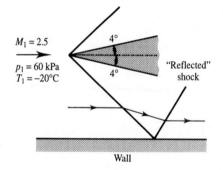

the oblique shock chart or the software gives

$$\beta = 26.6°$$

Hence $$M_{N1} = M_1 \sin \beta = 2.5 \sin 26.6 = 1.12$$

This value is also directly given by the software.

Next, using normal shock relations or tables or software for an upstream Mach number of 1.12 ($M_{N1}$) the following are obtained:

$$M_{N2} = 0.897, \qquad \frac{p_2}{p_1} = 1.336, \qquad \frac{T_2}{T_1} = 1.087$$

But: $$M_{N2} = M_2 \sin(\beta - \delta)$$

Hence: $$M_2 = \frac{0.897}{\sin(26.6 - 4)} = 2.334$$

Conditions behind the reflected wave, i.e., in region 3 will next be derived by considering the changes from region 2 to region 3 as shown in Fig. E6.2b.

Now for:

$$M_2 = 2.334, \qquad \delta = 4°$$

the oblique shock chart or the software gives:

$$\beta_2 = 28.5°$$

Hence: $$M_{N2} = M_2 \sin \beta_2 = 2.334 \sin 28.5 = 1.113$$

This value is also directly given by the software.

Next, using normal shock relations or tables or software for an upstream Mach number of 1.113 ($M_{N1}$), the following are obtained:

$$M_{N3} = 0.901\ 76, \qquad \frac{p_3}{p_2} = 1.297, \qquad \frac{T_3}{T_2} = 1.078$$

But: $$M_{N3} = M_3 \sin(\beta - \delta)$$

Hence: $$M_3 = \frac{0.9017}{\sin(28.5 - 4°)} = 2.17$$

Also: $$p_3 = \left( \frac{p_3}{p_2} \times \frac{p_2}{p_1} \right) p_1 = 1.297 \times 1.336 \times 60 = 104.0 \text{ kPa}$$

$$T_3 = \left( \frac{T_3}{T_2} \times \frac{T_2}{T_1} \right) T_1 = 1.078 \times 1.087 \times 253 \text{ K} = 296 \text{ K}$$

**FIGURE E6.2b**

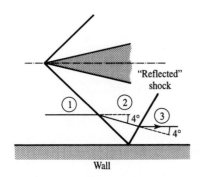

$$a_3 = \sqrt{\gamma R T_3} = \sqrt{1.4 \times 287 \times 296} = 345 \, \text{m/s}$$

From these results it follows that:

$$V_3 = M_3 a_3 = 2.17 \times 345 = 749 \, \text{m/s}$$

Therefore, after the reflection the pressure and velocity are 104 kPa and 749 m/s respectively.

The pressure acting on the wall near an oblique shock reflection is, if viscous effects are ignored, as shown in Fig. 6.13.

In real fluids, a boundary layer exists on the wall in which the velocity drops from its freestream value to zero at the wall. This means that the flow adjacent to the wall is subsonic and cannot sustain the pressure discontinuities associated with shock waves. Due to the presence of the boundary layer there is, therefore, a "spreading out" of the pressure distribution which may therefore resemble that shown in Fig. 6.14.

The actual form of the pressure distribution will depend on the type of boundary layer flow, i.e., laminar or turbulent, the thickness of the boundary layer and on the shock strength. The interaction can also cause a local separation bubble in the boundary layer as shown schematically in Fig. 6.15. The complexity of the reflection that can occur is also shown in Fig. 6.16.

Another point concerning the reflection of oblique shock waves should be noted. It was previously indicated that for a given initial Mach number, there is a minimum angle through which an oblique shock wave can turn a flow and

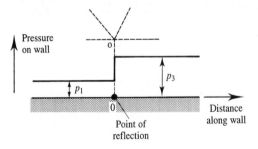

**FIGURE 6.13**
Wall pressure distribution near point of oblique shock wave reflection in ideal case.

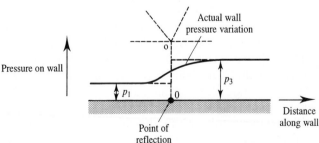

**FIGURE 6.14**
Wall pressure distribution near point of oblique shock wave reflection in real case.

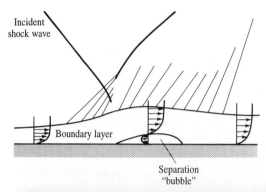

**FIGURE 6.15**
Boundary layer separation during shock wave–boundary layer interaction.

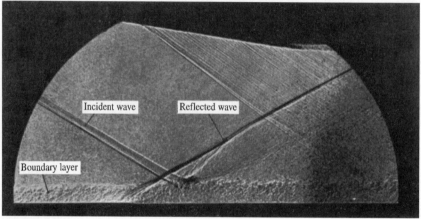

**FIGURE 6.16**
Reflection of oblique shock wave from wall with boundary layer.

that this angle decreases with decreasing Mach number. Therefore, considering the oblique shock wave reflection previously shown, it is possible for a situation to arise in which the maximum possible turning angle corresponding to $M_2$, i.e., to the flow behind the initial shock waves, is less than $\delta$, the angle required to bring the flow parallel to the wall. In this case, a so-called "Mach reflection" occurs, this being illustrated in Fig. 6.17 and shown schematically in Fig. 6.18. Here, a curved strong shock, behind which the flow is subsonic, forms near the wall. Since this subsonic flow need not all be parallel, the flow above the wall shock layer does not have to be parallel to the wall. The flow behind the curved wall shock is divided from the flow behind the "reflected" oblique shock by a slipline across which there are, of course, changes in velocity, temperature and entropy. These sliplines develop in the downstream direction into thin regions across which the changes in flow properties occur, these regions being called slipstreams.

   The details of the flow are influenced by downstream conditions due to the presence of the subsonic region, and the Mach reflection is difficult to treat analytically.

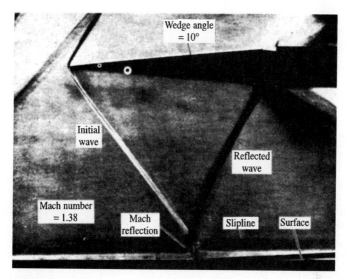

**FIGURE 6.17**
Mach reflection.

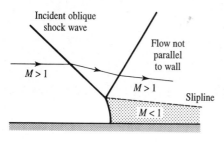

**FIGURE 6.18**
Schematic representation of a Mach reflection.

**EXAMPLE 6.3**

Air flows at a Mach number of 2.5 over a wedge, leading to the generation of an oblique shock wave. This oblique shock wave impinges on a flat wall, which is parallel to the flow upstream of the wedge, and is "reflected" from it. If various wedge angles have to be considered, what is the largest turning angle that a wedge could produce without a Mach reflection being produced at the wall?

***Solution***

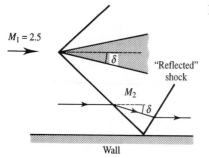

**FIGURE E6.3**

**TABLE E6.3**

| $\delta$ (guess) | $\beta$ | $M_{N1}$ | $M_{N2}$ | $M_2$ | $\delta_{max}$ | $\Delta\delta$ |
|---|---|---|---|---|---|---|
| 4° | 26.6° | 1.120 | 0.8967 | 2.333 | 27.87° | 23.87° |
| 12° | 33.8° | 1.391 | 0.7436 | 2.002 | 23.01° | 11.01° |
| 15° | 36.9° | 1.503 | 0.7002 | 1.877 | 20.73° | 5.73° |
| 18° | 40.4° | 1.620 | 0.6625 | 1.739 | 17.88° | −0.12° |
| 20° | 42.9° | 1.701 | 0.6402 | 1.645 | 15.74° | −4.26° |

A Mach reflection occurs when the maximum turning angle corresponding to the Mach number downstream of the incident wave, i.e., $M_2$, is not enough to bring the flow in region 3 parallel to the wall, i.e., is less than the turning angle produced by the wedge.

A simple iterative approach will be adopted here. A turning angle will be assumed. The maximum turning angle that can be produced by the reflected wave will be calculated, and the difference between this maximum value and the required turning angle will be calculated. This is termed $\Delta\delta$ in Table E6.3. When $\Delta\delta$ is negative, a Mach reflection will exist. The turning angle that makes $\Delta\delta = 0$ can then be deduced.

To illustrate how the results in the table are derived, consider the case where $\delta$ is assumed to be 12°. Now for:

$$M_1 = 2.5, \qquad \delta = 12°$$

the oblique shock chart or the software gives:

$$\beta = 33.8°$$

Hence: $\qquad M_{N1} = M1 \sin\beta = 2.5 \sin 33.8 = 1.391$

This value is also directly given by the software.

Next, using normal shock relations or tables or software for an upstream Mach number of 1.391 ($M_{N1}$) the following is obtained:

$$M_{N2} = 0.7436$$

But: $\qquad\qquad M_{N2} = M_2 \sin(\beta - \delta)$

Hence: $\qquad\qquad M_2 = \dfrac{0.7436}{\sin(33.8 - 12)} = 2.002$

Now for $M = 2.002$ the oblique shock chart or the software gives $\delta_{max} = 23.01°$. Hence:

$$\Delta\delta = \delta_{max} - \delta = 23.01 - 12 = 11.01°$$

Using this approach, the values in Table E6.3 have been derived.

Interpolating between the results given in the above table then indicates that $\Delta\delta = 0$ when $\delta = 17.9°$. Therefore, if the turning angle produced by the wedge is more than 17.9°, a Mach reflection will occur.

In the above discussion of the reflection of oblique shock waves, it was assumed that the shock was reflected off a flat wall. In some circumstances, the system involved may be designed so that the wall changes direction sharply at

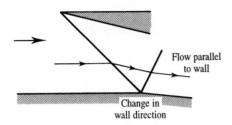

**FIGURE 6.19**
Change of wall direction at point of
shock impingement.

the point of shock impingement. This is shown in Fig. 6.19. The flow downstream of the reflected shock must be parallel to the wall downstream of the
reflection. Hence, the turning angle produced by the reflected wave must be
$\delta_1 - \delta_w$, the wall angle $\delta_w$ being measured in the direction shown in Fig. 6.19,
i.e., a convex corner is being taken as positive. For such a convex corner, then,
the reflected wave will be weaker than the wave that would be reflected from a
flat wall. If $\delta_w = \delta_1$, no reflected wave will occur, i.e., the incident oblique
shock wave will be "cancelled" by the turning of the wall. If $\delta_w > \delta_1$, an
expansion wave of the type discussed in the next chapter will be generated.

**EXAMPLE 6.4**
Air at a pressure of 60 kPa and a temperature of $-20°C$ flows at a Mach number
of 2.5 over a wedge, leading to the generation of an oblique shock wave. This
oblique shock wave impinges on a wall which "turns away from the flow" by 4°
exactly at the point where the oblique shock wave impinges on it, i.e. the wall has
a sharp convex 4° corner at the point where the incident shock impinges on it.
Sketch the flow pattern. If the leading edge of the wedge is 1 m above the wall,
how far behind this leading edge would the change in wall angle have to occur?

***Solution***
Since the change in wall direction is exactly equal to the turning angle produced
by the incident shock, the flow behind this incident shock is parallel to the wall.
There will, therefore, be no reflected wave, i.e., the shock is cancelled by the
corner. The flow pattern will therefore be as shown in Fig. E6.4.

Now, as indicated in the previous two examples, for:

$$M_1 = 2.5, \qquad \delta = 4°$$

the oblique shock chart or the software gives:

**FIGURE E6.4**

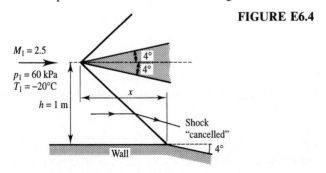

$$\beta = 26.6°$$

Hence, if $x$ is the distance of the point of impingement behind the leading edge of the wedge, it can be seen from Fig. E6.4 that:

$$\frac{h}{x} = \tan 26.6°$$

Since $h = 1$ m, this equation gives:

$$x = 1.99 \, \text{m}$$

Therefore, the sharp corner in the wall would have to be 1.99 m behind the leading edge of the wedge.

## 6.4
## INTERACTION OF OBLIQUE SHOCK WAVES

It will be noted from the results and discussion given of the properties of oblique shock waves that:

- An oblique shock wave always decreases the Mach number, i.e., $M_2 \leq M_1$.
- Considering only the non-strong shock solution, the shock angle, $\beta$, for a given turning angle, $\delta$, increases with decreasing Mach number.

Hence, if the flow around a concave wall consisting of several angular changes of equal magnitude is considered, the oblique shock waves generated at each step will tend to converge and coalesce into a single oblique shock wave which is stronger than any of the initial waves. This is illustrated in Fig. 6.20.

Now, the pressure and flow direction must be the same for all streamlines downstream of the last wave. But two or more weaker waves cannot produce the same changes as a single stronger wave and for this reason the "reflected"

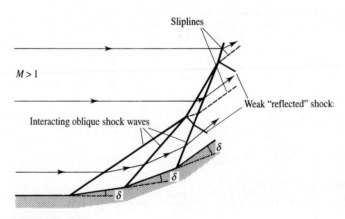

**FIGURE 6.20**
Interaction of oblique shock waves.

shocks shown must be generated. These waves are much weaker than the initial waves. While these "reflected" waves equalized the pressure and flow direction they cannot equalize the velocity, density, and entropy. For this reason, the sliplines shown in the diagram exist across which there is a jump in these properties. In theory, these sliplines are planes of discontinuity but in reality they grow into thin regions over which the changes in the properties occur.

A curved wall can be thought of as consisting of a series of small segments each producing a small fraction of the total change in flow direction. Each of these segments produces a weak wave and these weak waves interact in the manner described above to form a single oblique shock wave as shown in Figs. 6.21 and 6.22. In this case, instead of individual sliplines existing there is a region of variable density, velocity, and entropy. The reflected waves discussed above will, in this case, be negligible for most purposes because the initial waves are so weak. The final oblique shock resulting from the interaction must be that corresponding to the initial Mach number and the total turning angle, $\delta_{total}$.

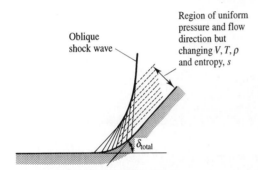

**FIGURE 6.21**
Interaction of oblique shock waves.

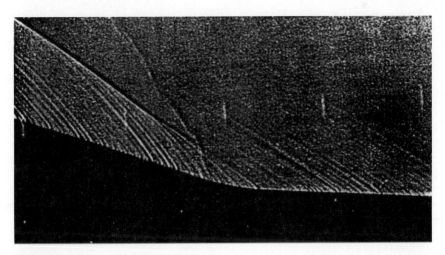

**FIGURE 6.22**
Interaction of oblique shock waves.

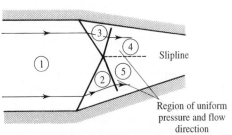

**FIGURE 6.23**
Intersection of oblique shock waves.

**FIGURE 6.24**
Intersection of oblique shock waves.

Lastly, consider what happens when oblique shock waves of differing strength generated by different surfaces interact as shown in Figs. 6.23 and 6.24. The flows in regions 4 and 5 shown in Fig. 6.23 must, of course, be parallel to each other. Therefore, conservation of momentum applied in a direction normal to the flows in these two regions indicates that the pressures in regions 4 and 5 must be the same. The initial waves separating regions 1 and 2 and regions 1 and 3 are, of course, determined by the Mach number in region 1 and the turning angles, $\theta$ and $\phi$. The properties of the "transmitted" waves are then determined from the condition that the pressures and flow directions in regions 4 and 5 must be the same. The density, velocity, and entropy will then be different in these two regions and the slip-stream shown must, therefore, exist. Of course, when $\phi = \theta$ the initial waves are both of the same strength as are the transmitted waves. No slipline then exists.

**EXAMPLE 6.5**
Air is flowing at a Mach number of 3 with a pressure and a temperature of 30 kPa and $-10°C$ respectively down a wide channel. The upper wall of this channel turns

through an angle of 4° "towards the flow" while the lower wall turns through an angle of 3° "towards the flow" leading to the generation of two oblique shock waves. These two oblique shock waves intersect each other. Find the pressure and flow direction downstream of the shock intersection.

### Solution
The flow situation being considered is shown in Fig. E6.5a. The conditions behind the two initial shock waves will first be considered, i.e., the conditions in the regions 2 and 3 shown in Fig. E6.5a will first be derived.

First consider region 2. Using the software or the oblique shock wave chart for $M = 3$ and $\delta = 4°$ gives $\beta = 22.3°$ and therefore:

$$M_{N1} = M_1 \sin \beta = 3 \sin 22.3° = 1.140$$

This is directly given by the software. Using the normal shock tables for $M = 1.14$ gives:

$$M_{N2} = 0.882$$

which is also directly given by the software, and

$$\frac{p_2}{p_1} = 1.350$$

Hence:    $$M_2 = \frac{M_{N2}}{\sin(\beta - \delta)} = \frac{0.882}{\sin(22.3 - 4)} = 2.799$$

Next consider region 3. Using the software or the oblique shock wave chart for $M = 3$ and $\delta = 3°$ gives $\beta = 21.6°$ and therefore:

$$M_{N1} = M_1 \sin \beta = 3 \sin 21.6° = 1.104$$

This is directly given by the software. Using the normal shock tables for $M = 1.104$ gives:

$$M_{N3} = 0.908$$

which is also directly given by the software, and

$$\frac{p_3}{p_1} = 1.255$$

Hence:    $$M_3 = \frac{M_{N3}}{\sin(\beta - \delta)} = \frac{0.908}{\sin(21.6 - 3)} = 2.848$$

Next consider region 4. The strengths of the shock waves after the intersection must be such that the pressure and the flow direction is the same throughout this region. It is convenient to consider two parts of region 4: Region 42 which is downstream of region 2 and region 43 which is downstream of region 3. The

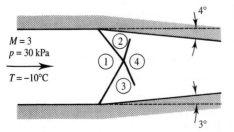

**FIGURE E6.5a**

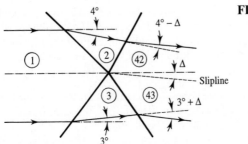

**FIGURE E6.5b**

solution must be such that the flow directions and pressures in regions 42 and 43 are the same. While there are elegant ways of obtaining the solution, a very simple approach will be adopted here. A flow direction that is the same in both regions 42 and 43 will be assumed. This direction will be specified by the angle $\Delta$ which is the flow direction relative to the flow upstream of the initial shock waves as defined in Fig. E6.5b.

The turning angle produced by the oblique shock wave between regions 2 and 42 is given by:

$$\delta = 4° - \Delta$$

Similarly, the turning angle produced by the oblique shock wave between regions 3 and 43 is given by:

$$\delta = 3° + \Delta$$

For this chosen flow direction, the pressures in the two regions, i.e., $p_{42}$ and $p_{43}$, can then be calculated. These two pressures will not in general be the same because the value of $\Delta$ has been guessed. The procedure is repeated for several different values of $\Delta$ and the value of $\Delta$ that makes $p_{42} = p_{43}$ can then be deduced.

To illustrate the procedure, consider the case where the guessed value of $\Delta$ is 1°. In this case the turning angle between region 2 and region 42 is $\delta = 3°$ and, it will be recalled, $M_2 = 2.799$. Using the software or the oblique shock wave chart for this value of $M$ and $\delta$ gives $\beta = 23.1°$ and therefore:

$$M_{N2} = M_2 \sin \beta = 2.799 \sin 23.1 = 1.098$$

This is directly given by the software. Using the normal shock tables for $M = 1.098$ gives:

$$\frac{p_{42}}{p_2} = 1.242$$

Now consider the change between region 3 and region 43. For $\Delta = 1°$, the turning angle between region 3 and region 43 is $\delta = 4°$ and, it will be recalled, $M_2 = 2.848$. Using the software or the oblique shock wave chart for this value of $M$ and $\delta$ gives $\beta = 23.5°$ and therefore:

$$M_{N3} = M_3 \sin \beta = 2.848 \sin 23.5 = 1.13$$

This is directly given by the software. Using the normal shock tables for $M = 1.17$ gives:

$$\frac{p_{43}}{p_3} = 1.335$$

TABLE E6.5

| $\Delta$ (chosen) | $p_{42}/p_1$ | $p_{43}/p_1$ | $p_{42}$ (kPa) | $p_{43}$ (kPa) |
|---|---|---|---|---|
| $+1°$ | 1.677 | 1.675 | 50.2 | 50.3 |
| $-1°$ | 1.920 | 1.453 | 57.6 | 43.6 |
| $0°$ | 1.793 | 1.561 | 53.8 | 46.8 |

Using these results gives:

$$p_{42} = \frac{p_{42}}{p_2}\frac{p_2}{p_1}p_1 = 1.242 \times 1.350 \times 30 = 50.2 \text{ kPa}$$

and:

$$p_{43} = \frac{p_{43}}{p_3}\frac{p_3}{p_1}p_1 = 1.335 \times 1.255 \times 30 = 50.3 \text{ kPa}$$

This procedure is repeated for several other values of $\Delta$ giving the results shown in Table E6.5.

Interpolating between the results given in the above table then indicates that $p_{42} = p_{43}$ approximately when $\Delta = +0.5°$ and $p_4 = 52$ kPa. The flow direction is, therefore, as indicated in Fig. E6.5c.

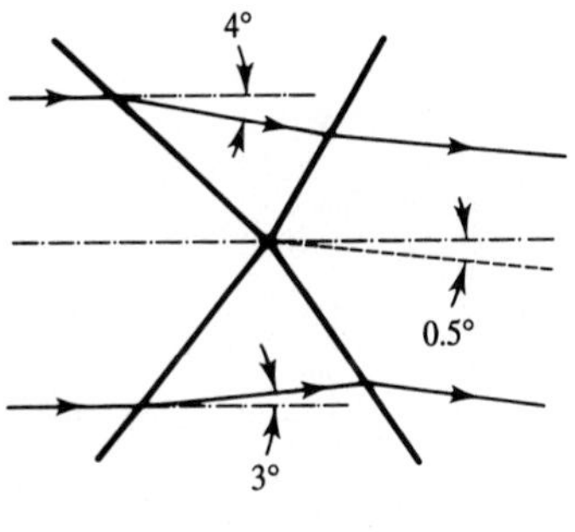

**FIGURE E6.5c**

## 6.5
## CONICAL SHOCK WAVES

The discussion given thus far in this chapter has been concerned with two-dimensional plane flows of the type shown in Fig. 6.25.

In such flows, the flow before and after the oblique shock wave is uniform as shown in Fig. 6.26.

A related but more complex flow is that associated with axisymmetric supersonic flow over a cone as shown in Fig. 6.27. In this situation, because the flow area increases with increasing distance from the centerline, as shown in Fig. 6.28, conservation of mass will require that the flow downstream of a conical shock wave be curved, as indicated in Fig. 6.29. The flow between the cone surface and the conical shock wave is therefore two-dimensional and cannot be analyzed using the simple procedures adopted in dealing with

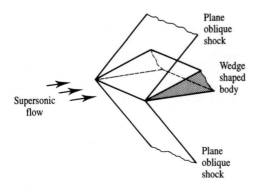

**FIGURE 6.25**
Oblique shock wave on wedge.

Plane
oblique
shock

Wedge
shaped
body

Supersonic
flow

Plane
oblique
shock

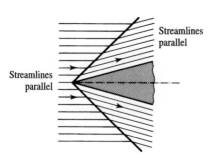

Streamlines
parallel

Streamlines
parallel

**FIGURE 6.26**
Flow direction upstream and downstream
of oblique shock wave on wedge.

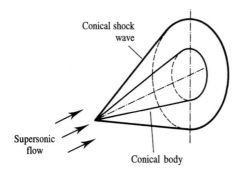

Conical shock
wave

Supersonic
flow

Conical body

**FIGURE 6.27**
Conical shock wave.

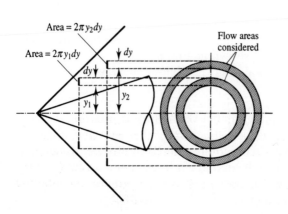

Area = $2\pi y_2 dy$

Area = $2\pi y_1 dy$

$dy$

$dy$

$y_1$

$y_2$

Flow areas
considered

**FIGURE 6.28**
Flow area between cone surface
and conical shock.

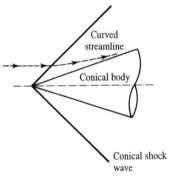

**FIGURE 6.29**
Curved streamline downstream of conical shock wave.

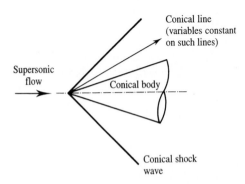

**FIGURE 6.30**
Conical flow downstream of conical shock wave.

plane oblique shock waves. However, as long as the conical shock is attached to the vertex of the cone, the flow behind a conical shock wave will be "conical," i.e., the flow variables are constant along conical surfaces originating at the vertex of the cone as shown in Fig. 6.30. By using this fact, a relatively simple ordinary differential equation can be derived to describe the flow behind an attached conical shock wave. This equation has to be numerically integrated.

## 6.6
## CONCLUDING REMARKS

When a supersonic flow is turned through a positive angle, i.e., is "towards itself" an oblique shock wave develops. The relation between the initial Mach number, the turning angle, and the oblique shock angle has been considered in this chapter. The interaction and reflection of oblique shock waves has also been considered. The difference between conical shock waves and oblique shock waves has also been discussed.

## PROBLEMS

**6.1.** Air is flowing over a flat wall. The Mach number, pressure, and temperature in the air-stream are 3, 50 kPa, and −20°C respectively. If the wall turns through an angle of 4° leading to the formation of an oblique shock wave, find the Mach number, the pressure and the temperature in the flow behind the shock wave.

**6.2.** An air flow in which the Mach number is 2.5 passes over a wedge with a half-angle of 10°, the wedge being symmetrically placed in the flow. Find the ratio of the stagnation pressures before and after the oblique shock wave generated at the leading edge of the wedge.

**6.3.** A symmetrical wedge with a 12° included angle is placed in an air flow in which the Mach number is 2.3 and the pressure is 60 kPa. If the centerline of the wedge is at an angle of 4° to the direction of flow, find the pressure difference between the two surfaces of the wedge.

**6.4.** Air flows over a wall at a supersonic velocity. The wall turns towards the flow generating an oblique shock wave. This wave is found to be at an angle of 50° to the initial flow direction. A scratch on the wall upstream of the shock wave is found to generate a very weak wave that is at an angle of 30° to the flow. Find the angle through which the wall turned.

**6.5.** Air flows over a plane wall at a Mach number of 3.5, the pressure in the flow being 100 kPa. The wall turns through an angle leading to the generation of an oblique shock wave whose strength is such that the pressure downstream of the corner is 548 kPa. Find the turning angle of the corner.

**6.6.** Air, flowing down a plane walled duct at a Mach number of 3, passes over a wedge. What is the largest included angle that this wedge can have if the oblique shock wave that is generated is attached to the wedge? Sketch the flow pattern that will exist if the wedge angle is greater than this maximum value.

**6.7.** A uniform air flow at a Mach number of 2.5 passes around a sharp concave corner in the wall which turns the flow through an angle of 10° and leads to the generation of an oblique shock wave. The pressure and temperature in the flow upstream of the corner are 70 kPa and 10°C respectively. Find the Mach number, the pressure, the temperature, and the stagnation pressure downstream of the oblique shock wave. How large would the corner angle have to be before the shock became detached from the corner?

**6.8.** Find the minimum values of the Mach number for which the oblique shock generated at the leading edge of a wedge placed in a supersonic air flow remains attached to the wedge for deflection angles of 15°, 25°, and 40°.

**6.9.** A wedge symmetrically placed in a supersonic air-stream is to be used to determine the Mach number in the flow. This will be done by using optical methods to measure the angle that the oblique shock wave attached to the leading edge of the wedge makes to the upstream flow. If the total included angle of the wedge that is to be used is 45°, find the Mach number range over which this method can be used.

**6.10.** Air flowing at a Mach number of 2.5 passes over a wedge that turns the flow through an angle of 5°. Find the pressure ratio across the oblique shock that is generated. If this oblique shock is reflected off a plane surface, find the overall pressure ratio.

**6.11.** An oblique shock wave with wave angle of 26° in an air-stream in which the Mach number is 2.7, the pressure is 100 kPa, and the temperature is 30° impinges on a straight wall. Find the Mach number, pressure, temperature, and stagnation pressure downstream of the reflected wave.

**6.12.** Air flows down a duct at a Mach number of 1.5. The top wall of the duct turns towards the flow leading to the generation of an oblique shock wave which strikes the flat lower wall of the duct and is reflected from it. What is the smallest turning angle that would give a Mach reflection off the lower wall?

**6.13.** A two-dimensional wedge with an included angle of 10° is placed in a wind tunnel which has parallel walls. If the Mach number in the freestream ahead of the wedge is 2 and if the axis of the wedge is inclined at an angle of 2° to the direction of the air flow, find the Mach numbers upstream and downstream of the oblique shock waves after they are reflected off the upper and lower walls of the tunnel. Also sketch the flow pattern marking the angles the waves make to the tunnel walls.

**6.14.** Air in which the pressure is 60 kPa is flowing down a plane walled duct at a Mach number of 2.5. The air-stream passes over a wedge with an included angle of 10°. The oblique shock wave that is generated by the wedge is reflected off the flat wall of the duct. Find the pressure and Mach number after the reflection.

**6.15.** Air at a pressure and temperature of 40 kPa and −30°C flows at a Mach number of 3 down a wide duct. The upper wall of the duct turns sharply through an angle of 5° leading to the formation of an oblique shock wave as shown in Fig. P6.15. Find the Mach number, temperature, and pressure behind this shock wave. As shown in the figure, this shock wave strikes the lower wall of the duct exactly at a point where the lower wall turns away from the flow

**FIGURE P6.15**

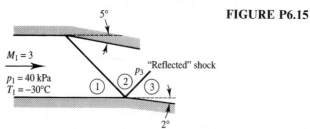

$M_1 = 3$

$p_1 = 40$ kPa
$T_1 = -30°C$

through an angle of 2°. Find the Mach number, pressure, and temperature behind the "reflected" wave.

**6.16.** Find the pressure ratio $p_3/p_1$ for the flow situation shown in Fig. P6.16.

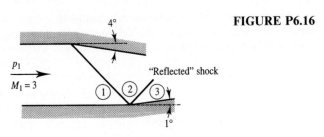

**FIGURE P6.16**

**6.17.** Find the pressure ratio $p_4/p_1$ for the flow situation shown in Fig. P6.17.

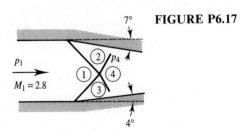

**FIGURE P6.17**

**6.18.** If the Mach number and pressure ahead of the oblique shock wave system shown in Fig. P6.18 are 3 and 50 kPa respectively, find the pressure in the region 4 downstream of the wave intersection.

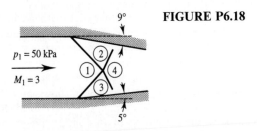

**FIGURE P6.18**

**6.19.** Air is flowing at a Mach number of 2 and a pressure of 70 kPa in a two-dimensional channel. The upper wall turns towards the flow through an angle of 5° and the lower wall turns towards the flow through an angle of 3°, two oblique shock waves thus being generated. These two shock waves intersect each other. Find the pressure in the region just downstream of the shock intersection.

**6.20.** An air-stream in which the Mach number is 3 and the pressure is 80 kPa flows between two parallel walls. The upper wall turns sharply through an angle of 18° and the lower wall turns sharply through an angle of 12° leading to the generation of two oblique shock waves which intersect each other. Sketch the flow pattern and find the flow direction, the Mach number and the pressure immediately downstream of the shock intersection.

**6.21.** Consider an air-stream flowing at a Mach number of 3.2 with a pressure of 60 kPa. Consider two cases. In the first case, the stream passes through a single normal shock wave and is then isentropically decelerated to a very low velocity. In the second case, the flow first passes through an oblique shock that turns the flow through 25° and then passes through a normal shock wave before being isentropically decelerated to a very low velocity. Compare the pressures attained in the two cases. Do the results indicate which arrangement should be used in decelerating a flow from supersonic to subsonic velocities at the inlet to a turbo jet engine in a supersonic aircraft?

**6.22.** A ram-jet engine is fitted to a small aircraft that cruises at a Mach number of 4 at an altitude where the pressure is 30 kPa and the temperature is −45°C. The air entering the engine is slowed to subsonic velocities by passing it through two oblique shock waves each of which turn the flow through 15° and by then passing it through a normal shock wave. Following the normal shock, the flow is isentropically decelerated to a Mach number of 0.1 before it enters the combustion zone. Find the values of the pressure and the temperature at the inlet to the combustion zone. What values would have been attained if initial deceleration had been through a single normal shock wave instead of through the combination of oblique shocks and a normal shock?

# CHAPTER 7

# Expansion Waves: Prandtl–Meyer Flow

## 7.1
## INTRODUCTION

The discussion given in Chapters 5 and 6 was concerned with waves that involved an increase in pressure, i.e., with shock waves. In this chapter, attention will be given to the types of waves that are generated when there is a decrease in pressure. For example, the type of wave that is generated when a supersonic flow passes over a convex corner and the type of wave that is generated when the end of a duct containing a gas at a pressure that is higher than that in the surrounding air is suddenly opened will be discussed in this chapter. These two situations are illustrated in Fig. 7.1.

Steady supersonic flows around convex corners will first be addressed in this chapter. Attention will then be given to unsteady flows.

## 7.2
## PRANDTL–MEYER FLOW

In the preceding chapter, supersonic flow over a concave corner, i.e., a corner involving a positive angular change in flow direction, was considered. It was indicated there that the flow over such a corner was associated with an oblique shock wave, this shock wave originating from the corner when it is sharp. Consider, now, the flow around a convex corner as shown in Fig. 7.2. To determine whether an oblique shock wave also occurs in this case, it is assumed that it does occur, a sharp corner being considered for simplicity as shown in Fig. 7.2.

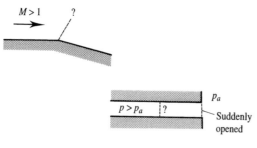

**FIGURE 7.1**
Flows involving a pressure decrease.

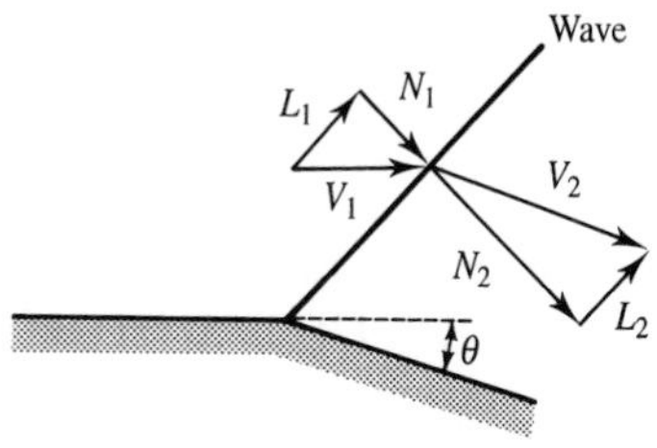

**FIGURE 7.2**
Assumed flow around convex corner.

Consider the velocity components indicated in Fig. 7.2. For the reasons given in the previous chapter, $L_1 = L_2$ and since $V_2$ must be parallel to the downstream wall, geometrical considerations show that $N_2 > N_1$. But $N_2$ and $N_1$ must be related by the normal shock wave relations and in dealing with normal shock waves, it was shown that an expansive shock was not possible since it would violate the second law of thermodynamics. It is, therefore, not possible for $N_2$ to be greater than $N_1$ and the flow over a convex corner cannot, therefore, take place through an oblique shock.

In order to understand the actual flow that occurs when a supersonic flow passes about a convex corner, consider what happens, in general, when the flow is turned through a differentially small angle, $d\theta$, this producing differentially small changes $dp$, $d\rho$, and $dT$ in the pressure, density, and temperature respectively. The present analysis applies whether $d\theta$ is positive or negative, i.e., whether the corner is concave or convex, the changes through the differentially weak Mach wave produced being isentropic (see later). By the reasoning previously given, the velocity component parallel to the wave, $L$, is unchanged by the wave. Hence, considering unit area of the wave, as indicated in Fig. 7.3, the equations of continuity and momentum give:

$$\rho N = (\rho + d\rho)(N + dN)$$

i.e.,
$$\rho\, dN + N\, d\rho = 0 \tag{7.1}$$

higher order terms having been neglected, and

$$p - (p + dp) = \rho N[(N + dN) - N]$$

i.e.,
$$-dp = \rho N\, dN \tag{7.2}$$

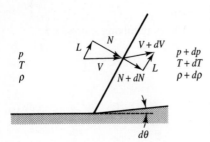

**FIGURE 7.3**
Changes produced by a weak wave.

Substituting for $dN$ from eq. (7.2) into eq. (7.1) gives:

$$N^2 = \frac{dp}{d\rho} \tag{7.3}$$

Now, in the limiting case of a very weak wave being considered, $(dp/d\rho)$ will be equal to the square of the upstream speed of sound, i.e.,

$$N^2 = a^2 \quad \text{or} \quad N = a \tag{7.4}$$

This is indicated in Fig. 7.4. Further, since $L$ is unchanged by the presence of the disturbance, it follows that:

$$(V + dV)\cos(\alpha - d\theta) = V \cos \alpha$$

i.e.,    $(V + dV)(\cos \alpha \cos d\theta + \sin \alpha \sin d\theta) = V \cos \alpha$

Expanding and ignoring higher order terms then gives:

$$V \cos \alpha + V \sin \alpha\, d\theta + dV \cos \alpha = V \cos \alpha$$

Therefore:    $$\frac{dV}{V} = -\tan \alpha\, d\theta$$

$$= \frac{-d\theta}{\sqrt{M^2 - 1}} \tag{7.5}$$

Further, since the energy equation gives:

$$\left(\frac{2\gamma}{\gamma - 1}\right)\left(\frac{p}{\rho}\right) + V^2 = \left(\frac{2\gamma}{\gamma - 1}\right)\left(\frac{p + dp}{\rho + d\rho}\right) + (V + dV)^2$$

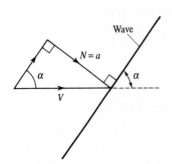

**FIGURE 7.4**
Velocity components near weak wave.
$\sin \alpha = (N/V) = (a/V) = 1/M.$

i.e., ignoring higher order terms:

$$\left(\frac{2\gamma}{\gamma-1}\right)\left(\frac{p}{\rho}\right) + V^2 = \left(\frac{2\gamma}{\gamma-1}\right)\left(\frac{p}{\rho}\right)\left[1 + \frac{dp}{p} - \frac{d\rho}{\rho}\right] + V^2 + 2V\,dV$$

so that:

$$\left(\frac{2\gamma}{\gamma-1}\right)\left(\frac{p}{\rho}\right)\left[\frac{dp}{p} - \frac{dp\,p\,d\rho}{p\,\rho\,dp}\right] = -2V\,dV \tag{7.6}$$

But by the previously made assumptions:

$$\frac{\gamma p}{\rho} = \frac{dp}{d\rho} = a^2$$

so eq. (7.6) becomes:

$$\left(\frac{2a^2}{\gamma-1}\right)\frac{dp}{p}\left[1 - \frac{1}{\gamma}\right] = -2V\,dV$$

i.e.,

$$\frac{dp}{p} = -\frac{\gamma V\,dV}{a^2} = -\gamma M^2 \frac{dV}{V} \tag{7.7}$$

or using eq. (7.5):

$$\frac{dp}{p} = \frac{\gamma M^2}{\sqrt{M^2-1}}\,d\theta \tag{7.8}$$

Further, since:

$$\frac{d\rho}{\rho} = \frac{d\rho\,dp\,p}{dp\,p\,\rho} = \frac{1}{a^2}\frac{dp}{p}\frac{a^2}{\gamma} = \frac{1}{\gamma}\frac{dp}{p}$$

it follows, using eq. (7.8), that:

$$\frac{d\rho}{\rho} = \frac{M^2}{\sqrt{M^2-1}}\,d\theta \tag{7.9}$$

Similarly:

$$\begin{aligned}
\frac{ds}{R} &= \left(\frac{1}{\gamma-1}\right)\ln\left(\frac{p_2}{p_1}\right) - \left(\frac{\gamma}{\gamma-1}\right)\ln\left(\frac{p_2}{\rho_1}\right) \\
&= \left(\frac{1}{\gamma-1}\right)\ln\left(1 + \frac{dp}{p}\right) - \left(\frac{\gamma}{\gamma-1}\right)\ln\left(1 + \frac{d\rho}{\rho}\right) \\
&= \left(\frac{1}{\gamma-1}\right)\frac{\gamma M^2\,d\theta}{\sqrt{M^2-1}} - \left(\frac{\gamma}{\gamma-1}\right)\frac{M^2\,d\theta}{\sqrt{M^2-1}} = 0 \tag{7.10}
\end{aligned}$$

Lastly, since:

$$M^2 = \frac{V^2}{a^2} = \frac{V^2\rho}{\gamma p}$$

the differential change in $M$ is given by:

$$2M \, dM = 2V \, dV \left( \frac{\rho}{\gamma p} \right) + \left( \frac{V}{\gamma p} \right) d\rho - \left( \frac{V\rho}{\gamma p^2} \right) dp$$

so that:

$$2\frac{dM}{M} = 2\frac{dV}{V} + \frac{d\rho}{\rho} - \frac{dp}{p}$$

$$= [-2 + M^2 - \gamma M^2] \frac{d\theta}{\sqrt{M^2 - 1}}$$

Hence:

$$\frac{dM}{M} = \left[ 1 + \frac{\gamma - 1}{2} M^2 \right] \frac{(-d\theta)}{\sqrt{M^2 - 1}} \qquad (7.11)$$

Thus a differentially small change in flow direction produces an isentropic disturbance such that:

$$dV \propto -d\theta$$

$$dp \propto d\theta$$

$$d\rho \propto d\theta$$

$$dM \propto -d\theta$$

$$ds = 0$$

these changes also being indicated in Fig. 7.5.

Thus, if flow around a corner, which may be considered to consist of an infinite number of differentially small angular changes, as shown in Fig. 7.6, is considered, it follows from the preceding results that for positive angular changes, the Mach number decreases and the waves converge to form an oblique shock wave while for a negative angular change the waves diverge. Thus, for negative changes in wall angle a region consisting of Mach waves is generated and the flow remains isentropic throughout. Such flows are called Prandtl–Meyer flows and their form is as shown in Fig. 7.7. In order to analytically determine the changes produced by such a flow, it is noted that eq. (7.11) must apply locally at all points within the expansion fan. Therefore,

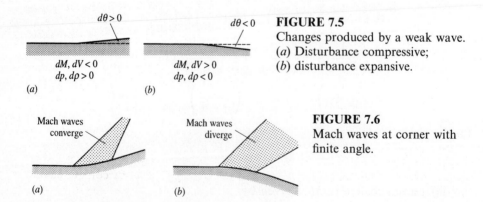

**FIGURE 7.5**
Changes produced by a weak wave.
(a) Disturbance compressive;
(b) disturbance expansive.

**FIGURE 7.6**
Mach waves at corner with finite angle.

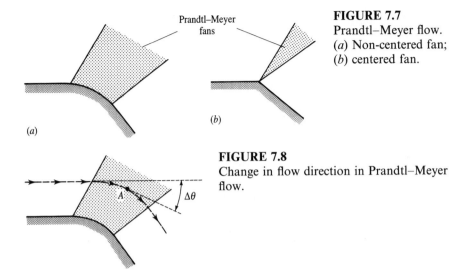

**FIGURE 7.7**
Prandtl–Meyer flow.
(a) Non-centered fan;
(b) centered fan.

**FIGURE 7.8**
Change in flow direction in Prandtl–Meyer flow.

if this equation is integrated it will give the relation between the flow properties and the change in flow direction at any point in the flow as indicated in Fig. 7.8.

Integration of eq. (7.11) gives:

$$\int -d\theta = \int \frac{\sqrt{M^2 - 1}}{1 + \frac{\gamma - 1}{2} M^2} \frac{dM}{M} + \text{constant} \qquad (7.12)$$

The right hand side of eq. (7.12) is easily integrated using standard techniques. Before giving the result, however, there are a couple of points to be noted. Firstly, since the change in flow direction must be negative for Prandtl–Meyer flow to exist, it is convenient to drop the negative sign on the left hand side. Secondly, to express the results in as convenient a form as possible, some standard condition is used to evaluate the constant. The initial boundary condition is, therefore, arbitrarily taken as:

$$\theta = 0 \quad \text{when} \quad M = 1, \quad \text{i.e.,} \quad \theta = 0: \ V = a = a^* \qquad (7.13)$$

The application of the results to flows in which the Mach number ahead of the corner is not 1 will be discussed later.

Using eq. (7.13) in eq. (7.12) then gives:

$$\theta = \sqrt{\frac{\gamma + 1}{\gamma - 1}} \tan^{-1} \sqrt{\frac{\gamma - 1}{\gamma + 1}(M^2 - 1)} - \tan^{-1} \sqrt{M^2 - 1} \qquad (7.14)$$

The relation between $\theta$ (usually expressed in degrees rather than radians as in eq. (7.14)) and $M$ as given by eq. (7.14) is usually listed in isentropic tables and is given in the isentropic flow table in Appendix B. $\theta$, of course, has no meaning when $M$ is less than 1.

Before discussing the application of eq. (7.14) it is worth considering the limiting case of $M \to \infty$. In this case since $\tan^{-1} \phi \to \pi/2$ as $\phi \to \infty$ it follows that as $M \to \infty$,

$$\theta \to \sqrt{\frac{\gamma + 1}{\gamma - 1}} \frac{\pi}{2} - \frac{\pi}{2}$$

$$= \frac{\pi}{2} \left\{ \sqrt{\frac{\gamma + 1}{\gamma - 1}} - 1 \right\} \tag{7.15}$$

For the case of $\gamma = 1.4$ this gives the limiting value of $\theta$ as:

$$\theta_{max} = \frac{\pi}{2} \{ \sqrt{6} - 1 \} = 130.5° \tag{7.16}$$

Thus, if a flow at a Mach number of 1 is turned through an angle of 130.5°, an infinite Mach number is generated and the pressure falls to zero. Expansion through a greater angle would, according to the present theory, lead to a vacuum adjacent to the wall as indicated in Fig. 7.9. Of course, in reality, the continuum and ideal gas assumptions cease to be valid long before this situation is reached.

Next, consider the application of eq. (7.14) to the calculation of the flow changes produced by a Prandtl–Meyer expansion. Referring to Fig. 7.10, the procedure is as follows.

1. From tables or graphs or from the equation find the value of $\theta$ corresponding to $M_1$, i.e., $\theta_1$. This is equivalent to assuming that the initial flow was generated by an expansion around a hypothetical corner from a Mach number of 1, the reference Mach number in the tables, to a Mach number of $M_1$.

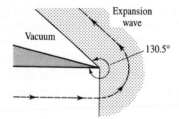

**FIGURE 7.9**
Expansion to zero pressure.

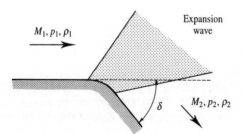

**FIGURE 7.10**
Flow changes through a Prandtl–Meyer wave.

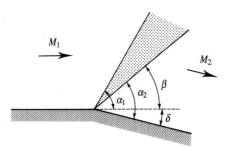

**FIGURE 7.11**
Angles associated with centered expansion wave.

2. Calculate the $\theta$ for the flow downstream of the corner. This will be given by:

$$\theta_2 = \theta_1 + \delta$$

3. Find, using tables or graphs or the equation, the downstream Mach number, $M_2$, corresponding to this value of $\theta_2$.
4. Any other required property of the downstream flow is obtained by noting that the expansion is isentropic and that the following relations, therefore, apply:

$$\frac{T_2}{T_1} = \frac{1 + \left(\dfrac{\gamma - 1}{2}\right) M_1^2}{1 + \left(\dfrac{\gamma - 1}{2}\right) M_2^2}$$

$$\frac{p_2}{p_1} = \left(\frac{T_2}{T_1}\right)^{\frac{\gamma}{\gamma-1}}, \qquad \frac{\rho_2}{\rho_1} = \left(\frac{T_2}{T_1}\right)^{\frac{1}{\gamma-1}}$$

Alternatively, where available, isentropic flow tables can be used, it being noted that the stagnation pressure remains constant across the wave.

5. If necessary, calculate the boundaries of the expansion wave. This is done by noting that they are the Mach lines corresponding to the upstream and downstream Mach numbers.

The various angles defined in Fig. 7.11 are given by:

$$\sin \alpha_1 = \frac{1}{M_1}, \qquad \sin \alpha_2 = \frac{1}{M_2}, \qquad \beta = \alpha_2 - \delta$$

**EXAMPLE 7.1**
Air flows at a Mach number of 1.8 with a pressure of 90 kPa and a temperature of 15°C down a wide channel. The upper wall of this channel turns through an angle of 5° "away from the flow" leading to the generation of an expansion wave. Find the pressure, Mach number, and temperature behind this expansion wave.

**Solution**

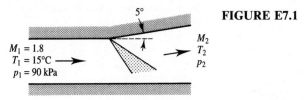

**FIGURE E7.1**

For $M_1 = 1.8$ isentropic flow relations, or tables or software give:

$$\theta_1 = 20.73°, \qquad \frac{p_{01}}{p_1} = 5.746, \qquad \frac{T_{01}}{T_1} = 1.648$$

After the expansion wave:

$$\theta_2 = \theta_1 + 5 = 20.73 + 5 = 25.73°$$

For this value of $\theta$, isentropic flow relations or tables or software give:

$$M_2 = 1.98, \qquad \frac{p_{02}}{p_2} = 7.585, \qquad \frac{T_{02}}{T_2} = 1.784$$

Therefore, because the flow through the expansion wave is isentropic so that $p_{02} = p_{01}$ and $T_{02} = T_{01}$, it follows that:

$$T_2 = \frac{T_{01}}{T_1}\frac{T_2}{T_{02}} T_1 = \frac{1.648}{1.784} \times (273 + 15) = 266.0 \text{ K} = -7°\text{C}$$

and

$$p_2 = \frac{p_{01}}{p_1}\frac{p_2}{p_{02}} p_1 = \frac{5.746}{7.585} \times 90 = 68.2 \text{ kPa}$$

Hence, the pressure, Mach number, and temperature downstream of the expansion wave are 68.2 kPa, 1.98, and $-7°$C respectively.

## 7.3
## REFLECTION AND INTERACTION OF EXPANSION WAVES

Just as with oblique shock waves, expansion waves can undergo reflection and can interact with each other. Consider, first, the reflection of an expansion wave from a straight wall as shown in Fig. 7.12. If $\theta_1$ is the Prandtl–Meyer angle corresponding to the initial flow conditions, i.e., to conditions in region 1, then the Prandtl–Meyer angle for the flow in the intermediate region 2 is, of

**FIGURE 7.12**
Reflection of an expansion wave from a flat wall.

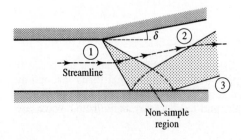

course, given by:

$$\theta_2 = \theta_1 + \delta$$

Since the end flow, i.e., the flow in region 3, must again be parallel to the wall, the reflected wave must also turn the flow through an angle of $\delta$ so that:

$$\theta_3 = \theta_2 + \delta = \theta_1 + 2\delta$$

Once this angle is determined, the Mach number in region 3 can be found and since the whole flow is isentropic, the conditions in region 3 can then be found in terms of those in region 1 using isentropic relations.

Inside the region of interaction of the incident and reflected waves, the relation between $\theta$ and $M$ previously derived cannot be directly applied. This region is known as a non-simple region. The flow is, of course, isentropic throughout.

### EXAMPLE 7.2

Air is flowing down a duct at a Mach number of 2.0 with a pressure of 90 kPa. The upper wall of the duct turns "away" from the flow through an angle of 10° leading to the formation of an expansion wave. This wave is reflected from the flat lower wall of the channel. Find the pressure after the reflection.

### Solution

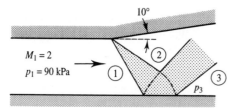

FIGURE E7.2

The initial wave and the reflected wave both turn the flow through 10°. Now, in the initial flow at $M_1 = 2$, isentropic relations or tables or the software give:

$$\frac{p_{01}}{p_1} = 7.83, \qquad \theta_1 = 26.38°$$

Hence:

$$\theta_3 = \theta_1 + 10 + 10 = 46.38°$$

Isentropic relations or tables or software then give for this value of $\theta$:

$$M_3 = 2.83, \qquad \frac{p_{03}}{p_3} = 28.41$$

But the entire flow is isentropic so:

$$p_{03} = p_{01}$$

and so:

$$p_3 = \frac{p_3}{p_{03}} \frac{p_{01}}{p_1} p_1 = \frac{7.83}{28.41} \times 90 = 24.81 \text{ kPa}$$

Therefore, the pressure and Mach number behind the reflected wave are 24.81 kPa and 2.83 respectively.

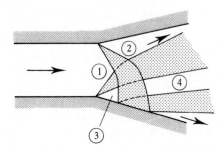

**FIGURE 7.13**
Interaction of expansion waves.

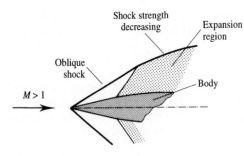

**FIGURE 7.14**
Interaction of an expansion wave
with an oblique shock wave.

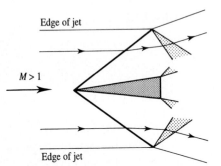

**FIGURE 7.15**
Reflection of an oblique shock wave from
the edge of a jet flow.

When expansion waves interact, the flow resembles that shown in Fig. 7.13. Since the whole flow is isentropic, region 4 must be a region of uniform properties, no slipstreams being generated when expansion waves interact as was the case when shock waves interacted.

In flows over bodies, expansion waves often interact with a shock wave, the shock being attenuated (weakened) by the interaction. An example of such an interaction is shown in Fig. 7.14. The interaction is complicated by the generation of a series of reflected waves.

An expansion wave can also be generated by the "reflection" of an oblique shock wave off a constant pressure boundary. To see how this can happen, consider a wedge shaped body placed in a two-dimensional supersonic jet flow as shown in Fig. 7.15. The edges of the jet are exposed to the stagnant surrounding gas and must, therefore, remain at the pressure that exists in this ambient gas. The flow ahead of the body, because it is a parallel flow, must also all be at the ambient pressure as indicated in Fig. 7.15. Oblique shock waves are generated at the leading edge of the body, these shock waves increasing the pressure. Expansion waves are therefore generated at the edges of the jet at the points at which the shock impinges on these edges, the expansion waves being of such a strength that they drop the pressure back to that in the ambient fluid as shown in Fig. 7.15.

**EXAMPLE 7.3**
A wedge shaped body with an included angle of 20° is placed symmetrically in a plane jet of air, the Mach number in the jet being 2.5. The ambient air pressure is 100 kPa. The oblique shock waves generated at the leading edge of the wedge are reflected off the edges of the jet. Find the Mach number and pressure in the flow downstream of the reflected waves.

**Solution**

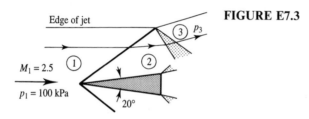

**FIGURE E7.3**

First consider the oblique shock wave. For $M = 2.5$ and $\delta = 10°$, normal shock relations or tables or the software give as discussed in Chapter 6:

$$M_2 = 2.086, \qquad M_{N1} = 1.319$$

For this value of $M_{N1}$, normal shock relations or tables or software give:

$$\frac{p_2}{p_1} = 1.866$$

while isentropic relations or tables or software give for $M_2 = 2.086$,

$$\frac{p_{20}}{p_2} = 8.945, \qquad \theta_2 = 28.72°$$

The strength of the "reflected" expansion waves must be such that $p_3 = p_1$, i.e., such that:

$$\frac{p_{03}}{p_3} = \frac{p_{03}}{p_{02}} \frac{p_{02}}{p_2} \frac{p_2}{p_1}$$

i.e., because the flow through the expansion wave is isentropic, $p_{03} = p_{02}$:

$$\frac{p_{03}}{p_3} = 8.95 \times 1.866 = 16.70$$

For this value of $p_{03}/p_3$ isentropic relations or tables or software give $M_3 = 2.485$.
    The Mach number downstream of the "reflected" expansion waves is therefore 2.485. The pressure behind these waves is, as assumed in the above calculations, the same as that in the initial flow, i.e., 100 kPa.

## 7.4
## BOUNDARY LAYER EFFECTS ON EXPANSION WAVES

For the same reasons that the presence of a boundary layer causes a "spreading out" of the pressure change when a shock wave impinges on a wall, the presence of a boundary layer modifies an expansion wave in the vicinity of a wall. This is illustrated in Fig. 7.16. The extent of the interaction again depends on the thickness and type of boundary layer.

## 7.5
## FLOW OVER BODIES INVOLVING SHOCK AND EXPANSION WAVES

Many bodies over which an effectively two-dimensional supersonic flow occurs in practice can be assumed to consist of a series of flat surfaces. For example, the type of body shown in Fig. 7.17 is similar to the cross-sectional shape of the control surfaces used on some supersonic vehicles. The flow over such a body can be calculated by noting that a series of oblique shock waves and expansion waves occur that cause the flow to be locally parallel to each of

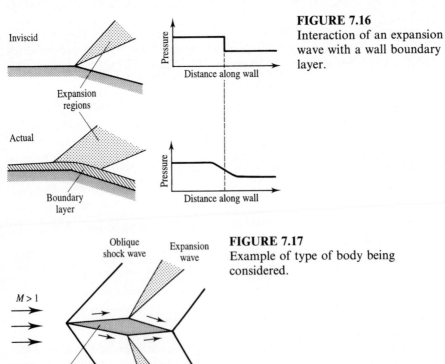

**FIGURE 7.16**
Interaction of an expansion wave with a wall boundary layer.

**FIGURE 7.17**
Example of type of body being considered.

the surfaces as illustrated in Fig. 7.17. Provided that the body is "slender," any secondary waves generated as a result of the interaction of the shock waves and expansion waves generated by the body will not impinge on the body and their presence will not effect the flow over the body. The flow over the body and the pressure acting on the surfaces over the body can then be calculated by separately using the oblique shock and the expansion wave results. Once the pressures on the surfaces of the body are found, the net force on the body can be found. This procedure is illustrated in the following examples.

### EXAMPLE 7.4
A simple wing may be modeled as a 0.25 m wide flat plate set at an angle of 3° to an air flow at a Mach number of 2.5, the pressure in this flow being 60 kPa. Assuming that the flow over the wing is two-dimensional, estimate the lift and drag force per meter span due to the wave formation on the wing. What other factor causes drag on the wing?

### Solution
The flow situation being considered is shown in Fig. E7.4. An expansion wave forms on the upper surface at the leading edge. This wave turns the flow parallel to the upper surface of the plate. Similarly, an oblique shock wave forms on the lower surface at the leading edge. This wave turns the flow parallel to the lower surface of the plate. Waves also form at the trailing edge of the plate but these waves have no effect on the pressures on the surfaces of the plate and they will not be considered here.

First consider the expansion wave which turns the flow parallel to the upper surface, the region adjacent to the upper surface being designated as 2 as indicated in Fig. E7.4. Now, in the freestream, i.e., in region 1, where the Mach number, $M_1$, is 2.5, isentropic relations or tables or the software give:

$$\frac{p_{01}}{p_1} = 17.09, \qquad \theta_1 = 39.13°$$

Hence since the flow is turned through 3° by the expansion wave, it follows that:

$$\theta_2 = 39.13 + 3° = 42.13°$$

Using this value of $\theta_2$, isentropic relations or tables or the software give:

$$M_2 = 2.63, \qquad \frac{p_{02}}{p_2} = 20.92$$

**FIGURE E7.4**

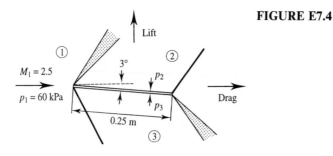

It therefore follows that, since the flow through the expansion wave is isentropic which means that $p_{02} = p_{01}$:

$$p_2 = \frac{p_2}{p_{02}}\frac{p_{01}}{p_1}p_1 = \frac{17.09}{20.92} \times 60 = 49.02 \text{ kPa}$$

Therefore, the pressure acting on the upper surface of the plate is 49 kPa.

Next, consider the oblique shock wave which turns the flow parallel to the lower surface, the region adjacent to the lower surface being designated as 3 as indicated in the Figure. Now, since $M_1$ is 2.5 and the turning angle $\delta$ produced by the oblique shock wave is 3°, oblique shock relations or charts or the software give:

$$\beta = 26°, \qquad M_{N1} = M_1 \sin \beta = 2.5 \sin 26 = 1.096$$

the latter values being given directly by the software.

Normal shock relations or tables or software then give for a Mach number of 1.096:

$$\frac{p_3}{p_1} = 1.23$$

From this it follows that:

$$p_3 = \frac{p_3}{p_1}p_1 = 1.23 \times 60 = 74 \text{ kPa}$$

Therefore, the pressure acting on the lower surface of the plate is 74 kPa.

The lift is the net force acting on the plate normal to the direction of initial flow while the drag is the net force parallel to the direction of initial flow. Therefore, since the plate area per meter span is 0.25 m², it follows that:

Lift per meter span $= (p_3 - p_2)A \cos 3$

$$= (74 - 49) \times 0.25 \times 0.999 = 6.23 \text{ kN/m span}$$

Drag per meter span $= (p_3 - p_2)A \sin 3$

$$= (74 - 49) \times 0.25 \times 0.0523 = 0.33 \text{ kN/m span}$$

Therefore, the lift and drag per meter span are 6.23 and 0.33 N respectively. This drag is that due to the pressure variation about the plate. It is termed the "wave drag." The skin friction, i.e., the force on the plate due to viscous forces, will also contribute to the drag.

**EXAMPLE 7.5**

Find the lift per meter span for the wedge shaped airfoil shown in Fig. E7.5a. Also sketch the flow pattern about the airfoil. The Mach number and the pressure ahead of the airfoil are 2.6 and 40 kPa respectively.

*Solution*

Consider the angles shown in Fig. E7.5b. It will be seen that:

$$0.4 \tan 2° = 0.3 \tan \psi$$

This gives $\psi = 2.67°$. It then follows that:

$$\phi = 2 + \psi = 2 + 2.67 = 4.67°$$

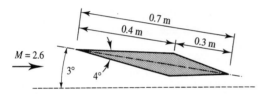

**FIGURE E7.5a**

**FIGURE E7.5b**

**FIGURE E7.5c**

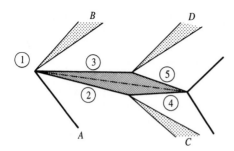

The wave pattern about the airfoil is then as shown in Fig. E7.5c. The waves that occur at the trailing edge do not effect the pressures on the surfaces of the airfoil and will not be analyzed here. The angles of turning produced by the waves shown in Fig. E7.5c are as follows:

$$\text{Shock wave } A - \text{Angle of turn} = 5°$$

$$\text{Expansion wave } B - \text{Angle of turn} = 1°$$

$$\text{Expansion wave } C - \text{Angle of turn} = 4.67°$$

$$\text{Expansion wave } D - \text{Angle of turn} = 4.67°$$

First consider shock wave $A$ which separates regions 1 and 2. Since $M_1$ is 2.6 and $\delta$ is 5°, oblique shock relations or charts or the software give:

$$\beta = 26.5°, \qquad M_{N1} = M_1 \sin \beta = 2.6 \sin 26.5 = 1.16$$

The latter value is given directly by the software.

Normal shock relations or tables or software then give for a Mach number of 1.16:

$$\frac{p_2}{p_1} = 1.403, \qquad M_{N2} = 0.868$$

From this it follows that:

$$p_2 = \frac{p_2}{p_1} p_1 = 1.403 \times 40 = 56.1 \text{ kPa}$$

and

$$M_2 = \frac{M_{N2}}{\tan(26.5 - 5)} = 2.37$$

Next consider expansion wave $B$ which separates regions 1 and 3. Now, in the freestream, i.e., in region 1, where the Mach number, $M_1$, is 2.6, isentropic relations or tables or the software give:

$$\frac{p_{01}}{p_1} = 19.95, \qquad \theta_1 = 41.41°$$

Hence since the flow is turned through 1° by the expansion wave, it follows that:

$$\theta_3 = 41.41 + 1 = 42.41°$$

Using this value of $\theta_3$, isentropic relations or tables or the software give:

$$M_3 = 2.64, \qquad \frac{p_{03}}{p_3} = 21.41$$

It therefore follows that, since the flow through the expansion wave is isentropic which means that $p_{03} = p_{01}$:

$$p_3 = \frac{p_3}{p_{03}} \frac{p_{01}}{p_1} p_1 = \frac{19.95}{21.41} \times 40 = 37.27 \text{ kPa}$$

Next consider expansion wave $C$ which separates regions 2 and 4. Now, ahead of the wave, i.e., in region 2, where the Mach number, $M_2$, is 2.37, isentropic relations or tables or the software give:

$$\frac{p_{02}}{p_1} = 13.95, \qquad \theta_2 = 36.02°$$

Hence since the flow is turned through 4.67° by the expansion wave, it follows that:

$$\theta_4 = 36.02 + 4.67 = 40.69°$$

Using this value of $\theta_4$, isentropic relations or tables or the software give:

$$M_4 = 2.57, \qquad \frac{p_{04}}{p_4} = 19.05$$

It therefore follows that, since the flow through the expansion wave is isentropic which means that $p_{04} = p_{02}$:

$$p_4 = \frac{p_4}{p_{04}} \frac{p_{02}}{p_2} p_2 = \frac{13.95}{19.05} \times 56.1 = 41.08 \text{ kPa}$$

Next consider expansion wave $D$ which separates regions 3 and 5. Now, ahead of the wave, i.e., in region 3, where the Mach number, $M_3$, is 2.64, isentropic relations or tables or the software give:

$$\frac{p_{03}}{p_3} = 21.23, \qquad \theta_3 = 42.41°$$

Hence since the flow is turned through 4.67° by the expansion wave, it follows that:

$$\theta_5 = 42.41 + 4.67 = 47.08°$$

Using this value of $\theta_5$, isentropic relations or tables or the software give:

$$M_5 = 2.87, \qquad \frac{p_{05}}{p_5} = 30.19$$

It therefore follows that, since the flow through the expansion wave is isentropic which means that $p_{05} = p_{03}$:

$$p_5 = \frac{p_5}{p_{05}} \frac{p_{03}}{p_3} p_3 = \frac{21.23}{30.19} \times 37.27 = 26.2 \text{ kPa}$$

Hence:

$$p_2 = 56.1 \text{ kPa}, \qquad p_3 = 37.3 \text{ kPa}, \qquad p_4 = 41.1 \text{ kPa}, \qquad p_5 = 26.2 \text{ kPa}$$

and therefore since the areas of the various surfaces per meter span are:

$$A_2 = 0.4/\cos 2 = 0.400 \text{ m}^2 = A_3$$

$$A_2 = 0.3/\cos 2.67 = 0.300 \text{ m}^2 = A_5$$

the lift and drag are given by

$$\text{Lift} = 56.1 \times 0.4 \times \cos 5 - 37.3 \times 0.4 \times \cos 1$$
$$+ 41.1 \times 0.3 \times \cos 0.33 - 26.2 \times 0.3 \times \cos 5.67 = 11.95 \text{ kN}$$

Hence, the lift per meter span is approximately 12 kN.

The waves that form at the trailing edge of the airfoil were not considered in the above two examples. Their strengths are, however, easily found by noting that the direction of flow and pressure must be the same in the entire region downstream of the airfoil.

## 7.6
## UNSTEADY EXPANSION WAVES

A type of flow that is related to the Prandtl–Meyer flow discussed above is the unsteady expansion wave. To understand the basic characteristics of this flow consider a piston at the end of a long duct as shown in Fig. 7.18.

If the piston is suddenly given a velocity, $dV$, in the direction of withdrawing it from the duct, the gas adjacent to the piston must have a velocity, $dV$, in the same direction as the piston and a weak wave (a sound wave) must propagate into the stationary gas at the local speed of sound, $a$. Across this wave there will be decreases in the speed of sound and pressure of magnitudes $da$ and $dp$ respectively. If the piston is then given another sudden acceleration to a new velocity, $2\,dV$, as shown in Fig. 7.19, then another weak wave will be generated which will move relative to the fluid at velocity $(a - da)$, i.e., the local velocity of sound. However, since the fluid ahead of this second wave already has velocity $dV$ in the opposite direction to that of wave propagation,

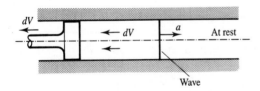

**FIGURE 7.18**
Generation of first wave.

the actual speed at which the wave propagates relative to the duct is $(a - da - dV)$. This is shown in Fig. 7.19.

If the piston had, in fact, been smoothly accelerated up to a velocity $V_p$ then the acceleration can be thought of as consisting of a series of differentially small jumps in velocity each of which produces an expansion wave which propagates down the duct at a lower velocity than its predecessor, thereby leading to an ever-widening expansion region. The process is conveniently shown on an $x$–$t$ diagram where $x$ is the distance along the duct and $t$ is the time. Such a diagram for the process being considered is shown in Fig. 7.20. The initial wave propagates at a speed, $a_1$, into the fluid so that the position of the head of the expansion is given by:

$$\text{Head: } \quad x = a_1 t \quad\quad (7.17)$$

The tail of the wave propagates at speed, $a_2$, relative to the fluid in region 2 and, therefore, since the fluid in this region has velocity, $u_p$, in the direction of piston motion, the velocity of the tail of the wave relative to the walls of the duct is:

$$\text{Tail: } \quad x - x_f = (a_2 - V_p)(t - t_f) \quad\quad (7.18)$$

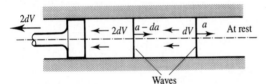

**FIGURE 7.19**

Generation of second weak wave.

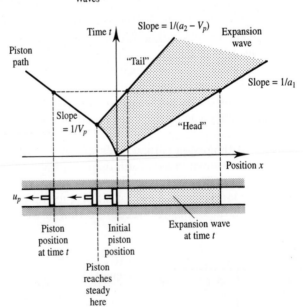

**FIGURE 7.20**

$x$–$t$ diagram for wave system generated by accelerating piston.

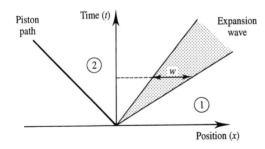

**FIGURE 7.21**
$x$–$t$ diagram for instantly accelerated piston case.

If the piston is instantly accelerated, the $x$–$t$ diagram takes the form shown in Fig. 7.21. Here, the path of the tail of the expansion wave is given by:

$$\text{Tail:} \quad x = (a_2 - V_p)t \tag{7.19}$$

and the width of the expansion region at any instant of time is given by:

$$w = [a_1 - (a_2 - V_p)]t \tag{7.20}$$

Unsteady expansion waves can be generated in other ways, notably by the rupture of diaphragms separating regions of high and low pressure. These will be discussed later.

In order to apply the equations derived above, $a_2$ has to be known. To find this consider the wave to be split, as previously explained, into a series of wavelets each produced by a differentially small jump in piston velocity and each producing a differentially small change in velocity, pressure, etc. If the local gas velocity in the wave is $V$, the wavelet is propagated with velocity $(a - V)$ relative to the tube walls as shown in Fig. 7.22. Consider the flow relative to the wave as indicated in Fig. 7.23. Applying the momentum equation to the flow across the wave gives:

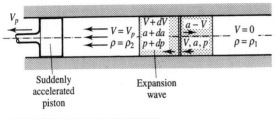

**FIGURE 7.22**
Wavelet considered in analysis of unsteady expansion wave.

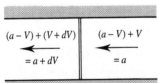

**FIGURE 7.23**
Flow relative to wavelet considered.

$$p - (p + dp) = \rho a[(a + dV) - a]$$

i.e.,
$$-dp = \rho a \, dV \tag{7.21}$$

But
$$\frac{dp}{d\rho} = a^2$$

so eq. (7.21) can be written as:

$$-\frac{d\rho}{\rho} = \frac{dV}{a} \tag{7.22}$$

Integrating this equation across the complete wave gives:

$$\int_0^{V_p} dV = -\int_{\rho_1}^{\rho_2} a \frac{d\rho}{\rho} \tag{7.23}$$

But since the entire flow is isentropic, $a$ is related to $\rho$ by:

$$\frac{a}{a_1} = \left(\frac{\rho}{\rho_1}\right)^{\frac{\gamma-1}{2}}$$

so the right hand side of eq. (7.23) becomes:

$$\int_{\rho_1}^{\rho_2} a \frac{d\rho}{\rho} = \int_{\rho_1}^{\rho_2} a_1 \left(\frac{\rho}{\rho_1}\right)^{\frac{\gamma-1}{2}} \frac{d\rho}{\rho}$$

$$= \left(\frac{2}{\gamma-1}\right) \frac{a_1}{\rho_1^{\frac{\gamma-1}{2}}} [\rho_2^{\frac{\gamma-1}{2}} - \rho_1^{\frac{\gamma-1}{2}}]$$

$$= \left(\frac{2}{\gamma-1}\right) a_1 \left[\left(\frac{\rho_2}{\rho_1}\right)^{\frac{\gamma-1}{2}} - 1\right]$$

$$= \left(\frac{2}{\gamma-1}\right) a_1 \left[\left(\frac{a_2}{a_1}\right) - 1\right]$$

$$= \frac{2a_2}{\gamma-1} - \frac{2a_1}{\gamma-1} \tag{7.24}$$

Equation (7.23) can, therefore, be written as:

$$V_p = -\frac{2a_2}{\gamma-1} + \frac{2a_1}{\gamma-1}$$

i.e.,
$$\frac{a_2}{a_1} = 1 - \left(\frac{\gamma-1}{2}\right)\left(\frac{V_p}{a_1}\right) \tag{7.25}$$

This is the basic equation of unsteady expansion wave flow. Once $(a_2/a_1)$ is known, the other changes across the wave can be calculated by noting that the flow through the wave is isentropic so that:

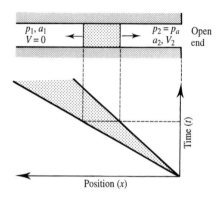

**FIGURE 7.24**

Unsteady expansion wave produced by sudden opening of the end of a tube.

$$\frac{p_2}{p_1} = \left(\frac{a_2}{a_1}\right)^{\frac{2\gamma}{\gamma-1}}, \qquad \frac{\rho_2}{\rho_1} = \left(\frac{a_2}{a_1}\right)^{\frac{2}{\gamma-1}}$$

In the above discussion, unsteady expansion waves associated with the motion of pistons were considered. As was noted earlier, however, such waves can be generated in other ways. As an example, consider a tube which is initially sealed at both ends and contains a gas at a pressure above atmospheric pressure. If one end of the tube is suddenly opened, the pressure at this end of the tube will drop to atmospheric and an unsteady expansion wave will propagate down the tube inducing flow out of the tube. This situation is shown in Fig. 7.24. The flow is equivalent to that which would have been generated by the instantaneous acceleration of a piston to velocity, $V_2$. From the previous work, it follows that across the expansion wave:

$$\frac{a_2}{a_1} = 1 - \left(\frac{\gamma-1}{2}\right)\frac{V_2}{a_1} \tag{7.26}$$

But the expansion is isentropic so that:

$$\left(\frac{p_2}{p_1}\right) = \left(\frac{a_2}{a_1}\right)^{\frac{2\gamma}{\gamma-1}}$$

Combining these equations and noting that $p_2 = p_a$, the atmospheric pressure, gives:

$$\left(\frac{p_a}{p_1}\right)^{\frac{\gamma-1}{2\gamma}} = 1 - \left(\frac{\gamma-1}{2}\right)\frac{V_2}{a_1}$$

i.e.,

$$V_2 = \left(\frac{2a_1}{\gamma-1}\right)\left\{1 - \left(\frac{p_a}{p_1}\right)^{\frac{\gamma-1}{2\gamma}}\right\} \tag{7.27}$$

which gives the velocity at which the gas will be discharged from the tube. It is interesting to note that no matter how large the initial pressure, $p_1$, is, there is a maximum velocity at which the gas will be discharged, this being:

$$V_{2\text{max}} = \frac{2a_1}{\gamma - 1} \tag{7.28}$$

This is, then, the maximum velocity which can be generated by an unsteady expansion wave propagating into a gas at rest. It should not be confused with the maximum escape velocity previously discussed which was the maximum velocity that could be generated by a steady isentropic expansion.

### EXAMPLE 7.6

A diaphragm at the end of 4 m long pipe containing air at a pressure of 200 kPa and a temperature of 30°C suddenly ruptures causing an expansion wave to propagate down the pipe. Find the velocity at which the air is discharged from the pipe if the ambient air pressure is 103 kPa. Also find the velocity of the front and the back of the wave and hence find the time taken for the front of the wave to reach the end of the pipe.

### Solution

The flow is shown in Fig. E7.6. Now it was shown above that:

$$V_2 = \left(\frac{2a_1}{\gamma - 1}\right)\left[1 - \left(\frac{p_a}{p_1}\right)^{\frac{\gamma-1}{2\gamma}}\right]$$

In the present case since $T_1 = 30°C = 303$ K, it follows that:

$$a_1 = \sqrt{\gamma R T_1} = \sqrt{1.4 \times 287 \times 303} = 348.9 \text{ m/s}$$

Hence:   $$V_2 = \left(\frac{2 \times 348.9 \text{ m/s}}{1.4 - 1}\right)\left[1 - \left(\frac{103}{200}\right)^{(1.4-1)/(2\times1.4)}\right] = 157.8 \text{ m/s}$$

Therefore, the velocity at which the air is discharged from the pipe is 158 m/s.

The front of the wave propagates at the local speed of sound in the undisturbed air, $a_1$, i.e., at 348.9 m/s. The tail of wave propagates at the local speed of sound behind the wave, $a_2$, relative to gas behind the wave, i.e., at $a_2 - V_2$ relative to the pipe. But, as shown above:

$$\frac{a_2}{a_1} = 1 - \left(\frac{\gamma - 1}{2}\right)\left(\frac{V_2}{a_1}\right)$$

i.e.,   $$\frac{a_2}{a_1} = 1 - \left(\frac{1.4 - 1}{2}\right)\left(\frac{157.8}{348.9}\right) = 0.91$$

Hence:   $$a_2 = 0.91 a_1 = 0.91 \times 348.9 = 317.5 \text{ m/s}$$

Therefore, the velocity of the tail of the wave relative to the walls of the pipe is $348.9 - 157.8 = 191.1$ m/s.

Because the front of the wave is moving at a velocity of 348.9 m/s, the time taken for the front of the wave to reach the end of the pipe is given by:

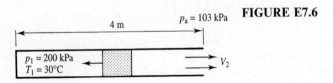

**FIGURE E7.6**

4 m   $p_a = 103$ kPa

$p_1 = 200$ kPa
$T_1 = 30°C$   $V_2$

$$t = \frac{4}{348.9} = 0.0015 \text{ s}$$

i.e., the time for the head of the wave to reach the end of the pipe is 0.0115 s.

An unsteady expansion wave is also generated when a moving shock wave reaches the end of an open duct as indicated in Figs. 7.25 and 7.26. Situations arise in a number of practical situations in which unsteady shock waves and expansion waves are simultaneously generated. Perhaps the simplest example of this is the flow that occurs in a so-called "shock tube." In its simplest form, this consists of a long tube of constant area divided into two sections by a diaphragm which is typically made from a thin sheet of metal which often has grooves cut into it to ensure that it can be easily and cleanly broken. The tube contains a high pressure gas on one side of the diaphragm and a low pressure gas on the other side of the diaphragm, as shown in Fig. 7.27. When the diaphragm is broken either by mechanical means or by increasing the pressure on the high pressure side of the diaphragm, a shock wave propagates into the

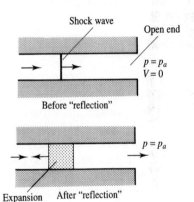

**FIGURE 7.25**
Generation of expansion wave by reflection of a moving shock wave from the open end of a tube.

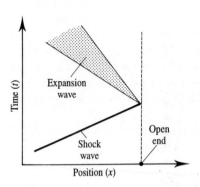

**FIGURE 7.26**
$x$–$t$ diagram for flow associated with the reflection of a moving shock wave from the open end of a tube.

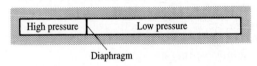

**FIGURE 7.27**
Arrangement of shock tube.

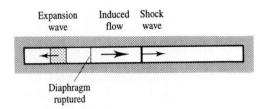

Expansion    Induced  Shock
wave         flow     wave

Diaphragm
ruptured

**FIGURE 7.28**
Waves generated in shock tube
following rupture of diaphragm.

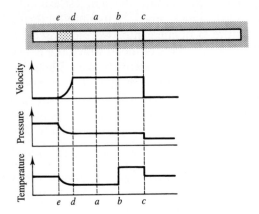

**FIGURE 7.29**
Velocity, pressure, and temperature
variations in shock tube: *aa*, position
of diaphragm; *bb*, position of
interface; *cc*, position of shock wave;
*dd–ee*, expansion wave.

low pressure section and an expansion wave propagates into the high pressure
section as illustrated in Fig. 7.28. Between the shock wave and the expansion
wave a region of uniform velocity is generated that can be used for many
different types of experimental studies. For example, a body can be placed in
the flow and the forces on it can be measured. The flow in a shock tube only
lasts for a short period of time, of course, because the waves are reflected off
the ends of the tube. However, this device has been widely used in many
studies of compressible flows.

The velocity that is generated in a shock tube is determined by noting that
the velocity and pressure behind the shock wave must be equal to the velocity
and pressure behind the expansion wave as indicated in Fig. 7.29. The shock
wave increases the temperature of the gas whereas the expansion wave
decreases the temperature of the gas. If the temperatures in the high and
low pressure sections of the tube are initially the same, it follows that there
will not be a uniform temperature between the shock wave and the expansion
wave, this being shown in Fig. 7.29.

The way in which the flow generated in a shock tube can be analyzed is
illustrated in a very basic way in the following example.

### EXAMPLE 7.7
A shock tube essentially consists of a long tube containing air and separated into
two sections by a diaphragm. The pressures on the two sides of the diaphragm are
300 kPa and 30 kPa and the temperature is 15°C in both sections. If the diaphragm is suddenly ruptured, find the velocity of the air between the moving
shock wave and the moving expansion wave that are generated.

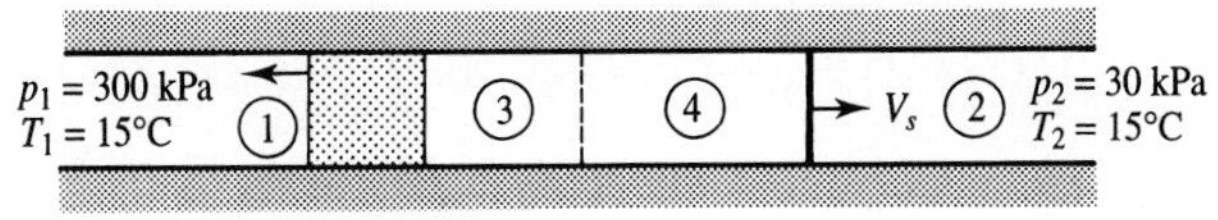

**FIGURE E7.7a**

### Solution
The flow that is generated by the rupturing of the diaphragm is shown in Fig. E7.7a. The speed of sound in the undisturbed air, i.e., in sections 1 and 2 is given by:

$$a_1 = a_2 = \sqrt{\gamma R T} = \sqrt{1.4 \times 287 \times 288} = 340.2 \,\text{m/s}$$

The strengths of the shock wave and expansion wave must be such that the pressure and velocity in the region between the shock wave and the expansion wave is everywhere the same, i.e., the strengths must be such that $p_3 = p_4$ and $V_3 = V_4$.

There are many procedures for obtaining the solution but a very simple trial-and-error type approach will be adopted here. In this approach the pressure between the two waves will be guessed, i.e., the value of $p_3 = p_4$ will be guessed. The air velocity behind the shock and behind the expansion wave will then be separately calculated. Because the value of the pressure is guessed, these two values will not in general be equal. Calculations will then be undertaken with different values of the guessed pressure and then the pressure that makes the air velocity behind the shock and behind the expansion wave equal will be deduced.

To illustrate the procedure, assume that:

$$p_3 = p_4 = 150 \,\text{kPa}$$

First consider the expansion wave. Across this wave:

$$\frac{p_3}{p_1} = \left[ 1 - \frac{\gamma - 1}{2} \frac{V_3}{a_1} \right]^{2\gamma/\gamma - 1}$$

For the specified conditions:

$$p_1 = 300 \,\text{kPa}, \qquad a_1 = 340.2 \,\text{m/s}$$

Hence for $p_3 = 150 \,\text{kPa}$, i.e., for:

$$\frac{p_3}{p_1} = \frac{150}{300} = 0.5$$

this equation gives:

$$0.5 = \left[ 1 - \frac{0.4}{0.2} \frac{V_3}{340.2} \right]^{2.8/0.4}$$

Solving for $V_3$ gives:

$$V_3 = 160.4 \,\text{m/s}$$

Next consider the flow across the shock wave. The flow relative to the shock is shown in Fig. E7.7b.

Since for $p_4 = 150 \,\text{kPa}$:

$$\frac{p_4}{p_2} = \frac{150}{30} = 5$$

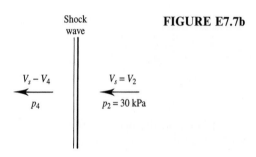

**FIGURE E7.7b**

For this pressure ratio, normal shock relations or tables or the software gives:

$$M_2 = 2.11, \qquad M_4 = 0.5598, \qquad \frac{T_4}{T_2} = 1.7789$$

But $M_2$ is the shock Mach number, i.e.,

$$M_2 = \frac{V_s}{a_2} = \frac{V_s}{340.2 \text{ m/s}}$$

Hence for the guessed pressure:

$$V_s = 340.2 \times 2.11 = 717.8 \text{ m/s}$$

Also:

$$T_4 = 1.7789 \times 288 = 512 \text{ K}$$

$$a_4 = \sqrt{1.4 \times 287 \times 512} = 454 \text{ m/s}$$

But, as discussed in Chapter 6:

$$M_4 = \frac{V_s - V_4}{a_4}$$

from which it follows that:

$$V_4 = V_s - M_4 a_4 = 717.8 - 0.5598 \times 454 = 463.7 \text{ m/s}$$

Hence, when it is guessed that $p_3 = p_4 = 150$ kPa, it is found that $V_3 = 160.4$ m/s and $V_4 = 463.7$ m/s.

Calculations of this type have been carried out for a number of other values of the guessed pressures, the results of some of these calculations being given in Table E7.7. By interpolation between these results it can be deduced that $V_3 = V_4$ when $p_3 = p_4 = 86$ kPa and that at this pressure $V_3 = V_4 = 279$ m/s, i.e., the velocity of air between the moving shock wave and the moving expansion wave is 279 m/s.

**TABLE E7.7**

| $p_3 = p_4$ (kPa) | $p_4/p_2$ | $M_2$ | $M_4$ | $a_4/a_1$ | $a_4$ (m/s) | $V_s$ (m/s) | $V_4$ (m/s) | $V_3$ (m/s) | $V_4 - V_3$ (m/s) |
|---|---|---|---|---|---|---|---|---|---|
| 80 | 2.67 | 1.56 | 0.6809 | 1.1664 | 397 | 531 | 261 | 293 | −32 |
| 85 | 2.83 | 1.60 | 0.6684 | 1.1781 | 401 | 544 | 276 | 280 | −4 |
| 90 | 3 | 1.65 | 0.6540 | 1.1928 | 405 | 561 | 296 | 269 | 27 |
| 100 | 3.3 | 1.73 | 0.6330 | 1.2166 | 414 | 589 | 327 | 247 | 80 |

# 7.7
# CONCLUDING REMARKS

Oblique shock waves are associated with a rise in pressure over a very thin region, the flow through such a wave being non-isentropic. These waves are associated with a turning of the flow "towards itself," i.e., with concave corners. When the flow is turned "away from itself" an expansion wave is generated, i.e., an expansion wave is generated at convex corners. Such a wave, which is termed a Prandtl–Meyer wave, is not thin, in general, and the flow through it is isentropic. The characteristics of such waves were discussed in this chapter. A related type of flow, an unsteady expansion wave, was also considered.


# PROBLEMS

**7.1.** Air is flowing over a flat wall. The Mach number, pressure, and temperature in the air-stream are 3, 50 kPa, and −20°C respectively. If the wall turns "away" from the flow through an angle of 10° leading to the formation of an expansion wave, what will be the Mach number, pressure, and temperature in the flow behind the wave?

**7.2.** Air flows along a flat wall at a Mach number of 3.5 and a pressure of 100 kPa. The wall turns towards the flow through an angle of 25° leading to the formation of an oblique shock wave. A short distance downstream of this, the wall turns away from the flow through an angle of 25° leading to the generation of an expansion wave causing the flow to be parallel to its original direction. Find the Mach number and pressure downstream of the expansion wave.

**7.3.** Air is flowing at a Mach number of 2 at a temperature and pressure of 100 kPa and 0°C down a duct. One wall of this duct turns through an angle of 5° away from the flow leading to the formation of an expansion wave. This expansion wave is reflected off the flat opposite wall of the duct. Find the pressure and temperature behind the reflected wave.

**7.4.** Air is flowing through a wide channel at a Mach number of 1.5, the pressure being 120 kPa. The upper wall of the channel turns through an angle of 4° "away" from the flow leading to the generation of an expansion wave. This expansion wave "reflects" off the flat lower surface of the channel. Find the Mach number and pressure after this reflection.

**7.5.** Air flowing at a Mach number of 3 is turned through an angle that leads to the generation of an expansion wave across which the pressure decreases by 60%. Find the angle that the upstream and downstream ends of the expansion wave make to the initial flow direction.

**7.6.** An air-stream flowing at a Mach number of 4 is expanded around a concave corner with an angle of 15° leading to the generation of an expansion wave.

Some distance downstream of this the air flows around a concave corner leading to the generation of an oblique shock wave and returning the flow to its original direction. If the pressure in the initial flow is 80 kPa, find the pressure downstream of the oblique shock wave.

**7.7.** An oblique shock wave occurs in an air flow in which the Mach number is 2.5, this shock wave turning the flow through 10°. The shock wave impinges on a free boundary along which the pressure is constant and equal to that existing upstream of the shock wave. The shock is "reflected" from this boundary as an expansion wave. Find the Mach number and flow direction downstream of this expansion wave.

**7.8.** Air is flowing through a wide channel at a Mach number of 2 at a pressure of 140 kPa. The upper wall of the channel turns through an angle of 8° "away" from the flow leading to the generation of an expansion wave, while the lower wall of the channel turns through an angle of 6° "away" from the flow also leading to the generation of an expansion wave. The two expansion waves interact and "pass through" each other. Find the Mach number, flow direction, and pressure just downstream of this interaction.

**7.9.** A symmetrical double-wedge shaped body with an included angle of 15° is aligned with an air flow in which the Mach number is 3 and the pressure is 20 kPa. The flow situation is, therefore, as shown in Fig. P7.9. Find the pressures acting on the surfaces of the body.

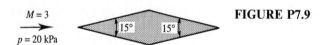

$M = 3$

$p = 20$ kPa

**FIGURE P7.9**

**7.10.** A simple wing may be modeled as a 0.25 m wide flat plate set at an angle of 3° to an airflow at a Mach number of 2.5, the pressure in this flow being 60 kPa. Assuming that the flow over the wing is two-dimensional, estimate the lift and drag force per meter span due to the wave formation on the wing. What other factor causes drag on the wing?

**7.11.** Consider two-dimensional flow over the double-wedge airfoil shown in Fig. P7.11. Find the lift and drag per meter span acting on the airfoil and sketch the flow pattern. How does the pressure vary over the surface of the airfoil?

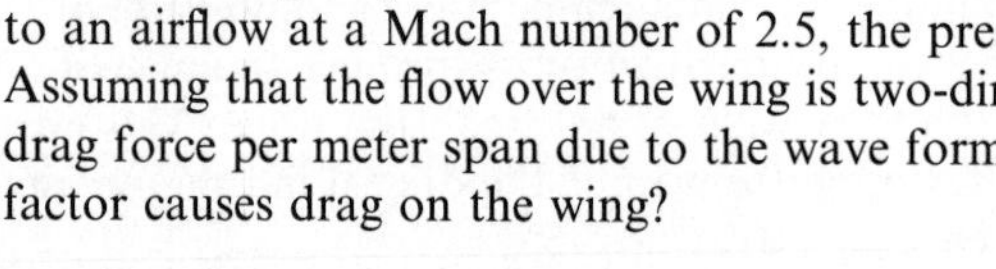

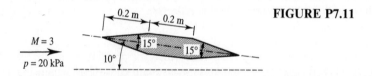

0.2 m    0.2 m

$M = 3$

$p = 20$ kPa    10°    15°    15°

**FIGURE P7.11**

**7.12.** For the double wedge airfoil shown in Fig. P7.12, find the lift per meter span if the Mach number and pressure in the uniform air flow ahead of the airfoil are 3 and 40 kPa respectively.

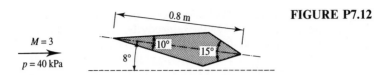

**FIGURE P7.12**

**7.13.** A safety diaphragm at the end of a 3 m long pipe containing air at a pressure of 200 kPa and a temperature of 10°C suddenly ruptures, causing an expansion wave to propagate down the pipe. Find the velocity at which the air is discharged from the pipe, the velocity of the front and the back of the wave, and the time taken for the front of the wave to reach the end of the pipe. Assume that the ambient pressure of the air surrounding the pipe is 100 kPa.

**7.14.** An unsteady expansion wave propagates down a duct containing air at rest at a pressure of 800 kPa and a temperature of 2000°C. The pressure behind the wave is 300 kPa. Find the velocity and the Mach number in the flow that is induced behind the wave in the duct.

**7.15.** Air is flowing through a long pipe at a velocity of 50 m/s at a pressure and temperature of 150 kPa and 40°C respectively. Valves at the inlet and exit to this pipe are suddenly and simultaneously closed. Discuss the waves that are generated in the pipe following valve closure and find the pressures acting on each valve immediately following the valve closure.

**7.16.** A long pipe is conveying air at a pressure and temperature of 150 kPa and 100°C respectively at a velocity of 160 m/s. Valves are fitted at both the inlet and the exit to the pipe. Discuss what waves will be developed and what pressure will act on the valve if (1) the inlet valve is suddenly closed, (2) the exit valve is suddenly closed.

**7.17.** A closed tube contains air at a pressure and temperature of 200 kPa and 30°C respectively. One end of the tube is suddenly opened to the surrounding atmosphere. At what velocity does the air leave the open end of the tube if the ambient pressure is 100 kPa?

**7.18.** A long pipe containing air is separated into two sections by a diaphragm. The pressure on one side of the diaphragm is 500 kPa and the pressure on the other side of the diaphragm is 100 kPa, the air temperature being 20°C in both sections. If the diaphragm suddenly ruptures causing a shock wave to move into the low pressure section, and an expansion wave to move into the high pressure section, find the pressure and air velocity in the region between the two waves.

**7.19.** A shock tube containing air has initial pressures on the two sides of the diaphragm of 400 kPa and 10 kPa, the temperature of the air being 25°C in both sections. If the diaphragm separating the two sections is suddenly ruptured, find the velocity, pressure, and temperature of the air between the moving shock wave and the moving expansion wave that are generated.

**7.20.** The air pressure in the high and low pressure sections of a constant diameter shock tube are 600 kPa and 20 kPa respectively. The temperatures in both sections are 30°C. After the diaphram that separates the two sections is ruptured, a shock wave propagates into the low pressure section and an expansion wave propagates into the high pressure section. Find the air velocity and temperatures between the two waves and the velocity of the shock wave.

# CHAPTER 8

# Variable Area Flow

## 8.1
## INTRODUCTION

Steady flow of a gas through a duct or streamtube which has a varying cross-sectional area will be considered in this chapter. Such flows, i.e., compressible gas flows through a duct whose cross-sectional area is varying, occur in many engineering devices, e.g., in the nozzle of a rocket engine and in the blade passages in turbo machines.

It will be assumed throughout this chapter that the flow can be adequately modeled by assuming it to be one-dimensional at all sections of the duct, i.e., quasi-one-dimensional flow will be assumed in this chapter. This means, by virtue of the discussion given in Chapter 2, that the rate of change of cross-sectional area with distance along the duct is not very large. It will also be assumed in this chapter in studying the effects of changes in area on the flow that the flow is isentropic everywhere except through any shock waves that may occur in the flow. This assumption is usually quite adequate since the effects of friction and heat transfer are usually restricted to a thin boundary layer in the types of flows here being considered and their effects can often be ignored or be adequately accounted for by introducing empirical constants. The presence of shock waves will have to be accounted for in the work of this chapter and the flow through these waves is, as discussed before, not isentropic.

## 8.2
## EFFECTS OF AREA CHANGES ON FLOW

Consider, first, the general effects of a change in area on the isentropic flow through a channel. The situation considered is shown in Fig. 8.1, i.e., the

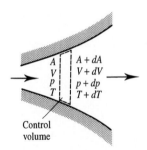

**FIGURE 8.1**
Flow changes considered in variable area channel.

Control
volume

effects of a differentially small change in area, $dA$, on the other variables, i.e., $V$, $p$, $T$, and $\rho$, are considered. The effects of $dA$ on the changes in pressure, density, velocity, etc., i.e., on $dp$, $d\rho$, $dV$, will be derived using the governing equations discussed earlier. The analysis presented here is an extension of some of the analyses given earlier and there is some overlap with earlier work.

Firstly, it is recalled that the continuity equation gives:

$$\rho A V = \text{mass flow rate} = \text{constant} \tag{8.1}$$

where $A$ is the cross-sectional area of the duct at any point. Applying this to the flow being considered gives:

$$\rho A V = (\rho + d\rho)(A + dA)(V + dV)$$

Since $dp$, $d\rho$, $dV$, and $dA$ are, by assumption, all small, this equation becomes, to first order accuracy (i.e., terms involving the products and squares of the differentially small quantities such as $d\rho \times dA$ are ignored):

$$\rho A V = A V \, d\rho + \rho V \, dA + \rho A \, dV$$

i.e., dividing through by $\rho A V$:

$$\frac{d\rho}{\rho} + \frac{dA}{A} + \frac{dV}{V} = 0 \tag{8.2}$$

Next, it is recalled that the energy equation gives:

$$c_p T + \frac{V^2}{2} = \text{constant}$$

which gives for the situation being considered:

$$c_p T + \frac{V^2}{2} = c_p(T + dT) + \frac{(V + dV)^2}{2}$$

i.e., to first order accuracy:

$$c_p \, dT + V \, dV = 0 \tag{8.3}$$

Further, the equation of state, gives:

$$p = \rho R T \quad \text{and} \quad p + dp = (\rho + d\rho)R(T + dT)$$

Subtracting these two equations and dividing the result by the first of the two equations gives to first order accuracy:

$$\frac{dp}{p} = \frac{d\rho}{\rho} + \frac{dT}{T}$$

(8.4)

Lastly, since the flow being considered is, by assumption, isentropic, it follows that:

$$\frac{p}{\rho^\gamma} = \text{constant} \quad \text{and} \quad \frac{p + dp}{(\rho + d\rho)^\gamma} = \text{constant}$$

(8.5)

Because $dp/p$ and $d\rho/\rho$ are by assumption small, the second of the above two equations gives to first order accuracy:

$$\frac{p}{\rho^\gamma} \frac{\left[1 + \dfrac{dp}{p}\right]}{\left[1 + \dfrac{d\rho}{\rho}\right]^\gamma} = \text{constant}$$

i.e.,

$$\frac{p}{\rho^\gamma} \frac{\left[1 + \dfrac{dp}{p}\right]}{\left[1 + \gamma\dfrac{d\rho}{\rho}\right]} = \text{constant}$$

i.e.,

$$\frac{p}{\rho^\gamma} \left[1 + \frac{dp}{p} - \gamma\frac{d\rho}{\rho}\right] = \text{constant}$$

Combining this with the first equation then gives to first order accuracy:

$$\frac{dp}{p} = \gamma \frac{d\rho}{\rho}$$

(8.6)

Equations (8.2), (8.3), (8.4), and (8.6) together are sufficient to determine the required results, i.e., to determine the relationship between the four variables $dp/p$, $dV/V$, $dT/T$, and $d\rho/\rho$ and the fractional area change $dA/A$. First combining eqs. (8.4) and (8.6) gives:

$$\frac{dT}{T} = (\gamma - 1)\frac{d\rho}{\rho}$$

(8.7)

which can be substituted into eq. (8.3) to give:

$$(\gamma - 1)\frac{d\rho}{\rho} + \frac{V^2}{c_p T}\frac{dV}{V} = 0$$

(8.8)

Now:

$$\frac{V^2}{c_p T} = \frac{V^2}{c_p a^2}\gamma R = \gamma\left(1 - \frac{1}{\gamma}\right)M^2 = (\gamma - 1)M^2$$

so eq. (8.8) can be written as:

$$\frac{d\rho}{\rho} = -M^2 \frac{dV}{V} \tag{8.9}$$

Substituting this into eq. (8.2) then gives:

$$\frac{dA}{A} = (M^2 - 1)\frac{dV}{V} \tag{8.10}$$

which may alternatively be written as:

$$\frac{dA}{dV} = (M^2 - 1)\frac{A}{V} \tag{8.11}$$

Because $A$ and $V$ are positive, it may be concluded from the above two equations that:

1. If $M < 1$, i.e., if the flow is subsonic, then $dA$ has the opposite sign to $dV$, i.e., decreasing the area increases the velocity and vice versa.
2. If $M > 1$, i.e., if the flow is supersonic, then $dA$ has the same sign as $dV$, i.e., decreasing the area decreases the velocity and vice versa.
3. If $M = 1$ then $dA/dV = 0$ and $A$ reaches an extremum. From (1) and (2) it follows that when $M = 1$, $A$ must be a minimum.

Further important conclusions regarding the effects of varying area on the flow variables can be obtained by writing eq. (8.10) as:

$$\frac{dA}{A} = (M^2 - 1)\frac{dM}{M}\frac{1}{a}\frac{dV}{dM} \tag{8.12}$$

However, by noting that $V = Ma$ it follows that:

$$\frac{dV}{V} = \frac{dM}{M} + \frac{da}{a} \tag{8.13}$$

Further:      $a = \sqrt{\gamma R T} \quad \text{and} \quad a + da = \sqrt{\gamma R (T + dT)}$

The second of these two equations gives to first order accuracy:

$$a + da = \sqrt{\gamma R T}\left(1 + \frac{dT}{2T}\right)$$

Hence dividing this result by the first of these equations gives:

$$\frac{da}{a} = \frac{1}{2}\frac{dT}{T}$$

and since the energy eq. (8.3) gives:

$$\frac{dT}{T} = -\frac{V^2}{c_p T}\frac{dV}{V} = -(\gamma - 1)M^2\frac{dV}{V}$$

it follows that:

$$\frac{da}{a} = -\left(\frac{\gamma - 1}{2}\right)M^2\frac{dV}{V}$$

(8.14)

Substituting this result into eq. (8.13) then gives:

$$\frac{dM}{M} = \frac{dV}{V}\left[1 + \left(\frac{\gamma - 1}{2}\right)M^2\right]$$

(8.15)

Substituting this, in turn, into eq. (8.12) gives:

$$\frac{dA}{A} = \frac{(M^2 - 1)\dfrac{1}{a}\dfrac{V}{M}}{1 + \left(\dfrac{\gamma - 1}{2}\right)M^2}\frac{dM}{M}$$

i.e.,

$$\frac{dA}{A} = \frac{(M^2 - 1)}{1 + \left(\dfrac{\gamma - 1}{2}\right)M^2}\frac{dM}{M}$$

( 8.16)

From this equation it follows that:

1. When $M < 1$, $dA$ has the opposite sign to $dM$, e.g., when $A$ increases, $M$ decreases.
2. When $M > 1$, $dA$ has the same sign as $dM$, e.g., when $A$ increases, $M$ increases.
3. When $M = 1$, $dA = 0$, i.e., $A$ is a minimum when $M = 1$. $dA$ can also be zero when $dM$ is zero, i.e., a minimum in the flow area can also be associated with a maximum or minimum in the Mach number.

The results derived above concerning the effect of area changes on the Mach number and the velocity are summarized in Fig. 8.2. From these results, it follows that if a subsonic flow is to be accelerated to a supersonic velocity it must be passed through a convergent–divergent passage or nozzle. The convergent portion accelerates the flow up to a Mach number of 1 and the divergent portion then accelerates the flow to supersonic velocity. At the throat, since $dA = 0$, the Mach number is equal to 1. This is summarized in Fig. 8.3. Of course, in such a nozzle, the pressure will decrease continuously throughout the nozzle. If the end pressure is not low enough, the flow will remain subsonic throughout as in a Venturi meter, i.e., supersonic flow will not be generated. In this situation, since $dA$ is still zero at the throat, $dV$ and $dM$ must be zero at the throat, i.e., the velocity and Mach number both reach a maximum at the throat (see later for a detailed discussion of the actual operating characteristics of such nozzles).

In a similar way, if the flow entering the nozzle is supersonic, two possibilities exist. Either the Mach number will decrease to 1 at the throat and then continue to decrease to subsonic values in the divergent portion of the nozzle, or the flow will remain supersonic throughout the nozzle, the Mach number

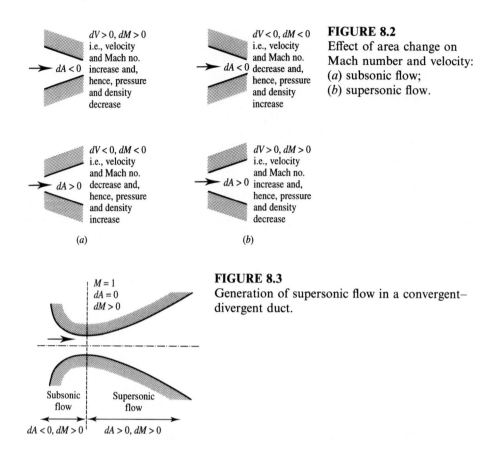

FIGURE 8.2

Effect of area change on Mach number and velocity: (a) subsonic flow; (b) supersonic flow.

FIGURE 8.3

Generation of supersonic flow in a convergent–divergent duct.

and velocity, in this case, decreasing in the convergent portion of the nozzle, reaching minimum although still supersonic values at the throat, and then increasing again in the divergent portion of the nozzle.

## 8.3
## EQUATIONS FOR VARIABLE AREA FLOW

Attention was given in the previous section to the changes in the flow variables produced by a differentially small change in area. Equations for the changes in the flow variables produced by finite changes in the flow area will be derived in the present section. The presence of shock waves in the flow will, for the present, be ignored. The flow is then, as previously discussed, analyzed using the assumption that it is one-dimensional at all sections and also that it is isentropic. The required relations could have been obtained by integrating the differential relations given in the previous section. It is easier, however, to derive the relations by directly applying the full energy and continuity equations.

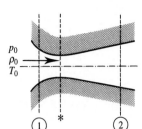

**FIGURE 8.4**
Flow through a convergent–divergent duct (section 1 can be anywhere in the duct).

Consider the flow of a gas from a large reservoir through some duct system as shown in Fig. 8.4. Because the reservoir is large, the stagnation conditions exist in the reservoir, i.e., the pressure, density, and temperature in the reservoir are $p_0$, $\rho_0$, and $T_0$ respectively. The equations governing the flow at some arbitrary section 1 are the continuity equation:

$$\rho_1 V_1 A_1 = \text{constant} = \text{mass flow rate, } \dot{m} \qquad (8.17)$$

and the energy equation:

$$V_1^2 + \frac{2}{(\gamma - 1)} a_1^2 = \frac{2}{(\gamma - 1)} a_0^2 \qquad (8.18)$$

The momentum equation could have been used instead of the energy equation since, as discussed earlier, they lead to the same results in isentropic flow.

In addition to eqs. (8.17) and (8.18), the isentropic relations apply at all points, i.e.,

$$\frac{a_1}{a_0} = \left(\frac{T_1}{T_0}\right)^{\frac{1}{2}} = \left(\frac{p_1}{p_0}\right)^{\frac{\gamma-1}{2\gamma}} = \left(\frac{\rho_1}{\rho_0}\right)^{\frac{\gamma-1}{2}}$$

An attempt will first be made to relate the conditions existing at the arbitrary section 1 of the duct to the pressure ratio, $p_1/p_0$, existing at this section. To do this, it is noted that the energy equation (8.18) gives:

$$V_1^2 = \left(\frac{2}{\gamma - 1}\right) a_0^2 \left[1 - \left(\frac{a_1}{a_2}\right)^2\right]$$

which becomes, using the isentropic relations:

$$V_1 = \left\{\left(\frac{2a_0^2}{\gamma - 1}\right) 1 - \left(\frac{p_1}{p_0}\right)^{\frac{\gamma-1}{\gamma}}\right\}^{\frac{1}{2}}$$

i.e.,

$$V_1 = \left\{\left(\frac{2\gamma}{\gamma - 1}\right)\left(\frac{p_0}{\rho_0}\right)\left[1 - \left(\frac{p_1}{p_0}\right)^{\frac{\gamma-1}{\gamma}}\right]\right\}^{\frac{1}{2}} \qquad (8.19)$$

This equation can also be written as:

$$V_1 = \left\{ (2c_p T_0) \left[ 1 - \left( \frac{p_1}{p_0} \right)^{\frac{\gamma-1}{\gamma}} \right] \right\}^{\frac{1}{2}}$$

Thus, the velocity at any section can be determined if the pressure at this section and the stagnation conditions are known.

Substituting eq. (8.19) into the continuity equation (8.17) and then using the isentropic relations gives:

$$\dot{m} = \rho_0 V_1 A_1 \frac{\rho_1}{\rho_0}$$

$$= \rho_0 A_1 \left( \frac{p_1}{p_0} \right)^{\frac{1}{\gamma}} \left\{ \left( \frac{2\gamma}{\gamma-1} \right) \left( \frac{p_0}{\rho_0} \right) \left[ 1 - \left( \frac{p_1}{p_0} \right)^{\frac{\gamma-1}{\gamma}} \right] \right\}^{\frac{1}{2}} \qquad (8.20)$$

Since $\dot{m}$ is a constant, i.e., the mass flow rate along the duct does not change because the flow is assumed to be steady, this equation can be used to relate the pressure at any point in the duct to the area, i.e., if subscript 2 refers to conditions at some other section of the duct, then eq. (8.20) gives:

$$\dot{m} = \rho_0 A_2 \left( \frac{p_2}{p_0} \right)^{\frac{1}{\gamma}} \left\{ \left( \frac{2\gamma}{\gamma-1} \right) \left( \frac{p_0}{\rho_0} \right) \left[ 1 - \left( \frac{p_2}{p_1} \right)^{\frac{\gamma-1}{\gamma}} \right] \right\}^{\frac{1}{2}} \qquad (8.21)$$

Dividing eq. (8.21) by eq. (8.20) then gives, on rearrangement:

$$\frac{A_2}{A_1} = \left( \frac{p_1}{p_2} \right)^{\frac{1}{\gamma}} \left[ \frac{1 - \left( \frac{p_1}{p_0} \right)^{\frac{\gamma-1}{\gamma}}}{1 - \left( \frac{p_2}{p_0} \right)^{\frac{\gamma-1}{\gamma}}} \right]^{\frac{1}{2}} \qquad (8.22)$$

This equation relates the pressures at any two sections of the duct to the areas of these sections.

In presenting the equations for flow in a variable area duct, it is convenient to choose, for reference, conditions at some specific point. A convenient point to use for this purpose is the point in the flow at which the Mach number is equal to 1. Of course, there may not actually be a real point in the flow at which the Mach number is equal to 1, but the conditions at such a point whether or not it really exists in the flow are convenient to use for reference purposes. As discussed before, the conditions at the point where $M = 1$ are known as the critical conditions and expressions for the pressure and velocity here have previously been derived. They will, however, be repeated here for reference. The energy equation gives the critical velocity, $V^*$, as:

$$V^{*2} = \frac{2}{\gamma + 1} a_0^2 \tag{8.23}$$

Since $V^* = a^*$, another way of writing this equation is:

$$\left(\frac{a^*}{a_0}\right)^2 = \frac{T^*}{T_0} = \frac{2}{\gamma + 1} \tag{8.24}$$

Using the isentropic relations allows this to be written as:

$$\frac{p^*}{p_0} = \left(\frac{2}{\gamma + 1}\right)^{\frac{\gamma}{\gamma - 1}} \tag{8.25}$$

Substituting this into eq. (8.20) gives the following equation that can be used to find the area of the duct where the critical conditions exist:

$$\dot{m} = \rho_0 A^* \left(\frac{2}{\gamma + 1}\right)^{\frac{1}{\gamma - 1}} \left\{\left(\frac{2\gamma}{\gamma - 1}\right) \frac{p_0}{\rho_0} \left(\frac{\gamma - 1}{\gamma + 1}\right)\right\}^{\frac{1}{2}}$$

This can be rearranged to give:

$$A^* = \frac{\dot{m}}{\sqrt{\gamma p_0 \rho_0}} \left(\frac{2}{\gamma + 1}\right)^{-\frac{\gamma + 1}{2(\gamma - 1)}} \tag{8.26}$$

It must be stressed, as already mentioned, that the critical conditions may not exist in the real flow, e.g., if the flow remains subsonic throughout, critical conditions will not exist at any point in the flow. Even in such cases, however, the critical conditions are used for reference.

The area at any point of the duct, expressed in terms of the critical area, $A^*$, can be related to the pressure by substituting eq. (8.25) into eq. (8.22) giving:

$$\frac{A}{A^*} = \left(\frac{2}{\gamma + 1}\right)^{\frac{1}{\gamma - 1}} \frac{1}{\left(\frac{p}{p_0}\right)^{\frac{1}{\gamma}}} \left\{\frac{1 - \frac{2}{\gamma + 1}}{1 - \left(\frac{p}{p_0}\right)^{\frac{\gamma - 1}{\gamma}}}\right\}^{\frac{1}{2}}$$

i.e.,

$$\frac{A}{A^*} = \frac{\left(\frac{2}{\gamma + 1}\right)^{\frac{\gamma + 1}{2(\gamma - 1)}} \left(\frac{\gamma - 1}{2}\right)^{\frac{1}{2}}}{\left\{\left(\frac{p}{p_0}\right)^{\frac{2}{\gamma}} - \left(\frac{p}{p_0}\right)^{\frac{\gamma + 1}{\gamma}}\right\}^{\frac{1}{2}}} \tag{8.27}$$

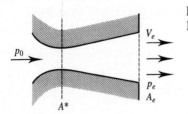

**FIGURE 8.5**
Exit plane conditions.

$p_0$

$V_e$

$p_e$
$A_e$

$A^*$

If a nozzle is to be designed for a given mass flow rate, $\dot{m}$, with a given overall pressure ratio, $(p_e/p_0)$ (see Fig. 8.5) then applying eqs. (8.19), (8.20), and (8.27) gives:

$$V_e = \left\{ \left( \frac{2\gamma}{\gamma - 1} \right) \left( \frac{p_0}{\rho_0} \right) \left[ 1 - \left( \frac{p_e}{p_0} \right)^{\frac{\gamma-1}{\gamma}} \right] \right\}^{\frac{1}{2}} \tag{8.28}$$

$$\dot{m} = \rho_0 A_e \left( \frac{p_e}{p_0} \right)^{\frac{1}{\gamma}} \left\{ \left( \frac{2\gamma}{\gamma - 1} \right) \left( \frac{p_0}{\rho_0} \right) \left[ 1 - \left( \frac{p_e}{p_0} \right)^{\frac{\gamma-1}{\gamma}} \right] \right\}^{\frac{1}{2}} \tag{8.29}$$

$$\frac{A_e}{A^*} = \frac{\left( \dfrac{2}{\gamma + 1} \right)^{\frac{\gamma+1}{2(\gamma-1)}} \left( \dfrac{\gamma - 1}{2} \right)^{\frac{1}{2}}}{\left\{ \left( \dfrac{p_e}{p_0} \right)^{\frac{2}{\gamma}} - \left( \dfrac{p_e}{p_0} \right)^{\frac{\gamma+1}{\gamma}} \right\}^{\frac{1}{2}}} \tag{8.30}$$

For a given set of stagnation conditions and given values of $\dot{m}$ and $(p_e/p_0)$, eq. (8.29) allows the exit area, $A_e$, to be found while eq. (8.30) allows the throat area, $A^*$, to be found. Further, for the given pressure ratio, $(p_e/p_0)$, eq. (8.28) allows the discharge velocity, $V_e$, to be found. The present analysis cannot be used to make any predictions concerning the optimum shape of the nozzle since it is based on the assumption that the flow is one-dimensional at all sections.

It is often convenient to express the variable area relations in terms of the Mach number, $M$, existing at any section. To do this, it is first noted, as before, that the energy equation:

$$V^2 + \left( \frac{2}{\gamma - 1} \right) a_1^2 = \left( \frac{2}{\gamma - 1} \right) a_0^2$$

gives:

$$V = Ma_0 \left\{ 1 + \left( \frac{\gamma - 1}{2} \right) M^2 \right\}^{-\frac{1}{2}} \tag{8.31}$$

The energy equation also gives:

$$\left( \frac{a}{a_0} \right)^2 = \left\{ 1 + \left( \frac{\gamma - 1}{2} \right) M^2 \right\}^{-1} \tag{8.32}$$

This in turn gives, using the isentropic relations:

$$\left(\frac{\rho}{\rho_0}\right) = \left\{1 + \left(\frac{\gamma - 1}{2}\right)M^2\right\}^{-\frac{1}{\gamma-1}} \tag{8.33}$$

Hence, since:

$$\dot{m} = \rho V A$$

it follows by using eqs. (8.31), (8.32), and (8.33) that:

$$\frac{\dot{m}}{A} = \frac{\rho_0 a_0 M}{\left\{1 + \left(\frac{\gamma - 1}{2}\right)M^2\right\}^{\frac{\gamma+1}{2(\gamma-1)}}} \tag{8.34}$$

Applying this equation between any two sections 1 and 2 of a duct then gives, since $\dot{m}$ must be the same at these two points:

$$\frac{A_2}{A_1} = \left(\frac{M_1}{M_2}\right)\left\{\frac{1 + \left(\frac{\gamma - 1}{2}\right)M_2^2}{1 + \left(\frac{\gamma - 1}{2}\right)M_1^2}\right\}^{\frac{\gamma+1}{2(\gamma-1)}} \tag{8.35}$$

Thus, if the ratio of the areas at the two sections is known, the Mach numbers at these sections can be related by this equation. It is again convenient to write this equation in terms of the duct area, $A^*$, at which the critical conditions exist. Since $M^*$ is by definition equal to 1 at this section, the Mach number, $M$, at some other section where the area is $A$ is then given by:

$$\frac{A}{A^*} = \left(\frac{1}{M}\right)\left\{\frac{1 + \left(\frac{\gamma - 1}{2}\right)M^2}{1 + \left(\frac{\gamma - 1}{2}\right)}\right\}^{\frac{\gamma+1}{2(\gamma-1)}}$$

$$= \frac{1}{M}\left\{\left(\frac{2}{\gamma + 1}\right)\left[1 + \left(\frac{\gamma - 1}{2}\right)M^2\right]\right\}^{\frac{\gamma+1}{2(\gamma-1)}} \tag{8.36}$$

Typical variations of $A/A^*$ with $M$ as given by this equation are shown in Fig. 8.6. This relationship is usually listed in isentropic tables (see Appendix B) and is given by the software that supports this book.

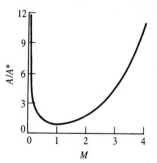

**FIGURE 8.6**
Variation of area ratio with Mach number for $\gamma = 1.4$.

It should be clearly understood that eq. (8.36) indicates that the Mach number is uniquely related to the cross-sectional area for a given value of $A^*$. This point will be considered further in the following discussion of nozzle characteristics.

### EXAMPLE 8.1

Hydrogen flows from a large reservoir through a convergent–divergent nozzle, the pressure and temperature in the reservoir being 600 kPa and 40°C respectively. The throat area of the nozzle is $10^{-4}$ m$^2$ and the pressure on the nozzle exit plane is 130 kPa. Assuming that the flow throughout the nozzle is isentropic and that the flow is steady and one-dimensional, find the mass flow rate through the nozzle and the exit area of the nozzle.

### Solution
The flow situation being considered is shown in Fig. E8.1. Because:

$$\frac{p^*}{p_0} = \left(\frac{2}{\gamma+1}\right)^{\frac{\gamma}{\gamma-1}}$$

it follows that since for hydrogen $\gamma = 1.407$:

$$\frac{p^*}{p_0} = \left(\frac{2}{2.407}\right)^{\frac{1.407}{0.407}} = 0.5271$$

This could also have been obtained using the software that supports this book.

But, in the situation being considered $p_e/p_0 = 130/600 = 0.2167 \ (< 0.5271)$. Hence, in this situation the flow is being accelerated to a supersonic velocity and at the throat $M = 1$ so:

$$\dot{m} = \rho_0 A^* \left(\frac{2}{\gamma+1}\right)^{\frac{1}{\gamma-1}} \left\{\left(\frac{2\gamma}{\gamma-1}\right)\frac{p_0}{\rho_0}\left(\frac{\gamma-1}{\gamma+1}\right)\right\}^{\frac{1}{2}}$$

so:

$$\dot{m} = \rho_0 A^* \left(\frac{2}{2.407}\right)^{\frac{1}{0.407}} \left\{\left(\frac{2.814}{0.407}\right)\frac{p_0}{\rho_0}\left(\frac{0.407}{2.407}\right)\right\}^{\frac{1}{2}}$$

But $\rho_0 = p_0/RT_0$ so, since $R = 8314.3/2.016 = 4124.2$ it follows that $\rho_0 = 600\,000/(4124.2 \times 313) = 0.4648$ kg/m$^3$ so:

$$\dot{m} = 0.4648 \times 0.0001 \times \left(\frac{2}{2.407}\right)^{\frac{1}{0.407}} \left\{\left(\frac{2.814}{0.407}\right)\frac{600\,000}{0.4648}\left(\frac{0.407}{2.407}\right)\right\}^{\frac{1}{2}}$$

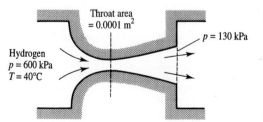

Throat area
= 0.0001 m$^2$

$p = 130$ kPa

Hydrogen
$p = 600$ kPa
$T = 40°C$

**FIGURE E8.1**

hence:                                      $\dot{m} = 0.036\,22$ kg/s

so the mass flow rate through the nozzle is $0.036\,22$ kg/s.
    The nozzle exit area can be found using:

$$\frac{A_e}{A^*} = \frac{\left(\dfrac{2}{\gamma+1}\right)^{\frac{\gamma+1}{2(\gamma-1)}}\left(\dfrac{\gamma-1}{2}\right)^{\frac{1}{2}}}{\left\{\left(\dfrac{p_e}{p_0}\right)^{\frac{2}{\gamma}}-\left(\dfrac{p_e}{p_0}\right)^{\frac{\gamma+1}{\gamma}}\right\}^{\frac{1}{2}}}$$

which gives:

$$\frac{A_e}{0.0001} = \frac{\left(\dfrac{2}{2.407}\right)^{\frac{2.407}{2\times0.407}}\left(\dfrac{0.407}{2}\right)^{\frac{1}{2}}}{\left\{\left(\dfrac{130\,000}{600\,000}\right)^{\frac{2}{1.407}}-\left(\dfrac{130\,000}{600\,000}\right)^{\frac{2.407}{1.407}}\right\}^{\frac{1}{2}}}$$

Hence:                                      $A_e = 0.000\,6416$ m$^2$

So the nozzle exit area is $6.416 \times 10^{-4}$ m$^2$.

## 8.4
## OPERATING CHARACTERISTICS OF NOZZLES

The concern here is with the effect of changes in the upstream and downstream pressures on the nature of the flow in and on the mass flow rate through a nozzle, i.e., through a variable area passage designed to accelerate a gas flow. In the present discussion of the operating characteristics of nozzles it will be assumed that the nozzle is connected to an upstream chamber in which the conditions, i.e., the upstream stagnation conditions, are kept constant while the conditions in the downstream chamber into which the nozzle discharges are varied. The pressure in the downstream chamber is termed the back pressure. The situation considered is, therefore, as shown in Fig. 8.7. The characteristics of convergent and convergent–divergent nozzles will be separately discussed.

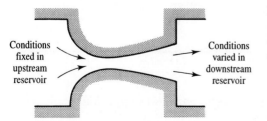

Conditions
fixed in
upstream
reservoir

Conditions
varied in
downstream
reservoir

**FIGURE 8.7**
Flow situation considered in discussing nozzle operating characteristics.

**Convergent Nozzle**

In this section, a nozzle of the type shown in Fig. 8.8 will be considered. If the back pressure, $p_b$, is initially equal to the supply pressure, $p_0$, there will be no flow through the nozzle. As the back pressure, $p_b$, is decreased, flow commences, this flow initially being subsonic throughout the nozzle. Under these circumstances, the pressure on the exit plane of the nozzle, $p_e$, remains equal to the back pressure, $p_b$, and the Mach number on the exit plane is less than 1. In this region of operation, a reduction in $p_b$ produces an increase in the mass flow rate, $\dot{m}$. This type of flow continues to exist until $p_b$ is reduced to the critical pressure corresponding to $p_0$, i.e., until:

$$p_b = p^* = p_0 \left( \frac{2}{\gamma + 1} \right)^{\frac{\gamma}{\gamma - 1}}$$

When $p_b$ has been decreased to this value, the Mach number on the exit plane becomes equal to 1. Further reductions in $p_b$ have no effect on the flow in the nozzle, i.e., $p_e$ remains equal to $p^*$, the mass flow rate remains constant and the Mach number in the exit plane remains equal to 1. Since $p_b < p_e$ in this state, the expansion from $p_e$ to $p_b$ takes place outside the nozzle through a series of expansion waves, the flow then resembling that shown in Fig. 8.9.

The reason that reductions in the back pressure have no effect on the flow in the nozzle once the Mach number at the nozzle exit plane has reached 1 can be understood by realizing that, as previously discussed, the effect of changes in the back pressure are propagated into the nozzle at the speed of sound

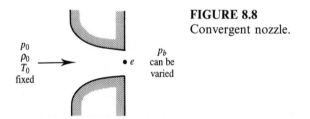

**FIGURE 8.8**
Convergent nozzle.

$p_0$
$p_0$
$T_0$
fixed

$\bullet \, e$

$p_b$
can be
varied

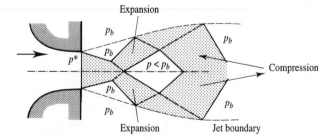

**FIGURE 8.9**
Flow near the exit to a convergent nozzle when $p_b$ is less than the critical pressure.

relative to the fluid. The speed of propagation of the effects of the changes in the back pressure up the nozzle relative to the nozzle will, therefore, be equal to the speed of sound minus the local velocity of the gas. Hence, once the gas velocity on the exit plane becomes equal to the local speed of sound, i.e., once the exit plane Mach number becomes equal to 1, the effect of changes in the back pressure cannot be propagated up the nozzle. As a result, once the exit plane Mach number has reached a value of 1, further reductions in back pressure can have no influence on the flow in the nozzle and cannot therefore effect the mass flow rate through the nozzle.

The following relations therefore apply to the flow through a convergent nozzle:

When $p_b > p^*$:

$$p_e = p_b$$

$$V_e = \left\{ \left( \frac{2\gamma}{\gamma - 1} \right) \left( \frac{p_0}{\rho_0} \right) \left[ 1 - \left( \frac{p_b}{p_0} \right)^{\frac{\gamma - 1}{\gamma}} \right] \right\}^{\frac{1}{2}}$$

$$\dot{m} = \rho_0 A_e \left( \frac{p_b}{p_0} \right)^{\frac{1}{\gamma}} \left\{ \left( \frac{2\gamma}{\gamma - 1} \right) \left( \frac{p_0}{\rho_0} \right) \left[ 1 - \left( \frac{p_b}{p_0} \right)^{\frac{\gamma - 1}{\gamma}} \right] \right\}^{\frac{1}{2}}$$

When $p_b \leq p^*$:

$$\frac{p_e}{p_0} = \left( \frac{2}{\gamma + 1} \right)^{\frac{\gamma}{\gamma - 1}}$$

$$V_e = \sqrt{ \left( \frac{2\gamma}{\gamma + 1} \right) \frac{p_0}{\rho_0} }$$

$$\dot{m} = \sqrt{\gamma p_0 \rho_0} A_e \left( \frac{2}{\gamma + 1} \right)^{\frac{\gamma + 1}{2(\gamma - 1)}}$$

all of these quantities being constants.

The variation of $p_e$ and $\dot{m}$ with $p_b$ for a convergent nozzle is therefore as shown in Fig. 8.10. While $p_b/p_0 > p^*/p_0$, $p_e/p_0 = p_b/p_0$ and $\dot{m}$ increases with

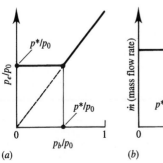

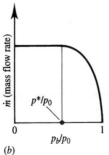

**FIGURE 8.10**
Effect of back pressure on (a) the exit plane pressure and (b) the mass flow rate, $\dot{m}$, through a convergent nozzle.

decreasing $p_b/p_0$. However, when $p_b/p_0 < p^*/p_0$, $p_e/p_0 = p^*/p_0$ and $\dot{m}$ is un-affected by changes in $p_b/p_0$. Under these circumstances, when changes in the back pressure have no effect on the mass flow rate through the nozzle, the nozzle is said to be "choked."

**EXAMPLE 8.2**
Consider the flow of air out of a large vessel through a convergent nozzle to the atmosphere, the atmospheric pressure being 101.1 kPa. The temperature of the air in the vessel is 40°C and the pressure in this vessel is varied. Show how the mass flow rate out of the nozzle per unit exit section area varies with the pressure in the vessel. Assume steady, one-dimensional, isentropic flow in the nozzle.

*Solution*

**FIGURE E8.2a**

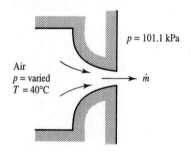

$p = 101.1$ kPa

Air
$p$ = varied
$T = 40°C$

$\dot{m}$

The velocity on the exit plane is given by:

$$V_e = M_e a_e = M_e a_0 \frac{a_e}{a_0}$$

The mass flow rate is given by:

$$\dot{m} = \rho_e V_e A_e = \frac{\rho_e}{\rho_0} \rho_0 V_e A_e$$

Hence:

$$\frac{\dot{m}}{A_e} = \frac{\rho_e}{\rho_0} \rho_0 M_e a_0 \frac{a_e}{a_0} = M_e a_0 \rho_0 \frac{\rho_e}{\rho_0} \frac{a_e}{a_0}$$

But:

$$\rho_0 = \frac{p_0}{RT_0} = \frac{p_0}{287 \times 313} = \frac{p_0}{89\,831}$$

and:

$$a_0 = \sqrt{\gamma RT_0} = \sqrt{1.4 \times 287 \times 313} = 354.6 \text{ m/s}$$

Hence:

$$\frac{\dot{m}}{A_e} = M_e \times 354.6 \times \frac{p_0}{89\,831} \frac{\rho_e}{\rho_0} \frac{a_e}{a_0}$$

Because the flow in the nozzle is isentropic, isentropic relations or tables or the software provided give the values of $\rho_e/\rho_0$, $M_e$, and $a_e/a_0$ for any value of $p_e/p_0$ and, therefore, for any value of $101\,100/p_0$, $p_0$ being in pascals. This will apply until $p_0$ has risen to a value that causes $M_e$ to reach a value of 1 which, according to isentropic relations or tables or the software provided, occurs when $p_0/p_e = 1.8929$. Once $p_0$ has risen to this value, i.e., 191.37 kPa, $p_0/p_e$ remains equal to 1.8929 and:

**TABLE E8.2**

| $p_0$ (kPa) | $p_0/p_e$ | $M_e$ | $a_0/a_e$ | $\rho_0/\rho_e$ | $\dot{m}/A_e$ |
|---|---|---|---|---|---|
| 101.1 | 1.000 | 0.0000 | 1.000 | 1.000 | 0.000 |
| 120 | 1.187 | 0.5009 | 1.130 | 1.025 | 204.9 |
| 140 | 1.385 | 0.6981 | 1.262 | 1.047 | 292.0 |
| 160 | 1.583 | 0.8371 | 1.388 | 1.068 | 356.7 |
| 180 | 1.780 | 0.9465 | 1.510 | 1.086 | 410.1 |
| 191.4 | 1.893 | 1.0000 | 1.577 | 1.096 | 437.1 |
| 200 | 1.893 | 1.0000 | 1.577 | 1.096 | 456.8 |
| 250 | 1.893 | 1.0000 | 1.577 | 1.096 | 571.0 |
| 300 | 1.893 | 1.0000 | 1.577 | 1.096 | 685.2 |
| 400 | 1.893 | 1.0000 | 1.577 | 1.096 | 913.6 |
| 500 | 1.893 | 1.0000 | 1.577 | 1.096 | 1141.9 |

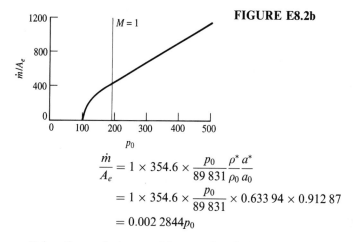

**FIGURE E8.2b**

$$\frac{\dot{m}}{A_e} = 1 \times 354.6 \times \frac{p_0}{89\,831} \frac{\rho^* a^*}{\rho_0 a_0}$$

$$= 1 \times 354.6 \times \frac{p_0}{89\,831} \times 0.633\,94 \times 0.912\,87$$

$$= 0.002\,2844 p_0$$

Using these relations and isentropic relations or tables or the software provided, Table E8.2 can be constructed. The variation is shown in Fig. E8.2b.

### Convergent–Divergent Nozzle

In this section, a nozzle of the type shown in Fig. 8.11 will be considered. As $p_b$ varies, four more or less separate flow regimes can be identified. The nature of the flow that exists in these four regimes can again be explained by considering how the flow in the nozzle changes as the back pressure $p_b$ is decreased. The four flow regimes are then as follows:

1. When $p_b$ is very nearly the same as $p_0$ the flow remains subsonic throughout. The flow in the nozzle is then similar to that in a venturi, the pressure dropping from $p_0$ to a minimum value which is greater than $p^*$ at the throat and then increasing again to $p_e = p_b$ at the exit. In this flow regime the pressure distribution in the nozzle therefore resembles that shown in Fig.

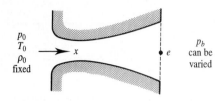

**FIGURE 8.11**
Convergent–divergent nozzle.

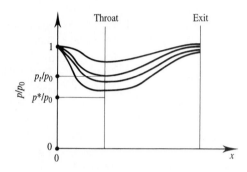

**FIGURE 8.12**
Pressure distribution with subsonic flow in a convergent–divergent nozzle.

8.12. With this type of flow, the mass flow rate through the nozzle increases as $p_b$ is decreased.

Since eq. (8.22) applies between any two points in the nozzle, it can be applied between the throat and the nozzle exit in order to relate the pressure at the throat, $p_t$, to the back pressure, $p_b$, which, with subsonic flow throughout the nozzle, is equal to $p_e$. This procedure gives:

$$\frac{A_e}{A^*} = \left(\frac{p_t}{p_b}\right)^{\frac{1}{\gamma}} \left[ \frac{1 - \left(\frac{p_t}{p_0}\right)^{\frac{\gamma-1}{\gamma}}}{1 - \left(\frac{p_b}{p_0}\right)^{\frac{\gamma-1}{\gamma}}} \right]^{\frac{1}{2}}$$

i.e.,   $\left(\dfrac{p_t}{p_0}\right)^{\frac{1}{\gamma}} \left\{ 1 - \left(\dfrac{p_t}{p_0}\right)^{\frac{\gamma-1}{\gamma}} \right\}^{\frac{1}{2}} = \left(\dfrac{A_e}{A^*}\right) \left(\dfrac{p_b}{p_0}\right)^{\frac{1}{\gamma}} \left\{ 1 - \left(\dfrac{p_b}{p_0}\right)^{\frac{\gamma-1}{\gamma}} \right\}^{\frac{1}{2}}$   (8.37)

2. In the first flow regime, discussed above, as the back pressure decreases, the throat pressure, which is lower than the back pressure, also decreases. This continues until $p_b$ has dropped to a value at which the throat pressure becomes equal to the critical pressure and the Mach number at the throat becomes equal to 1. The back pressure at which this occurs is obtained by substituting:

$$p_t = p^* = p_0 \left(\frac{2}{\gamma+1}\right)^{\frac{\gamma}{\gamma-1}}$$

into eq. (8.37) giving:

$$\left(\frac{A_e}{A^*}\right)\left(\frac{p_{b\,\mathrm{crit}}}{p_0}\right)^{\frac{1}{\gamma}}\left\{1-\left(\frac{p_{b\,\mathrm{crit}}}{p_0}\right)^{\frac{\gamma-1}{\gamma}}\right\}=\left(\frac{\gamma-1}{\gamma+1}\right)\left(\frac{2}{\gamma+1}\right)^{\frac{1}{\gamma-1}} \quad (8.38)$$

Here $p_{b\,\mathrm{crit}}$ is the back pressure at which the throat pressure first drops to the critical value.

Once a Mach number of 1 has been reached at the throat, further reductions in the back pressure cannot effect conditions upstream of the throat and cannot, therefore, alter the mass flow rate through the nozzle. The nozzle is, therefore, choked once the back pressure is decreased to $p_{b\,\mathrm{crit}}$ and the second flow regime is entered when $p_b$ has decreased to $p_{b\,\mathrm{crit}}$.

As the back pressure is reduced below $p_{b\,\mathrm{crit}}$, a region of supersonic flow develops just downstream of the throat. This region of supersonic flow is terminated by a normal shock wave. The shock wave increases the pressure and reduces the velocity to a subsonic value. The flow then decelerates subsonically until the pressure on the exit plane $p_e$ is equal to the back pressure, $p_b$. The flow in the nozzle under these circumstances is as shown in Fig. 8.13. In the above discussion, the supersonic portion of the flow is assumed to be terminated by a normal shock wave. In real flows, if the nozzle is of relatively small size and the boundary layer consequently relatively thick, a complex wave system can actually occur near the end of the supersonic flow region as a result of the interaction of the shock wave with the boundary layer. However, even in such cases, the characteristics of the flow can often be adequately modeled by assuming a normal shock wave.

As the back pressure is further reduced, the extent of the supersonic flow region increases, the shock wave moving further down the divergent portion of the nozzle. The nozzle pressure distribution in this regime, therefore, resembles that shown in Fig. 8.14.

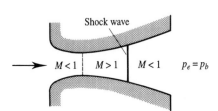

**FIGURE 8.13**
Flow in a convergent–divergent nozzle in regime 2.

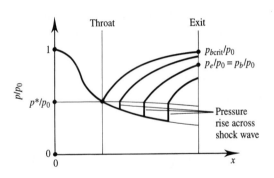

**FIGURE 8.14**
Pressure distribution in a convergent–divergent nozzle in regime 2.

3. As noted above, in flow regime 2, as the back pressure decreases the shock wave moves down the divergent portion of the nozzle towards the exit plane. Eventually, $p_b$ will drop to a value at which the shock wave is on the exit plane of the nozzle. It is when this occurs that flow regime 3 is entered. The back pressure at which this occurs can be estimated by assuming that the shock wave, which at the beginning of this regime lies in the exit plane of the nozzle, can be adequately modeled as a normal shock. The flow at the exit plane of the nozzle is then as shown in Fig. 8.15.

Since the flow in the nozzle is now isentropic throughout, the pressure, $p_e$, and Mach number, $M_e$, on the exit plane ahead of the shock can be found using isentropic relations. They are, therefore, given by the following two equations:

$$\left(\frac{p_e}{p_0}\right)^{\frac{2}{\gamma}} - \left(\frac{p_e}{p_0}\right)^{\frac{\gamma+1}{\gamma}} = \left(\frac{2}{\gamma+1}\right)^{\frac{\gamma+1}{\gamma-1}}\left(\frac{\gamma-1}{2}\right)\left(\frac{A^*}{A_e}\right)^{\frac{1}{2}}$$

$$\frac{1}{M_e}\left\{\left(\frac{2}{\gamma+1}\right)\left[1+\left(\frac{\gamma-1}{2}\right)M_e^2\right]^{\frac{\gamma+1}{2(\gamma-1)}}\right\} = \frac{A_e}{A^*}$$

these being obtained by using eqs. (8.27) and (8.36) respectively. Thus, since $(A_e/A^*)$ is known, $(p_e/p_0)$ and $M_e$ can be calculated. They will, of course, be equal to the design pressure ratio and the design exit Mach number because the flow in the nozzle is isentropic throughout.

Next, it is noted that since the flow behind a normal shock wave is subsonic, the static pressure behind the shock wave must be the back pressure, $p_{b3}$. Therefore, the normal shock wave relations given in Chapter 5 give:

$$\frac{p_{b3}}{p_e} = \frac{2\gamma M_e^2 - (\gamma-1)}{(\gamma+1)}$$

Hence, since $M_e$ is known, $p_{b3}/p_e$ can be calculated. $p_{b3}/p_0$ is then given by noting that:

$$\frac{p_{b3}}{p_0} = \left(\frac{p_{b3}}{p_e}\right)\left(\frac{p_e}{p_0}\right)$$

As $p_b$ is decreased below $p_{b3}$, conditions at all sections of the nozzle remain unchanged and the pressure on the exit plane, $p_e$, remains unchanged. With decreasing $p_b$, however, the shock wave moves outside

**FIGURE 8.15**
Flow at exit of convergent–divergent nozzle at beginning of regime 3.

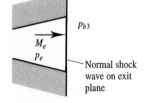

$p_{b3}$

$M_e$

$p_e$

Normal shock wave on exit plane

the nozzle and the compression from $p_e$ to $p_b$ takes place through a series of oblique shock waves outside the nozzle, the flow pattern resembling that shown in Fig. 8.16.

In this state the nozzle is said to be "over-expanded" because the exit plane pressure is less than the back pressure. The pressure distribution in the nozzle in this regime resembles that shown in Fig. 8.17.

4. In flow regime 3, as $p_b$ is decreased the oblique shock waves in the discharge flow become weaker and weaker and the difference between $p_e$ and $p_b$ becomes smaller and smaller. Eventually, a point is reached at which $p_b$ is just equal to the exit plane pressure, $p_e$. The nozzle is then operating at its "design" pressure ratio and there are no waves inside or outside the nozzle. A further reduction in $p_b$ moves the flow into regime 4. The design back pressure, i.e., the back pressure at which regime 4 begins, is given by eq. (8.30) as:

$$\left(\frac{p_{b4}}{p_0}\right)^{\frac{2}{\gamma}} - \left(\frac{p_{b4}}{p_0}\right)^{\frac{\gamma+1}{\gamma}} = \left(\frac{2}{\gamma+1}\right)^{\frac{\gamma+1}{(\gamma-1)}}\left(\frac{\gamma-1}{2}\right)\left(\frac{A^*}{A_e}\right)^{\frac{1}{2}}$$

If $p_b$ is further reduced it becomes less than the exit plane pressure, $p_e$, which remains constant, of course, from the beginning of regime 3, and the expansion from $p_e$ to $p_b$ takes place through a series of expansion waves outside the nozzle as shown in Fig. 8.18.

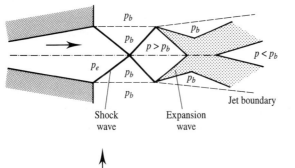

Shock
wave

Expansion
wave

Jet boundary

**FIGURE 8.16**
Flow near exit of over-expanded convergent–divergent nozzle.

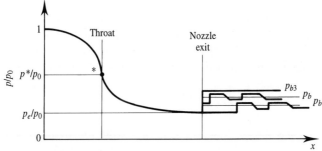

**FIGURE 8.17**
Pressure distribution in a convergent–divergent nozzle in regime 3.

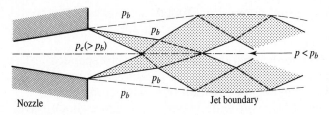

**FIGURE 8.18**
Flow near exit of under-expanded convergent–divergent nozzle.

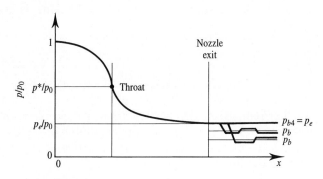

**FIGURE 8.19**
Pressure distribution in a convergent–divergent nozzle in regime 4.

In this state the nozzle is said to be "under-expanded" because the pressure in the exit plane is greater than the back pressure. The nozzle pressure distribution in this regime resembles that shown in Fig. 8.19.

The characteristics of a convergent–divergent nozzle are summarized in Figs. 8.20 and 8.21. Some of the features of flow through a convergent–divergent nozzle discussed above can be seen in the set of photographs of the flow near the exit to a two-dimensional nozzle shown in Fig. 8.22.

Instead of using the equations given above to determine the limits of the various flow regimes they can be determined using isentropic and normal shock tables or the software. This is illustrated in the examples given below.

The above discussion was concerned with the situation where the supply pressure $p_0$ is kept constant and the back pressure $p_b$ is varied. The flow changes that occur in other situations, e.g., when there are variations in $p_0$ for a fixed value of $p_b$, can easily be deduced from the above discussion.

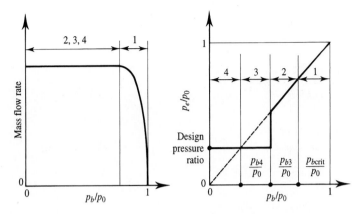

**FIGURE 8.20**
Effect of back pressure on mass flow rate and exit plane pressure in a convergent–divergent nozzle.

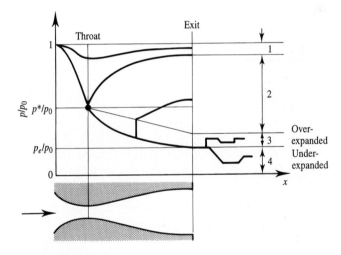

**FIGURE 8.21**
Pressure distribution in a convergent–divergent nozzle.

**EXAMPLE 8.3**
Air is expanded through a convergent–divergent nozzle from a large reservoir in which the pressure and temperature are 600 kPa and 40°C respectively. The design back pressure is 100 kPa.
   Find:

1. The ratio of the nozzle exit area to the nozzle throat area
2. The discharge velocity from the nozzle under design considerations
3. At what back pressure will there be a normal shock at the exit plane of the nozzle?

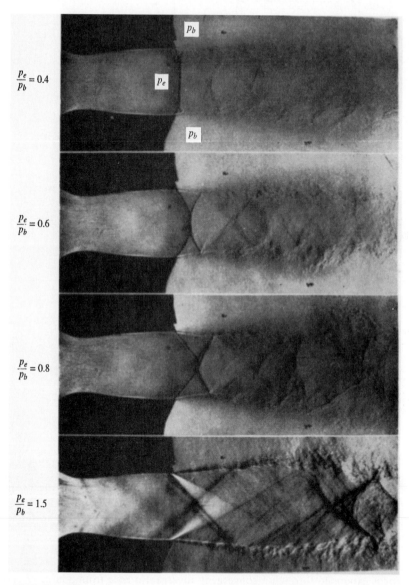

**FIGURE 8.22**
Flow near the exit of a convergent–divergent nozzle.

## Solution

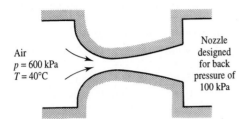

**FIGURE E8.3**

Here $p_0 = 600$ kPa and $T_0 = 40°C$ and the design back pressure is $p_b = 100$ kPa.

1. When operating at the design conditions $p_e = p_b$ so:

$$\frac{p_e}{p_0} = \frac{100}{600} = 0.1667$$

Hence using isentropic relations or tables or the software gives:

$$M_e = 1.83$$

and

$$\frac{A^*}{A_e} = 0.6792$$

Hence, the ratio of the nozzle exit area to the nozzle throat area is:

$$\frac{A_e}{A^*} = \frac{1}{0.6792} = 1.472$$

2. Since $M_e = 1.83$, using isentropic relations or tables or the software gives:

$$\frac{a_e}{a_0} = 0.7739$$

But:     $a_0 = \sqrt{\gamma R T_0} = \sqrt{1.4(286.8)(40 + 273)} = 345$ m/s

Hence:     $V_e = \dfrac{V_e}{a_e}\dfrac{a_e}{a_0} a_0 = 1.83 \times 0.7739 \times 354.5 = 502.1$ m/s

Hence, the nozzle discharge velocity under design conditions is 502.1 m/s.

3. When there is a normal shock wave on the exit plane of the nozzle the design conditions will exist upstream of the shock. Hence, using $M_1 = 1.83$, normal shock wave relations or tables or the software give:

$$M_2 = 0.6099$$

and

$$\frac{p_2}{p_1} = 3.740$$

Hence:     $p_2 = \dfrac{p_2}{p_1} p_1 = 3.740 \times 100 = 374.0$ kPa

Therefore, there will be a normal shock wave on the exit plane of the nozzle when $p_b = p_2 = 374.0$ kPa.

### EXAMPLE 8.4

A convergent–divergent nozzle is designed to expand air from a chamber in which the pressure is 800 kPa and the temperature is 40°C to give a Mach number of 2.7. The throat area of the nozzle is 0.08 m². Find:

1. The exit area of the nozzle
2. The mass flow rate through the nozzle when operating under design conditions
3. The design back pressure
4. The lowest back pressure for which there is only subsonic flow in the nozzle
5. The back pressure at which there will be a normal shock wave on the exit plane of the nozzle
6. The back pressure below which there are no shock waves in the nozzle
7. The range of back pressures over which there are oblique shock waves in the exhaust from the nozzle
8. The range of back pressures over which there are expansion waves in the exhaust from the nozzle.

**Solution**

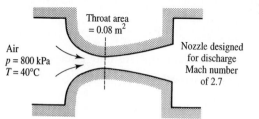

**FIGURE E8.4**

1. For $M = 2.7$, isentropic flow relations or tables or software give:

$$\frac{A}{A^*} = 3.183$$

Hence: $$A_e = 3.183 \times 0.08 = 0.255 \, \text{m}^2$$

2. Now $\dot{m} = \rho^* V^* A^*$. But:

$$\rho_0 = \frac{p_0}{RT_0} = \frac{800 \times 10^3}{287 \times 313} = 8.91 \, \text{kg/m}^3$$

Also, isentropic flow relations or tables or software give for $M = 1$:

$$\frac{\rho^*}{\rho_0} = 0.633\,94, \qquad \frac{T^*}{T_0} = 0.830\,55$$

From which it follows that:

$$T^* = 0.830\,55 \times 313 = 260 \, \text{K}$$

Also: $$V^* = a^* = \sqrt{\gamma R T^*} = \sqrt{1.4 \times 287 \times 260} = 323.2 \, \text{m/s}$$

and: $$\rho^* = 0.633\,94 \times 8.91 = 5.65 \, \text{kg/m}^3$$

Hence: $$\dot{m} = \rho^* V^* A^* = 5.65 \times 323.2 \times 0.8 = 146 \, \text{kg/s}$$

3. Isentropic flow relations or tables or software give for $M = 2.7$:

$$\frac{p_0}{p} = 23.283, \quad \text{hence} \quad p_{\text{design}} = 800/23.283 = 34.36 \, \text{kPa}$$

4. Subsonic isentropic flow relations or tables or software give $p_0/p = 1.025$ for $A/A^* = 3.183$. Hence the Mach number at the throat will reach 1 when the back pressure has dropped to $800/1.025 = 780.5 \, \text{kPa}$.

5. When there is a shock wave on the exit plane of the nozzle, the Mach number ahead of the shock is 2.7 and the pressure is 34.36 kPa. But for a Mach number of 2.7, normal shock relations or tables or software give

$$\frac{p_b}{p_{\text{design}}} = 8.338\,32$$

Hence the back pressure at which there is a shock wave on the nozzle exit plane is:

$$p_b = 8.338\,32 \times 34.36 = 286.5\,\text{kPa}$$

6. When the back pressure has dropped below that established in 5 above, the shock wave moves out of the nozzle. Hence, there are no shock waves in the nozzle when:

$$p_b < 286.5\,\text{kPa}$$

7. There are oblique shocks in the exhaust when the back pressure is below that established in 5 above and hence there are oblique shock waves in the exhaust when:

$$34.36\,\text{kPa} < p_b < 286.5\,\text{kPa}$$

8. Expansion waves will occur when $p_b < p_{\text{design}}$, i.e., when $p_b < 34.36\,\text{kPa}$.

## 8.5
## CONVERGENT–DIVERGENT SUPERSONIC DIFFUSERS

With most present day air-breathing aircraft engines, it is necessary to decelerate the air to subsonic velocity before passing it to the engine. This deceleration, which also increases the pressure of the air, is carried out in the diffuser. From the discussion given in the earlier part of this chapter it follows that this deceleration can be accomplished by passing the air through a convergent–divergent diffuser, the deceleration potentially being shockless at the design flight Mach number of the aircraft. In such a case the flow through the diffuser will be as shown in Fig. 8.23.

The present discussion will be concerned with how this state is achieved and how the diffuser performs when the aircraft is flying at Mach numbers different from the design value. Initially, the concern will be with fixed diffusers, i.e., diffusers in which the throat and inlet areas are fixed.

Now, in the ideal case, the diffuser will, of course, be designed so that at the design Mach number, the ratio of throat to inlet area is equal to the value of $A^*/A$ corresponding to the design Mach number. However, since $A^*/A$ increases as $M$ decreases, it is obvious that when the aircraft is accelerating up to the design Mach number the diffuser will not be able to "swallow" all the air flowing towards the intake. Under these conditions, therefore, a

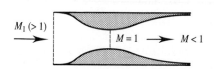

$M_1 (> 1)$    $M = 1 \rightarrow M < 1$

**FIGURE 8.23**
Ideal flow through a convergent–divergent supersonic diffuser.

normal shock stands ahead of the diffuser and decreases the velocity to a subsonic velocity. The air is then able to "spill" over the intake. This is illustrated in Fig. 8.24.

As the flight Mach number is increased and the throat area becomes closer to that required for the Mach number, the shock moves towards the intake. However, even when the design Mach number is reached, there is still a normal shock ahead of the diffuser and spillage still occurs. It is only when some Mach number that is greater than the design Mach number is reached that the shock moves up to the intake and all the air reaching the diffuser is ingested. This situation is illustrated in Fig. 8.25. The Mach number, $M_m$, at which this occurs is found by noting that this set of circumstances is reached because the Mach number behind the shock wave is now sufficiently small for the throat to intake area ratio to be equal to the value $A^*/A$ that corresponds to the Mach number in subsonic flow immediately behind the shock wave. $M_m$ is thus found by using subsonic relations or tables or software to give the subsonic Mach number corresponding to the diffuser area ratio and then using normal shock relations or tables or software to obtain the supersonic Mach number which gives this downstream Mach number value.

Any further slight increase in the Mach number will cause the shock wave to be "swallowed" by the diffuser and the shock wave then settles in the divergent portion of the diffuser as shown in Fig. 8.26.

Thus, in order for the fixed area diffuser to swallow the shock, it is necessary to use "over-speeding," i.e., to accelerate the aircraft to a Mach number that is greater than the design Mach number.

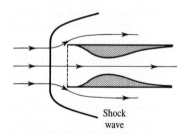

**FIGURE 8.24**
Flow near intake to fixed supersonic diffuser when $M < M_{\text{design}}$.

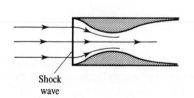

**FIGURE 8.25**
Flow near intake to fixed supersonic diffuser just before shock is swallowed.

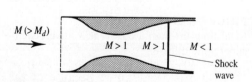

**FIGURE 8.26**
Flow after shock wave is swallowed.

Once the shock has been swallowed, the flight Mach number can be decreased and as it is decreased the shock wave moves up the divergent portion of the diffuser towards the throat, decreasing in strength as it does so because the Mach number upstream of the shock decreases as the shock moves towards the throat. At the design Mach number, the shock wave reaches the throat and, because the upstream Mach number is then 1, it disappears. The originally discussed shockless design condition then exists. In reality, it would not be practical to operate under exactly these conditions because the slightest decrease in Mach number would cause the shock to be "disgorged," i.e., for the shock to move right out of the diffuser again and the whole "over-speeding" process required to swallow the shock would have to be repeated.

### EXAMPLE 8.5

A small jet aircraft which is designed to cruise at a Mach number of 1.7 has a convergent–divergent intake diffuser with a fixed area ratio. Find the ideal area ratio for this diffuser and the Mach number to which the aircraft must be taken in order to swallow the normal shock wave if the diffuser has this ideal area ratio.

### Solution

Just before the shock is swallowed, the flow is as shown in Fig. E8.5. From isentropic relations or tables or software at $M_1 = 1.7$, i.e., with no shock:

$$\frac{A_i}{A^*} = 1.338$$

From isentropic relations or tables or software for $A/A^* = 1.338$ and subsonic flow:

$$M_2 = 0.5$$

Hence from normal shock relations or tables or software for $M_2 = 0.5$:

$$M_1 = 2.64$$

**FIGURE E8.5**

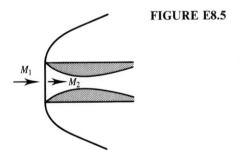

One criterion that can be used to describe the performance of a diffuser is the stagnation pressure recovery ratio, i.e., the ratio of the pressure that would be obtained if the air were brought to rest by the diffuser to the stagnation pressure corresponding to the freestream flow. With the fixed area ratio diffuser discussed above under design conditions when no shock waves exist, this ratio is 1, but under other circumstances there is a stagnation pressure loss across the shock and the ratio is less than 1. Because of the

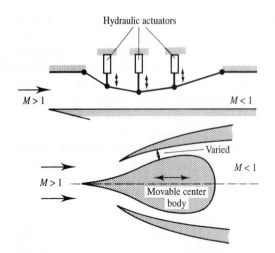

**FIGURE 8.27**
Simple variable area supersonic diffusers.

poor performance at off-design conditions and because of the over-speeding required to swallow the shock, simple fixed convergent–divergent diffusers are seldom used. These disadvantages can, to some extent, be overcome by using a diffuser with a variable area throat but this, of course, adds mechanical complexities. The operation of some forms of variable area convergent–divergent diffuser are illustrated in a very schematic way in Fig. 8.27. With a variable throat area diffuser, the throat can be opened to allow the shock to be swallowed and then the throat area can be decreased until the shock disappears at all Mach numbers, i.e., the throat area is adjusted to match the actual Mach number and no over-speeding is required. This is illustrated in the following example.

### EXAMPLE 8.6

A small jet aircraft designed to cruise at a Mach number of 2.5 has an intake diffuser with a variable area ratio. Find the ratio of the throat area under these cruise conditions to the throat area required when the aircraft is flying at a Mach number of 1.3. Assume the diffuser intake area does not change. If the aircraft when flying at cruise conditions is suddenly slowed down without altering the diffuser area ratio, sketch the diffuser flow pattern that will then exist.

### *Solution*

Using isentropic relations or tables or software gives $A/A^* = 2.6367$ for a Mach number of 2.5 and $A/A^* = 1.0663$ for a Mach number of 1.3. Hence:

For $$M = 2.5: \quad \frac{A_e}{A_{\text{throat}}} = 2.6367$$

For $$M = 1.3: \quad \frac{A_e}{A_{\text{throat}}} = 1.0663$$

Since the intake area $A_e$ is the same at the two Mach numbers, it follows that:

$$\frac{A_{\text{throat}} \text{ at } M = 2.5}{A_{\text{throat}} \text{ at } M = 1.3} = \frac{1/2.6367}{1/1.0663} = 0.404$$

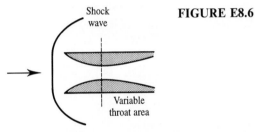

**FIGURE E8.6**

Hence the throat area required under cruise conditions is 40.4 percent of that required at a Mach number of 1.3.

If the aircraft is slowed down without increasing the throat area the diffuser will not be able to "swallow" the flow, and a shock wave will form ahead of the diffuser as shown in Fig. E8.6.

Because of the difficulties associated with convergent–divergent diffusers, oblique shock diffusers are commonly used. In such diffusers, much of the deceleration is done through oblique shocks which incur less stagnation pressure loss than normal shock waves that tend to occur in convergent–divergent diffusers. Usually a normal shock wave will exist at some point in the system but because the Mach number upstream of this normal shock has been decreased through the oblique shocks, the losses are less than would exist if the normal shock occurred at the freestream Mach number. A typical oblique shock diffuser is shown very schematically in Fig. 8.28.

In some cases, a curved compressive corner instead of a series of distinct steps each generating an oblique shock wave is used. The corner flow is still followed by a normal shock wave. This is shown schematically in Fig. 8.29 and a photograph of the flow near an intake that utilizes this system is shown in Fig. 8.30. The situation shown actually involves an axisymmetric (conical) system.

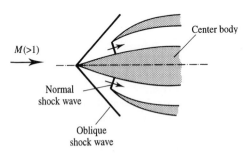

**FIGURE 8.28**
Simple oblique shock diffuser.

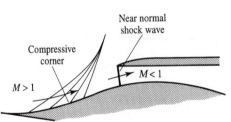

**FIGURE 8.29**
Diffuser with curved compressive corner.

**FIGURE 8.30**
Flow into diffuser with curved compressive corner.


**EXAMPLE 8.7**
An aircraft is to cruise at a Mach number of 3. The stagnation pressure in the flow
ahead of the aircraft is 400 kPa. Consider three possible intake scenarios:

1. An intake that involves a normal shock in the freestream ahead of the intake
   followed by an isentropic deceleration of the subsonic flow behind the shock
   wave to an essentially zero velocity
2. An oblique shock wave diffuser in which the air flows through an oblique
   shock wave, which causes the flow to turn through 15°, followed by a normal
   shock and then an isentropic deceleration of the subsonic flow behind the
   normal shock wave to an essentially zero velocity
3. An ideal shockless convergent–divergent diffuser in which the air is isentropically brought to an essentially zero velocity.

Find the pressures at the exits of each of these diffusers.

***Solution***
The three situations considered are shown in Fig. E8.7.

Now for a normal shock at a Mach number 3, normal shock relations
or tables or the software gives $p_{02}/p_{01} = 0.3283$. Hence in Case 1,
$p_{02} = 0.3283 \times 400 = 131.3$ kPa.

For an oblique shock with an upstream Mach number of 3 and a turning
angle of 15°, oblique shock relations or charts or the software give $M_{N1} = 1.600$
and $M_2 = 2.255$. Normal shock relations or tables or the software give
$p_{02}/p_{01} = 0.8952$ for a Mach number of 1.600 and $p_{03}/p_{02} = 0.6168$ for a Mach
number of 2.255. Hence in Case 2, $p_{03} = 0.8952 \times 0.6168 \times 400 = 220.9$ kPa.

In Case 3 there is no loss in stagnation pressure since there are no shock waves
so in this case $p_{02} = 40$ kPa.

Therefore, the pressures at the exit of the diffuser in the three cases are
131.3 kPa, 220.9 kPa, and 400 kPa respectively.

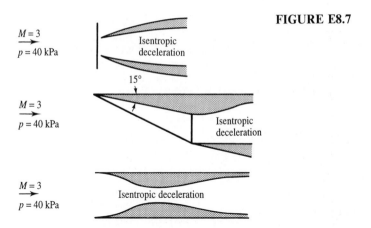

**FIGURE E8.7**

The above discussion was concerned with supersonic diffusers for aircraft. Similar problems often arise in machines that involve compressible gas flow. As an illustration of this, consider a supersonic wind-tunnel. In such a tunnel, the air is accelerated to the desired supersonic Mach number in the working section of the tunnel by passing it through a convergent–divergent nozzle. The air must then be decelerated back to a subsonic Mach number after it leaves the working section. This is usually done by fitting a convergent–divergent diffuser to the tunnel. Under ideal circumstances, the flow will be accelerated to a supersonic Mach number in the nozzle and then back to a subsonic Mach number in the diffuser, the throat area of the diffuser under these circumstances ideally then being equal to the throat area of the nozzle. This flow situation is shown in Fig. 8.31.

The tunnel is started, however, by either increasing the pressure ahead of the nozzle or decreasing the pressure behind the diffuser. If operating characteristics of a nozzle discussed earlier are considered, it will be seen that a normal shock wave must exist in the nozzle during the start-up process. This shock will get stronger and stronger as it moves down the divergent section of the nozzle from the throat towards the working section and will be strongest when it lies at the end of the nozzle. This flow situation is shown in Fig. 8.32.

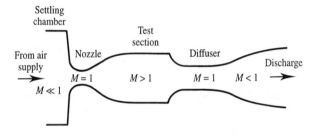

**FIGURE 8.31**
Ideal flow in a supersonic wind-tunnel.

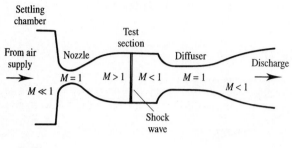

**FIGURE 8.32**
Flow in a supersonic
wind-tunnel during
starting.

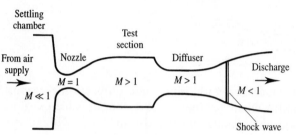

**FIGURE 8.33**
Flow in a supersonic
wind-tunnel after the
shock has been
swallowed.

Because there is a loss of stagnation pressure through the shock wave, the
area of the throat of the diffuser will have to be bigger than that of the nozzle
if the diffuser is to be able to "swallow" this starting shock wave. If the
diffuser throat area is sufficiently large, once the normal shock has reached
the working section it will jump through the diffuser and settle down some-
where in the diverging portion of the diffuser. The flow that then exists is
shown in Fig. 8.33.

If the wind-tunnel was fitted with a diffuser that had a variable throat
area, the diffuser throat area could then be decreased until the shock wave
disappeared and the ideal flow shown in Fig. 8.31 is obtained.

### EXAMPLE 8.8
A wind-tunnel designed for a test section Mach number of 3 is fitted with a
variable area diffuser. Find the ratio of the diffuser throat area when operating
under ideal running conditions to the diffuser throat area during starting when
there is a shock wave in the working section.

### *Solution*
When, as shown in Fig. E8.8, the shock is about to be swallowed, normal shock
relations or tables or the software give for $M_1 = 3$:

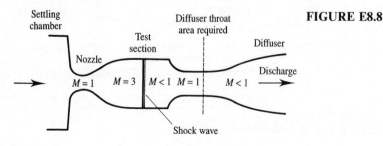

**FIGURE E8.8**

$$M_2 = 0.475$$

Isentropic subsonic relations or tables or the software then give for $M = 0.475$:

$$\frac{A_{ts}}{A_i} = \frac{A^*}{A} = 0.72$$

When the shock has been swallowed and the throat area decreased to eliminate the shock, supersonic isentropic subsonic relations or tables or the software give for $M = 3$:

$$\frac{A_t}{A_i} = \frac{A^*}{A} = 0.236$$

Therefore, the ratio of the diffuser throat area is:

$$\frac{A_t}{A_{ts}} = \frac{A_t/A_i}{A_{ts}/A_i}$$

$$= \frac{0.236}{0.72}$$

$$= 0.328$$

## 8.6
## TRANSONIC FLOW OVER A BODY

In the flows discussed above, the flow was, in general, subsonic in part of the system and supersonic in other parts of the system, the supersonic flow region usually being terminated by a normal shock, e.g., consider the nozzle flow shown in Fig. 8.13. Another situation in which the same general type of flow occurs will be discussed in the present section in a very qualitative manner. Consider flow over a body such as that shown in Fig. 8.34.

The velocity and, hence, the Mach number is increased as the flow passes over the surface. As a result, as the freestream velocity is increased, a point will be reached at which, even when the flow in the freestream ahead of the body is subsonic, the Mach number at some point in the flow over the body reaches a value of 1. The freestream Mach number at which this occurs, i.e., at which the Mach number at some point in the flow over the body first reaches a value of 1, is called the critical Mach number, $M_{crit}$. As the freestream Mach number is increased above the critical value, regions of supersonic flow develop in the flow over the body, these regions being terminated by normal shock waves as indicated in Fig. 8.35.

Because of the sharp pressure rise across the shock wave, the boundary layer on the body tends to separate from the surface at the shock wave as indicated in Fig. 8.36.

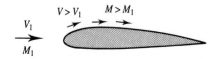

**FIGURE 8.34**
Velocity changes in external flow over a body.

$M_1 < 1$

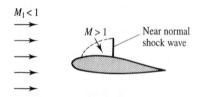

**FIGURE 8.35**
Transonic flow over a body.

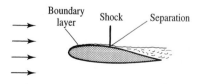

**FIGURE 8.36**
Shock induced boundary layer separation in transonic flow.

As a result of this boundary layer separation, the drag on the body rises as the freestream Mach number rises above the critical Mach number, the form of the variation being as shown in Fig. 8.37. The drag force on the body rises to a maximum near a Mach number of 1 and then decreases with further increase in the freestream Mach number as shown in Fig. 8.37. In Fig. 8.37, the drag has been expressed in terms of the dimensionless drag coefficient, $C_D$, which is defined by:

$$C_D = \frac{D}{\rho_1 V_1^2 A/2}$$

where $D$ is the drag on the body, $\rho_1$ and $V_1$ are the density and velocity in the freestream ahead of the body, and $A$ is the characteristic area of the body. The drag only starts to increase significantly at a freestream Mach number that is higher than the critical Mach number, the value of the freestream Mach number at which a significant drag rise first occurs is called the divergence Mach number, $M_{\text{div}}$. With further increase in the freestream Mach number, the extent of the supersonic flow regions increase and the normal shocks move towards the trailing edge of the body. This continues until the freestream Mach number ahead of the body reaches a value of 1. A shock wave then forms ahead of the body and the normal shock waves on the body move to the trailing edge of the body. The changes in the flow are shown in Fig. 8.38.

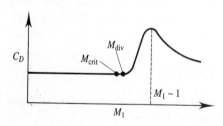

**FIGURE 8.37**
Variation of coefficient of drag with Mach number in transonic flow.

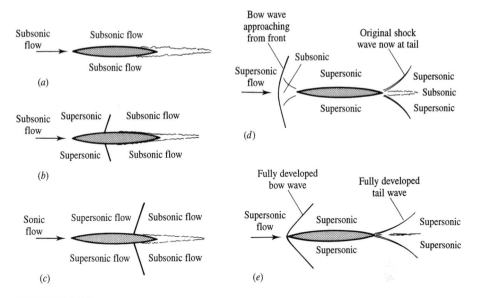

**FIGURE 8.38**
Variation in the flow over a body with Mach number in transonic flow.
(a) $M = 0.6$; (b) $M = 0.8$; (c) $M = 1.0$; (d) $M = 1.1$; (e) $M = 2.0$.

Because the thrust generated by the engines of an aircraft must balance the drag in steady horizontal flight, many aircraft will not have enough thrust to allow them to pass through the region of drag rise in the transonic region. For this reason, the drag rise near a freestream Mach number of 1 is termed the "sound barrier." When flying in the transonic region, the position of the shock waves and the extent of the boundary layer separation induced by the shock waves is usually fluctuating with time, and this together with the unsteady forces produced by the impingement of the unsteady separated flow on other components of the aircraft gives rise to the buffeting often associated with "breaking the sound barrier."

**EXAMPLE 8.9**
An aircraft has a wing area of 30 m² and a mass of 9000 kg. The turbojet engine which propels this aircraft has a maximum thrust of 50 kN. The drag coefficient, based on wing area, for this aircraft has a value of 0.02 at low Mach numbers. Near a Mach number of 1, this drag coefficient rises to a maximum value of 0.045. Determine:

1. What is the highest speed at which this aircraft could fly at an altitude of 5000 m if there was no increase in drag near a Mach number of 1 due to compressibility?
2. Whether this aircraft can "break the sound barrier" in steady horizontal flight and in a steady vertical dive at an altitude of 5000 m.

**Solution**

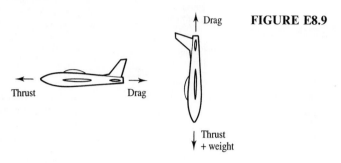

**FIGURE E8.9**

1. In steady horizontal flight, the thrust is equal to the drag, i.e.,

$$T = D$$

But:

$$D = C_D \tfrac{1}{2} \rho V^2 A_w$$

At an altitude of 5000 m in the standard atmosphere the density is 0.737 kg/m³ so if $C_D = 0.02$ the maximum speed at which the aircraft can fly is determined by:

$$50\,000 = 0.02 \times \tfrac{1}{2} \times 0.737 \times V^2 \times 30$$

Solving this gives $V = 475$ m/s. But the speed of sound at an altitude of 5000 m in the standard atmosphere is 320 m/s, so if there was no drag rise due to compressibility, the aircraft would be able to fly at a maximum Mach number of $475/320 = 1.48$.

2. To determine whether the aircraft can "break the sound barrier" the drag on the aircraft when flying at a Mach number of 1 with a drag coefficient of 0.045 is calculated. Since when flying at $M = 1$ at the specified altitude, $V = a = 320$ m/s, the drag on the aircraft is:

$$D = 0.045 \times \tfrac{1}{2} \times 0.737 \times 320^2 \times 30 = 50\,941 \text{ N}$$

This is greater than the maximum thrust of the engine, so the aircraft will not be able to reach a Mach number of 1 in horizontal flight. However, in a vertical dive, the force available to overcome the drag is the sum of the thrust and the gravitational force on the aircraft, i.e., $50\,000 + (9000 \times 9.81)$, i.e., 138 290 N, so the aircraft will be able to reach supersonic velocities in a dive.

**8.7**
## CONCLUDING REMARKS

In order to accelerate a gas flow from subsonic to supersonic velocities in a nozzle, it is necessary to use a convergent–divergent nozzle, the Mach number being 1 at the throat. If conditions upstream of the nozzle are kept constant, there is a limit to the mass flow rate through the nozzle that can be achieved by lowering the back pressure. If the back pressure is low enough for this maximum mass flow rate to be reached, the nozzle is said to be choked.

Convergent–divergent systems designed to decelerate a supersonic flow to a subsonic velocity have also been discussed.


# PROBLEMS

**8.1.** Air is discharged from a large reservoir, in which the pressure and temperature are 0.8 MPa and 25°C respectively, through a convergent nozzle with an exit diameter of 5 cm. The nozzle discharges to the atmosphere. Find the mass flow rate through the nozzle and the pressure and temperature on the nozzle exit plane.

**8.2.** A supersonic nozzle possessing an area ratio of 3.0 is supplied from a large reservoir and is allowed to exhaust to atmospheric pressure. Determine the range of reservoir pressures over which a normal shock will appear in the nozzle. For what value of reservoir pressure will the nozzle be perfectly expanded, with supersonic flow at the exit plane? Find the minimum reservoir pressure to produce sonic flow at the nozzle throat. Assume isentropic flow except for shocks with $\gamma = 1.4$.

**8.3.** A converging–diverging nozzle is designed to operate with an exit Mach number of 1.75. The nozzle is supplied from an air reservoir at 1000 psia. Assuming one-dimensional flow, calculate:

1. Maximum back pressure to choke the nozzle
2. Range of back pressures over which a normal shock will appear in the nozzle.
3. Back pressure for the nozzle to be perfectly expanded to the design Mach number
4. Range of back pressures for supersonic flow at the nozzle exit plane

**8.4.** A convergent–divergent nozzle is designed to expand air from a chamber in which the pressure is 700 kPa and the temperature is 35°C to give a Mach number of 1.6. The mass flow rate through the nozzle under design conditions is 0.012 kg/s. Find:

1. The throat and exit areas of the nozzle
2. The design back pressure and the temperature of the air leaving the nozzle with this back pressure
3. The lowest back pressure for which there will be no supersonic flow in the nozzle
4. The back pressure below which there are no shock waves in the nozzle

**8.5.** A converging–diverging nozzle is designed to generate an exit Mach number of 2. The nozzle is supplied with air from a large reservoir in which the pressure is kept at 6.5 MPa. Assuming one-dimensional isentropic flow find:

1. The maximum back pressure at which the nozzle will be choked
2. The range of back pressures over which there will be a shock in the nozzle
3. The design back pressure

4. The range of back pressures over which there is supersonic flow on the nozzle exit plane

**8.6.** A convergent–divergent nozzle is designed to expand air from a chamber in which the pressure is 800 kPa and the temperature is 40°C to give a Mach number of 2.5. The throat area of the nozzle is 0.0025 m². Find:

1. The flow rate through the nozzle under design conditions
2. The exit area of the nozzle
3. The design back pressure and the temperature of the air leaving the nozzle with this back pressure
4. The lowest back pressure for which there is only subsonic flow in the nozzle
5. The back pressure at which there is a normal shock wave on the exit plane of the nozzle
6. The back pressure below which there are no shock waves in the nozzle
7. The range of back pressures over which there are oblique shock waves in the exhaust from the nozzle
8. The range of back pressures over which there are expansion waves in the exhaust from the nozzle
9. The back pressure at which a normal shock wave occurs in the divergent section of the nozzle at a point where the nozzle area is half way between the throat and the exit plane areas

**8.7.** A variable area diffuser is fitted to an aircraft designed to operate at a Mach number of 3.5. If the shock wave is "swallowed" at this Mach number, find the ratio of the throat area for shockless operation to the throat area at which the shock wave is swallowed.

**8.8.** A nozzle is designed to expand air from a chamber in which the pressure and temperature are 800 kPa and 40°C respectively to a Mach number of 2.5. The throat area of this nozzle is to be 0.1 m². Find:

1. The exit area of the nozzle
2. The mass flow rate through the nozzle when operating at design conditions
3. The back pressure at which there will be a normal shock wave on the exit plane of the nozzle
4. The range of back pressures over which expansion waves will occur outside the nozzle.

**8.9.** A rocket nozzle is designed to operate supersonically with a chamber pressure of 500 psia and an ambient pressure of 14.7 psia. Find the ratio between the thrust at sea level to the thrust in space (0 psia). Assume a constant chamber pressure, with a chamber temperature of 2500 R. Assume the rocket exhaust gases behave as a perfect gas with $\gamma = 1.4$ and $R = 20$ ft-lbf/lbm R.

**8.10.** Air flows through a convergent–divergent nozzle that has an inlet area of 0.0025 m². The inlet temperature and pressure are 50°C and 550 kPa respectively, and the velocity at the inlet is 80 m/s. If the flow is assumed to be isentropic, and if the exit pressure is 120 kPa, find the throat and exit areas and the exit velocity.

**8.11.** Air flows through a convergent–divergent passage. The passage inlet area is 5 cm$^2$, the minimum area is 3 cm$^2$ and the exit area is 4 cm$^2$. The air velocity at the inlet to the passage is 120 m/s. The pressure is 700 kPa and the temperature is 40°C. Assuming that the flow is isentropic, find the mass flow rate through the passage, the Mach number at the minimum area section, and the velocity and pressure at the exit section.

**8.12.** Air flows through a convergent–divergent nozzle from a large reservoir in which the pressure and temperature are maintained at 700 kPa and 60°C respectively. The rate of air flow through the nozzle is 1 kg/s. On the exit plane of the nozzle the stagnation pressure is 550 kPa and the static pressure is 500 kPa. A shock wave occurs in the nozzle and the flow can be assumed to be isentropic everywhere except through the shock wave. Find the nozzle throat area, the Mach numbers before and after the shock, the nozzle areas at the point where the shock occurs and on the exit plane, and the air density on the exit plane of the nozzle.

**8.13.** Air flows through a convergent–divergent nozzle. The air has a Mach number of 0.50 and a pressure and a temperature of 280 kPa and 10°C respectively at the inlet to the nozzle. The nozzle throat area is $6.5 \times 10^{-4}$ m$^2$ and the ratio of the exit area to the throat area is 4. If the pressure on the exit plane of the nozzle is 170 kPa, find the Mach number and the temperature on the nozzle exit plane and the nozzle area at the point in the nozzle at which the normal shock wave occurs.

**8.14.** Air is supplied to a convergent–divergent nozzle from a large tank in which the pressure and temperature are kept at 700 kPa and 40°C respectively. If the nozzle has an exit area that is 1.6 times the throat area and if a normal shock occurs in the nozzle at a section where the area is 1.2 times the throat area, find the pressure, temperature, and Mach number at the nozzle exit. Assume one-dimensional, isentropic flow.

**8.15.** Air enters a convergent–divergent nozzle at Mach number 0.2. The stagnation pressure is 700 kPa and the stagnation temperature is 5°C. The throat area of the nozzle is $46 \times 10^{-4}$ m$^2$ and the exit area is $230 \times 10^{-4}$ m$^2$. If the pressure at the exit to the nozzle is 500 kPa, determine if there is a shock in the divergent portion of the nozzle. If there is a shock wave, determine the nozzle area at which the shock occurs and the Mach number and pressure just before and just after the shock wave.

**8.16.** Air at a temperature of 20°C and a pressure of 101 kPa flows through a convergent–divergent nozzle at the rate of 0.5 kg/s. The exit area of the nozzle is 1.355 times the inlet area. If the air leaves the nozzle at a static temperature of 20°C and a stagnation temperature of 30°C, calculate the inlet and exit Mach numbers, the increase in entropy (if any) between the inlet and the outlet, the area at which the shock (if any) occurs and the stagnation pressure at the exit.

**8.17.** Air flows from a large reservoir in which the temperature and pressure are 80°C and 780 kPa through a convergent–divergent nozzle which has a throat diameter of 2.5 cm. When the back pressure is 560 kPa, a shock wave is found to occur at a location in the nozzle where the static pressure is 210 kPa. Find the exit area, exit temperature, the exit Mach number, the area at which the shock wave occurs, and the pressure ratio across the shock.

**8.18.** Air flows from a reservoir in which the pressure is kept at 124 kPa through a convergent–divergent nozzle and exhausts to the atmosphere where the pressure is 101.3 kPa. Under these conditions the nozzle is choked and the flow is subsonic on both sides of the throat. To what value must the pressure in the reservoir be raised so that there is a normal shock on the nozzle exit plane?

**8.19.** Air flows through a convergent–divergent nozzle. The nozzle exit and throat areas are 0.5 and 0.25 $m^2$ respectively. If the inlet stagnation pressure is 200 kPa and the back pressure is 120 kPa, determine the nozzle area at which the normal shock wave is located. What is the increase in entropy across the shock? At what back pressure will the shock wave be located on the nozzle exit plane?

**8.20.** Air at a pressure of 350 kPa, a temperature of 80°C, and a velocity of 180 m/s enters a convergent–divergent nozzle. A normal shock occurs in the nozzle at a location where the Mach number is 2. If the air mass flow rate through the nozzle is 0.7 kg/s and if the pressure on the nozzle exit plane is 260 kPa, find the nozzle throat area, the nozzle exit area, the temperatures upstream and downstream of the shock wave, and the change in entropy through the nozzle.

**8.21.** Air with a stagnation pressure and temperature of 100 kPa and 150°C is expanded through a convergent–divergent nozzle that is designed to give an exit Mach number of 2. The nozzle exit plane area is 30 $cm^2$. Find the mass rate of flow through the nozzle when operating at design conditions and the exit plane pressure under these design conditions. Also find the exit plane pressure if a normal shock wave occurs in the divergent portion of the nozzle at a section where the area is half way between the throat and the exit plane areas.

**8.22.** Air flows through a convergent–divergent nozzle with an exit area to throat area ratio of 4.0. If a normal shock wave occurs in the nozzle at a location where area is 2.5 the throat area, find the Mach number on the exit plane of the nozzle.

**8.23.** Air flows through a convergent–divergent nozzle. The stagnation temperature of supply air is 200°C and the nozzle has an exit area of 2 $m^2$ and a throat area of 1 $m^2$. If the air from the nozzle is discharged to an ambient pressure of 70 kPa, find the minimum supply stagnation pressure required to produce choking in the nozzle and the mass flow rate through the nozzle when it is choked. Also find the supply stagnation pressure that exists if a normal shock wave occurs in the divergent portion of the nozzle at a section where the area is 1.5 $m^2$.

**8.24.** Air flows from a large reservoir in which the pressure is 450 kPa through a convergent–divergent nozzle. A normal shock wave occurs in the diverging portion of the nozzle at a point where the nozzle area is twice the throat area. Find the Mach numbers on each side of this shock wave. If the Mach number on the exit plane of the nozzle is 0.2, find the back pressure required to maintain the shock at this location.

**8.25.** The stagnation pressure and temperature of the inlet to a supersonic wind-tunnel are 100 kPa and 30°C respectively. The Mach number in the test section of the tunnel is 2. If the cross-sectional area of the test section is 1.2 m$^2$, find the throat areas of the nozzle and diffuser.

**8.26.** Air flows through a convergent nozzle. At a section within this nozzle at which this cross-sectional area is 0.01 m$^2$, the pressure is 300 kPa and the temperature is 30°C. If the velocity at this section of the nozzle is 150 m/s, find the Mach number at this section, the stagnation temperature and pressure, and the mass flow rate through the nozzle. If the nozzle is choked, find the area, pressure, and temperature at the exit of the nozzle.

**8.27.** Carbon dioxide flows through an 8 cm inside diameter pipe. In order to determine the mass flow rate, a venturi meter with a throat diameter of 5 cm is installed in the pipe. The pressure and temperature just upstream of the venturi meter are 600 kPa and 40°C, respectively. The difference between the pressure just upstream of the venturi meter and the pressure at the throat of the venturi meter is 15 kPa. Find the mass flow rate of carbon dioxide.

**8.28.** A large rocket engine designed to propel a satellite launcher has a thrust of $10^6$ lbf when operating at sea-level, the exit plane pressure being equal to the ambient pressure under these conditions. The combustion chamber pressure and temperature are 500 psia and 4500°F respectively. If the products of combustion can be assumed to have the properties of air and if the flow through the nozzle can be assumed to be isentropic, find the throat and exit diameters of the nozzle.

**8.29.** Air, at a pressure of 700 kPa and a temperature of 80°C, flows through a convergent–divergent nozzle. The inlet area is 0.005 m$^2$ and the pressure on the exit plane is 40 kPa. If the mass flow rate through the nozzle is 1 kg/s, find, assuming one-dimensional isentropic flow, the Mach number, temperature, and velocity of the air at the discharge plane.

**8.30.** A small jet aircraft designed to cruise at a Mach number of 1.5 has an intake diffuser with a fixed area ratio. Find the ideal area ratio for this diffuser and the Mach number to which the aircraft must be taken in order to swallow the normal shock wave if the diffuser has this ideal area ratio.

**8.31.** A fixed supersonic convergent–divergent diffuser is designed to operate at a Mach number of 1.7. To what Mach number would the inlet have to be accelerated in order to swallow the shock during start-up?

**8.32.** A jet aircraft is designed to fly at a Mach number of 1.9. It is fitted with a variable-area diffuser. If the diffuser just "swallows" the shock wave at the design Mach number and the throat area is then reduced to give a "shockless" flow, find the percentage reduction in diffuser throat area that is required.

**8.33.** A small jet aircraft designed to cruise at a Mach number of 3 has an intake diffuser with a variable area ratio. Find the ratio of the throat area under these cruise conditions to the throat area required when the aircraft is flying at a Mach number of 1.5. Assume the diffuser intake area does not change.

**8.34.** A convergent–divergent supersonic diffuser is to be used at Mach 3.0. The diffuser is to use a variable throat area so as to swallow the starting shock. What percentage increase in throat area will be necessary?

**8.35.** A wind-tunnel, designed for a test section Mach number of 4, is fitted with a variable area diffuser. Find the ratio of the diffuser throat area when operating under ideal running conditions to the diffuser throat area during starting when there is a shock wave in the working section, the Mach number ahead of this shock being 4.

**8.36.** Air flows from a tank in which the pressure is kept at 750 kPa and the temperature is kept at 30°C through a converging nozzle which discharges the air to the atmosphere. If the throat area of this nozzle is $0.6 \, cm^2$, find the rate at which the air is discharged from the tank in kg/s.

**8.37.** In transonic wind-tunnel testing, the small area decrease caused by placing the model in the test section, i.e., by the model blockage, can cause relatively large changes in the flow in the test section. To illustrate this effect, consider a tunnel that has an empty test section Mach number of 1.08. The test section has an area of $1 \, m^2$ and the stagnation temperature of the air flowing through the test section is 25°C. If a model with a cross-sectional area of $0.005 \, m^2$ is placed in this test section, find the percentage change in test section velocity. Assume one-dimensional isentropic flow.

**8.38.** Consider one-dimensional isentropic flow through a convergent–divergent nozzle that has a throat area of $10 \, cm^2$. The pressure at the throat is 310 kPa and the flow goes from subsonic to supersonic velocities in the nozzle. Find the pressures and Mach numbers at points in the nozzle upstream and downstream of the throat where the nozzle cross-sectional area is $29 \, cm^2$.

**8.39.** A moving piston forces air from a well-insulated 15 cm diameter pipe through a convergent nozzle fitted to the end of the pipe. The nozzle has an exit diameter of 4 mm and the air is discharged to the atmosphere. If the force on the piston is 3700 N and the air temperature is 30°C, estimate the velocity on the nozzle exit plane, the piston velocity, and the mass flow rate at which the air is discharged from the nozzle.

**8.40.** A convergent–divergent nozzle with an exit area to throat area ratio of 3 is supplied with air from a reservoir in which the pressure is 350 kPa. The air from the nozzle is discharged into another large reservoir. It is found that the flow leaving the nozzle exit is directed inwards at an angle of 4° to the nozzle centerline. The velocity on the nozzle exit plane is supersonic. What is the pressure in the second reservoir?

**8.41.** A convergent–divergent nozzle has an exit area to throat area ratio of 4. It is supplied with air from a large reservoir in which the pressure is kept at 500 kPa and it discharges into another large reservoir in which the pressure is kept at 10 kPa. Expansion waves form at the exit edges of the nozzle causing the discharge flow to be directed outwards. Find the angle that the edge of the discharge flow makes to the axis of the nozzle.

**8.42.** A small meteorite punches a 3 cm diameter hole in the skin of an orbiting space laboratory. The pressure and temperature in the laboratory are 80 kPa and 20°C respectively. Estimate the initial rate at which air flows out of the laboratory. State the assumptions you make in arriving at your estimate.

**8.43.** A jet engine is running on a test bed. The stagnation pressure and stagnation temperature just upstream of the convergent nozzle fitted to the rear of this engine are found to be 700 kPa and 700°C respectively. The exit diameter of the nozzle is 0.5 m. If the test is being run at an ambient pressure of 101 kPa, find the mass flow rate through the engine, the jet exit velocity and the thrust that is developed by the engine. Assume the gases have the properties of air.

**8.44.** Air flows through a convergent–divergent nozzle from a large reservoir in which the pressure is 300 kPa and the temperature is 100°C. The nozzle has a throat area of 1 cm$^2$ and an exit area of 4 cm$^2$. The nozzle discharges into another large reservoir in which the pressure can be varied. For what range of back pressures will the mass flow rate through the nozzle be constant and what will the mass flow rate be under these circumstances?

**8.45.** Air flows through a convergent–divergent nozzle, the flow becoming supersonic in the divergent section of the nozzle. A normal shock wave occurs in the divergent section. If the static pressure behind this shock wave is equal to the static pressure at the throat of the nozzle, find the ratio of the nozzle area at which the shock wave occurs to the nozzle throat area.

**8.46.** A nozzle is designed to expand air from a chamber in which the pressure and temperature are 800 kPa and 40°C respectively to a Mach number of 2.5. The throat area of this nozzle is to be 0.05 m$^2$. Find:

1. The exit area of the nozzle
2. The mass flow rate through the nozzle when operating at design conditions
3. The back pressure at which there will be a normal shock wave on the exit plane of the nozzle

4. The range of back pressures over which there will be a normal shock wave in the nozzle
5. The range of back pressures over which oblique shock waves will occur outside the nozzle
6. The range of back pressures over which expansion waves will occur outside the nozzle

# CHAPTER 9

# Adiabatic Flow in a Duct with Friction

## 9.1
## INTRODUCTION

In the discussion of flow through ducts given in the preceding chapters, it was assumed, in almost all cases, that the effects of viscosity were negligible. This is often an adequate assumption when dealing with flow through nozzles or short ducts. For long ducts, however, the effects of viscosity, i.e., the effects of fluid friction at the walls, can in fact be dominant. This is illustrated in Fig. 9.1 which shows typical Mach number and pressure variations in a constant area duct with and without friction. In incompressible flow through a duct of constant cross-sectional area, the friction only affects the pressure, which drops in the direction of flow. The velocity in such a situation remains constant along the duct. In compressible flow, however, friction effects all of the flow variables, i.e., the changes in pressure cause changes in density which lead to changes in velocity.

In some cases, the effects of viscosity may be negligible over part of the flow but then be very important in other parts of the flow. This is illustrated in Fig. 9.2.

In the present chapter, consideration will be given to the effects of viscosity on steady gas flows through ducts under such conditions that compressibility effects are important. In the analyses given in the present chapter it will be assumed that the flow is adiabatic, i.e., that the duct is well insulated (a discussion of the effects of friction in the presence of heat exchange is given in the next two chapters). Attention will, in the present chapter, mainly be restricted to flow in a constant area duct although a brief discussion of the effects of area change will be given at the end of the chapter.

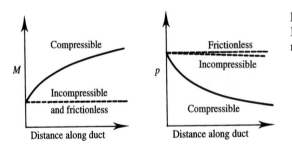

**FIGURE 9.1**
Effect of friction on Mach number in a duct.

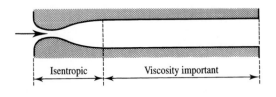

**FIGURE 9.2**
Effect of friction in different portions of flow.

## 9.2
## FLOW IN A CONSTANT AREA DUCT

Attention will here be given to the effects of wall friction on adiabatic flow through a duct whose cross-sectional area does not change. This type of flow, i.e., compressible adiabatic flow in a constant area duct with frictional effects, is known as "Fanno" flow.

Consider the momentum balance for the small portion of the duct shown in Fig. 9.3. This gives, since steady flow is being considered:

Net pressure force − Force due to wall shear stress

$$= \text{Mass flow rate} \times (\text{Velocity out} - \text{Velocity in})$$

i.e., since the force due to the wall shear stress is equal to the product of the shear stress and the surface area of the portion of the duct being considered:

$$pA - (p + dp)A - \tau_w \, (\text{perimeter}) \, dx = \rho V A (V + dV - V)$$

where $\tau_w$ is the wall shear stress. Therefore:

$$-dp - \tau_w \frac{P}{A} dx = \rho V \, dV \tag{9.1}$$

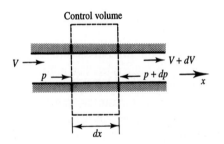

Control volume

**FIGURE 9.3**
Control volume used in analysis of frictional flow in a duct.

where $P$ is the perimeter of the duct and $A$ is its cross-sectional area. In the case of a circular duct, $P = \pi D$ and $A = \pi D^2/4$ so $P/A = 4/D$. For this reason, even for non-circular ducts, the ratio $P/A$ is usually expressed in terms of an equivalent diameter, called the "hydraulic diameter," $D_H$, defined by:

$$D_H = \frac{4 \, (\text{Area})}{\text{Perimeter}} = \frac{4A}{P} \tag{9.2}$$

Dividing eq. (9.1) by $\rho V^2$ then gives:

$$-\frac{dp}{\rho V^2} - \frac{\tau_w}{\rho V^2} \frac{P}{A} dx = \frac{dV}{V} \tag{9.3}$$

Next consider continuity of mass. This gives:

$$\rho A V = \text{constant}$$

But, in the present case since flow in a constant area duct is being considered, this reduces to:

$$\rho V = \text{constant}$$

Hence, for the portion of the duct being considered:

$$\rho V = (\rho + d\rho)(V + dV)$$

From this it follows that to first order of accuracy, i.e., by assuming that the term $d\rho \times dV$ is negligible because $d\rho$, $dV$, etc. are very small:

$$\rho \, dV + V \, d\rho = 0 \tag{9.4}$$

Dividing this equation through by $\rho V$ then gives:

$$\frac{dV}{V} + \frac{d\rho}{\rho} = 0 \tag{9.5}$$

Next, consider the energy equation:

$$c_p T + \frac{V^2}{2} = \text{constant} = c_p(T + dT) + \frac{(V + dV)^2}{2}$$

which gives to first order of accuracy:

$$c_p \, dT + V \, dV = 0 \tag{9.6}$$

Also, the equation of state gives:

$$\frac{p}{\rho} = RT, \quad \text{i.e.,} \quad p = \rho RT$$

and $\quad \dfrac{p + dp}{\rho + d\rho} = R(T + dT), \quad$ i.e., $\quad p + dp = (\rho + d\rho)R(T + dT)$

Hence to first order of accuracy:

$$dp = RT \, d\rho + \rho R \, dT \tag{9.7}$$

Further, since:

$$M^2 = \frac{V^2}{a^2} = \frac{V^2}{\gamma RT}$$

and

$$(M + dM)^2 = \frac{(V + dV)^2}{(a + da)^2} = \frac{(V + dV)^2}{\gamma R(T + dT)}$$

i.e.,

$$M^2 \left(1 + \frac{dM}{M}\right)^2 = \frac{V^2(1 + dV/V)^2}{\gamma RT(1 + dT/T)}$$

i.e., to first order of accuracy:

$$M^2 \left(1 + 2\frac{dM}{M}\right) = \frac{V^2}{\gamma RT} \left(1 + 2\frac{dV}{V}\right) \left(1 - \frac{dT}{T}\right)$$

i.e.,

$$M^2 \left(1 + 2\frac{dM}{M}\right) = \frac{V^2}{\gamma RT} \left(1 + 2\frac{dV}{V} - \frac{dT}{T}\right)$$

Hence, it follows that to first order of accuracy:

$$\frac{dM}{M} = \frac{dV}{V} - \frac{dT}{2T} \tag{9.8}$$

The above equations represent a set of five equations in five unknowns $dM$, $dV$, $dp$, $dT$, and $d\rho$ which can be solved to give expressions for $dM$, $dV$, etc. as illustrated below.

Equation (9.6) gives:

$$\frac{dT}{T} + \frac{V}{c_p T}\frac{dV}{} = 0 \tag{9.9}$$

But since:

$$M^2 = \frac{V^2}{\gamma RT}$$

and

$$(M + dM)^2 = \frac{(V + dV)^2}{\gamma R(T + dT)}$$

it follows that:

$$2M \, dM = \frac{2V \, dV}{\gamma RT} - \frac{V^2}{\gamma RT^2} dT$$

which can be rearranged to give:

$$V \, dV = \gamma RTM \, dM + \frac{V^2}{2}\frac{dT}{T}$$

Hence, eq. (9.9) becomes:

$$\frac{dT}{T} + \frac{\gamma R}{c_p} M \, dM + \frac{V^2}{2c_p T}\frac{dT}{T} = 0 \tag{9.10}$$

Now, recalling that:

$$R = c_p - c_v, \quad \text{i.e.,} \quad c_p = \gamma R/(\gamma - 1)$$

eq. (9.10) becomes:

$$\frac{dT}{T} + (\gamma - 1)M\,dM + \frac{\gamma - 1}{2}M^2\frac{dT}{T} = 0$$

which can be rearranged to give:

$$\frac{dT}{T} = -\frac{(\gamma - 1)M^2}{[1 + (\gamma - 1)M^2/2]}\frac{dM}{M} \tag{9.11}$$

Now eq. (9.7) gives, on dividing through by $p = \rho RT$,

$$\frac{dp}{p} = \frac{d\rho}{\rho} + \frac{dT}{T} \tag{9.12}$$

Eliminating $d\rho/\rho$ between this equation and eq. (9.5) then gives:

$$\frac{dp}{p} = -\frac{dV}{V} + \frac{dT}{T} \tag{9.13}$$

Hence, since eq. (9.8) gives:

$$\frac{dV}{V} = \frac{dM}{M} + \frac{dT}{2T} \tag{9.14}$$

eq. (9.13) gives:

$$\frac{dp}{p} = -\frac{dM}{M} + \frac{dT}{2T} \tag{9.15}$$

Using eq. (9.11) then gives:

$$\frac{dp}{p} = -\left[1 + \frac{(\gamma - 1)M^2/2}{1 + (\gamma - 1)M^2/2}\right]\frac{dM}{M} \tag{9.16}$$

Next, it is noted that eq. (9.3) gives by using $a^2 = \gamma p/\rho$:

$$\frac{dV}{V} + \frac{1}{\gamma M^2}\frac{dp}{p} + \frac{\tau_w}{\rho V^2}\frac{P}{A}dx = 0 \tag{9.17}$$

Substituting for $dV/V$ from eq. (9.8) allows this equation to be written as:

$$\frac{dM}{M} + \frac{1}{2}\frac{dT}{T} + \frac{1}{\gamma M^2}\frac{dp}{p} - \frac{\tau_w}{\rho V^2}\frac{P}{A}dx = 0 \tag{9.18}$$

Substituting eqs. (9.11) and (9.16) into this equation then gives:

$$\frac{dM}{M} - \frac{(\gamma - 1)M^2/2}{[1 + (\gamma - 1)M^2/2]}\frac{dM}{M} - \frac{1 + (\gamma - 1)M^2}{[1 + (\gamma - 1)M^2/2]}\frac{dM}{M} + \frac{\tau_w}{\rho V^2}\frac{P}{A}dx = 0$$

This is easily rearranged to give:

$$\frac{dM}{M} = \frac{\gamma M^2[1 + (\gamma - 1)/2M^2]}{(1 - M^2)}\left[\frac{\tau_w}{\rho V^2}\frac{P}{A}dx\right] \tag{9.19}$$

Substituting this equation back into eq. (9.11) then gives:

$$\frac{dT}{T} = -\frac{\gamma(\gamma-1)M^4}{(1-M^2)}\left[\frac{\tau_w}{\rho V^2}\frac{P}{A}\,dx\right] \tag{9.20}$$

Similarly, substituting eq. (9.19) into eq. (9.16) gives:

$$\frac{dp}{p} = -\frac{\gamma M^2[1+(\gamma-1)M^2]}{(1-M^2)}\left[\frac{\tau_w}{\rho V^2}\frac{P}{A}\,dx\right] \tag{9.21}$$

Since the wall shear stress, the velocity and the Mach number are always positive, eq. (9.19) indicates that the sign of $dM$ depends on the sign of $(1-M^2)$. This equation, therefore, shows that if the Mach number is less than 1, friction causes the Mach number to increase, whereas if the Mach number is greater than 1, friction causes the Mach number to decrease. Viscosity, therefore, always causes the Mach number to tend towards 1. Since once a Mach number of 1 is attained, changes in the downstream conditions cannot affect the upstream flow, it follows that "choking" can occur as a result of friction.

In the same way, eqs. (9.20) and (9.21) show that if $M$ is less than 1, $dT$ and $dp$ are negative whereas if $M$ is greater than 1, $dT$ and $dp$ are positive.

Lastly, consider the entropy change caused by friction. Since:

$$ds = c_p\frac{dT}{T} - R\frac{dp}{p}$$

i.e., since:

$$\frac{ds}{c_p} = \frac{dT}{T} - \frac{\gamma-1}{\gamma}\frac{dp}{p}$$

it follows, using eqs. (9.20) and (9.21), that:

$$\frac{ds}{c_p} = (\gamma-1)M^2\left[\frac{\tau_w}{\rho V^2}\frac{P}{A}\,dx\right] \tag{9.22}$$

This shows that, as is required, the entropy always increases.

The above relations, therefore, together indicate that in constant area compressible duct flow with friction the flow variables change as indicated in Table 9.1. These results are also summarized in Fig. 9.4.

The above equations contain the wall shear stress, $\tau_w$. This is expressed in terms of a dimensionless wall shear stress, $f$, which is defined by:

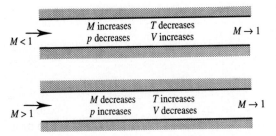

$M<1$   M increases   T decreases   $M\rightarrow 1$
p decreases   V increases

$M>1$   M decreases   T increases   $M\rightarrow 1$
p increases   V decreases

FIGURE 9.4
Effect of friction on flow variables in an adiabatic flow in a constant area duct.

**TABLE 9.1**
**Changes in flow variables produced by friction**

|       | $dM$ | $dV$ | $dP$ | $dT$ | $ds$ |
|-------|:----:|:----:|:----:|:----:|:----:|
| $M < 1$ |  +   |  +   |  −   |  −   |  +   |
| $M > 1$ |  −   |  −   |  +   |  +   |  +   |

+ means a quantity is increasing, − means it is decreasing.

$$f = \frac{\tau_w}{\frac{1}{2}\rho V^2} \tag{9.23}$$

The dimensionless wall shear stress, $f$, is termed the "Fanning friction factor." Another friction factor, $f_D$, termed the "Darcy friction factor," is commonly used in the analysis of incompressible fluid flows in ducts. The two friction factors are related by:

$$f = \frac{f_D}{4}$$

In general:

$$f = \text{function } (Re,\ \epsilon/D_H,\ M)$$

where $Re$ is the Reynolds number based on the mean velocity and the hydraulic diameter and $\epsilon$ is a measure of the mean height of the wall roughness. The effect of $M$ has, however, been found to be small. Therefore, for most purposes, it is adequate to assume that:

$$f = \text{function } (Re,\ \epsilon/D_H)$$

Thus, $f$ will be given by the same equations or charts that apply in low speed duct flow. A Moody chart that can be used to give the Darcy friction factor is given in Fig. 9.5.

Alternatively, some easy to use expressions for the friction factor have been proposed, e.g.:

Laminar flow:

$$f = \frac{16}{Re}$$

Turbulent flow:

$$f = 0.0625 \left/ \left[\log\left(\frac{\epsilon}{3.7D_H} + \frac{5.74}{Re^{0.9}}\right)\right]^2 \right.$$

Transition from laminar to turbulent flow can in most cases be assumed to occur at a Reynolds number of about 2300. Most compressible gas flows in ducts will involve turbulent flow.

The value of the wall roughness $\epsilon$ depends on the type of material from which the duct is made. For drawn tubing it has a value of approximately 0.0015 mm and for commercial steel it has a value of approximately 0.045 mm.

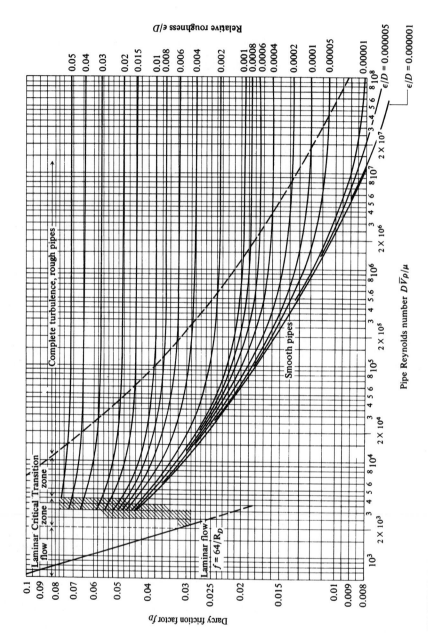

**FIGURE 9.5**
Moody chart.

In terms of $f$, eq. (9.19) can be written:

$$\frac{4f\,dx}{D_H} = \frac{2(1-M^2)}{\gamma M^2(1+\frac{1}{2}(\gamma-1)M^2)}\frac{dM}{M} \tag{9.24}$$

where, for convenience, the "hydraulic diameter," which was discussed before and which is defined by $D_H = 4A/P$, has been introduced.

Integration of eq. (9.24) then gives:

$$\int_0^l \frac{4f\,dx}{D_H} = \int_{M_1}^{M_2} \frac{2(1-M^2)}{\gamma M^2(1+\frac{1}{2}(\gamma-1)M^2)}\frac{dM}{M}$$

i.e.,

$$\frac{4\bar{f}}{D_H}l = \int_{M_1}^{M_2} \frac{2(1-M^2)}{\gamma M^2(1+\frac{1}{2}(\gamma-1)M^2)}\frac{dM}{M} \tag{9.25}$$

where $\bar{f}$ is the "mean friction factor" over the length, $l$, of the duct. As will be discussed below, the changes in the friction factor along the duct are usually small and $f$ can often be assumed to be constant. Hence, in what follows because the variations in $f$ are small, $f$ will be used in place of $\bar{f}$.

The integral in the above equation can be evaluated by applying the method of partial fractions. This gives:

$$\frac{4f}{D_H}l = \frac{1}{\gamma}\left(\frac{1}{M_1^2}-\frac{1}{M_2^2}\right) + \frac{\gamma+1}{2\gamma}\ln\frac{M_1^2(1+\frac{1}{2}(\gamma-1)M_2^2)}{M_2^2(1+\frac{1}{2}(\gamma-1)M_1^2)} \tag{9.26}$$

This equation allows the change in Mach number over a given length of duct to be found. In order to describe the results given by this equation, it is convenient to select a reference value for $M_2$. Since, when friction is important, the Mach number always tends to 1, $M_2$ is conventionally set equal to 1 and the length of duct required to give this value of $M$, which is usually given the symbol $l^*$ or $l_{\max}$, is introduced, i.e., in presenting the values given by this equation, the duct length required to give a Mach number of 1, $l^*$, is introduced. With $M_2$ set equal to 1, the above equation gives:

$$\frac{4f}{D_H}l^* = \left(\frac{1-M^2}{\gamma M^2}\right) + \frac{\gamma+1}{2\gamma}\ln\frac{(\gamma+1)M^2}{2(1+\frac{1}{2}(\gamma-1)M^2)} \tag{9.27}$$

Similarly, integrating eq. (9.16) between a point in the duct at which the pressure is $p$ and the real or imaginary point in the duct at which $M = 1$ and at which $p = p^*$ gives:

$$\int_p^{p^*}\frac{dp}{p} = \int_M^1\left[1 + \frac{(\gamma-1)M^2/2}{(1+\frac{1}{2}(\gamma-1)M^2)}\right]\frac{dM}{M}$$

which gives:

$$\frac{p}{p^*} = \frac{1}{M}\left[\frac{(\gamma+1)/2}{1+(\gamma+1)M^2/2}\right]^{1/2} \tag{9.28}$$

Similarly, integrating eq. (9.11) gives:

$$\int_T^{T^*} \frac{dT}{T} = -\int_M^1 \frac{(\gamma - 1)M}{1 + (\gamma - 1)M^2/2} dM$$

From this it follows that:

$$\frac{T}{T^*} = \frac{(\gamma + 1)/2}{1 + (\gamma - 1)M^2/2} \tag{9.29}$$

where, in these equations, $T^*$ is the temperature at the real or imaginary point at which the Mach number is 1.

Using these results and the definitions of the stagnation pressure and temperature, relations for $p_0/p_0^*$ and $T_0/T_0^*$ can be found. For example, since:

$$\frac{p_0}{p} = \left[ 1 + \frac{\gamma - 1}{2} M^2 \right]^{\frac{\gamma}{\gamma - 1}}$$

it follows that:

$$\frac{p_0}{p_0^*} = \frac{p}{p^*} \left[ \frac{1 + (\gamma - 1)M^2/2}{(\gamma + 1)/2} \right]^{\frac{\gamma}{\gamma - 1}}$$

Hence, using eq. (9.28) gives:

$$\frac{p_0}{p_0^*} = \frac{1}{M} \left[ \frac{1 + (\gamma - 1)M^2/2}{(\gamma + 1)/2} \right]^{\frac{\gamma + 1}{2(\gamma - 1)}} \tag{9.30}$$

The above equations have the form:

$$\frac{4fl^*}{D} = \text{function} (M)$$

$$\frac{p}{p^*} = \text{function} (M)$$

$$\frac{T}{T^*} = \text{function} (M)$$

etc.

Values of these functions are available in graphs and in tables (see Appendix D) and are given by the software that supports this book. A typical chart, this being for $\gamma = 1.4$, is given in Fig. 9.6.

Values of the variables are also easily found by using a programmable calculator. The procedure used is illustrated by the listing in Appendix L of a very simple BASIC program that allows the value of $M$ to be entered and then finds the value of the other quantities.

The way in which the charts, tables, or software are used to find the changes in the values of the flow variables in ducts whose length is less than that required to give a Mach number of 1 is illustrated in the following examples. Basically, the procedure uses the fact that if $l_{12}$ is the actual duct length and if $M_1$ and $M_2$ are the Mach numbers at the beginning and end of the duct then $l_{12} = l_1^* - l_2^*$, similar equations applying for the changes in the other flow variables.

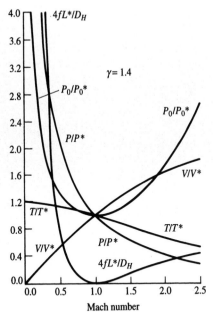

**FIGURE 9.6**
Variation of flow variables in adiabatic constant area duct flow with friction.

**EXAMPLE 9.1**
Air flows in a 5 cm diameter pipe. The air enters at $M = 2.5$ and is to leave at $M = 1.5$. What length of pipe is required? What length of pipe would give $M = 1$ at the exit? Assume that $\bar{f} = 0.002$ and that the flow is adiabatic.

*Solution*

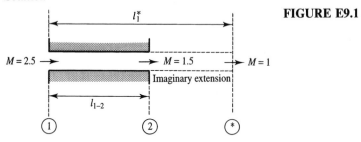

**FIGURE E9.1**

For Mach numbers of 2.5 and 1.5, tables or charts or the software give $4f\,l^*/D = 0.432$ and $0.136$ respectively. Hence, it follows that:

$$\frac{4f\,l_{1-2}}{D} = \frac{4f\,l_1^*}{D} - \frac{4f\,l_2^*}{D} = 0.432 - 0.136$$

from which it follows that:

$$l_{1-2} = \frac{0.296 \times D}{4f} = \frac{0.296 \times 0.05}{4 \times 0.002} = 1.85 \text{ m}$$

Hence the length of the pipe is 1.85 m.

To reach a Mach number of 1 at the exit it is necessary that the length of the pipe be equal to $l_1^*$, i.e., be given by:

$$l_1^* = \frac{4fl_1^*}{D} \frac{D}{4f} = \frac{0.432 \times 0.05}{4 \times 0.002} = 2.7 \text{ m}$$

Hence, to achieve a Mach number of 1 at the exit of the pipe its length must be 2.7 m. The pipe length required to achieve a Mach number of 1 is thus approximately 54 diameters. In supersonic flow, therefore, a Mach number of 1 is achieved with a relatively short duct length.

### EXAMPLE 9.2

Air flows out of a pipe with a diameter of 0.3 m at a rate of 1000 m³ per minute at a pressure and temperature of 150 kPa and 293 K respectively. If the pipe is 50 m long, find assuming that $f = 0.005$, the Mach number at the exit, the inlet pressure, and the inlet temperature.

### *Solution*

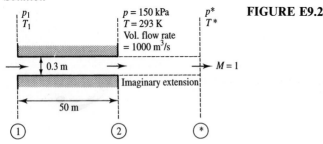

**FIGURE E9.2**

At the exit, i.e., at point 2

$$V_2 = \frac{Q}{A_2} = \frac{(1000/60)}{(\pi/4)(0.3)^2} = 236 \text{ m/s}$$

But:

$$a_2 = \sqrt{\gamma R T_2} = \sqrt{1.4 \times 287 \times 293} = 343 \text{ m/s}$$

Hence:

$$M_2 = \frac{V_2}{a_2} = \frac{236}{343} = 0.688$$

At point 2, therefore, the relations derived above or tables or the software give:

$$\frac{4\bar{f}l_2^*}{D} = 0.228$$

$$\frac{p_2}{p^*} = 1.54, \qquad \frac{T_2}{T^*} = 1.10$$

Now:

$$l_{1-2} = l_1^* - l_2^*$$

hence:

$$\frac{4f l_{1-2}}{D} = \frac{4f l_1^*}{D} - \frac{4f l_2^*}{D}$$

From which it follows that:

$$\frac{(4)(0.005)(50)}{0.3} = \frac{4f l_1^*}{D} - 0.228$$

which gives:

$$\frac{4f l_1^*}{D} = 3.6$$

From the relations derived above or from tables or using the software for $4fl_1^*/D = 3.6$, the following is obtained:

$$M_1 = 0.345, \qquad p_1/p^* = 3.14, \qquad T_1/T^* = 1.17$$

Hence:
$$p_1 = \frac{p_1/p^*}{p_2/p^*} \times p_2 = \frac{3.14}{1.54} 150 = 306 \text{ kPa}$$

and:
$$T_1 = \frac{T_1/T^*}{T_2/T^*} T_2 = \frac{1.17}{1.10}(293) = 312 \text{ K} = 39°\text{C}$$

Therefore the Mach number, pressure, and temperature at the inlet are 0.345, 306 kPa, and 39°C respectively.

## 9.3
## FRICTION FACTOR VARIATIONS

It was assumed in the above analysis that the friction factor could be treated as constant and in the examples given above its value was assumed to be known. Now, as discussed before:

$$f = \text{function } (Re, \epsilon/D_H) \quad \text{where } Re = \frac{\rho V D_H}{\mu}$$

Since the mass flow rate through the duct is given by $\dot{m} = \rho V A$ it follows that the Reynolds number is given by:

$$Re = \frac{\dot{m} D_H}{A\mu} = \frac{4\dot{m}}{P\mu}$$

In a given situation, $\dot{m} = $ constant and $P = $ constant. Therefore:

$$Re \propto \frac{1}{\mu}$$

The value of the coefficient of viscosity, $\mu$, varies somewhat with temperature, its variation for gases being approximately described by:

$$\frac{\mu_1}{\mu_2} = \left(\frac{T_1}{T_2}\right)^n$$

where $T$ is the absolute temperature. The index $n$ in this equation is roughly between 0.5 and 0.8 for common gases. Therefore, the changes in $Re$ along a duct flow are usually small. Furthermore, since the flow is usually turbulent, $f$ is only weakly dependent on $Re$. Hence, it is usually adequate to treat $f$ as constant and to evaluate it using inlet conditions.

### EXAMPLE 9.3
Air flows through a 5 cm diameter stainless steel pipe. The air enters the pipe at a Mach number of 0.3 with a pressure of 150 kPa and a temperature of 40°C. Using the friction factor evaluated at the inlet conditions, determine the Mach number, temperature, and pressure at distances of 4, 8, 12, 16, 18, and 19 m from the inlet.

Using these results, find the actual friction factors at these positions in the pipe and compare these values with the assumed constant value.

**Solution**

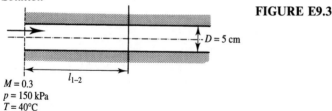

**FIGURE E9.3**

$D = 5$ cm

$l_{1-2}$

$M = 0.3$
$p = 150$ kPa
$T = 40°C$

Considering the inlet conditions:

$$a_1 = \sqrt{\gamma RT_1} = \sqrt{1.4 \times 287 \times 313} = 354.6 \text{ m/s}$$

Hence:

$$V_1 = M_1 \times a_1 = 0.3 \times 354.6 = 106.4 \text{ m/s}$$

The inlet density is given by:

$$\rho_1 = \frac{p_1}{RT_1} = \frac{150\,000}{287 \times 313} = 1.67 \text{ kg/m}^3$$

The kinematic viscosity $\mu$ depends only on temperature over the range of pressures here being considered and is given approximately by using Sutherland's law which for air gives:

$$\mu = 0.000\,017\,16 \left(\frac{T}{273.1}\right)^{1.5} \frac{384.1}{T + 111} \text{ N} \cdot \text{s/m}^2$$

Hence at the inlet conditions:

$$\mu = 0.000\,017\,16 \left(\frac{313}{273.1}\right)^{1.5} \frac{384.1}{313 + 111} = 0.000\,019\,07 \text{ N} \cdot \text{s/m}^2$$

The Reynolds number based on the inlet conditions is, therefore, given by:

$$Re = \frac{\rho VD}{\mu} = \frac{1.67 \times 106.4 \times 0.05}{0.000\,019\,07} = 467\,600$$

The flow is thus turbulent. For stainless steel tubing, the wall roughness $\epsilon$ is 0.0015 mm. Therefore, the roughness ratio $\epsilon/D$ for the pipe being considered is $0.0015/50 = 0.000\,03$. Using the equation given before for the friction factor, i.e.,

$$f = 0.0625 \bigg/ \left[\log\left(\frac{\epsilon}{3.7D_H} + \frac{5.74}{Re^{0.9}}\right)\right]^2$$

then gives:

$$f = 0.0625 \bigg/ \left[\log\left(\frac{0.000\,03}{3.7} + \frac{5.74}{467\,600^{0.9}}\right)\right]^2 = 0.003\,43$$

Now at the inlet Mach number of 0.3 the relations derived above or tables or the software give:

$$\frac{4fl_1^*}{D} = 5.299, \qquad \frac{p_1}{p^*} = 3.619, \qquad \frac{T_1}{T^*} = 1.179$$

At any other point in the flow distance $l_{1-2}$ from the inlet:

$$l_2^* = l_1^* - l_{1-2}$$

hence:

$$\frac{4f l_2^*}{D} = \frac{4f l_1^*}{D} - \frac{4f l_{1-2}}{D}$$

i.e., using the values derived above:

$$\frac{4f l_2^*}{D} = 5.299 - \frac{4 \times 0.003\,43 \times l_{1-2}}{0.05} = 5.299 - 0.2736 \times l_{1-2}$$

Using the values of $4f l_2^*/D$ so found, the values of $M_2$, $p_2/p^*$, and $T_2/T^*$ can be found. Then since $p_1$ and $T_1$ are known, the values of $p_2$ and $T_2$ can be found using:

$$p_2 = \frac{p_2/p^*}{p_1/p^*} \times p_1$$

and:

$$T_2 = \frac{T_2/T^*}{T_1/T^*} T_1$$

Using the value of $T_2$ so found, the speed of sound $a_2$ can be found and the velocity can be found by setting $V_2 = M_2 a_2$. The viscosity and density can also be found using the derived values of temperature and pressure using:

$$\mu_2 = 0.000\,017\,16 \left(\frac{T_2}{273.1}\right)^{1.5} \frac{384.1}{T_2 + 111} \ \text{N} \cdot \text{s/m}^2$$

and:

$$\rho_2 = \frac{p_2}{R T_2}$$

With the density, viscosity, and velocity determined, the Reynolds number can be found and the friction factor can then be calculated using the assumed value of $\epsilon$ and the specified value of $D$:

$$f_2 = 0.0625 \left/ \left[ \log \left( 0.000\,0081 + \frac{5.74}{Re_2^{0.9}} \right) \right] \right.$$

Using this procedure in conjunction with the relations derived above or the tables or the software gives the data shown in Table E9.3 for the prescribed values of $l_{1-2}$. From these results it will be seen that the friction factor varies by little more than 1% and, therefore, that the use of a constant value of $f$ is justified.

**TABLE E9.3**

| $l_{1-2}$ | $4f l_2^*/D$ | $M_2$ | $p_2/p^*$ | $T_2/T^*$ | $p_2$ (kPa) | $T_2$ (K) | $V_2$ (m/s) | $Re_2$ | $f_2$ |
|---|---|---|---|---|---|---|---|---|---|
| 4 | 4.205 | 0.326 | 3.325 | 1.175 | 137.8 | 311.9 | 115.4 | 466 904 | 0.003 43 |
| 8 | 3.110 | 0.361 | 2.996 | 1.170 | 124.2 | 310.6 | 127.5 | 468 380 | 0.003 42 |
| 12 | 2.016 | 0.419 | 2.570 | 1.159 | 106.5 | 307.7 | 147.3 | 471 944 | 0.003 42 |
| 16 | 0.921 | 0.518 | 2.060 | 1.139 | 85.38 | 302.4 | 180.5 | 478 098 | 0.003 41 |
| 18 | 0.374 | 0.633 | 1.665 | 1.111 | 69.01 | 295.0 | 217.9 | 487 390 | 0.003 40 |
| 19 | 0.101 | 0.771 | 1.343 | 1.073 | 55.76 | 284.9 | 260.8 | 500 597 | 0.003 39 |

In cases where the hydraulic diameter of the duct or the flow rate through the duct has to be found, the friction factor cannot initially be directly found and an iterative approach has to be used. The inlet Mach number is usually initially guessed and the unknown diameter or flow rate is found. This value is used to find an improved value of the Mach number and the calculation is repeated. This is continued until a converged result is obtained. This procedure is illustrated in the following example. Although sophisticated iteration procedures can be used, a straightforward procedure that involves obtaining solutions for a series of guessed values of the initial Mach number and then deducing the correct initial Mach number from the results so obtained will be used here.

### EXAMPLE 9.4

A worker in a protective suit in a hazardous area is supplied with oxygen at a rate of 0.06 kg/s through an "umbilical cord" that has a length of 8 m. The pressure and temperature of the oxygen at the inlet to the umbilical cord are maintained at 250 kPa and 10°C respectively and the pressure at the outlet end of the cord is to be 50 kPa. Find the diameter of the umbilical cord if it is made from a material that gives an effective wall roughness, $\epsilon$, of 0.005 mm.

### Solution

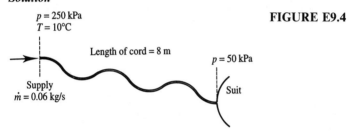

**FIGURE E9.4**

The situation being considered is shown schematically in Fig. E9.4. Since the inlet conditions are kept constant and since the gas involved is oxygen:

$$a_1 = \sqrt{\gamma R T_1} = \sqrt{1.4 \times 260 \times 283} = 320.8 \text{ m/s}$$

$$\rho_1 = \frac{p_1}{R T_1} = \frac{250\,000}{260 \times 283} = 3.4 \text{ kg/m}^3$$

Sutherland's equation will again be used to calculate $\mu$, i.e.,

$$\mu_1 = 0.000\,019\,19 \left(\frac{T}{273.1}\right)^{1.5} \frac{412.1}{T + 139} \text{ N} \cdot \text{s/m}^2$$

$$= 0.000\,019\,19 \left(\frac{283}{273.1}\right)^{1.5} \frac{412.1}{283 + 139} = 0.000\,019\,77 \text{ N} \cdot \text{s/m}^2$$

the constants appropriate to oxygen having been used in Sutherland's equation for $\mu$.

The procedure is then as follows:

1. A value for the inlet Mach number, $M_1$, is guessed
2. The value of $p_1/p^*$ corresponding to this value $M_1$ is obtained using the relations derived above or tables or the software
3. The inlet velocity is calculated using:

$$V_1 = M_1 \times a_1 = M_1 \times 320.8$$

4. The diameter is calculated using:

$$\rho_1 \left( \frac{\pi D^2}{4} \right) V_1 = \dot{m}$$

i.e.,
$$D = \sqrt{\frac{4 \times 0.06}{\pi \times 3.4 \times V_1}}$$

5. The Reynolds number is calculated using:

$$Re = \frac{\rho_1 V_1 D}{\mu_1} = \frac{3.4 V_1 D}{0.000\,0197}$$

6. The friction factor is found using:

$$f = 0.0625 \Big/ \left[ \log\left( \frac{\epsilon}{3.7D} + \frac{5.74}{Re^{0.9}} \right) \right]^2$$

$\epsilon$ being set equal to 0.000 005 m
7. The value of $4fl_1^*/D$ corresponding to the chosen value of $M_1$ is obtained using the relations derived above or tables or the software
8. The value of $4fl_2^*/D$ is found using:

$$\frac{4fl_2^*}{D} = \frac{4fl_1^*}{D} - \frac{4fl_{1-2}}{D}$$

where $l_{1-2}$ is the length of the cord, i.e., 8 m
9. The value of $p_2/p^*$ corresponding to this value of $4fl_2^*/D$ is found using the relations derived above or tables or the software
10. The value of $p_2$ is then found using:

$$p_2 = \frac{p_2/p^*}{p_1/p^*} \times p_1 = \frac{p_2/p^*}{p_1/p^*} \times 250 \text{ kPa}$$

11. The value of $p_2$ so obtained is compared to the required value of 50 kPa.

Of course, if too large a value of $M_1$ is selected, a negative value of $4fl_2^*/D$ is obtained indicating that the assumed inlet Mach number is not possible. A typical set of results is shown in Table E9.4. From these results, it can be deduced that an umbilical cord with a diameter of 0.0169 m (16.9 mm) will give the correct discharge pressure of 50 kPa. This diameter corresponds to an inlet Mach number of 0.242.

**TABLE E9.4**

| $M_1$ | $V_1$ (m/s) | $D$ | $Re$ | $f$ | $4fl_{1-2}/D$ | $l_{1-2}$ | $4fl_1/D$ * | $4fl_2/D$ * | $p_2$ (kPa) |
|-------|-------------|------|------|------|---------------|-----------|-------------|-------------|-------------|
| 0.1 | 32.08 | 0.0265 | 146 009 | 0.004 19 | 5.065 | 66.92 | 61.86 | 10.54 | 240.77 |
| 0.2 | 64.16 | 0.0187 | 206 488 | 0.003 93 | 6.717 | 14.53 | 7.816 | 3.58 | 81.78 |
| 0.3 | 96.24 | 0.0153 | 252 895 | 0.003 79 | 7.936 | 5.299 | −2.64 | — | — |
| 0.24 | 76.99 | 0.0171 | 226 196 | 0.003 87 | 7.239 | 9.387 | 2.148 | 2.27 | 51.86 |
| 0.25 | 80.20 | 0.0167 | 230 860 | 0.003 85 | 7.362 | 8.483 | 1.121 | 1.90 | 43.40 |
| 0.26 | 83.41 | 0.0164 | 235 432 | 0.003 84 | 7.482 | 7.688 | 0.207 | 1.36 | 31.44 |
| 0.27 | 86.62 | 0.0161 | 239 917 | 0.003 83 | 7.599 | 6.983 | −0.62 | — | — |

## 9.4
## THE FANNO LINE

The Fanno line has, in the past, been used extensively in describing the changes that occur in an adiabatic flow in a duct with friction, such flow, as noted before, being called Fanno flow. The Fanno line shows the flow process on a $T$–$s$ or $h$–$s$ diagram, it being noted that, since the flow of a perfect gas is being considered, $h = c_p T$ and $c_p$ is a constant. Now, as noted before:

$$\frac{ds}{c_p} = \frac{dT}{T} - \frac{(\gamma - 1)}{\gamma}\frac{dp}{p}$$

Hence, since eq. (9.13) gives:

$$\frac{dp}{p} = \frac{dT}{T} - \frac{dV}{V}$$

and since eq. (9.9) gives:

$$\frac{dT}{T} = -\frac{V\,dV}{c_p T}$$

it follows that:

$$\frac{ds}{c_p} = \frac{dT}{T} - \frac{(\gamma - 1)}{\gamma}\frac{dT}{T}\left(1 + \frac{c_p T}{V^2}\right) \tag{9.31}$$

But the energy equation gives, since adiabatic flow is being considered:

$$V^2 = 2c_p(T_0 - T)$$

Combining this result with eq. (9.31) allows the variation of $s$ with $T$ for a given $T_0$ to be found, since the above two equations together give:

$$\frac{ds}{c_p} = \frac{dT}{T} - \frac{(\gamma - 1)}{\gamma}\frac{dT}{T}\left(1 + \frac{T}{2(T_0 - T)}\right)$$

i.e.,

$$\frac{ds}{c_p} = \frac{1}{\gamma}\frac{dT}{T} - \frac{\gamma - 1}{2\gamma}\frac{dT}{T_0 - T}$$

This equation can be integrated to give the variation of $s$ with $T$. To do this, some arbitrary temperature $T_1$ at which the entropy is taken to have an arbitrary datum value of $s_1$ (usually taken as 0) is introduced. Integration of the above equation then gives for a given $T_0$:

$$\frac{s - s_1}{c_p} = \frac{1}{\gamma}\int_{T_1}^{T}\frac{dT}{T} - \frac{\gamma - 1}{2\gamma}\int_{T_1}^{T}\frac{dT}{T_0 - T}$$

Carrying out the integration then gives:

$$\frac{s - s_1}{c_p} = \ln\left[\left(\frac{T}{T_1}\right)^{\frac{1}{\gamma}}\left(\frac{T_0 - T}{T_0 - T_1}\right)^{\frac{\gamma - 1}{2\gamma}}\right] \tag{9.32}$$

This then allows the variation of $s$ with $T$ for a given $T_0$ and arbitrarily selected $T_1$ to be found. The variation resembles that shown in Fig. 9.7.

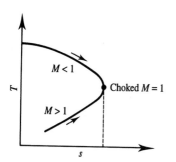

**FIGURE 9.7**
The Fanno line.

The line describing the variation of $s$ with $T$ is called a "Fanno line" and adiabatic flow in a constant area duct with significant friction effects is, as mentioned before, termed "Fanno flow." The Fanno line shows all possible combinations of entropy and temperature that can exist in an adiabatic flow in a constant area duct at a given stagnation temperature, the stagnation temperature, of course, remaining constant in an adiabatic flow. The maximum entropy point on the Fanno line is, for the reasons previously discussed, the point at which the Mach number is 1. The upper portion of the curve, which is associated with higher values of $T$, applies in subsonic flow whereas the lower portion of the curve, which is associated with lower values of $T$, applies in supersonic flow. Since, as discussed before, the entropy always increases, this again shows how the Mach number always moves towards 1.

**EXAMPLE 9.5**
Consider an adiabatic air flow in a constant area duct. If the stagnation temperature in the flow is 750 K and if entropy is arbitrarily taken as being equal to zero when the temperature is 0°C, plot the variation of entropy with temperature in the flow. Find the temperature at which the Mach number is 1 and show this on the plot.

**Solution**
Here $T_0 = 750$ K, $T_1 = 273$ K, and $s_1 = 0$ and, since air flow is being considered, $c_p = 1006$ J/kg·K and $\gamma = 1.4$. Equation (9.32) therefore gives:

$$\frac{s}{1006} = \ln\left[\left(\frac{T}{273}\right)^{\frac{1}{1.4}}\left(\frac{750 - T}{750 - 273}\right)^{\frac{1.4-1}{2.8}}\right]$$

For any chosen value of $T$ this allows the value of $s$ to be found. Since $T$ must be less than 750 K, i.e., less than 477°C, results will be obtained for temperatures between 0 and 470°C. Some values of $s$ and $T$ given by the above equation are shown in Table E9.5.

It has been shown several times that:

$$\frac{T_0}{T} = 1 + \frac{\gamma - 1}{2}M^2$$

Hence, in the present case when the Mach number is equal to 1:

$$\frac{750}{T} = 1 + \frac{1.4 - 1}{2}, \quad \text{i.e.,} \quad T = \frac{750}{1.2} = 625 \text{ K}$$

**TABLE E9.5**

| $T$ (°C) | $T$ (K) | $s$ (J/kg·K) |
|---|---|---|
| 0 | 273 | 0 |
| 100 | 373 | 82.7 |
| 200 | 473 | 137.6 |
| 250 | 523 | 156.5 |
| 300 | 573 | 169.5 |
| 325 | 598 | 173.3 |
| 350 | 623 | 175.0 |
| 375 | 648 | 173.5 |
| 400 | 673 | 167.8 |
| 450 | 723 | 124.7 |
| 470 | 743 | 49.0 |

Therefore, the temperature is 352°C when $M = 1$. This, as will be seen from the above table, corresponds to the maximum entropy temperature.

The variation of entropy with temperature is plotted in Fig. E9.5.

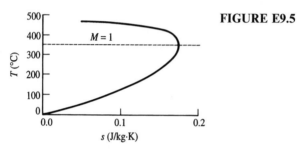

**FIGURE E9.5**

## 9.5
## FRICTIONAL FLOW IN A DUCT PRECEDED BY AN ISENTROPIC NOZZLE

Consider the case where a constant area duct is supplied with gas which flows into the duct through a nozzle from a large chamber. The duct discharges into another large chamber. As discussed before, it is usually adequate to assume that friction effects are negligible in the nozzle. This is because the nozzle is usually relatively short and because the flow is accelerating through the nozzle. To illustrate what happens in this type of flow, it will be assumed that the conditions in the first large chamber, i.e., the stagnation conditions upstream of the nozzle, are kept constant and that the back pressure in the large chamber into which the duct discharges is varied.

The case where the duct is preceded by a converging nozzle will first be considered. The flow situation considered in this case is shown in Fig. 9.8.

When the back pressure $p_b$ is equal to the supply chamber pressure there is, of course, no flow through the duct. As the back pressure is decreased, the mass flow rate through the duct and the Mach number at the duct exit increase, the exit plane pressure $p_e$ being equal to the back pressure $p_b$. This continues until $M_e$ reaches a value of 1. Further decreases in the back pressure

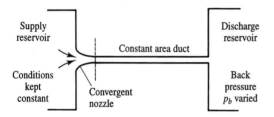

**FIGURE 9.8**
Convergent nozzle flow situation.

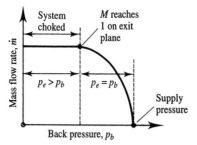

**FIGURE 9.9**
Variation of mass flow rate for convergent nozzle flow situation.

have no effect on the flow in the duct, the adjustment from $p_e$ to $p_b$ in this situation taking place through expansion waves outside the duct. The mass flow rate is thus limited, i.e., the flow is "choked," as a result of friction. The variation of the mass flow rate with back pressure is illustrated in Fig. 9.9. Since once the flow is choked the Mach number always has a value of 1 at the exit of the duct, if the length of the duct is changed, the Mach number at the discharge of the nozzle and the mass flow rate will change.

Next, consider the case where the constant area duct is preceded by a convergent–divergent nozzle. Again when the back pressure $p_b$ is equal to the supply chamber pressure there is, of course, no flow through the duct. As the back pressure is then initially decreased, the Mach number increases at the nozzle throat but then decreases again in the divergent portion of the nozzle, the flow remaining subsonic throughout the nozzle. The Mach number then increases in the constant area duct as a result of friction. As the back pressure is further decreased, one possibility is that the Mach number will reach a value of 1.0 at the duct exit, i.e., that the flow will choke at the end of the duct as a result of friction and that the flow will remain subsonic in the nozzle no matter how low the back pressure gets. This situation is exactly the same as that which occurs with a convergent nozzle, i.e., the same as that discussed above. A more likely situation is, however, that the system will be sized so that as the back pressure is decreased the Mach number at the nozzle throat will reach a value of 1.0 before the Mach number at the duct exit reaches a value of 1, i.e., the nozzle will choke before choking occurs in the duct. Once the Mach number has reached a value of 1.0 at the nozzle throat, a region of supersonic flow develops downstream of the throat with further decrease in back pressure. The region of supersonic flow is terminated by a normal shock wave. Because there is a significant region of subsonic flow near

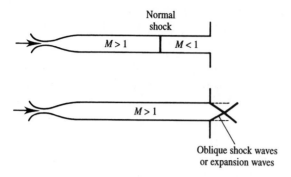

**FIGURE 9.10**
Variation of flow pattern with back pressure for convergent–divergent nozzle flow situation.

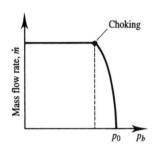

**FIGURE 9.11**
Variation of mass flow rate for convergent–divergent nozzle flow situation.

the wall, the effects of this shock wave can be spread out over a significant length of the duct in small ducts. However, the flow can usually be adequately analyzed by assuming that a conventional normal shock wave occurs. As the back pressure is further decreased, the shock wave moves towards the duct exit, eventually reaching the exit. Once the back pressure has been reduced to a value at which the shock wave is at the exit plane of the duct, further reductions in the back pressure have no effect on the flow in the duct system, the adjustment from the exit plane pressure to the back pressure taking place through oblique shock waves or expansion waves outside the duct. The changes in the flow and the mass flow rate with back pressure are illustrated in Figs. 9.10 and 9.11.

The effects of back pressure on the flow are illustrated in the following two examples.

**EXAMPLE 9.6**
Air, in a large reservoir, at a pressure of 200 kPa and a temperature of 30°C is expanded through a convergent nozzle. The air then flows down a pipe with a diameter of 25 mm. If the Mach number at the exit of the nozzle, i.e., at the inlet to pipe is 0.2 and the Mach number at the end of this pipe is 0.8, find, assuming that the flow in the nozzle is isentropic and the flow in the pipe adiabatic, the length of the pipe and the pressure at the exit of the pipe. Also find the pressure in the reservoir into which the pipe discharges at which choking first occurs and the inlet Mach number under these conditions. Also plot a graph of pipe inlet and outlet Mach number against discharge reservoir pressure. It can be assumed in all calculations that $f = 0.005$.

**Solution**

Considering the flow across the nozzle, isentropic relations or tables or the software give for $M_1 = 0.2$, $p_0/p = 1.0283$.

Next, consider the flow in the pipe. At the inlet for $M_1 = 0.2$, the relations given earlier in this chapter, or tables, or the software give $4f l_1^*/D = 14.533$ and $p_1/p^* = 5.4555$. Similarly, at the outlet for $M_2 = 0.8$, the relations given earlier in this chapter, or tables or the software give $4f l_2^*/D = 0.072\,29$ and $p_2/p^* = 1.2892$. But, as discussed before:

$$\frac{4f l_{1-2}}{D} = \frac{4f l_1^*}{D} - \frac{4f l_2^*}{D}$$

Therefore:

$$\frac{4f l_{1-2}}{D} = 14.533 - 0.072\,29 = 14.4607$$

which gives:

$$l_{1-2} = \frac{14.4607 \times 0.025}{4 \times 0.005} = 18.1 \text{ m}$$

Therefore the length of the pipe is 18.1 m. Also, using the results given above:

$$p_2 = \frac{p_2}{p^*}\frac{p^*}{p_1}\frac{p_1}{p_0}p_0 = \frac{1.2892}{5.4555} \times 0.9725 \times 200 = 45.96 \text{ kPa}$$

Therefore the pressure at the exit is 45.96 kPa.

Choking occurs when the exit Mach number reaches a value of 1. Now, when $M_2 = 1$, $4f l_2^*/D = 0.0$, and $p_2/p^* = 1$. Hence, since:

$$\frac{4f l_2^*}{D} = \frac{4f l_1^*}{D} - \frac{4f l_{1-2}}{D} = \frac{4f l_1^*}{D} - 14.4607$$

It follows that when choking occurs, $4f l_1^*/D = 14.4607$. The relations given earlier in this chapter, or tables or the software indicate that this value of $4f l_1^*/D$ corresponds to $M_1 = 0.2005$ and $p_1/p^* = 5.455$. Isentropic relations or tables or the software give for this value of $M_1$, $p_0/p_1 = 1.028$. Hence, since when the exit Mach number is 1, $p_2 = p^*$, it follows that:

$$p_2 = \frac{p^*}{p_1}\frac{p_1}{p_0}p_0 = \frac{200}{1.028 \times 5.455} = 35.66 \text{ kPa}$$

It follows that an exit Mach number of 1 will be obtained when

$$p_2 = p^* = 35.66 \text{ kPa}$$

To determine the Mach number variation with back pressure it is recalled that, because the friction factor is assumed to remain the same, the following still applies:

$$\frac{4f l_{1-2}}{D} = 14.4607$$

Hence:

$$\frac{4f l_2^*}{D} = \frac{4f l_1^*}{D} - \frac{4f l_{1-2}}{D} = \frac{4f l_1^*}{D} - 14.4607$$

Also as before:

$$p_2 = \frac{p_2}{p^*}\frac{p^*}{p_1}\frac{p_1}{p_0}p_0 = \frac{p_2}{p^*}\frac{p^*}{p_1}\frac{p_1}{p_0} \times 200$$

For any chosen value of the nozzle exit Mach number $M_1$ the values of $p_0/p_1$, $4f l_1^*/D$, and $p_1/p^*$ can be found using isentropic and Fanno relations or tables or

**TABLE E9.6**

| $M_1$ | $p_0/p_1$ | $4fl_1/D$ | $p_1/p^*$ | $4fl_2/D$ | $p_2/p^*$ | $p_2$ (kPa) | $M_2$ |
|------|-----------|-----------|-----------|-----------|-----------|-------------|-------|
| 0.00 | 1.000 | — | — | — | — | 200.0 | 0.000 |
| 0.12 | 1.010 | 45.408 | 9.1156 | 30.801 | 7.60 | 165.1 | 0.144 |
| 0.15 | 1.016 | 27.932 | 7.2866 | 13.471 | 5.27 | 142.4 | 0.205 |
| 0.17 | 1.020 | 21.115 | 6.4253 | 6.6543 | 3.95 | 120.5 | 0.275 |
| 0.19 | 1.026 | 16.375 | 5.7448 | 1.9143 | 2.54 | 86.19 | 0.423 |
| 0.20 | 1.028 | 14.533 | 5.4555 | 0.0723 | 1.29 | 45.96 | 0.800 |
| 0.2005 | 1.028 | 14.607 | 5.4554 | 0.0000 | 1.00 | 35.66 | 1.000 |

the software. The value of $4fl_2^*/D$ can then be found, and then using Fanno relations or tables or the software the values of $M_2$ and $p_2/p^*$ can be obtained.

A typical set of results obtained using the above procedure is shown in Table E9.6.

The variations of $M_1$ and $M_2$ with $p_2$ is shown in Fig. E9.6.

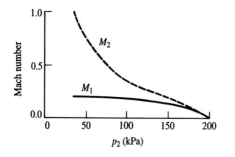

**FIGURE E9.6**

**EXAMPLE 9.7**

Air flows steadily from a large reservoir through a convergent–divergent nozzle into a 0.3 m diameter pipe with a length of 3.5 m. The conditions in the reservoir are such that the Mach number and the pressure at the inlet to the pipe are 2 and 101.3 kPa respectively. The average friction factor, $f$, for the flow in the pipe is estimated to be 0.005.

1. If no shocks occur, find $M$ and $p$ at the exit of the pipe
2. If there is a normal shock at the exit of the pipe, find the back pressure in the chamber into which the pipe is discharging
3. Find the back pressure in the chamber into which the pipe is discharging when there is a shock halfway down the pipe

**Solution**

1. The flow situation considered is shown in Fig. E9.7a.

For $M_1 = 2$, the relations for Fanno flow derived above or tables or the software give:

$$\frac{4fl_1^*}{D} = 0.3049, \qquad \frac{p_1}{p^*} = 0.4083$$

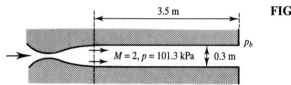

**FIGURE E9.7a**

Now, at the exit of the pipe:

$$\frac{4f\,l_2^*}{D} = \frac{4f\,l_1^*}{D} - \frac{4f\,l_{1-2}}{D}$$

$$= 0.3049 - \frac{4 \times 0.005}{0.3} \times 3.5 = 0.0717$$

For this value of $4f\,l_2^*/D$ the relations for Fanno flow or tables or the software give $M_2 = 1.32$ and $p_2/p^* = 0.715$. Hence:

$$p_2 = \frac{p_2}{p^*} \times \frac{p^*}{p_1} \times p_1 = \frac{0.715}{0.4083} \times 101.3 = 177.3 \text{ kPa}$$

Hence, when there are no shocks in the pipe, the pressure and the Mach number at the exit of the pipe are 177.3 kPa and 1.32 respectively.

2. When there is a shock wave on the exit plane of the pipe, the pressure and the Mach number just upstream of the shock will be 177.3 kPa and 1.32 respectively. Now for $M = 1.32$, normal shock relations or tables or software give:

$$\frac{p_2}{p_1} = 1.866$$

Hence the back pressure is given by:

$$p_{back} = 1.866 \times 177.3 = 331 \text{ kPa}$$

3. The situation being considered is shown in Fig. E9.7b. At a point halfway down the pipe:

$$\frac{4f\,l_2^*}{D} = \frac{4f\,l_1^*}{D} - \frac{4f(l_1^* - l_2^*)}{D}$$

$$= 0.3049 - \frac{4 \times 0.005}{0.3}\left(\frac{3.5}{2}\right) = 0.19$$

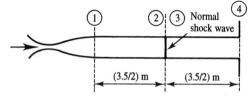

**FIGURE E9.7b**

For this value of $4f\,l_2^*/D$ the relations for Fanno flow or tables or the software give $M_2 = 1.65$ and $p_2/p^* = 0.534$. Hence:

$$p_2 = \frac{p_2}{p^*} \times \frac{p^*}{p_1} \times p_1 = \frac{0.534}{0.408} \times 101.3 = 132.6 \text{ kPa}$$

Now for $M = 1.65$, normal shock relations or tables or software give:

$$\frac{p_3}{p_2} = 3.010, \qquad M_3 = 0.654$$

Lastly, consider the subsonic flow downstream of the shock. Now for $M = 0.655$, the relations for Fanno flow derived above or tables or the software give:

$$\frac{4f\,l_3^*}{D} = 0.31, \qquad \frac{p_3}{p^*} = 1.6$$

Hence, at the exit plane:

$$\frac{4f\,l_4^*}{D} = \frac{4f\,l_3^*}{D} - \frac{4f(l_3 - l_4)}{D}$$

$$= 0.31 - \left(\frac{4 \times .005}{0.3}\right)\left(\frac{3.5}{2}\right) = 0.194$$

For this value of $4f l^*/D$ the relations for Fanno flow or tables or the software give $M_4 = 0.71$ and $p_4/p^* = 1.48$. Hence:

$$p_4 = p_{\text{back}} = \frac{p_4}{p^*} \times \frac{p^*}{p_3} \times \frac{p_3}{p_2} \times p_2$$

$$= \frac{1.48}{1.6} \times 3.01 \times 132.6 = 369 \text{ kPa}$$

Hence when the back pressure is 369 kPa there is a shock halfway down the pipe.

The effects of the pressure on the flow are summarized in Fig. E9.7c.

**FIGURE E9.7c**

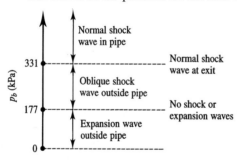

For this value of $4f l^*/D$ the relations for Fanno flow or tables or the

## 9.6
## THE EFFECTS OF FRICTION ON VARIABLE AREA FLOW

In the discussion of the effects of friction on flow in a duct given above, it was assumed that the area of the duct remained constant. In some cases, however, flows occur in which the effects of both friction and area changes are important. This will be considered in the present section.

Consider the flow through the control volume shown in Fig. 9.12. For this control volume, the continuity equation gives, as before:

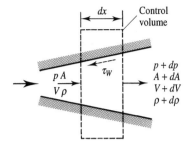

**FIGURE 9.12**
Control volume used in analysis of variable area flow.

$$\frac{d\rho}{\rho} + \frac{dA}{A} + \frac{dV}{V} = 0 \tag{9.33}$$

while the momentum equation gives for this control volume:

$$pA + \left(p + \frac{dp}{2}\right) dA - (p + dp)(A + dA) - \tau_w A_s \cos\theta = \rho A V \, dV \tag{9.34}$$

The second term on the left hand side represents the force due to the pressure on the curved wall. The fourth term on the left hand side is the shear stress term, $A_s$ being the wall area, i.e., the surface area, and $\theta$ the angle of inclination of the wall. Neglecting second order terms and expressing the shear stress in terms of the friction factor then gives:

$$dp + \frac{\rho V^2}{2} \frac{4f \, dx}{D_H} + \rho V \, dV = 0 \tag{9.35}$$

which in turn can be rewritten as:

$$\frac{dp}{p} + \frac{\gamma M^2}{2} \frac{4f \, dx}{D_H} + \gamma M^2 \frac{dV}{V} = 0 \tag{9.36}$$

Next, using the perfect gas law and combining it with the continuity equation result gives the following:

$$\frac{dp}{p} = \frac{d\rho}{\rho} + \frac{dT}{T} = -\frac{dA}{A} - \frac{dV}{V} + \frac{dT}{T} \tag{9.37}$$

Substituting this into the momentum equation then gives:

$$-\frac{dA}{A} - \frac{dV}{V} + \frac{dT}{T} + \frac{\gamma M^2}{2} \frac{4f \, dx}{D_H} + \gamma M^2 \frac{dV}{V} = 0 \tag{9.38}$$

But, as discussed before, the definition of the Mach number gives:

$$\frac{dV}{V} = \frac{1}{2} \frac{dT}{T} + \frac{dM}{M} \tag{9.39}$$

Substituting this into eq. (9.38) then gives:

$$-\frac{dA}{A} - \frac{dM}{M} + \frac{1}{2} \frac{dT}{T} + \frac{\gamma M^2}{2} \frac{4f \, dx}{D_H} + \frac{\gamma M^2}{2} \frac{dT}{T} + \gamma M^2 \frac{dM}{M} = 0 \tag{9.40}$$

Now, if the derivation of eq. (9.11) is considered it will be seen that it is based only on the use of the energy equation and the definition of the Mach number. Equation (9.11) therefore applies whether or not the area is changing. Hence, here again:

$$\frac{dT}{T} = -\frac{(\gamma-1)M^2}{[1+(\gamma-1)M^2/2]}\frac{dM}{M} \tag{9.41}$$

But eq. (9.40) can be written as:

$$-\frac{dA}{A} + (\gamma M^2 - 1)\frac{dM}{M} + \frac{(\gamma M^2 + 1)}{2}\frac{dT}{T} + \frac{\gamma M^2}{2}\frac{4f\,dx}{D_H} = 0 \tag{9.42}$$

Hence, using eq. (9.41) and rearranging gives:

$$-\frac{dA}{A} - \frac{(M^2-1)}{[1+(\gamma-1)M^2/2]}\frac{dM}{M} + \frac{\gamma M^2}{2}\frac{4f\,dx}{D_H} = 0$$

which can be rearranged to give:

$$\frac{dM}{M} = -\left[\frac{1+(\gamma-1)M^2/2}{1-M^2}\right]\frac{dA}{A} + \left[\frac{1+(\gamma-1)M^2/2}{1-M^2}\right]\frac{\gamma M^2}{2}\frac{4f\,dx}{D_H} \tag{9.43}$$

The above equation indicates that the changes in Mach number along the duct are due to both area changes (the first term on the right hand side) and to viscosity effects (the second term on the right hand side). When the area is constant the first term on the right hand side is zero and the Fanno line equation is obtained. When friction effects are negligible, the second term on the right hand side can be neglected and the same equation as previously used in the analysis of isentropic variable area flow is obtained. It should also be noted that it is possible for the Mach number to remain constant even when the area is changing provided the viscosity effects just balance the area change effects. In such a case:

$$\frac{dA}{dx} = \frac{2f\gamma M^2 A}{D_H}$$

Hence, in all situations, if $M$ stays constant, the area, $A$, must increase.

It will be noted that both terms in eq. (9.43) contain the term $1 - M^2$ in the denominator and the equation therefore has a singularity when $M = 1$. Writing eq. (9.43) as:

$$(1-M^2)\frac{dM}{M} = -\left(1+\frac{\gamma-1}{2}M^2\right)\frac{dA}{A} + \left(1+\frac{\gamma-1}{2}M^2\right)\frac{\gamma M^2}{2}\frac{4f\,dx}{D_H} \tag{9.44}$$

If the Mach number in the flow is to change from a subsonic to a supersonic value or from a supersonic to a subsonic value without a shock wave in the flow, i.e., if a sonic point occurs in the flow, eq. (9.44) shows that this sonic point must occur at the point in the flow where, since $M = 1$ at this point:

$$-\left(1 + \frac{\gamma - 1}{2}\right)\frac{dA}{A} + \left(1 + \frac{\gamma - 1}{2}\right)\frac{\gamma\,4f\;dx}{2\;D_H} = 0$$

i.e.,
$$-\frac{\gamma + 1}{2}\frac{dA}{A} + \frac{\gamma(\gamma + 1)}{4}\frac{4f\;dx}{D_H} = 0 \qquad (9.45)$$

If friction effects are negligible, i.e., if the second term in this equation is negligible, the sonic point will occur when $dA$ is zero as discussed before. However, when friction effects are significant, because the second term in eq. (9.45) always has a positive value, the sonic point must occur at a point in the flow at which $dA$ is positive, i.e., in a divergent portion of the flow.

In most cases it is necessary to integrate eq. (9.40) by numerical methods simultaneously with the equations for $p$, $T$, and $\rho$. If the changes in Reynolds number due to area changes are large, it may be necessary to allow for the changes in the friction factor in this calculation. This procedure will be discussed in a more general way in Chapter 11. The basic aspects of the procedure are, however, illustrated in the following example.

**EXAMPLE 9.8**
Consider steady, adiabatic air flow through a duct that has a circular cross-sectional shape. The inlet diameter of the duct is 6 cm and the duct has a length of 1.5 m. The air enters the duct with a Mach number of 0.6, a pressure of 150 kPa and a temperature of 40°C. Determine the Mach number variation along the duct if:

1. Its cross-sectional area increases linearly by 50%
2. Its cross-sectional area decreases linearly by 5%
3. Its cross-sectional area remains constant.

Assume that the friction factor can be treated as a constant and so can be evaluated using the inlet conditions. Assume that the duct walls are smooth.

**Solution**
The cross-sectional area of the duct will be given by:

$$A = A_i\left(1 + K\frac{x}{L}\right)$$

where $x$ is the distance along the duct from the inlet, $L$ is the length of the duct, i.e., 3 m, and $A_i$ is the inlet area. $K$ is a constant that is equal to $+0.5$ in Case 1, $-0.05$ in Case 2, and 0 in Case 3. With the area specified in this way, the diameter at any $x$ can be found using:

$$D = \sqrt{\frac{4}{\pi}A}$$

At 40°C, the coefficient of viscosity $\mu$ of air is $0.000\,019\,07\ \mathrm{N \cdot s/m^2}$ and the speed of sound in the air is 354.6 m/s which means that the inlet velocity is 212.8 m/s. At a pressure of 150 kPa and a temperature of 40°C, the density of the air $\rho$ is $1.67\ \mathrm{kg/m^3}$. (The above properties of air were derived in Example 9.3.) The Reynolds number based on the inlet conditions is, therefore, given by:

$$Re = \frac{\rho V D}{\mu} = \frac{1.67 \times 212.8 \times 0.06}{0.000\,019\,07} = 1\,122\,240$$

The flow is thus turbulent. Since the wall roughness $\epsilon$ is assumed to be 0, using the equation given before for the friction factor, i.e.,

$$f = 0.0625 \Big/ \left[\log\left(\frac{\epsilon}{3.7D_H} + \frac{5.74}{Re^{0.9}}\right)\right]^2$$

then gives:

$$f = 0.0625 \Big/ \left[\log\left(0 + \frac{5.74}{1\,122\,240^{0.9}}\right)\right]^2 = 0.002\,85$$

Therefore, since $\gamma$ is equal to 1.4, eq. (9.43) gives for the present situation:

$$dM = -\frac{1+0.2M^2}{1-M^2} M \frac{dA}{A} + \frac{1+0.2M^2}{1-M^2} 0.7M^3 \frac{0.0114\,dx}{D}$$

This equation will be integrated to give the variation of $M$ with $x$. This will be done using a very simple procedure. The duct will be broken down into a series of $N$ sections of equal length, i.e.,

$$dx = \frac{L}{N}$$

and since the duct area changes linearly:

$$dA = \frac{A_{\text{exit}} - A_i}{N}$$

The procedure involves:

1. Set $M = M_i$, $A = A_i$, and $D = D_i$ and calculate the change in the Mach number across the first section using:

$$dM = M\left[-\frac{1+0.2M^2}{1-M^2}\frac{dA}{A} + \frac{1+0.2M^2}{1-M^2}0.7M^3\frac{0.0114\,dx}{D}\right]$$

2. Find the Mach number at the end of the first section using $M = M + dM$ and the area at this section using $A = A + dA$. Then find the diameter at this section using:

$$D = \sqrt{\frac{4}{\pi}A}$$

3. Repeat the above sets to find the Mach number at the end of the next step and so on, continuing until the end of the duct is reached.

The above procedure can easily be carried out using a spreadsheet program. Alternatively, a simple BASIC procedure that implements the procedure is listed in Program E9.8:

**Program E9.8**

```
10  CLS
20  PRINT" "
30  PRINT"VARIABLE AREA FLOW WITH FRICTION"
40  PRINT" ":PRINT" "
50  INPUT" NUMBER OF STEPS ";N
60  INPUT" AREA VARIATION COEFF. ";K
70  PRINT" "
```

```
 80   PI = 3.1416
 90   L = 1.5
100   MI = 0.6
110   DI = 0.06
120   AI = PI*DI*DI/4
130   AO = (1 + K)*AI
140   DA = (AO-AI)/N
150   DX = L/N
160   PRINT " X"," A"," M"
170   PRINT" "
180   X = 0.0
190   M = MI
200   A = AI
210   PRINT USING "####.########";X,A,M
220   FOR I = 1 TO N
230   D = (4*A/PI)^0.5
240   F = (1 + 0.2*M*M)*M/(1-M*M)
250   DM = F*(-DA/A + 0.0114*0.7*M*M*DX/D)
260   M = M + DM
270   A = A + DA
280   X = X + DX
290   PRINT USING "####.########";X,A,M
300   NEXT I
310   STOP
320   END
```

Results obtained using this program are listed in Table E9.8. These were obtained using $N = 60$.

**TABLE E9.8**

| x (m) | M (Case 1) | M (Case 2) | M (Case 3) |
|-------|-----------|-----------|-----------|
| 0.0 | 0.6 | 0.6 | 0.6 |
| 0.2 | 0.546 | 0.617 | 0.610 |
| 0.4 | 0.506 | 0.637 | 0.621 |
| 0.6 | 0.471 | 0.658 | 0.632 |
| 0.8 | 0.441 | 0.683 | 0.645 |
| 1.0 | 0.416 | 0.712 | 0.659 |
| 1.2 | 0.393 | 0.748 | 0.674 |
| 1.4 | 0.373 | 0.796 | 0.691 |
| 1.5 | 0.364 | 0.829 | 0.700 |

# 9.7
# CONCLUDING REMARKS

It has been shown that viscous effects, i.e., the effects of wall friction, in a constant area, adiabatic duct flow cause the Mach number to tend towards 1. As a result, it is possible for choking to occur due to viscous effects. An expression for the duct length required to produce choking was derived and relations for the ratio of local values of the flow variables to the values that these variables would have at the real or hypothetical point at which $M = 1$ were derived. The wall shear stress was expressed in terms of the friction factor and the way in which the friction factor is obtained was discussed. The effects of viscous friction on adiabatic flow in a variable area duct were also briefly discussed and it was shown that in such flows the sonic point is not, in general, at a point of minimum area.

## PROBLEMS

**9.1.** Air flows through a duct with a constant cross-sectional area. The pressure, temperature, and Mach number at the inlet to the duct are 180 kPa, 30°C, and 0.25 respectively. If the Mach number at the exit of the duct has risen to 0.75 as a result of friction, determine the pressure, temperature, and velocity at the exit. Assume that the flow is adiabatic.

**9.2.** Air flows through a well insulated 4 in diameter pipe at the rate of 500 lbm/min. The pressure drops from 50 psia at the inlet to the pipe to a value of 40 psia at the exit. If the temperature at the inlet is 200°F, find the Mach number at the exit of the pipe.

**9.3.** Consider compressible flow through a long, well-insulated duct. At the inlet to the duct the pressure and temperature are 100 kPa and 30°C respectively. Assuming that the flow is adiabatic and that the pipe is sufficiently long to ensure that the flow is choked at the exit, find the velocity and temperature at the pipe exit.

**9.4.** Air flows through a 5 cm diameter pipe. Measurements indicate that at the inlet to the pipe the velocity is 70 m/s, the temperature is 80°C, and the pressure 1 MPa. Find the temperature, the pressure, and the Mach number at the exit to the pipe if the pipe is 25 m long. Assume that the flow is adiabatic and that the mean friction factor is 0.005.

**9.5.** Air flows down a pipe with a diameter of 0.15 m. At the inlet to the pipe, the Mach number is 0.1, the pressure is 70 kPa, and the temperature is 35°C. If the flow can be assumed to be adiabatic and if the mean friction factor is 0.005, determine the length of the pipe if the Mach number at the exit is 0.6. Also find the pressure and temperature at the exit to the pipe.

**9.6.** Air flows from a large tank through a well-insulated 12 mm diameter pipe. If the air enters the pipe at a Mach number of 0.2 and leaves at a Mach number of 0.6, find the length of the pipe. Assume a mean friction factor of 0.005. How much longer must the pipe be if the exit Mach number is 1? If the pipe is 75 cm longer than this latter value and if the same conditions exist in the supply chamber, what reduction in the flow rate will occur?

**9.7.** Air flows from a large tank, in which the pressure and temperature are 100 kPa and 30°C respectively, through a 1.6 m long pipe with a diameter of 2.5 cm. The pipe is connected to a short convergent nozzle with an exit diameter of 2.1 cm. The air from this nozzle is discharged into a large tank in which the pressure is maintained at 35 kPa. Assuming that the friction factor is equal to 0.002, find the mass flow rate through the system. The flow in the nozzle can be assumed to be isentropic and the pipe can be assumed to be heavily insulated.

**9.8.** Air at an inlet temperature of 60°C flows with a subsonic velocity through an insulated pipe having an inside diameter of 5 cm and a length of 5 m. The pressure at the exit to the pipe is 101 kPa and the flow is choked at the end of the pipe. If the average friction factor is 0.005, determine the inlet and exit Mach numbers, the mass flow rate and the change in temperature and pressure through the pipe.

**9.9.** Hydrogen flows through a 50 mm diameter pipe. The inlet pressure is 400 kPa, the inlet velocity is 300 m/s and the inlet temperature is 30°C. How long is the pipe if the flow is choked at the exit end? Assume a mean friction factor of 0.0058 and that the flow is adiabatic.

**9.10.** Air flows through a 0.15 m × 0.25 m rectangular duct. The Mach number, pressure, and temperature at a certain section of the duct are found to be 2, 75 kPa, and 5°C respectively. Assuming the mean friction factor of 0.006, find the maximum length of duct that can be installed downstream of this section if no shock wave is to occur in the duct. Also find the exit pressure and temperature that will exist with this maximum length of duct.

**9.11.** Air is stored in a tank at a pressure and temperature of 1.6 MPa and 20°C respectively. What is the maximum possible mass rate of flow from the tank through a pipe with a diameter of 1.2 cm and a length of 30 cm? The pipe discharges to the atmosphere and the atmospheric pressure is 101 kPa. The average friction factor can be assumed to be 0.006 and the flow in the pipe can be assumed to be subsonic and adiabatic.

**9.12.** Air flows through a 12 m long pipe which has a diameter of 25 mm. At the inlet to the pipe, the air velocity is 80 m/s, the pressure is 350 kPa and the temperature is 50°C. If the mean friction factor is 0.005, find the velocity, pressure, and temperature at the end of the pipe. Assume the flow to be adiabatic.

**9.13.** Air is expanded from a large reservoir in which the pressure and temperature are 200 kPa and 30°C respectively through a convergent nozzle which gives an exit Mach number of 0.2. The air then flows down a pipe with a diameter of 25 mm, the Mach number at the end of this pipe being 0.8. Assuming that the flow in the nozzle is isentropic and the flow in the pipe adiabatic, find the length of the pipe and the pressure at the exit of the pipe. The friction factor in the pipe can be assumed to be 0.005.

**9.14.** Air flows through a 4 cm diameter pipe. At the inlet to the pipe the stagnation pressure is 150 kPa, the stagnation temperature is 80°C, and the velocity is 120 m/s. If the mean friction factor is 0.006, and if the flow can be assumed to be adiabatic, find the maximum duct length before choking occurs.

**9.15.** Air flows through a 0.5 in diameter pipe at subsonic velocities. The pipe is 20 feet long and the pressure and temperature at the inlet to the pipe are 60 psia and 130°F. The pipe is discharged into a large vessel in which the pressure is kept at 20 psia. If the mean friction factor is assumed to be 0.0055 and if the flow is assumed to be adiabatic, find the mass flow rate through the pipe.

**9.16.** Air flows through a circular pipe at a rate of 8.3 kg/s. The Mach number at the inlet to the pipe is 0.15 and at the exit to the pipe is 0.5. The pressure and temperature at the inlet are 350 kPa and 38°C respectively. Assuming the flow to be adiabatic, and the mean friction factor to be 0.005, find the length and the diameter of the duct and the pressure and temperature at the exit of the duct.

**9.17.** Air is expanded from a large reservoir, in which the pressure and temperature are 250 kPa and 30°C respectively, through a convergent nozzle which gives an exit Mach number of 0.3. The air from the nozzle flows down a pipe having a diameter of 5 cm. The Mach number at the end of this pipe is 0.95. Find the length of the pipe and the pressure at the end of the pipe. If the actual pipe length was only 0.75 of this length, find the Mach number and the pressure that would exist at the end of the pipe. The flow in the nozzle can be assumed to be isentropic and the friction factor in the pipe can be assumed to be 0.005.

**9.18.** Air flows at a steady rate through a pipe with an internal diameter of 26 mm and a length of 15 m. The pressure and temperature in the air at the inlet to the pipe are 140 kPa and 120°C respectively. Assuming that the flow is adiabatic and using an average friction factor for the flow of 0.005, find the maximum possible mass flow rate through the pipe. Also find the temperature and pressure at the exit of the pipe when the Mach number at the exit is equal to 1.

**9.19.** Air flows down a 20 mm diameter pipe which has a length of 0.8 m. If the velocity at the inlet to the pipe is 200 m/s and its temperature is 30°C, find the average friction factor if the flow is choked at the exit to the pipe. Assume the flow to be adiabatic.

**9.20.** Air enters an insulated pipe with a diameter of 7.5 cm at a Mach number of 3.0. As a result of friction, the Mach diameter decreases to a value of 1.5 at the exit to the pipe. If the mean friction factor is equal to 0.002, find the length of the pipe.

**9.21.** A convergent–divergent nozzle supplies air to a well-insulated constant area duct. At the inlet to the duct the Mach number is 2, the pressure is 140 kPa, and the temperature is −100°C. If the Mach number is 1 at the exit to the duct, determine the pressure and temperature at the duct exit.

**9.22.** Air flows down a constant area pipe which has a diameter of 5 cm. The Mach number at the inlet to the pipe is 2 and the inlet pressure and temperature are 80 kPa and 20°C respectively. The flow in the pipe can be assumed to be adiabatic. If the pipe is 0.6 m long and the average friction factor is 0.005, find the Mach number, pressure, and temperature at the exit of the pipe. If, on leaving the pipe, the air flows through a convergent–divergent nozzle which has an exit area that is three times the throat area, and if the air stream leaves the nozzle at a subsonic velocity, find the pressure and the Mach number at the exit of the nozzle if the flow in the nozzle can be assumed to be isentropic.

**9.23.** Air with a stagnation pressure of 600 kPa and a stagnation temperature of 150°C flows through a convergent–divergent nozzle, the Mach number being greater than 1 at the nozzle exit. The throat area of the nozzle is 1 cm². The flow from the nozzle enters a duct which has a constant area of 3 cm². If the flow in the nozzle can be assumed to be isentropic and if the flow in the duct can be assumed to be adiabatic and if the mean friction factor is 0.004, find the temperature and the pressure on the exit plane of the duct.

**9.24.** Air enters a pipe having a diameter of 0.1 m and a length of 1 m with a Mach number of 2 and a pressure of 90 kPa. Assuming the flow to be adiabatic and the mean friction factor to be 0.005, plot a graph of the pressure variation along the length of the duct.

**9.25.** An air stream enters a 2.5 cm diameter pipe with a Mach number of 2.5 and a pressure and temperature of 30 kPa and −15°C respectively. The average friction factor can be assumed to be 0.005. Determine the maximum possible length of tube if there are to be no shock waves in the flow. Also find the values of the pressure and the temperature at the tube exit for this maximum length. Assume the flow to be adiabatic.

**9.26.** Air flows from a large reservoir in which the pressure and temperature are 1 MPa and 30°C respectively through a convergent–divergent nozzle and into a constant area duct. The ratio of the nozzle exit area to its throat area is 3.0 and the length-to-diameter ratio of the duct is 15. Assuming that the flow in the nozzle is isentropic, that the flow in the duct is adiabatic, and that the average friction factor is 0.005, find the back pressure for a normal shock to appear at the exit to the duct.

**9.27.** Air enters a pipe at a Mach number of 2.5, a temperature of 40°C, and a pressure of 70 kPa. The pipe has a diameter of 2.0 cm and the flow can be assumed to be adiabatic. A shock occurs in the pipe at a location where the Mach number is 2. If the Mach number at the exit from the pipe is 0.8 and if the average friction factor is 0.005, find the distance of the shock from the entrance to the pipe and the total length of the pipe. Also find the pressure at the exit of the pipe.

**9.28.** Air with a stagnation pressure of 700 kPa flows through a convergent–divergent nozzle with an exit-to-throat area ratio of 3. The flow in this nozzle can be assumed to be isentropic. The air from the nozzle enters a well-insulated duct with a length-to-diameter ratio of 20. The mean friction factor is 0.002. The air from the duct is discharged into a large reservoir in which the pressure is 100 kPa. Find the Mach numbers at the inlet and exit of the duct.

**9.29.** Air with a stagnation pressure of 300 kPa and a stagnation temperature of 30°C enters a constant area duct at Mach number of 3. The duct has a length-to-diameter ratio of 60. The flow can be assumed to be adiabatic and the average friction factor is 0.0025. If the pressure at the exit to the duct is 50 kPa, determine the Mach number at the exit of the duct and the location of the shock down the duct in diameters. (Hint: Apply an iterative solution using guessed values of the shock wave position.)

**9.30.** Air flows at a steady rate through a 0.08 m diameter 1.5 m long pipe. A convergent–divergent nozzle expands the air to a Mach number of 2.25 and a pressure of 40 kPa at the inlet to the pipe. The air from the pipe is discharged into a large chamber in which the pressure can be varied. Assuming a friction factor of 0.003, find the pressure in this chamber if:

1. There are no shock or expansion waves in the flow
2. There is a normal shock wave on the exit plane of the pipe.

**9.31.** Air is expanded from a large reservoir in which the pressure and temperature are 300 kPa and 20°C respectively through a convergent–divergent nozzle which gives an exit Mach number of 1.5. The air from the nozzle flows down a 5 cm diameter pipe to a reservoir in which the pressure is 140 kPa. A normal shock wave occurs at the end of the pipe. Find the length of the pipe. Discuss what will occur if the pressure in the reservoir at the discharge end of the pipe is decreased. The flow in the nozzle can be assumed to be isentropic and that in the pipe can be assumed to be adiabatic. The friction factor in the pipe can be assumed to be 0.002.

**9.32.** Air at an initial temperature of 45°C is to be transported through a 50 m long, well-insulated pipe. If the mean friction factor in the pipe can be assumed to be 0.025, find the minimum pipe diameter that can be used to carry the flow without choking occurring if the inlet air velocity is 50 m/s, 100 m/s, and 400 m/s.

**9.33.** Air flows from a supersonic nozzle with a throat diameter of 6.5 cm into a 13 cm diameter pipe. The stagnation pressure at the inlet to the pipe is 700 kPa. At distances of 5 diameters and 33 diameters from the inlet to the pipe, the static pressures are measured and found to be 24.5 kPa and 50 kPa, respectively. Determine the Mach numbers at these two sections and the mean friction factor between the two sections. The flow in the nozzle can be assumed to be isentropic and the flow in the pipe can be assumed to be adiabatic.

**9.34.** An apparatus that is used for determining friction factors with air flow through a pipe consists of a reservoir connected to a convergent–divergent nozzle which in turn is connected to the pipe. The nozzle has a throat diameter of 0.6 cm and the diameter of the pipe, which is well insulated, is 0.9 cm. In one experiment, the pressure and temperature in the reservoir are 1.7 MPa and 40°C respectively and the pressure at a distance of 0.15 m from the inlet to the pipe is 340 kPa. Calculate the average friction factor in the pipe. Assume that the flow in the nozzle is isentropic.

**9.35.** Air is expanded from a large reservoir in which the pressure and temperature are 600 kPa and 30°C respectively through a convergent–divergent nozzle which gives an exit Mach number of 2.0. The air from the nozzle flows down a 3.5 cm diameter pipe.

1. If the Mach number at the end of this pipe is 1.2, find the length of the pipe.
2. If the back pressure is changed until a normal shock wave occurs half way between the nozzle exit plane and the exit plane of the pipe, find the back pressure.

The flow in the nozzle can be assumed to be isentropic and the friction factor in the pipe can be assumed to be 0.005.

**9.36.** Air enters a linearly converging duct with a circular cross-section at a Mach number of 0.6. The inlet diameter of the duct is 10 cm and the wall makes an angle of 10° to the axis of the duct. Using numerical integration, produce a plot of Mach number along the duct up to the point where the Mach number reaches a value of 1. The mean friction factor can be assumed to be equal to 0.005 and the flow can be assumed to be adiabatic.

**9.37.** A high speed wind-tunnel is supplied with air from a collection of interconnected compressed air tanks situated outside of the laboratory. The air is delivered from the tanks to the tunnel through a long 100 cm diameter pipe. The pressure at the inlet to this supply pipe is 10 MPa and the air is to be delivered to the tunnel at a pressure of 1 MPa. How long can this supply pipe be if choking is not to occur? Assume adiabatic subsonic one-dimensional flow and a mean friction factor of 0.005.

# CHAPTER 10

# Flow with Heat Addition or Removal

## 10.1
## INTRODUCTION

Most of the previous chapters in this book have been devoted to flows in which the effects of viscosity could be neglected over most of the flow field and to flows that could be assumed to be adiabatic. The effects of viscosity (i.e., fluid friction) on quasi-one-dimensional flows were discussed in Chapter 9. In the present chapter, the effects of either the addition of heat to or the extraction of heat from the flow will be discussed. The heat addition or removal may result, for example, from the heating or cooling of the wall of the duct through which the gas is flowing or from chemical reactions that occur in the flow such as in a combustion chamber or due to evaporation of liquid droplets being carried in the flow.

In the first portion of this chapter, it will be assumed that the effects of viscosity on the flow are negligible compared to the effects of the heat exchange. This is usually an adequate assumption in processes in which relatively large amounts of heat are added to or removed from the flow such as when combustion is occurring. It will also be assumed in this chapter that the gas composition does not change and that the specific heat ratio, $\gamma$, therefore, does not change. This assumption may not be adequate in some combustion systems.

## 10.2
## ONE-DIMENSIONAL FLOW IN A CONSTANT AREA DUCT
## NEGLECTING VISCOSITY

The effects of heat exchange on one-dimensional flow in a constant area duct in which the effects of viscosity are negligible will first be examined. Consider

262

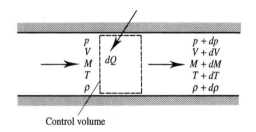

Control volume

**FIGURE 10.1**
Control volume used in analysis of
flow through a duct with heat
exchange.

the flow through the control volume shown in Fig. 10.1. Conservation of mass
gives, as discussed in earlier chapters:

$$\rho A V = \text{constant}$$

But, in the present case since flow in a constant area duct is being considered,
this reduces, as in the previous chapter, to:

$$\rho V = \text{constant}$$

Hence, for the portion of the duct being considered:

$$\rho V = (\rho + d\rho)(V + dV)$$

From this it follows that to first order of accuracy, i.e., by assuming that the
term $d\rho \times dV$ is negligible because $d\rho$, $dV$, etc. are very small:

$$\rho \, dV + V \, d\rho = 0$$

Dividing this equation through by $\rho V$ then gives:

$$\frac{dV}{V} + \frac{d\rho}{\rho} = 0 \tag{10.1}$$

Similarly, conservation of momentum applied to the control volume
shown in Fig. 10.1, gives:

$$pA - (p + dp)A = \rho V A(V + dV - V)$$

i.e.,
$$dp + \rho V \, dV = 0 \tag{10.2}$$

Lastly, conservation of energy gives, since no work is being done on the
system:

$$dQ = \dot{m}\left[c_p(T + dT) + \frac{(V + dV)^2}{2} - c_p T - \frac{V^2}{2}\right] \tag{10.3}$$

It should be noted that heat exchange is, thermodynamically, strictly the
energy transfer that results from temperature differences. In combustion systems, it would be more correct to write the energy equation as:

$$\dot{m}\Delta h_f = \dot{m}\left[c_p(T + dT) + \frac{(V + dV)^2}{2} - c_p T - \frac{V^2}{2}\right]$$

where $\Delta h_f$ is the difference between the heats of formation of the reactants
and the products. However, this can be quite adequately dealt with by treating

$\dot{m}\Delta h_f$ as the "heat produced as the result of the combustion" and there is, then, no reasons in most analyses to consider the physical origin of the "heat."

It is convenient to define $dq$ as the rate of heat addition per unit mass flow rate, i.e., to define:

$$dq = \frac{dQ}{\dot{m}}$$

The conservation of energy equation, eq. (10.3), can then be written as:

$$dq = c_p\, dT + V\, dV \qquad (10.4)$$

Alternatively, consider the definition of the stagnation temperature, $T_0$, given earlier, i.e.,

$$T_0 = T + \frac{V^2}{2c_p}$$

or:

$$T_0 + dT_0 = (T + dT) + \frac{(V + dV)^2}{2c_p}$$

These equations together give to first order accuracy:

$$dT_0 = dT + \frac{V\, dV}{c_p}$$

Hence, it will be seen that eq. (10.4) can be written as:

$$dq = c_p\, dT_0 \qquad (10.5)$$

The above equations can now be used to relate the changes in the flow variables to the rate of heat addition. Equation (10.4) gives:

$$\frac{dq}{dV} = c_p \frac{dT}{dV} + V \qquad (10.6)$$

But, the perfect gas law gives:

$$\frac{p}{\rho} = RT, \quad \text{i.e.,} \quad p = \rho RT$$

and:

$$\frac{p + dp}{\rho + d\rho} = R(T + dT), \quad \text{i.e.,} \quad p + dp = (\rho + d\rho)R(T + dT)$$

which can be used, as shown in earlier chapters, to give:

$$\frac{dp}{p} = \frac{d\rho}{\rho} + \frac{dT}{T} \qquad (10.7)$$

But eq. (10.1) gives:

$$\frac{d\rho}{\rho} = -\frac{dV}{V}$$

and eq. (10.2) gives:

$$-\frac{dp}{p} = -\frac{\rho V\, dV}{p}$$

so eq. (10.7) can be written as:

$$-\frac{\rho V \, dV}{p} = -\frac{dV}{V} + \frac{dT}{T}$$

Noting that the perfect gas law gives $p/\rho = RT$, this equation can be written:

$$-\frac{V \, dV}{RT} = -\frac{dV}{V} + \frac{dT}{T}$$

i.e.,
$$\frac{dT}{dV} = \frac{T}{V} - \frac{V}{R} \qquad\qquad (10.8)$$

Substituting this into eq. (10.6) then gives:

$$\frac{dq}{dV} = c_p \frac{T}{V} - V\left(\frac{c_p}{R} - 1\right)$$

Hence, since $R = c_p - c_v$ so that, as shown before, $R/c_p = (\gamma - 1)/\gamma$, it follows that:

$$\frac{dq}{dV} = c_p \frac{T}{V} - \frac{V}{\gamma - 1} \qquad\qquad (10.9)$$

The first term in this equation decreases with increasing $V$ whereas the second term increases with increasing $V$. At very low velocities, the first term dominates and, therefore, since $T$ and $V$ are positive, $dq/dV$ is positive, i.e., $dV$ has the same sign as $dq$. This means that at low velocities, adding heat increases the velocity while removing heat decreases the velocity. At high velocities, the second term in eq. (10.9) will dominate and $dq/dV$ will be negative. Thus, at high velocities, $dV$ has the opposite sign to $dq$. This means that at high velocities, adding heat decreases the velocity while removing heat increases the velocity. Transition from one form of behavior (i.e., $dq/dV$ being positive) to the other (i.e., $dq/dV$ being negative) will occur when $dq/dV$ is zero. From eq. (10.9) it follows that this will occur when:

$$c_p \frac{T}{V} = \frac{V}{\gamma - 1}$$

i.e., when:
$$V = \sqrt{(\gamma - 1)c_p T} = \sqrt{\gamma RT} = a \qquad\qquad (10.10)$$

This shows that when $M < 1$, $dq/dV$ is positive and when $M > 1$, $dq/dV$ is negative. When $M = 1$, $dq/dV$ is zero. These results are summarized in Table 10.1.

Now it will be recalled that since $V = Ma$ and $V + dV = (M + dM)(a + da) = Ma + M \, da + a \, dM$ it follows that:

$$\frac{dV}{V} = \frac{da}{a} + \frac{dM}{M}$$

TABLE 10.1
**Changes in flow variables produced by heat exchange**

|  | Heat addition | Heat removal |
| --- | --- | --- |
| $M < 1$ | $V$ increases | $V$ decreases |
| $M > 1$ | $V$ decreases | $V$ increases |

Therefore, since as shown above:

$$-\frac{\rho V\,dV}{p} = -\frac{dV}{V} + \frac{dT}{T}$$

which can be written as:

$$-\frac{\gamma V\,dV}{a^2} = -\frac{dV}{V} + \frac{dT}{T}$$

i.e.,

$$(1 - \gamma M^2)\frac{dV}{V} = \frac{dT}{T}$$

it follows that:

$$\frac{(1 + \gamma M^2)}{2}\frac{dV}{V} = \frac{dM}{M}$$

This shows that $dM$ has the same sign as $dV$. This means that adding heat will tend to move $M$ towards 1 whereas removing heat will tend to move $M$ away from 1.

It should also be noted that eq. (10.8) shows that $dT/dV$ will be positive at low velocities and negative at high velocities. $dT/dV$ will be zero when $T/V = V/R$, i.e., when $V^2 = RT$, i.e., when:

$$M = \frac{1}{\sqrt{\gamma}} \tag{10.11}$$

This shows that the maximum temperature in a flow will exist where $M = 1/\sqrt{\gamma}$.

Entropy changes associated with the heat addition or removal will next be considered. Since, as discussed elsewhere:

$$\frac{ds}{c_p} = \frac{dT}{T} - \frac{(\gamma - 1)}{\gamma}\frac{dp}{p} \tag{10.12}$$

it follows using eqs. (10.7) and (10.1) that:

$$\frac{ds}{c_p} = \frac{1}{\gamma}\frac{dT}{T} + \frac{(\gamma - 1)}{\gamma}\frac{dV}{V} \tag{10.13}$$

Substituting eq. (10.8) into this equation and rearranging then gives:

$$\frac{ds}{c_p} = (1 - M^2)\frac{dV}{V} \tag{10.14}$$

But, as shown above $dq/dV$ is positive if $M < 1$ and negative if $M > 1$. Equation (10.14), therefore, shows that heat addition always increases the entropy while heat removal always decreases the entropy. The addition of heat to a flow thus moves the entropy towards a maximum at $M = 1$. This is indicated in Table 10.2.

**TABLE 10.2**
**Changes in entropy produced by heat exchange**

|          | Heat addition | Heat removal |
|----------|---------------|--------------|
| $M < 1$  | $s$ increases | $s$ decreases |
| $M > 1$  | $s$ increases | $s$ decreases |

**EXAMPLE 10.1**
Air flowing through a constant area duct is being heated. The air temperature at a certain section of the duct is 200°C. If, over a short section of the duct, the stagnation temperature increases by 1%, estimate the percentage increases in $V$ and $M$ and the value of $ds/c_p$ for Mach numbers, at the section considered, of 0.4, 0.8, 1.2, and 1.6.

**Solution**
Equation (10.5) gives:

$$\frac{dq}{c_p T_0} = \frac{dT_0}{T_0}$$

But in the present case:

$$\frac{dT_0}{T_0} = 0.01$$

so $dq/c_p T_0 = 0.01$.
    Equation (10.9) gives:

$$\frac{dq/(c_p T_0)}{dV/V} = \frac{T}{T_0} - \frac{V^2}{(\gamma - 1)c_p T_0}$$

i.e.,

$$\frac{0.01}{dV/V} = \frac{T}{T_0} - \frac{V^2}{0.4 \times 1007 T_0}$$

i.e.,

$$\frac{dV}{V} = 0.01 \bigg/ \left(\frac{T}{T_0} - \frac{V^2}{402.8 T_0}\right)$$

But:

$$\frac{T_0}{T} = 1 + \left(\frac{\gamma - 1}{2}\right)M^2 = 1 + 0.2M^2$$

so:

$$T_0 = 473(1 + 0.2M^2)$$

Also:    $V = M\sqrt{\gamma R T} = M\sqrt{1.4 \times 287 \times 473} = 435.9M$ m/s

Hence, for any value of $M$, the values of $V$, $T_0/T$, and $T_0$ can be found. The value of $dV/V$ can then be calculated. The fractional change in the Mach number is then found using:

$$\frac{dM}{M} = \frac{(1 + \gamma M^2)}{2}\frac{dV}{V} = \frac{(1 + 1.4M^2)}{2}\frac{dV}{V}$$

while the entropy change is found using eq. (10.14), i.e., using:

$$\frac{ds}{c_p} = (1 - M^2)\frac{dV}{V}$$

Results obtained using this procedure for the specified values of Mach number are given in Table E10.1.

**TABLE E10.1**

| M | dV/V (%) | dM/M (%) | ds/c_p |
|---|---|---|---|
| 0.4 | 1.23 | 0.75 | 0.63 |
| 0.8 | 3.12 | 2.96 | 1.06 |
| 1.2 | −2.95 | −4.45 | 1.96 |
| 1.6 | −1.17 | −2.42 | 3.02 |

It is again convenient to express the changes that occur in the flow in terms of the Mach number. Consider two points in the flow as shown in Fig. 10.2.

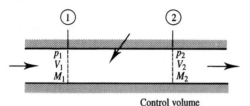

**FIGURE 10.2**
Points in flow considered.

Control volume

The conservation of momentum equation applied to the control volume shown gives:

$$p_1 + \rho_1 V_1^2 = p_2 + \rho_2 V_2^2$$

Since $a^2 = \gamma p/\rho$, this equation can be written as:

$$p_1(1 + \gamma M_1^2) = p_2(1 + \gamma M_2^2)$$

i.e.,
$$\frac{p_2}{p_1} = \frac{1 + \gamma M_1^2}{1 + \gamma M_2^2} \tag{10.15}$$

Now by definition, the stagnation pressure is given by:

$$\frac{p_0}{p} = \left(1 + \frac{\gamma - 1}{2} M^2\right)^{\frac{\gamma}{\gamma - 1}} \tag{10.16}$$

When there is heat addition or removal the stagnation pressure will change. From eqs. (10.15) and (10.16) it follows that:

$$\frac{p_{02}}{p_{01}} = \frac{p_2 \left[1 + (\gamma - 1)M_2^2/2\right]^{\gamma/\gamma-1}}{p_1 \left[1 + (\gamma - 1)M_1^2/2\right]^{\gamma/\gamma-1}}$$

i.e.,
$$\frac{p_{02}}{p_{01}} = \frac{\left[1 + \gamma M_1^2\right] \left[1 + (\gamma - 1)M_2^2/2\right]^{\gamma/\gamma-1}}{\left[1 + \gamma M_2^2\right] \left[1 + (\gamma - 1)M_1^2/2\right]^{\gamma/\gamma-1}} \tag{10.17}$$

Next, it is noted that the continuity equation gives:

$$\frac{\rho_2}{\rho_1} = \frac{V_1}{V_2} \tag{10.18}$$

while the perfect gas law gives:

$$\frac{\rho_2}{\rho_1} = \frac{p_2}{p_1}\frac{T_1}{T_2} \tag{10.19}$$

These two equations together then give:

$$\frac{T_2}{T_1} = \frac{p_2}{p_1}\frac{V_2}{V_1} \tag{10.20}$$

But:

$$\frac{V_2}{V_1} = \frac{M_2 a_2}{M_1 a_1} = \frac{M_2}{M_1}\sqrt{\frac{T_2}{T_1}} \tag{10.21}$$

Substituting this and eq. (10.15) into eq. (10.20) then gives:

$$\frac{T_2}{T_1} = \frac{M_2^2(1+\gamma M_1^2)^2}{M_1^2(1+\gamma M_2^2)^2} \tag{10.22}$$

Substituting this back into eq. (10.21) then gives:

$$\frac{V_2}{V_1} = \frac{M_2^2(1+\gamma M_1^2)}{M_1^2(1+\gamma M_2^2)} \tag{10.23}$$

Also substituting eqs. (10.23) and (10.15) into eq. (10.19) gives:

$$\frac{\rho_2}{\rho_1} = \frac{M_1^2(1+\gamma M_2^2)}{M_2^2(1+\gamma M_1^2)} \tag{10.24}$$

Lastly, it will be recalled that the stagnation temperature is given by:

$$\frac{T_0}{T} = 1 + \frac{(\gamma-1)}{2}M^2$$

from which it follows that:

$$\frac{T_{02}}{T_{01}} = \frac{T_2}{T_1}\frac{[1+(\gamma-1)M_2^2/2]}{[1+(\gamma-1)M_1^2/2]}$$

Substituting eq. (10.22) into this equation then gives:

$$\frac{T_{02}}{T_{01}} = \frac{M_2^2(1+\gamma M_1^2)^2}{M_1^2(1+\gamma M_2^2)^2}\frac{[1+(\gamma-1)M_2^2/2]}{[1+(\gamma-1)M_1^2/2]} \tag{10.25}$$

It should be noted that eq. (10.5) shows that the change in $T_0$ always has the same sign as $q$, i.e., adding heat always increases $T_0$ while removing heat always decreases $T_0$. Further, it was shown earlier that adding heat always moves $M$ towards 1 whereas removing heat always moves $M$ away from 1. Therefore, $T_0$ must have a maximum value when $M = 1$. The variation of $T_0$ with $M$ is, therefore, as illustrated in Fig. 10.3.

For any value of $M_1$, the right hand side of eq. (10.25) has a maximum when $M_2 = 1$, i.e.,

$$\frac{T_{0max}}{T_{01}} = \frac{1}{2(1+\gamma)M_1^2}\frac{(1+\gamma M_1^2)^2}{[1+(\gamma-1)M_1^2/2]} \tag{10.26}$$

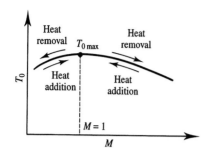

FIGURE 10.3
Variation of stagnation temperature with Mach number.

Lastly, the relation between the entropy change and Mach number will be determined. Since:

$$s_2 - s_1 = c_p \ln \frac{T_2}{T_1} - R \ln \frac{p_2}{p_1}$$

i.e.,

$$\frac{s_2 - s_1}{c_p} = \ln \frac{T_2}{T_1} - \frac{(\gamma - 1)}{\gamma} \ln \frac{p_2}{p_1}$$

i.e.,

$$\frac{s_2 - s_1}{c_p} = \ln \left[ \frac{T_2}{T_1} \left( \frac{p_1}{p_2} \right)^{\frac{(\gamma - 1)}{\gamma}} \right]$$

it follows, using eqs. (10.21) and (10.15), that:

$$\frac{s_2 - s_1}{c_p} = \ln \left[ \frac{M_2^2 (1 + \gamma M_1^2)^2}{M_1^2 (1 + \gamma M_2^2)^2} \left( \frac{1 + \gamma M_2^2}{1 + \gamma M_1^2} \right)^{\frac{(\gamma - 1)}{\gamma}} \right]$$

i.e.,

$$\frac{s_2 - s_1}{c_p} = \ln \left[ \frac{M_2^2}{M_1^2} \left( \frac{1 + \gamma M_1^2}{1 + \gamma M_2^2} \right)^{\frac{(\gamma + 1)}{\gamma}} \right] \tag{10.27}$$

For any specified initial conditions, eqs. (10.15), (10.17), (10.21), (10.23), (10.24), (10.25), and (10.27) allow the downstream conditions to be found for any specified heat addition or removal. To do this, it is recalled that eq. (10.5) gives:

$$q = c_p(T_{02} - T_{01})$$

i.e.,

$$\frac{q}{c_p T_{01}} = \frac{T_{02}}{T_{01}} - 1 \tag{10.28}$$

Hence, using eq. (10.25):

$$\frac{q}{c_p T_{01}} = \frac{M_2^2 [1 + \gamma M_1^2]^2 [1 + (\gamma - 1) M_2^2 / 2]}{M_1^2 [1 + \gamma M_2^2]^2 [1 + (\gamma - 1) M_1^2 / 2]} - 1 \tag{10.29}$$

For any specified values of $M_1$ and $q/c_p T_{01}$, this equation allows $M_2$ to be determined. The equations listed above then allow the changes in all other flow variables to be found. Since $q/c_p T_{01}$, is, by virtue of eq. (10.28), a unique function of $T_{02}/T_{01}$, it is usual to use the stagnation temperature ratio as the

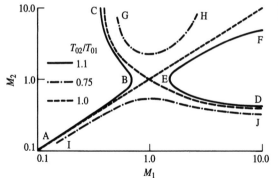

**FIGURE 10.4**
Variation of $M_2$ with $M_1$.

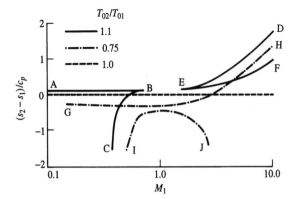

**FIGURE 10.5**
Variation of entropy change with $M_1$.

measure of the heat exchange. There is then a relationship between $M_2$, $M_1$, and $T_{02}/T_{01}$. The form of this relationship is illustrated in Fig. 10.4. When $T_{02}/T_{01} = 1$ the flow is adiabatic and in subsonic flow, in this case, $M_2 = M_1$. However, when the flow is supersonic, i.e., $M_1 > 1$, there are two possible values of $M_2$, i.e., either $M_2 = M_1$ or $M_2 < 1$. These, of course, correspond to the possibility of no normal shock or of a normal shock occurring. Furthermore, for $M_1 < 1$ and $T_{02}/T_{01} > 1$, i.e., for heat addition, it is possible to obtain a solution which involves $M_2 > 1$. Such solutions are, for reasons discussed above, not physically possible.

The variation of entropy change with $M_1$ as given by eqs. (10.27) and (10.29) is shown in Fig. 10.5. The curves in this figure are lettered to correspond to those in Fig. 10.4. As discussed above, if heat is added to the flow, the Mach number tends towards 1 while if heat is extracted from the flow, the Mach number moves away from 1. It is convenient, therefore, to write the above equations in terms of conditions that exist when $M_2 = 1$. In this case $p_2 = p^*$, $T_2 = T^*$, etc. The equations given above can in this case be used to give:

$$\frac{T_0}{T_0^*} = \frac{2(\gamma + 1)M^2[1 + (\gamma - 1)M^2/2]}{(1 + \gamma M^2)^2} \tag{10.30}$$

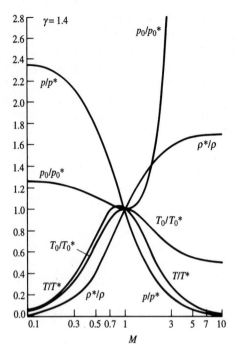

**FIGURE 10.6**
Variation of properties in constant
area duct flow with heat exchange.

$$\frac{T}{T^*} = \frac{(1+\gamma)^2 M^2}{(1+\gamma M^2)^2} \tag{10.31}$$

$$\frac{p}{p^*} = \frac{(1+\gamma)}{(1+\gamma M^2)} \tag{10.32}$$

$$\frac{p_0}{p_0^*} = \left(\frac{1+\gamma}{1+\gamma M^2}\right)\left\{\left(\frac{2}{\gamma+1}\right)\left[1+\frac{(\gamma-1)}{2}M^2\right]\right\}^{\frac{\gamma}{\gamma-1}} \tag{10.33}$$

$$\frac{V}{V^*} = \frac{(1+\gamma)M^2}{(1+\gamma M^2)} \tag{10.34}$$

$$\frac{\rho}{\rho^*} = \frac{(1+\gamma M^2)}{(1+\gamma)M^2} \tag{10.35}$$

$$\frac{s-s^*}{c_p} = \ln\left[M^2\left(\frac{1+\gamma}{1+\gamma M^2}\right)^{\frac{\gamma+1}{\gamma}}\right] \tag{10.36}$$

Thus the values of the quantities $T_0/T_0^*$, $T/T^*$, etc., can be determined
for any value of $M$. The variations are shown in Fig. 10.6. Values of these
quantities are also listed in the table given in Appendix E. Values of these
quantities are also given by the software provided. Values can also easily be
found using a programmable calculator, the simple BASIC program listed in
Appendix L illustrating the procedure used.

In most real situations, there will not actually be a point in the flow at which $M = 1$. However, even in such cases, it is convenient to use the conditions at the hypothetical point at which $M = 1$ for reference purposes.

It should be noted that, in the derivations of the above equations, no assumptions have been made as to how the heat has been added between the two sections of the flow considered. The heat could have been added over just a small part of the flow or it could have been added uniformly over the entire region of the flow between the two sections considered.

### EXAMPLE 10.2
Air flows through a constant area duct. The pressure and temperature of the air at the inlet to the duct are 100 kPa and 10°C respectively and the inlet Mach number is 2.8. Heat is transferred to the air as it flows through the duct and as a result the Mach number at the exit is 1.3. Find the pressure and temperature at the exit. If no shock waves occur in the flow, find the maximum amount of heat that can be transferred to the air per unit mass of air. Also find the exit pressure and temperature that would exist with this maximum heat transfer rate. Assume that the flow is steady, that the effects of wall friction can be neglected and that the air behaves as a perfect gas.

### Solution

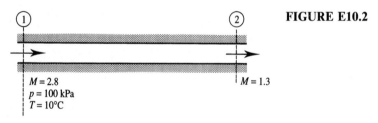

**FIGURE E10.2**

$M = 2.8$
$p = 100$ kPa
$T = 10°C$

$M = 1.3$

The flow situation under consideration is shown in Fig. E10.2. Now for a Mach number of 2.8, using:

$$\frac{T_0}{T} = 1 + \frac{\gamma - 1}{2} M^2 = 1 + 0.2M^2$$

or using the software or isentropic tables gives $T_{01}/T_1 = 2.568$. Hence:

$$T_{01} = 2.568 \times (273 + 10) = 726.7 \text{ K}$$

Next, using the relations given above or the tables or the software for frictionless flow in a constant area duct with heat exchange gives:

For $M = 2.8$:

$$\frac{p}{p^*} = 0.2004, \qquad \frac{T}{T^*} = 0.3149, \qquad \frac{T_0}{T_0^*} = 0.6738$$

For $M = 1.3$:

$$\frac{p}{p^*} = 0.7130, \qquad \frac{T}{T^*} = 0.8592, \qquad \frac{T_0}{T_0^*} = 0.9580$$

Using these values gives:

$$p_2 = \frac{p_2/p^*}{p_1/p^*} p_1 = \frac{0.7130}{0.2004} \times 100 = 355.8 \text{ kPa}$$

and:

$$T_2 = \frac{T_2/T^*}{T_1/T^*} T_1 = \frac{0.8592}{0.3149} \times 283 = 772.2 \text{ K} = 499.2°C$$

Therefore, the pressure and temperature of the air at the outlet to the duct are 355.8 kPa and 499.2°C respectively.

When the maximum amount of heat has been transferred, the Mach number at the exit is 1 and $T_{02} = T_0^*$. Hence, in this case:

$$T_{02} = \frac{T_{01}}{T_{01}/T_0^*} = \frac{726.7}{0.6738} = 1078.5 \text{ K}$$

But:

$$q = c_p(T_{02} - T_{01}) = 1.007(1078.5 - 726.7) = 354.3 \text{ kJ/kg}$$

it having been assumed that $c_p = 1.007 \text{ kJ/kg}$.

Also, when the maximum amount of heat has been transferred:

$$p_2 = p^* = \frac{p_1}{p_1/p^*} = \frac{100.7}{0.2004} = 499.0 \text{ kPa}$$

and:

$$T_2 = T^* = \frac{T_1}{T_1/T^*} = \frac{283}{0.3149} = 898.7 \text{ K} = 625.7°C$$

Therefore, when the maximum amount of heat is transferred, the heat added is 354.3 kJ for every kilogram of air flowing through the duct and under these circumstances the pressure and temperature of the air at the outlet to the duct are 499.0 kPa and 625.7°C respectively.

### EXAMPLE 10.3

Air flows through a constant area duct whose walls are kept at a low temperature. The air enters the pipe at a Mach number of 0.52, a pressure of 200 kPa, and a temperature of 350°C. The rate of heat transfer from the air to the walls of pipe is estimated to be 400 kJ/kg of air. Find the Mach number, temperature, and pressure at the exit of the pipe. Assume that the flow is steady, that the effects of wall friction are negligible, and that the air behaves as a perfect gas.

### Solution

For a Mach number of 0.52, using:

$$\frac{T_0}{T} = 1 + \frac{\gamma - 1}{2} M^2 = 1 + 0.2 M^2$$

or using the software or isentropic tables gives $T_{01}/T_1 = 1.054$. Hence:

$$T_{01} = 1.054 \times (350 + 10) = 656.6 \text{ K}$$

Next, using the relations given above or the tables or the software for frictionless flow in a constant area duct with heat exchange gives:

For $M = 0.52$:

$$\frac{p}{p^*} = 1.7414, \qquad \frac{T}{T^*} = 0.8196, \qquad \frac{T_0}{T_0^*} = 0.7199$$

But:

$$q = c_p(T_{02} - T_{01}) = 1.007(T_{02} - 656.6)$$

it again having been assumed that $c_p = 1.007\,\text{kJ/kg}$. Therefore, because $q = -400\,\text{kJ/kg}$ (heat is transferred from air to walls):

$$400 = 1.007(T_{02} - 656.6), \quad \text{i.e.,} \quad T_{02} = 259.4\,\text{K} = -13.6°\text{C}$$

Hence:
$$\frac{T_{02}}{T_0^*} = \frac{T_{02}}{T_{01}}\frac{T_{01}}{T_0^*} = \frac{259.4}{656.6}0.7199 = 50.2844$$

For this value of $T_{02}/T_0^*$, using the relations given above or the tables or the software for frictionless flow in a constant area duct with heat exchange gives:

$$M = 0.2656, \qquad \frac{p}{p^*} = 2.184, \qquad \frac{T}{T^*} = 0.3365$$

Hence:
$$p_2 = \frac{p_2/p^*}{p_1/p^*}p_1 = \frac{2.184}{1.741} \times 200 = 250.9\,\text{kPa}$$

and:
$$T_2 = \frac{T_2/T^*}{T_1/T^*}T_1 = \frac{0.3365}{0.8196} \times 623 = 256\,\text{K} = -17°\text{C}$$

Therefore, the Mach number, pressure, and temperature of the air at the outlet to the duct are 0.2656, 250.9 kPa, and $-17°\text{C}$ respectively.

**EXAMPLE 10.4**
Air enters a combustion chamber at a velocity of 80 m/s with a pressure and temperature of 180 kPa and 120°C. Find the maximum amount of heat that can be generated in the combustion chamber per unit mass of air. If the fuel has a heating value of 45 MJ/kg, find the air–fuel ratio. If the air–fuel ratio is adjusted until it is 90% of this value, find the reduction in the mass flow rate through the combustion chamber that must occur if the inlet stagnation pressure and stagnation temperature remain the same. Assume that the flow is steady, that the effects of wall friction can be neglected, that the effects of the mass of the fuel can be neglected, and that the air behaves as a perfect gas.

**Solution**
Using the prescribed inlet conditions gives:

$$M_1 = \frac{V_1}{\sqrt{\gamma R T_1}} = \frac{80}{\sqrt{1.4 \times 287 \times 393}} = 0.1981$$

For this value of $M_1$ using:

$$\frac{T_0}{T} = 1 + \frac{\gamma - 1}{2}M^2 = 1 + 0.2M^2$$

or using the software or isentropic tables gives $T_{01}/T_1 = 1.008$. Hence:

$$T_{01} = 1.008 \times 393 = 396.1\,\text{K}$$

Also, for this value of $M_1$ using the relations given above or the tables or the software for frictionless flow in a constant area duct with heat exchange gives:

$$\frac{p}{p^*} = 2.2754, \qquad \frac{T}{T^*} = 0.2029, \qquad \frac{T_0}{T_0^*} = 0.1704$$

Now, when the maximum amount of heat is generated, the value of $M_2$ will be equal to 1 and $T_{02} = T_0^*$. Hence, in this case:

$$T_{02} = \frac{T_0}{T_{01}/T_0^*} = \frac{396.1}{0.1704} = 2324.5 \text{ K}$$

But:     $q = c_p(T_{02} - T_{01}) = 1.007(2324.5 - 396.7) = 1942.2 \text{ kJ/kg}$

$$= 1.9422 \text{ MJ/kg}$$

it having been assumed that $c_p = 1.007 \text{ kJ/kg} \cdot \text{K}$.

Hence, the air–fuel ratio, i.e., the ratio of the mass of air to mass of fuel, when the Mach number is 1 at the exit with the specified inlet velocity is:

$$AF = \frac{\text{Heat generation per unit mass of fuel}}{\text{Heat generation per unit mass of air}} = \frac{45}{1.9422} = 23.17$$

The adjusted air–fuel ratio is $0.9 \times 23.17 = 20.85$. The adjusted heat generation rate per unit mass of air is therefore $1942.2/0.9 = 2158 \text{ kJ/kg}$. Hence, since the mass flow decreases until the Mach number at exit is again 1, it follows that:

$$q = c_p(T_{02} - T_{01}) = c_p(T_0^* - T_{01})$$

Hence since the inlet stagnation temperature is the same:

$$2158 = 1.007 \times (T_0^* - 396.1)$$

This gives $T_0^* = 2539.1$ K so $T_{01}/T_0^* = 396.1/2539.1 = 0.1560$.

This then gives, using the relations given above or the tables or the software for frictionless flow in a constant area duct with heat exchange:

$$M_1 = 0.1886$$

Now the mass flow rate is given by:

$$\dot{m} = \rho_1 V_1 A = \rho_0 a_0 A \frac{\rho_1 a_1}{\rho_0 a_0} M_1$$

and isentropic relations or tables or the software give, for $M = 0.198$:

$$\frac{\rho}{\rho_0} = 0.980, \qquad \frac{a}{a_0} = 0.996$$

and for $M = 0.1886$:

$$\frac{\rho}{\rho_0} = 0.982, \qquad \frac{a}{a_0} = 0.997$$

Hence, since the stagnation conditions are the same and the area is the same in the two cases:

$$\frac{\text{Adjusted mass flow rate}}{\text{Original mass flow rate}} = \frac{0.982}{0.980} \times \frac{0.997}{0.996} \times \frac{0.1886}{0.198} = 0.955$$

Therefore, the mass flow rate is decreased by 4.5%.

## 10.3
## ENTROPY–TEMPERATURE RELATIONS

Traditionally, the variation of entropy with temperature has been used in examining one-dimensional flows with heat addition. To establish the relation between these quantities, it is noted that for a given flow along a duct, as discussed above:

$$\frac{s - s^*}{c_p} = \ln\left[\frac{T}{T^*}\right] - \frac{(\gamma - 1)}{\gamma}\ln\left[\frac{p}{p^*}\right] \tag{10.37}$$

To establish the relation between $(s - s^*)/c_p$ and $T/T^*$ it is necessary to express $p/p^*$ in terms of $T/T^*$. Now, from eq. (10.32), the following is obtained:

$$M^2 = \frac{(1 + \gamma)}{\gamma}\left[\frac{p^*}{p}\right] - \frac{1}{\gamma}$$

while from eqs. (10.31) and (10.32) it will be seen that:

$$\frac{p}{p^*} = \frac{1}{M}\left(\frac{T}{T^*}\right)^{0.5}$$

Combining these two equations then gives:

$$\left(\frac{p}{p^*}\right)^2\left[(1 + \gamma)\frac{p^*}{p} - 1\right] = \gamma\frac{T}{T^*}$$

Rearranging this equation gives:

$$\left(\frac{p}{p^*}\right)^2 - (1 + \gamma)\frac{p}{p^*} + \gamma\frac{T}{T^*} = 0$$

Solving this quadratic equation for $p^*/p$ then gives:

$$\frac{p}{p^*} = \frac{(1 + \gamma)^2}{2} \pm \frac{\sqrt{(1 + \gamma)^2 - 4\gamma(T/T^*)}}{2} \tag{10.38}$$

Substituting this into eq. (10.37) gives the required relationship as:

$$\frac{s - s^*}{c_p} = \ln\left[\frac{T}{T^*}\right] - \left[\frac{\gamma - 1}{\gamma}\right]\ln\left\{\frac{(1 + \gamma)^2}{2} \pm \frac{\sqrt{(1 + \gamma)^2 - 4\gamma(T/T^*)}}{2}\right\} \tag{10.39}$$

This allows the variation of $T/T^*$ with $(s - s^*)/c_p$ to be found for any specified value of $\gamma$. The variation resembles that shown in Fig. 10.7. Such a curve, which basically gives the variation of $T$ with $s$, is termed the "Rayleigh line," and the type of flow here being considered, i.e., one-dimensional flow with heat addition and negligible friction, is often termed "Rayleigh flow." It will be seen from Fig. 10.7 that for any value of $(s - s^*)/c_p$, there are two possible values of $T/T^*$, one involving subsonic flow and the other involving supersonic flow, as indicated in the figure. The discussion given earlier—see the discussion of eq. (10.14)—indicated that $s$ will be a maximum when $M = 1$. The point A in Fig. 10.7 is, therefore, the point at which $M = 1$ and at which $s = s^*$ and $T = T^*$. Since heat addition moves the Mach number towards 1, while heat removal moves it away from 1, the effects of $q$ are as shown in the figure. Point B in Fig. 10.7 is the point at which $T$ is a maximum.

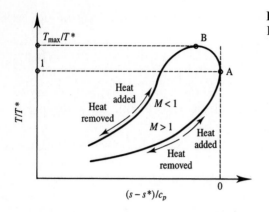

**FIGURE 10.7**
Rayleigh line.

The value of $M$ at which this occurs can be found by differentiating eq. (10.31) to give $dT/dM$ and then setting this equal to zero to give the value of $M$ at which $T$ is a maximum. The procedure gives:

$$\frac{2M(1+\gamma)^2}{(1+\gamma M^2)^2} - \frac{4\gamma M^3(1+\gamma)^2}{(1+\gamma M^2)^3} = 0$$

which, on rearrangement, gives the following for the Mach number at which the temperature is a maximum:

$$M = \sqrt{\frac{1}{\gamma}} \tag{10.40}$$

This result was previously given in eq. (10.11).

Substituting eq. (10.40) back into eq. (10.31) then gives the maximum value of $T$ as:

$$\frac{T_{max}}{T^*} = \frac{(1+\gamma)^2}{4\gamma} \tag{10.41}$$

The effects of the heat addition on the flow variables in Rayleigh flow are summarized in Table 10.3.

**TABLE 10.3**
**Changes in flow variables produced by heat exchange**

|       | $q$ | $T_0$ | $M$ | $p$ | $V$ | $s$ |
|-------|-----|-------|-----|-----|-----|-----|
| $M < 1$ | +   | +     | +   | −   | +   | +   |
| $M > 1$ | +   | +     | −   | +   | −   | +   |
| $M < 1$ | −   | −     | −   | +   | −   | −   |
| $M > 1$ | −   | −     | +   | −   | +   | −   |

+ means a quantity is increasing, − means it is decreasing.

## 10.4
## VARIABLE AREA FLOW WITH HEAT ADDITION

The analysis given in the previous section assumed that the flow area, $A$, was constant. In some real situations, the effects of both changing flow area and heat exchange are important. This situation will be considered in the present section, the effects of wall friction again being assumed to be negligible. The quasi-one-dimensional assumption will be used.

Consider the flow through the control volume shown in Fig. 10.8. Since, in the situation being considered, the flow area is changing, the continuity equation gives, as discussed in earlier chapters:

$$\frac{d\rho}{\rho} + \frac{dA}{A} + \frac{dV}{V} = 0 \tag{10.42}$$

while, as also discussed in earlier chapters, conservation of momentum gives:

$$pA + \left(p + \frac{dp}{2}\right) dA - (p + dp)(A + dA) = \rho A V \, dV$$

The second term on the left hand side represents the force due to the pressure on the curved wall. Neglecting second order terms then gives:

$$dp + \rho V \, dV = 0 \tag{10.43}$$

Also, as discussed before in this chapter, the equation of state gives:

$$\frac{dp}{p} = \frac{d\rho}{\rho} + \frac{dT}{T} \tag{10.44}$$

whereas, as also discussed earlier in this chapter, conservation of energy gives:

$$dq = c_p \, dT_0 \tag{10.45}$$

where:

$$dT_0 = dT + \frac{V \, dV}{c_p} \tag{10.46}$$

**FIGURE 10.8**
Control volume used in variable area analysis.

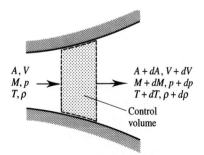

A, V
M, p
T, ρ

A + dA, V + dV
M + dM, p + dp
T + dT, ρ + dρ

Control
volume

Equations (10.43) and (10.44) together give:

$$\frac{d\rho}{\rho} + \frac{dT}{T} + \frac{\rho V \, dV}{p} = 0$$

Hence, since $a^2 = \gamma p/\rho$, it follows from this equation that:

$$\frac{d\rho}{\rho} + \frac{dT}{T} + \gamma M^2 \frac{dV}{V} = 0$$

Using eq. (10.42) with this equation then gives:

$$-\frac{dA}{A} + \frac{dT}{T} + (\gamma M^2 - 1)\frac{dV}{V} = 0 \qquad (10.47)$$

Now since $V^2 = M^2 a^2 = M^2 \gamma RT$ and similarly since $(V + dV)^2 = (M + dM)^2 \gamma R(T + dT)$, i.e., to first order accuracy:

$$V\left(1 + 2\frac{dV}{V}\right) = M^2 \gamma RT\left(1 + \frac{dT}{T} + 2\frac{dM}{M}\right)$$

From the above relations, it follows that:

$$\frac{dV}{V} = \frac{1}{2}\frac{dT}{T} + \frac{dM}{M}$$

Hence, eq. (10.47) can be written as:

$$-\frac{dA}{A} + \frac{(\gamma M^2 + 1)}{2}\frac{dT}{T} + (\gamma M^2 - 1)\frac{dM}{M} = 0 \qquad (10.48)$$

It is next, once again, noted that:

$$T_0 = T\left[1 + \frac{\gamma - 1}{2}M^2\right]$$

and that:

$$T_0 + dT_0 = (T + dT) + \left(\frac{\gamma - 1}{2}\right)(M + dM)^2$$

so that:

$$\frac{dT_0}{T} = \left[1 + \frac{(\gamma - 1)}{2}M^2\right]\frac{dT}{T} + (\gamma - 1)M^2\frac{dM}{M}$$

which, using eq. (10.45), gives:

$$\frac{dq}{c_p T} = \left[1 + \frac{(\gamma - 1)}{2}M^2\right]\frac{dT}{T} + (\gamma - 1)M^2\frac{dM}{M} \qquad (10.49)$$

This equation can then be combined with eq. (10.48) to give:

$$-\frac{dA}{A} + \frac{(\gamma M^2 + 1)/2}{[1 + (\gamma - 1)M^2/2]}\frac{dq}{c_p T} + \frac{(M^2 - 1)}{[1 + (\gamma - 1)M^2/2]}\frac{dM}{M} = 0 \qquad (10.50)$$

Alternatively, this equation can be written as:

$$-\frac{dA}{A} + \frac{(\gamma M^2 + 1)/2}{[1 + (\gamma - 1)M^2/2]}\frac{dT_0}{T} + \frac{(M^2 + 1)}{[1 + (\gamma - 1)M^2/2]}\frac{dM}{M} = 0$$

The first term in these equations effectively represents the effect of the area change on the Mach number, whereas the second term in the equations effectively represents the effect of the heat exchange on the Mach number. If the heat exchange is zero, i.e., if $dq = 0$, then, as discussed in Chapter 8, when $M = 1$ and $dM$ is non-zero, i.e., the Mach number is not a maximum or a minimum, $dA$ must be zero, i.e., the point at which $M = 1$ in a flow in which $M$ changes from a subsonic to a supersonic value or from a supersonic to a subsonic value corresponds to the point at which the area is a minimum. However, when there is heat addition or removal, the point at which $M = 1$ corresponds to the point at which:

$$-\frac{dA}{A} + \frac{(\gamma + 1)/2}{[1 + (\gamma - 1)/2]}\frac{dq}{c_p T} = 0 \tag{10.51}$$

Because $\gamma > 1$, the coefficient of $dq/c_p T$ in this equation is always positive. This means that if there is heat addition, i.e., if $dq > 0$, then $dA$ will be positive when $M = 1$ whereas if there is heat extraction, i.e., $dq < 0$, then $dA$ will be negative when $M = 1$. Therefore, when there is heat addition or removal, the section at which the Mach number is one does not coincide with the section of minimum flow area.

### EXAMPLE 10.5
Air enters a convergent duct at a Mach number of 0.8 with a pressure of 600 kPa and a temperature of 50°C. If the exit area of the duct is 60 percent of that at the inlet and if the shape of the duct is such that the Mach number remains constant in the duct, find the stagnation temperature and the temperature at the exit of the duct and the amount of heat being transferred to the air per unit mass of air. Assume that the flow is steady, that the effects of wall friction are negligible, and that the air behaves as a perfect gas.

*Solution*

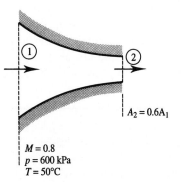

**FIGURE E10.5**

$A_2 = 0.6A_1$

$M = 0.8$
$p = 600$ kPa
$T = 50°C$

Using the prescribed inlet conditions:

$$T_{01} = T_1 \left(1 + \frac{\gamma - 1}{2} M^2\right) = 323(1 + 0.2 \times 0.8^2)$$

$$= 323 \times 1.128 = 364.3 \text{ K}$$

Now since the Mach number remains constant, i.e., since $dM = 0$, eq. (10.50) gives:

$$\frac{dA}{A} = \frac{(\gamma M^2 + 1)}{[1 + (\gamma - 1)M^2/2]} \frac{dT_0}{T}$$

but since:

$$\frac{T_0}{T} = 1 + \frac{(\gamma - 1)}{2} M^2$$

this equation can be written as:

$$\frac{dA}{A} = \frac{(\gamma M^2 + 1)}{2} \frac{dT_0}{T_0}$$

Integrating this equation between the inlet and the outlet gives, because $M$ is constant:

$$\ln\left(\frac{A_2}{A_1}\right) = \frac{(\gamma M^2 + 1)}{2} \ln\left(\frac{T_{02}}{T_{01}}\right)$$

Using the prescribed conditions then gives:

$$\ln(0.6) = \frac{(1.4 \times 0.8^2 + 1)}{2} \ln\left(\frac{T_{02}}{364.3}\right)$$

which can be solved to give the outlet stagnation temperature as $T_{02} = 212.5$ K. Using this value then gives:

$$q = c_p(T_{02} - T_{01}) = 1.007(212.5 - 364.3) = -152.9 \text{ kJ/kg}$$

So the amount of heat removed per unit mass of air is 152.9 kJ/kg. Since the Mach number does not change, the exit temperature is given by:

$$T_2 = \frac{T_{02}}{\left(1 + \frac{\gamma - 1}{2} M^2\right)} = \frac{212.5}{(1 + 0.2 \times 0.8^2)} = 188.4 \text{ K}$$

i.e., the air leaves at a temperature of 188.4 K.

The situation discussed in the above example is a very particular one and eq. (10.50) cannot, in general, be directly integrated to give the variation of $M$ with $A$ when $dq$ is nonzero. However, it is quite easy to integrate this equation numerically to give this variation. This procedure will be discussed in a more general way in Chapter 11. The basic aspects of the procedure are, however, illustrated in the following example.

**EXAMPLE 10.6**
Consider steady air flow through a duct that has a circular cross-sectional shape. The inlet diameter of the duct is 6 cm and the duct has a length of 1.5 m. The air enters the duct with a Mach number of 0.4 and a temperature of 40°C. Heat is

added to the flow in the duct at a uniform rate which is such that the stagnation temperature increases by 246 K in the duct. Determine the Mach number and temperature variations along the duct if

1. Its cross-sectional area increases linearly by 50 percent
2. Its cross-sectional area decreases linearly by 5 percent
3. Its cross-sectional area remains constant.

Assume that the air behaves as a perfect gas, that the flow is steady and that the effects of friction are negligible.

### Solution
The procedure used here is basically the same as that used in the previous chapter when dealing with the effects of area change on adiabatic flow with friction. The cross-sectional area of the duct will be given by:

$$A = A_i\left(1 + K\frac{x}{L}\right)$$

where $x$ is the distance along the duct from the inlet, $L$ is the length of the duct, i.e., 1.5 m, and $A_i$ is the inlet area. $K$ is a constant that is equal to $+0.5$ in the first case, $-0.05$ in the second case, and 0 in the third case. With the area specified in this way, the diameter at any $x$ can be found using:

$$D = \sqrt{\frac{4}{\pi}A}$$

Since $\gamma$ is equal to 1.4, eq. (10.50) gives for the present situation:

$$-\frac{dA}{A} + \frac{(1.4M^2 + 1)}{(1 + 0.2M^2)}\frac{dT_0}{T} + \frac{(M^2 - 1)}{(1 + 0.2M^2)}\frac{dM}{M} = 0$$

This equation will be integrated to give the variation of $M$ with $x$. This will be done using a very simple procedure. The duct will be broken down into a series of $N$ sections of equal length, i.e.,

$$dx = \frac{L}{N}$$

and since the duct area changes linearly:

$$dA = \frac{A_{\text{exit}} - A_i}{N}$$

and since the heat is added uniformly:

$$dT_0 = \frac{\text{Increase in } T_0}{N}$$

The procedure involves:

1. Set $M = M_i$, $A = A_i$, $T = T_i$, and $D = D_i$ and calculate the change in the Mach number across the first section using:

$$dM = -\frac{M[1 + 0.2M^2]}{(M^2 - 1)}\left\{-\frac{dA}{A} + \frac{(1.4M^2 - 1)\,dT_0}{[1 + 0.2M^2]\,T}\right\}$$

2. Find the Mach number at the end of the first section using $M = M + dM$, the area at this section using $A = A + dA$ and the stagnation temperature at the end of the section. Then find the diameter at this section using:

$$D = \sqrt{\frac{4}{\pi} A}$$

and the temperature at this section using:

$$T = T_0 \left/ \left[ 1 + \frac{\gamma - 1}{2} M^2 \right] \right.$$

3. Repeat the above sets to find the Mach number at the end of the next step and so on, continuing until the end of the duct is reached.

The above procedure can easily be carried out using a spreadsheet program. Alternatively, a simple BASIC program that implements the procedure is listed in Program E10.6.

**Program E10.6**

```
 10   CLS
 20   PRINT" "
 30   PRINT" VARIABLE AREA FLOW WITH HEAT EXCHANGE"
 40   PRINT" ":PRINT" "
 50   INPUT" NUMBER OF STEPS ";N
 60   INPUT" INCREASE IN STAGNATION TEMP. (K) ";DT
 70   INPUT" AREA VARIATION COEFF. ";K
 80   PRINT" "
 90   PI=3.1416
100   L=1.5
110   TI=313
120   MI=0.4
130   DI=0.06
140   T0=TI*(1.0+0.2*MI*MI)
150   AI=PI*DI*DI/4
160   AO=(1+K)*AI
170   DA=(AO-AI)/N
180   DS=DT/N
190   DX=L/N
200   PRINT "   X","   A","   M"
210   PRINT" "
220   X=0.0
230   M=MI
240   A=AI
250   T=TI
260   PRINT USING "####.########";X,A,M
270   FOR I=1 TO N
280   D=(4*A/PI)^0.5
290   F1=(1+0.2*M*M)*M/(M*M-1)
300   F2=(1.4*M*M+1)*M/(2.0*(M*M-1))
310   DM=F1*DA/A-F2*DS/T
320   M=M+DM
330   A=A+DA
340   X=X+DX
350   T0=T0+DS
360   T=T0/(1.0+0.2*M*M)
370   PRINT USING "####.########";X,A,M,T0,T
380   NEXT I
390   STOP
400   END
```

Results obtained using this program are shown in Figs. E10.6a and b.

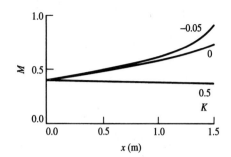

**FIGURE E10.6a**

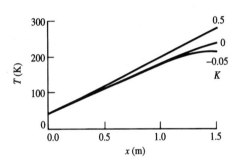

**FIGURE E10.6b**

## 10.5
## ONE-DIMENSIONAL CONSTANT AREA FLOW WITH BOTH HEAT EXCHANGE AND FRICTION

In some situations that are of practical importance, the effects of both heat exchange and the shear stress at the wall resulting from viscosity are important. To illustrate how such flows can be analyzed, attention will be given to one-dimensional flow in a duct of constant cross-sectional area. Consider, again, the flow through the control volume shown in Fig. 10.9. The form of the continuity equation is, of course, the same whether heat exchange or friction is considered. In this situation it therefore again gives:

$$\frac{dV}{V} + \frac{d\rho}{\rho} = 0 \tag{10.52}$$

Similarly, since the energy balance is not affected by the friction, it again gives:

$$dq = c_p \, dT + V \, dV = c_p \, dT_0 \tag{10.53}$$

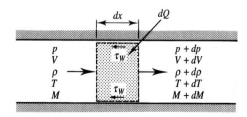

**FIGURE 10.9**
Control volume used in analysis of flow with heat exchange and friction.

Therefore, since:

$$\frac{T_0}{T} = 1 + \frac{(\gamma - 1)}{2} M^2$$

it can easily be shown, as was done in the previous section, that:

$$\frac{dT_0}{T} = \left[1 + \frac{\gamma - 1}{2} M^2\right] \frac{dT}{T} + (\gamma - 1)M^2 \frac{dM}{M}$$

which, when substituted into eq. (10.53) gives:

$$\frac{dq}{c_p T} = \left[1 + \frac{\gamma - 1}{2} M^2\right] \frac{dT}{T} + (\gamma - 1)M^2 \frac{dM}{M} \tag{10.54}$$

Because the viscous stresses at the wall are important, the conservation of momentum equation gives for the control volume being considered:

$$dp + \tau_w \frac{P}{A} dx + \rho V \, dV = 0 \tag{10.55}$$

where $P$ is again the perimeter of the duct and $A$ is its cross-sectional area.

The wall shear stress, $\tau_w$, is, as in the previous chapter, expressed in terms of the dimensionless friction factor, $f$, as follows:

$$\tau_w = f \tfrac{1}{2} \rho V^2 \tag{10.56}$$

and the ratio $P/A$ is, as before, expressed in terms of the hydraulic diameter, $D_H$, using:

$$D_H = \frac{4 \, (\text{Area})}{\text{Perimeter}} = \frac{4A}{P} \tag{10.57}$$

Equation (10.55) can then be written as:

$$dp + \tfrac{1}{2}\rho V^2 \frac{f}{D_H} dx + \rho V \, dV = 0 \tag{10.58}$$

Now as shown above—see discussion of eq. (10.48)—since:

$$V^2 = M^2 a^2 = M^2 \gamma RT$$

it follows that:

$$\frac{dV}{V} = \frac{dM}{M} + \frac{1}{2} \frac{dT}{T} \tag{10.59}$$

so eq. (10.58) gives, noting that $a^2 = \gamma p/\rho$:

$$dp + \frac{\gamma}{2} M^2 \frac{f}{D_H} dx + \gamma M^2 \frac{dM}{M} + \frac{\gamma M^2}{2} \frac{dT}{T} = 0 \tag{10.60}$$

Also, the equation of state for a perfect gas gives, as shown before:

$$\frac{dp}{p} = \frac{d\rho}{\rho} + \frac{dT}{T}$$

This can be combined with the continuity equation, i.e., eq. (10.52), to give:

$$\frac{dp}{p} = -\frac{dV}{V} + \frac{dT}{T} \tag{10.61}$$

Substituting eq. (10.59) into eq. (10.61) then gives:

$$\frac{dp}{p} = -\frac{dM}{M} + \frac{1}{2}\frac{dT}{T} \tag{10.62}$$

This, in turn, can be substituted into eq. (10.60) to give:

$$-\frac{dM}{M} + \frac{1}{2}\frac{dT}{T} + \frac{\gamma}{2}M^2\frac{f}{D_H}dx + \gamma M^2\frac{dM}{M} + \frac{\gamma M^2}{2}\frac{dT}{T} = 0$$

i.e.,

$$\frac{(1+\gamma M^2)}{2}\frac{dT}{T} + \frac{\gamma}{2}M^2\frac{f}{D_H}dx - (1-\gamma M^2)\frac{dM}{M} = 0 \tag{10.63}$$

Substituting the value of $dT/T$ given by eq. (10.54) into this equation then gives:

$$\left(\frac{1+\gamma M^2}{2}\right)\frac{dq}{c_pT[1+(\gamma-1)M^2/2]} + \frac{\gamma M^2}{2}\frac{f\,dx}{D_H}$$

$$= \left\{(1-\gamma M^2) + \frac{(1+\gamma M^2)}{2}\frac{(\gamma-1)M^2}{[1+(\gamma-1)M^2/2]}\right\}\frac{dM}{M} \tag{10.64}$$

For specified values of $dq/c_p$ and $f/D_H$, this equation together with eqs. (10.63) and (10.62), can be numerically integrated to give the variations of $M$, $T$, and $p$ along the duct.

The first term on the left hand side of eq. (10.64) determines the relative effect of the heat exchange on the Mach number variation whereas the second term on the left hand side determines the relative effect of the wall friction on this variation. As mentioned before, when the heat exchange is the result of combustion, the first term is usually far greater than the second term on the left hand side, i.e., the effect of friction is negligible compared to the effect of heat exchange. It is, therefore, usually only in cases where the heat exchange is the result of heat transfer from the walls of the duct to the gas that both the friction term and the heat exchange term are important. Now, it is usual to express the heat transfer rate from the duct wall to the gas in terms of a heat transfer coefficient, $h$, defined by:

$$dq_w = h\,dA_w(T_w - T) \tag{10.65}$$

where $dq_w$ is the rate of heat transfer through the portion of the wall of surface area $dA_w$ considered, $T_w$ is the wall temperature at the point considered and $T$ is the mean gas temperature at the section considered. These quantities are defined in Fig. 10.10. The gas is brought to rest at the wall so even if there is no heat transfer at the wall, i.e., if $dq_w$ is zero, the wall temperature will be higher than $T$. The wall temperature in this adiabatic flow case is termed the adiabatic wall temperature, $T_{wa}$. If the process that the gas particles undergo

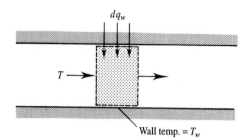

**FIGURE 10.10**
Wall section used in defining the heat transfer coefficient.

in being brought to rest at the wall of the duct is adiabatic, the wall temperature when $dq_w$ is zero, i.e., $T_{wa}$, will be the stagnation temperature $T_0$ which is, of course, given by:

$$\frac{T_0}{T} = 1 + \frac{\gamma - 1}{2} M^2$$

However, the actual process experienced by the fluid in being brought to rest at the wall is not adiabatic and the wall temperature when $dq_w$ is zero will be somewhat less than $T_0$, i.e.,

$$\frac{T_{wa} - T}{T_0 - T} = r(< 1) \qquad (10.66)$$

In this equation, $r$ is termed the "recovery factor." The difference between $T_{wa}$ and $T_0$ is, however, comparatively small for gases which have a Prandtl number near 1, i.e., for such gases, $r$ has a value near 1. It will, therefore, be assumed, here, that $T_{wa} = T_0$. In this case, the heat transfer can only be from the wall to the gas if $T_w > T_0$. If $T_w < T_0$, heat will be transferred from the gas to the wall. From this discussion it follows that the most appropriate temperature to take as the gas temperature in eq. (10.65) is $T_0$, i.e., to write eq. (10.65) as:

$$dq_w = h\, dA_w (T_w - T_0) \qquad (10.67)$$

Now, by definition:

$$dq_w = \frac{dq_w}{\dot{m}} = \frac{dq_w}{\rho A V} \qquad (10.68)$$

As shown in books on heat transfer, an analysis of the velocity and temperature fields in a flow with heat transfer indicates that there is a relationship between the heat transfer coefficient, $h$, and the friction factor, $f$. This relationship is termed the "Reynolds' analogy" and for gases that have a Prandtl number that is near 1, this analogy gives:

$$h = \frac{\rho V c_p V}{8f} \qquad (10.69)$$

Substituting eqs. (10.69) and (10.67) into eq. (10.68) then gives:

$$dq = \frac{c_p f}{8} \frac{dA_w}{A} (T_w - T_0) \qquad (10.70)$$

But:

$$dA_w = P\, dx$$

so using eq. (10.57), eq. (10.70) becomes:

$$\frac{dq}{c_p} = \frac{f\,dx}{2D_H}(T_w - T_0) \tag{10.71}$$

Substituting this into eq. (10.64) then gives:

$$\frac{(T_w - T_0)}{T_0}\left(\frac{1 + \gamma M^2}{2}\right)\frac{f\,dx}{D_H} + \frac{\gamma M^2 f\,dx}{2\,D_H}$$

$$= \left\{(1 - \gamma M^2) + \frac{(1 + \gamma M^2)}{2}\frac{(\gamma - 1)M^2}{[1 + (\gamma - 1)M^2/2]}\right\}\frac{dM}{M} \tag{10.72}$$

The ratio of the heat transfer effect, i.e., the first term on the left hand side of eq. (10.72), to the fluid friction effects, i.e., the second term on the left hand side of eq. (10.72), will be seen to depend on:

$$\frac{(T_w - T_0)}{T_0}\frac{(1 + \gamma M^2)}{2\gamma M^2}$$

Hence, for a given situation, i.e., for a given $M$, this ratio will depend on $(T_w - T_0)/T_0$. If $T_w$ is very much greater than $T_0$, heat transfer effects will predominate. However, if $T_w$ is approximately equal to $T_0$, friction effects will dominate.

When both heat exchange and friction effects are important, eq. (10.72) must be integrated numerically to give the variation of $M$ with $x$. Other equations given in this section can then be used to simultaneously determine the variations of the other variables, e.g., eq. (10.63) can be used to determine the changes in $T$, eq. (10.62) can be used to find the changes in $p$, and eq. (10.61) can be used to find the changes in $V$.

## 10.6
## ISOTHERMAL FLOW WITH FRICTION IN A CONSTANT AREA DUCT

When gases are transported through long pipelines that are not heavily insulated the gas temperature frequently remains approximately constant. Of course, this will normally require that heat is either transferred from the gas to the surroundings or that heat is transferred from the surroundings to the gas flowing through the pipe. Because the pipe is usually long in such cases and the heat transfer rates relatively low, the effects of friction cannot usually be neglected. Hence, in the present section, one-dimensional isothermal gas flow through a constant area duct with significant friction effects will be considered.

The analysis is, of course, again based on the use of the conservation of mass, momentum, and energy equations together with the perfect gas law. The same type of control volume as used in the previous section is again consid-

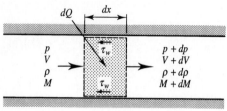

**FIGURE 10.11**
Control volume used in analysis of isothermal duct flow.

ered, this being shown in Fig. 10.11. For this control volume, the governing equations give since $T$ is constant:

*Conservation of mass:*

$$\frac{dV}{V} + \frac{d\rho}{\rho} = 0 \tag{10.73}$$

*Conservation of momentum:*

$$dp + \tau_w \frac{P}{A} dx + \rho V \, dV = 0 \tag{10.74}$$

where $P$ is again the perimeter of the duct and $A$ is its cross-sectional area.

*Conservation of energy:*

$$dq = V \, dV \tag{10.75}$$

*Perfect gas law:*

$$\frac{dp}{p} = \frac{d\rho}{\rho} \tag{10.76}$$

Further, from the definition of the Mach number, since when $T$ is constant, $a$ is constant:

$$\frac{dM}{M} = \frac{dV}{V} \tag{10.77}$$

The above equations represent a set of five equations involving, as variables, $dV$, $d\rho$, $dp$, $dq$, and $dM$. Before solving for these variables, the wall shear stress, $\tau_w$ is written in terms of the friction factor and the hydraulic diameter is again introduced, these being defined as discussed above by:

$$\tau_w = f \tfrac{1}{2}\rho V^2, \qquad D_H = \frac{4A}{P}$$

In terms of these, eq. (10.74) becomes:

$$dp + \rho V^2 \frac{2f}{D_H} dx + \rho V \, dV = 0 \tag{10.78}$$

which can be written:

$$\frac{dp}{p} + \frac{2f}{D_H} dx \frac{\rho V^2}{p} + \frac{\rho V}{p} dV = 0$$

i.e.,
$$\frac{dp}{p} + \frac{2f}{D_H} dx \gamma M^2 + \gamma M^2 \frac{dV}{V} = 0 \qquad (10.79)$$

But eqs. (10.73), (10.76), and (10.77) together give:

$$\frac{dV}{V} = -\frac{d\rho}{\rho} = -\frac{dp}{p} = \frac{dM}{M} \qquad (10.80)$$

Therefore, eq. (10.79) gives:

$$\frac{dp}{p}(1 - \gamma M^2) + \gamma M^2 \frac{2f}{D_H} dx = 0$$

i.e.,
$$\frac{dp}{p} + \left[\frac{\gamma M^2}{1 - \gamma M^2}\right] \frac{2f}{D_H} dx = 0 \qquad (10.81)$$

Hence:
$$\frac{dV}{V} = -\frac{d\rho}{\rho} = -\frac{dp}{p} = \frac{dM}{M} = \left[\frac{\gamma M^2}{1 - \gamma M^2}\right] \frac{2f}{D_H} dx \qquad (10.82)$$

Using this result, eq. (10.75) gives:

$$\frac{dq}{V^2} = \frac{dV}{V} = \left[\frac{\gamma M^2}{1 - \gamma M^2}\right] \frac{2f}{D_H} dx \qquad (10.83)$$

Also since:
$$\frac{ds}{c_p T} = \frac{dT}{T} - \left[\frac{\gamma - 1}{\gamma}\right] \frac{dp}{p} = -\left[\frac{\gamma - 1}{\gamma}\right] \frac{dp}{p}$$

it follows from eq. (10.83) that:

$$\frac{ds}{c_p T} = \left[\frac{\gamma - 1}{\gamma}\right] \left[\frac{\gamma M^2}{1 - \gamma M^2}\right] \frac{2f}{D_H} dx \qquad (10.84)$$

Each of eqs. (10.82), (10.83), and (10.84), for the changes in the variables through the control volume contains the term $1 - \gamma M^2$ in the denominator on the right hand side, each of the other terms on this right side always being positive. Therefore, the change in the variables through the control volume, i.e., $dp$, $dV$, etc., changes sign when $M = 1/\sqrt{\gamma}$. The signs of these changes above and below this value are shown in Table 10.4.

**TABLE 10.4**
**Changes in flow variables produced by friction in isothermal flow**

| | $dq$ | $dp$ | $d\rho$ | $dV$ | $dM$ | $ds$ |
|---|---|---|---|---|---|---|
| $M < 1/\sqrt{\gamma}$ | + | − | − | + | + | + |
| $M > 1/\sqrt{\gamma}$ | − | + | + | − | − | − |

+ means a quantity is increasing, − means it is decreasing.

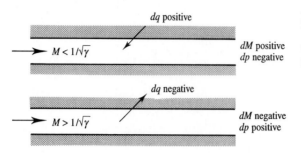

**FIGURE 10.12**
Variation of differential heat transfer rate with Mach number.

The form of the variation of $dq$ with $M$ is illustrated in Fig. 10.12. As $M$ tends to $1/\sqrt{\gamma}$, $dq$ tends to infinity, i.e., an infinite amount of heat would have to be transferred to or removed from the flow to keep the temperature of the gas constant if $M$ were to reach $1/\sqrt{\gamma}$. Therefore, $1/\sqrt{\gamma}$ is a limiting value for $M$. This can further be seen from the fact that when $M > 1/\sqrt{\gamma}$ the Mach number decreases in the flow direction whereas when $M < 1/\sqrt{\gamma}$, the Mach number increases. Hence, in all cases, $M$ tends to $1/\sqrt{\gamma}$.

The relationships given above basically determine the rate of change of the flow variables along the duct. They must be integrated to give the change in these flow variables over a specified length of the duct $l_{12}$ indicated in Fig. 10.13.

Equation (10.80) can be directly integrated to give:

$$\frac{V_2}{V_1} = \frac{\rho_1}{\rho_2} = \frac{p_1}{p_2} = \frac{M_2}{M_1} \tag{10.85}$$

To relate the changes in these variables to the length of the duct, it is noted that eq. (10.82) gives:

$$\frac{dM}{\gamma M^3} - \frac{dM}{M} = \frac{2f}{D_H} dx \tag{10.86}$$

This equation can be integrated between the points 1 and 2 being considered to give:

$$\left[\frac{1}{2\gamma M_1^2} - \frac{1}{2\gamma M_2^2}\right] - \ln\left[\frac{M_2}{M_1}\right] = \frac{2f l_{12}}{D_H} \tag{10.87}$$

For any value of $M_1$, this equation allows $M_2$ to be found. This determines $M_2/M_1$ and so allows $p_2/p_1$, $\rho_2/\rho_1$, and $V_2/V_1$ to be found using eq. (10.85).

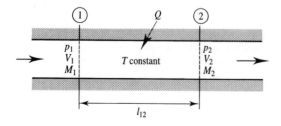

**FIGURE 10.13**
Changes over duct length considered.

Since the limiting value of $M_2$ is $1/\sqrt{\gamma}$, eq. (10.87) gives the following equation for the maximum duct length $l^*$ in terms of the initial Mach number $M$:

$$\left[\frac{1}{2\gamma M^2}-\frac{1}{2}\right]+\ln[\sqrt{\gamma}M]=\frac{2fl^*}{D_H}$$

i.e.,

$$\left[\frac{1-\gamma M^2}{\gamma M^2}\right]+\ln[\gamma M^2]=\frac{4fl^*}{D_H} \qquad (10.88)$$

For any value of $M$, this allows the value of $4fl^*/D_H$ to be found. Hence, since $M = M^* = 1/\sqrt{\gamma}$, when $l_{12} = l^*$ so that $M^*/M = 1/\sqrt{\gamma}M$, it follows from eq. (10.85) that:

$$\frac{\rho}{\rho^*}=\frac{p}{p^*}=\frac{V^*}{V}=\frac{M^*}{M}=\frac{1}{\sqrt{\gamma}M} \qquad (10.89)$$

Therefore, $4fl^*/D_H$, $\rho/\rho^*$, $p/p^*$, and $V/V^*$ are functions of $M$ for a given value of $\gamma$. The ratios of the stagnation properties can then be obtained by using the relation between the stagnation property and the Mach number, e.g., since:

$$\frac{T_0}{T}=1+\frac{\gamma-1}{2}M^2$$

it follows that:

$$\frac{T_0}{T_0^*}=\frac{1+(\gamma-1)M^2/2}{1+[(\gamma-1)/2](1/\gamma)}$$

i.e.,

$$\frac{T_0}{T_0^*}=\frac{2\gamma}{3\gamma-1}\left(1+\frac{\gamma-1}{2}M^2\right)$$

The values given by these relations are often presented in tables (see Appendix F) and charts. A typical set of values, these being for $\gamma = 1.4$, is given in Fig. 10.14. The results are also given by the software supplied with this book. They

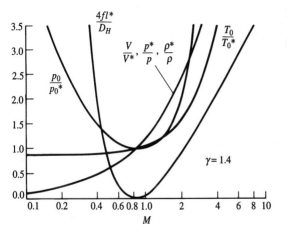

**FIGURE 10.14**
Variation of properties in isothermal flow for $\gamma = 1.4$.

can also be easily obtained using a programmable calculator, the procedure used in that case being illustrated by the simple BASIC program listed in Appendix L.

The way in which these results are used is illustrated in the following example.

**EXAMPLE 10.7**
Natural gas flows through a 0.075 m diameter pipeline which has a length of 750 m. The flow can be assumed to be isothermal with a temperature of 15°C. The Mach number and pressure at the inlet are 0.09 and 900 kPa. If the mean friction factor for the flow is 0.002, find the Mach number and the pressure at the exit to the pipe. Also find the maximum possible length of the pipe and the exit pressure with this length of pipe. Assume that the flow is steady and that the natural gas has a specific heat ratio at 1.3.

**Solution**

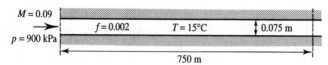

**FIGURE E10.7**

For the inlet Mach number, $M_1$, of 0.09, using the relations given above or the tables or the software for isothermal flow in a constant area duct with friction gives for $\gamma = 1.3$:

$$\frac{4fl^*}{D} = 89.41, \qquad \frac{p}{p^*} = 9.745$$

Since:

$$l_2^* = l_1^* - l_{12}$$

it follows that:

$$\frac{4fl_2^*}{D} = \frac{4fl_1^*}{D} - \frac{4fl_{12}}{D}$$

Hence:

$$\frac{4fl_2^*}{D} = 89.41 - \frac{4 \times 0.002 \times 750}{0.075} = 9.41$$

For this value of $4fl^*/D$ using the relations given above or the tables or the software for isothermal flow in a constant area duct with friction gives for $\gamma = 1.3$:

$$M = 0.2435, \qquad \frac{p}{p^*} = 3.6019$$

Using these values gives:

$$p_2 = \frac{p_2/p^*}{p_1/p^*}p_1 = \frac{3.6019}{9.745} \times 900 = 332.7 \text{ kPa}$$

Therefore, the pressure of the gas at the outlet to the pipe is 332.7 kPa. The maximum pipe length is $l_1^*$ which is given by:

$$\frac{4 \times 0.002 \times l_1^*}{0.075} = 89.41$$

which gives $l_1^* = 838.2$ m, i.e., the maximum pipe length is 838.2 m.

With the maximum pipe length, $p_2 = p^*$ so the exit pressure is given by:

$$p_2 = \frac{p^*}{p_1}p_1 = \frac{900}{9.745} = 92.4\,\text{kPa}$$

Therefore, the pressure of the gas at the outlet to the maximum length pipe is 92.4 kPa.

## 10.7
## COMBUSTION WAVES

The discussion given thus far in this chapter has, basically, been concerned with the effect of heat exchange on compressible flow through a duct. There are, however, other types of compressible flow in which heat exchange is important. One such case is the flow through a combustion wave. Here, the heat addition is the result of the "release of chemical energy" associated with the combustion. It is thus dependent on the difference between the heat of formation of the initial gas mixture, i.e., the reactants, and that of the products of combustion. A combustion wave can be modeled, basically, as a composite wave consisting of a shock wave sustained by the release of chemical energy in a combustion zone immediately following the shock wave, the "combustion wave" consisting of the shock and the combustion zone. The detailed structure of the wave will not, however, be considered here. Combustion waves are often so strong that the high-temperature effects discussed in Chapter 15 become important. However, most of the main features of such flows can be adequately described by assuming that the gas is perfect. It should also be noted that the combustion wave can either be stationary with the gas flowing through it or it can be propagating through the gas. The wave will here be analyzed using a coordinator system that is fixed relative to the wave, i.e., the wave will be at rest in the analysis presented below. The results of this analysis can then be applied to a moving wave using the same procedure as was adopted in Chapter 5 to deal with moving shock waves.

Consider the flow relative to a combustion wave, i.e., consider the situation shown in Fig. 10.15.

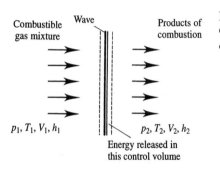

**FIGURE 10.15**
Control volume used in the analysis of a detonation wave.

The flow through the control volume shown in Fig. 10.15 is of course, as before, analyzed by applying conservation of mass, momentum, and energy. Conservation of mass and conservation of momentum are unaffected by the presence of the combustion and give, as with the conventional normal shock wave flow considered in Chapter 5:

$$\rho_1 V_1 = \rho_2 V_2 \ (= \dot{m}) \tag{10.90}$$

and:
$$p_1 + \rho_1 V_1^2 = p_2 + \rho_2 V_2^2 \tag{10.91}$$

unit frontal area of the wave having again been considered.

The above two equations can be combined by noting that eq. (10.91) gives:

$$p_1 + \frac{\dot{m}^2}{\rho_1} = p_2 + \frac{\dot{m}^2}{\rho_2}$$

i.e.,
$$\dot{m}^2 = \frac{p_2 - p_1}{(1/\rho_1) - (1/\rho_2)} \tag{10.92}$$

Hence, using eq. (10.90), it follows that, i.e.,

$$V_1 = \frac{1}{\rho_1}\sqrt{\frac{p_2 - p_1}{(1/\rho_1) - (1/\rho_2)}}, \qquad V_2 = \frac{1}{\rho_2}\sqrt{\frac{p_2 - p_1}{(1/\rho_1) - (1/\rho_2)}} \tag{10.93}$$

Conservation of energy applied to the flow across the wave is next considered. This gives, because of the heat generation:

$$c_p T_1 + \frac{V_1^2}{2} + q = c_p T_2 + \frac{V_2^2}{2} \tag{10.94}$$

The heat generation rate per unit mass, $q$, depends on the chemical reaction and is assumed to be known.

Now, eq. (10.93) gives:

$$V_2^2 - V_1^2 = \frac{p_2 - p_1}{(1/\rho_1) - (1/\rho_2)}\left[\frac{1}{\rho_2^2} - \frac{1}{\rho_1^2}\right]$$

i.e.,
$$V_2^2 - V_1^2 = (p_1 - p_2)\left[\frac{1}{\rho_2} + \frac{1}{\rho_1}\right] \tag{10.95}$$

Now, by utilizing the perfect gas law, eq. (10.94) can be written as:

$$V_2^2 - V_1^2 = \left[\frac{2\gamma}{\gamma - 1}\right]\left[\frac{p_1}{\rho_1} - \frac{p_2}{\rho_2}\right] + q \tag{10.96}$$

Equations (10.95) and (10.96) together then give:

$$(p_1 - p_2)\left[\frac{1}{\rho_2} + \frac{1}{\rho_1}\right] = \left[\frac{2\gamma}{\gamma - 1}\right]\left[\frac{p_1}{\rho_1} - \frac{p_2}{\rho_2}\right] + q$$

i.e.,
$$\left[1 - \frac{p_2}{p_1}\right]\left[\frac{\rho_1}{\rho_2} + 1\right] = \left[\frac{2\gamma}{\gamma - 1}\right]\left[1 - \left(\frac{p_2}{p_1}\right)\left(\frac{\rho_1}{\rho_2}\right)\right] + \frac{\rho_1 q}{p_1} \tag{10.97}$$

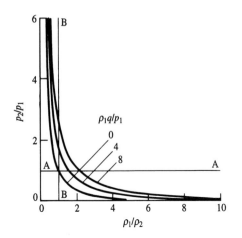

**FIGURE 10.16**
Relation between pressure and density ratios across a combustion wave.

This equation is the equivalent of eq. (5.13) which was derived for the flow through an adiabatic normal shock wave. The equivalents of the other adiabatic normal shock wave relations given in Chapter 5 can easily be derived for a detonation wave by using the same basic procedure as adopted in Chapter 5.

Equation (10.97) determines the variation of $p_2/p_1$ with $\rho_1/\rho_2$ for any value of $\rho_1 q/p_1$. The form of the variation is shown in Fig. 10.16. In Fig. 10.16, the curve corresponding to $q = 0$, i.e., to adiabatic flow, applies to conventional normal shock waves. The point on this $q = 0$ at which $p_2/p_1 = 1$ and $\rho_1/\rho_2 = 1$ represents conditions in the flow upstream of the combustion wave. It is not possible to determine from this graph what conditions will exist downstream of the wave for given upstream conditions and given $q$. Therefore, the limits on possible downstream solutions have to be considered. Consider eq. (10.93). Since the velocity both before and after the wave must be positive, it follows that $p_2 - p_1$ must have the same sign as $(1/\rho_1 - 1/\rho_2)$. Therefore, a combustion wave must be such that $p_2/p_1 > 1$ and $\rho_2/\rho_1 > 1$ or $p_2/p_1 < 1$ and $\rho_2/\rho_1 < 1$. Considering Fig. 10.16, it will be seen that this requires that only solutions to the left of BB or below AA are possible, i.e., only the solutions indicated in Fig. 10.17 are possible.

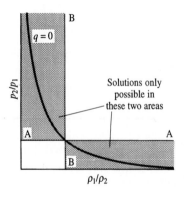

**FIGURE 10.17**
Possible detonation wave solutions.

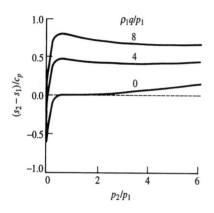

**FIGURE 10.18**
Entropy change across a combustion wave.

To decide whether there are further limits on the possible solution the entropy change across the wave is considered. This is given by:

$$\frac{s_2 - s_1}{c_p} = \ln\frac{T_2}{T_1} - \frac{(\gamma - 1)}{\gamma}\ln\frac{p_2}{p_1}$$

i.e., using the perfect gas law:

$$\frac{s_2 - s_1}{c_p} = \ln\left[\left(\frac{p_2}{p_1}\right)^{1/\gamma}\left(\frac{\rho_1}{\rho_2}\right)\right] \qquad (10.98)$$

This equation in conjunction with eq. (10.97) allows the variation of $s_2 - s_1/c_p$ with $p_2/p_1$ for any value of $\rho_1 q/p_1$ to be determined. The form of the variation is shown in Fig. 10.18. The curve corresponding to $q = 0$, i.e., to adiabatic flow, applies to conventional normal shock waves. It will be seen from Fig. 10.17 that for the $q = 0$ case the entropy increases across the wave when $p_2/p_1 > 1$ and decreases across the wave when $p_2/p_1 < 1$. When $q = 0$, therefore, only waves across which the density and, hence, the pressure increases are possible, i.e., only a compressive shock is possible. This result was previously derived in Chapter 5. However, when $q > 0$, it will be seen from Fig. 10.17 that an entropy increase can occur both when $p_2/p_1 > 1$ and when $p_2/p_1 < 1$. When there is heat addition, then, both compressive and expansive waves are possible. Combustion waves across which $p_2/p_1 > 1$ and therefore across which $\rho_1/\rho_2 < 1$ are termed "detonation" waves whereas combustion waves across which $p_2/p_1 < 1$, and therefore across which $\rho_1/\rho_2 > 1$ are termed "deflagration" waves. The propagation of deflagration waves is mainly governed by the rate of diffusion of heat and mass and their characteristics will not be considered further here. Therefore, attention will here be restricted to detonation waves, i.e., to waves across which $\rho_1/\rho_2 < 1$ and $p_2/p_1 > 1$, the conditions across such waves being determined by the curves given in Fig. 10.16. However, the actual pressure and density ratios across the detonation wave for any value of $q$ have still not been defined. Now it will be seen from Fig. 10.18 that for a detonation wave, the entropy increase passes through a minimum as $\rho_1/\rho_2$ decreases from 1 towards 0. It has been proposed and experimentally verified that the conditions behind detonation waves of the type here being considered correspond to those at this minimum entropy increase point. Such detonation waves are termed "Chapman–Jouguet waves." They occur, for example, when combustion is initiated at the closed end of a long duct containing a combustible gas and when the combustion results from the temperature rise across a normal shock propagating down a duct into a combustible mixture.

The fact that the entropy rise across such waves is a minimum allows the conditions behind a detonation wave to be determined. To do this, it is noted that eq. (10.98) can be written:

$$\frac{s_2 - s_1}{c_p} = \frac{1}{\gamma} \ln \left[ \frac{p_2}{p_1} \right] + \ln \left[ \frac{\rho_1}{\rho_2} \right] \tag{10.99}$$

The point of minimum entropy on the $(s_2 - s_1)/c_p$ against $\rho_1/\rho_2$ curve is required. To find this point it is noted that the above equation gives:

$$\frac{d(s/c_p)}{d(\rho_1/\rho_2)} = \frac{1}{\gamma(p_2/p_1)} \frac{d(p_2/p_1)}{d(\rho_1/\rho_2)} + \frac{1}{(\rho_1/\rho_2)} \tag{10.100}$$

From this it follows that:

$$\frac{d(s/c_p)}{d(\rho_1/\rho_2)} = 0$$

when:

$$\frac{d(p_2/p_1)}{d(\rho_1/\rho_2)} = \frac{\gamma(p_2/p_1)}{(\rho_1/\rho_2)} \tag{10.101}$$

But taking the derivative of eq. (10.97) with respect to $\rho_1/\rho_2$ for a given value of $\rho_1 q/p_1$ gives on rearrangement:

$$1 + \frac{\gamma + 1}{\gamma - 1}(p_2/p_1) = \left[ 1 - \frac{\gamma + 1}{\gamma - 1} \left( \frac{\rho_1}{\rho_2} \right) \right] \frac{d(p_2/p_1)}{d(\rho_1/\rho_2)}$$

Substituting for $d(p_2/p_1)/d(\rho_1/\rho_2)$ from eq. (10.101) then gives:

$$1 + \frac{\gamma + 1}{\gamma - 1}(p_2/p_1) = \left[ -\gamma \frac{(p_2/p_1)}{(\rho_1/\rho_2)} + \frac{\gamma(\gamma + 1)}{\gamma - 1} \left( \frac{p_2}{p_1} \right) \right]$$

This equation gives, on rearrangement:

$$\left( \frac{\rho_1}{\rho_2} \right) = \frac{\gamma(p_2/p_1)}{(\gamma + 1)(p_2/p_1) - 1} \tag{10.102}$$

In addition, since:

$$M_1^2 = \frac{V_1^2}{a_1^2} = \frac{V_1^2 \rho_1}{\gamma p_1}$$

eq. (10.93) gives:

$$M_1^2 = \left( \frac{1}{\gamma} \right) \frac{p_2/p_1 - 1}{1 - \rho_1/\rho_2} = \left( \frac{1}{\gamma} \right) \left[ (\gamma + 1) \left( \frac{p_2}{p_1} \right) - 1 \right]$$

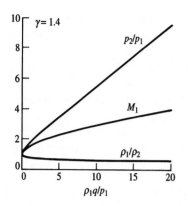

**FIGURE 10.19**
Pressure and density ratios across a detonation wave.

This equation in conjunction with eqs. (10.97) and (10.102) defines the values of $p_2/p_1$, $\rho_1/\rho_2$, and $M_1$ for a detonation wave for any specified value of $\rho_1 q/p_1$. The form of the pressure and density ratio variations with $\rho_1 q/p_1$ is illustrated in Fig. 10.19 which gives results for $\gamma = 1.4$. Very high pressure ratios across a detonation wave will be seen to be possible.

Equation (10.93) gives the velocity behind the detonation wave as:

$$V_2^2 = \left(\frac{p_1 \rho_1}{\rho_2^2}\right) \frac{p_2/p_1 - 1}{1 - \rho_1/\rho_2}$$

i.e.,

$$\frac{V_2^2}{\gamma p_2/\rho_2} = \frac{\rho_1/\rho_2}{\gamma(p_2/p_1)} \frac{p_2/p_1 - 1}{1 - \rho_1/\rho_2}$$

i.e.,

$$M_2^2 = \frac{\rho_1/\rho_2}{\gamma(p_2/p_1)} \frac{p_2/p_1 - 1}{1 - \rho_1/\rho_2}$$

Using eq. (10.102) to give the density ratio in terms of the pressure ratio then gives for a detonation wave, i.e., for a wave for which the entropy rise is a minimum:

$$M_2^2 = 1, \quad \text{i.e.,} \quad M_2 = 1 \tag{10.103}$$

Hence, the Mach number behind a Chapman–Jouguet wave is always equal to 1.

It should be noted for very strong detonation waves, i.e., for high values of $p_2/p_1$, the 1 in the denominator on the right hand side of eq. (10.102) will be negligible compared to the first term in the denominator. Therefore, for a very strong detonation wave, the density ratio reaches a limit of:

$$\frac{\rho_1}{\rho_2} = \frac{\gamma}{\gamma + 1} \tag{10.104}$$

Further, for strong detonation waves, eq. (10.97) can be approximated by:

$$-\left(\frac{p_2}{p_2}\right)\left(\frac{\rho_1}{\rho_2} + 1\right) = -\left(\frac{2\gamma}{\gamma - 1}\right)\left(\frac{p_2}{p_1}\right)\left(\frac{\rho_1}{\rho_2}\right) + \frac{\rho_1 q}{p_1}$$

Substituting the limiting value of this into $p_1/p_2$ as given by eq. (10.104) into this equation and rearranging gives the limiting pressure ratio for very strong detonation waves as:

$$\frac{p_2}{p_1} = (\gamma - 1)(\rho_1 q/p_1) \tag{10.105}$$

When the limiting values of the pressure and density ratios as given by eqs. (10.103) and (10.104) are compared with the results given in Fig. 10.19, it will be seen that the results do tend to these values at large values of $\rho_1 q/p_1$. The limiting velocity ahead of a very strong detonation wave, i.e., the limiting velocity of propagation of a strong moving detonation wave, is given by noting that when $p_2/p_1$ is large, eq. (10.93) gives:

$$V_1^2 = \frac{p_2/\rho_1}{1 - \rho_1/\rho_2}$$

which gives when eq. (10.104) is used:

$$V_1^2 = (\gamma + 1)\left(\frac{p_2}{p_1}\right) \tag{10.106}$$

This equation, together with eq. (10.105), gives the limiting value of $V_1$.

Whereas these limiting values for strong detonation waves give an indication of the characteristics of such waves, in many cases these waves will involve such large temperature increases that the assumption of a perfect gas on which these equations are based becomes invalid.

### EXAMPLE 10.8

A long insulated duct contains a combustible gas mixture at a pressure of 120 kPa and a temperature of 20°C. If a detonation wave propagates down the duct as a result of the ignition of the gas at one end of the duct, find the velocity at which the wave is moving down the duct and the pressure behind the wave. Also find the velocity behind the wave relative to the wave. The combustion causes a heat "release" of 2 MJ/kg of gas. Assume that the gas mixture has the properties of air.

### Solution

The equation of state gives:

$$\rho = \frac{p}{RT} = \frac{120\,000}{287 \times 293} = 1.427 \text{ kg/m}^3$$

Hence, here:

$$\frac{\rho_1 q}{p_1} = \frac{1.427 \times 2\,000\,000}{120\,000} = 23.78$$

Equation (10.97) therefore gives:

$$\left[1 - \frac{p_2}{p_1}\right]\left[\frac{\rho_1}{\rho_2} + 1\right] = 7\left[1 - \left(\frac{p_2}{p_1}\right)\left(\frac{\rho_1}{\rho_2}\right)\right] + 23.78$$

While eq. (10.102) gives:

$$\left(\frac{\rho_1}{\rho_2}\right) = \frac{1.4(p_2/p_1)}{2.4(p_2/p_1) - 1}$$

These two equations together determine the values of $\rho_1/\rho_2$ and $p_2/p_1$. Many algorithms and much software exists that allows the rapid determination of these values. The simplest approach is to guess a series of values of $p_2/p_1$ and then to use the second of the above equations to determine the corresponding value of $\rho_1/\rho_2$. Using these values, the left and right hand sides of the first of the above equations are found and these values compared and the value of $p_2/p_1$ that makes the two sides of the first equation equal can be deduced. Whatever procedure is adopted, the following results are obtained:

$$\frac{p_2}{p_1} = 7.73 \quad \text{and} \quad \frac{\rho_1}{\rho_2} = 0.617$$

so that:

$$p_2 = 7.73 \times 120 = 927.6 \text{ kPa}$$

The pressure behind the wave is, therefore, 927.6 kPa.

Then since:

$$M_1^2 = \left(\frac{1}{\gamma}\right) \frac{p_2/p_1 - 1}{1 - \rho_1/\rho_2}$$

the following is obtained:

$$M_1^2 = \left(\frac{1}{1.4}\right) \frac{7.73 - 1}{1 - 0.617}$$

From which it follows that:

$$M_1 = 3.53$$

and so:

$$V_1 = 3.53 \times \sqrt{1.4 \times 287 \times 293} = 1211.2 \text{ m/s}$$

This is the velocity relative to the wave upstream of the wave and is, therefore, equal to the velocity at which the wave is propagating, i.e., the wave is moving at a speed of 1211.2 m/s.

Now, the equation of state gives:

$$\frac{T_2}{T_1} = \left(\frac{p_2}{p_1}\right)\left(\frac{\rho_1}{\rho_2}\right)$$

so:

$$T_2 = 293 \times 7.73 \times 0.617 = 1397.4 \text{ K}$$

Then since the Mach number behind the wave is always equal to 1, i.e.,

$$M_2 = 1$$

it follows that:

$$V_1 = 1 \times \sqrt{1.4 \times 287 \times 1397.4} = 749.3 \text{ m/s}$$

Therefore, the velocity behind the wave relative to the wave is 749.3 m/s.

## 10.8
## CONDENSATION SHOCKS

A condensation shock is a region of relatively rapid change that occurs when there is vapor-to-liquid phase change in a supersaturated gas flow. To understand how a condensation shock can be formed, consider the flow of moist air, i.e., air containing water vapor, through a convergent–divergent nozzle. Now,

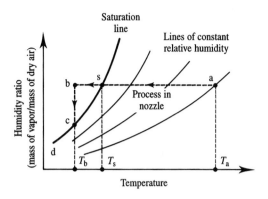

**FIGURE 10.20**
Humidity ratio: temperature variation during flow through a nozzle.

as the air flows down the nozzle, its temperature decreases. But the maximum amount of water vapor that air can contain per unit mass decreases with temperature. Thus, although the air is initially unsaturated, i.e., contains less than the maximum possible amount of water vapor, a point can be reached in the nozzle at which the air becomes saturated with water vapor. This is indicated schematically on a psychrometric chart in Fig. 10.20.

Up until the point at which condensation starts to occur, the mass of water vapor per unit mass of air, i.e., the humidity ratio, remains constant. Therefore, the process in the nozzle before condensation starts to occur can be represented by the horizontal line **asb** shown in Fig. 10.20. Once point s is reached, condensation would start to occur if the air–water vapor mixture remained in thermodynamic equilibrium, i.e., the process would be represented by line **sd** in Fig. 10.20. However, the growth of the liquid water drops in the flow that should occur once point s is reached is a relatively slow process and does not, in fact, start until point **b** in Fig. 10.20 is reached. Between points s and **b** in Fig. 10.20, the air is supersaturated because it contains more water vapor than would be possible if the air was in thermodynamic equilibrium at the same temperature. Once condensation starts, however, the process is fairly rapid and occurs over a relatively short distance. This region of condensation is termed the condensation "shock." The process through the shock is represented by the line **bc** in Fig. 10.20, the air being saturated and with a very low moisture ratio after it has reached point **c**, the rest of the flow process being along the line **cd**. The flow process is shown schematically in Fig. 10.21.

Because the degree of supersaturation that usually occurs with nozzle flows is high, almost all of the water initially in the air is condensed out in the condensation shock. The enthalpy change through a condensation shock is, therefore, approximately given by:

$$h_b - h_d = c_p(T_b - T_d) + \omega L$$

where $\omega$ is the initial humidity ratio and $L$ is the latent heat of evaporation. The specific heat $c_p$ can be taken as that of air. The term $\omega L$ behaves in the same way as a heat generation term, i.e., a condensation shock is similar to the

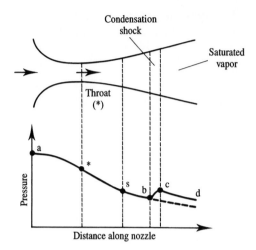

**FIGURE 10.21**
Formation of condensation shock in nozzle.

combustion waves discussed in the previous section. However, the conditions at which the condensation shock occurs are determined by the rate at which the condensation develops, not by entropy change considerations such as those associated with a detonation wave.

Although referred to as a condensation "shock," the nature of the process is really quite different from that which occurs in a conventional normal shock wave. The pressure change across a condensation shock is usually relatively small and the flow behind the "shock" is usually still supersonic. A condensation shock is also usually much thicker than a conventional shock wave.

Condensation shocks can also occur with steam flow through a nozzle. As the steam flows through the nozzle, the temperature and pressure decrease and the temperature eventually reaches the saturation temperature. However, condensation does not then normally begin, the flow continuing as a super-saturated vapor until a point is reached at which relative rapid condensation and a return to thermodynamic equilibrium occurs, i.e., a condensation shock occurs.

## 10.9
## CONCLUDING REMARKS

It was shown that, in the case of constant area flows when there is heat exchange but when the wall friction effects are negligible, the addition of heat to the flow will cause the Mach number to move towards 1, i.e., it will cause the Mach number to increase in a subsonic flow and will cause the Mach number to decrease in a supersonic flow. Heat addition can, therefore, cause the Mach number to reach 1, i.e., can cause choking to occur. It was also shown that flows in which there are changes in flow area and in which the effects of friction at the wall are not negligible can be analyzed using the same basic approach but that, in these cases, simple analytical expressions for the variations for the flow properties could not be derived. Consideration was also given to a combustion wave in which there is effectively a normal shock wave and an associated region of heat release. Two types of combustion wave were shown to be possible but detailed attention was only given to detonation waves which are normally associated with the release of large amounts of chemical energy. Expressions for the changes across such waves were derived and it was shown that the Mach number behind such waves is always 1.

## PROBLEMS

**10.1.** Air flows through a constant area duct. The air has a temperature of 20°C and a Mach number of 0.5 at the entrance to the duct. It is desired to transfer heat to the duct such that at the exit of the duct the stagnation temperature is 1180°C. Is this possible? If not, what adjustment must be made to Mach number at the entrance in order to give a discharge stagnation temperature of 1180°C? Ignore the effects of friction.

**10.2.** Air with a stagnation temperature of 430°C and a stagnation pressure of 1.6 MPa enters a constant area duct in which heat is transferred to the air. The Mach number at the inlet to the duct is 3 and the Mach number at the exit to the duct is 1. Determine the stagnation temperature and the stagnation pressure at the exit. Ignore the effects of friction.

**10.3.** Air at a temperature of 100°C with a pressure of 101 kPa enters a constant area combustion chamber at a velocity of 130 m/s. Determine the maximum amount of heat that can be transferred to the air flow per unit mass of air. Assume friction effects are negligible.

**10.4.** Heat is supplied to air flowing through a short tube causing the Mach number to increase from an initial value of 0.3 to a final value of 0.6. If the initial air temperature is 40°C and if the effects of friction are neglected, find the heat supplied per unit mass.

**10.5.** Air flows through a 10 cm diameter pipe at a rate of 0.18 kg/s. If the air enters the pipe at a temperature of 0°C and a pressure of 60 kPa, how much heat can be added to the air per kilogram without choking the flow? The effects of friction are negligible.

**10.6.** Air enters a constant area combustion chamber at a pressure of 101 kPa and a temperature of 70°C with a velocity of 130 m/s. By ignoring the effects of friction, determine the maximum amount of heat that can be transferred to the flow per unit mass of air.

**10.7.** Air flows through a constant diameter duct. At the inlet the velocity is 300 m/s, and the stagnation temperature is 90°C. If the Mach number at the exit is 0.3, determine the direction and the rate of heat transfer. For the same conditions at the inlet, determine the amount of heat that must be transferred to the system per unit mass of air if the flow is to be sonic at the exit of the duct.

**10.8.** Air flows through a constant area combustion chamber which has a diameter of 0.15 m. The inlet stagnation temperature is 335 K, the inlet stagnation pressure is 1.4 MPa, and the inlet Mach number is 0.55. Find the maximum rate at which heat can be added to the flow. Neglect the effects of friction.

**10.9.** Air enters a constant area duct at a Mach number of 0.15, a pressure of 200 kPa, and a temperature of 20°C. Heat is added to the air that flows through the duct at a rate of 60 kJ/kg of air. Assuming that the flow is steady and that the effects of wall friction can be ignored, find the temperature, pressure, and Mach number at which the air leaves the duct. Assume that the air behaves as a perfect gas.

**10.10.** At the inlet to a constant area combustion chamber the Mach number is 0.2 and the stagnation temperature is 120°C. What is the amount of heat transfer to the gas per unit mass if the Mach number is 0.7 at the exit of the chamber? What is the maximum possible amount of heat transfer? The gas can be assumed to have the properties of air.

**10.11.** Air flows through a constant area duct. At the inlet to the duct the stagnation pressure is 600 kPa and the stagnation temperature 200°C. If the Mach number at the inlet of the duct is 0.5 and if the flow is choked at the exit of the duct, determine the heat transfer per unit mass and the exit temperature. Assume that friction effects can be neglected.

**10.12.** Air enters a constant diameter duct at a pressure of 200 kPa. At the exit of the duct the pressure is 120 kPa, the Mach number is 0.75, and the stagnation temperature is 330°C. Determine the inlet Mach number and the heat transfer per unit mass of air.

**10.13.** Air flows through a 4 in diameter pipe. The pressure and temperature at the inlet to the pipe are 15 psia and 70°F. The velocity at the inlet is 200 ft/sec. If the temperature at the exit to the pipe is 1300°F, how much heat must be added to the air and what will be the exit pressure, velocity, and Mach number? Ignore the effects of friction.

**10.14.** Air flows through a constant area duct. The air enters the duct at a pressure of 1 MPa, a Mach number of 0.5, and stagnation temperature of 45°C. The Mach number at the duct exit is 0.9 and the stagnation temperature at this point is 160°C. Find the amount of heat transferred per unit mass to or from the air in the duct. Also find the pressure at the duct exit. Assume that the effects of friction are negligible.

**10.15.** Air enters a constant area duct at a pressure of 620 kPa and a temperature of 300°C, the velocity at the inlet being 100 m/s. If the velocity at the exit to the duct is 210 m/s, determine the pressure, temperature, stagnation pressure, and stagnation temperature at the exit to the duct. Also find the heat transfer per unit mass in the duct. Assume that the effects of friction are negligible.

**10.16.** Air enters a pipe at a pressure of 200 kPa. At the exit of the pipe, the pressure is 120 kPa, the Mach number is 0.75, and the stagnation temperature is 300°C. Determine the inlet Mach number and the heat transfer rate to the air assuming that the effects of friction can be ignored.

**10.17.** Air with a stagnation pressure of 600 kPa and a stagnation temperature of 200°C enters a constant area pipe with a diameter of 2 cm. Heat is transferred to the air as it flows through the duct at a rate of 100 kJ/kg. The air is then isentropically brought to rest in a large chamber. Plot the mass flow rate of air as a function of back pressure (i.e., the chamber pressure) for the range 0 to 400 kPa. Assume frictionless flow.

**10.18.** Air flows through a rectangular duct with a 10 cm × 16 cm cross-sectional area. The air velocity, pressure, and temperature at the inlet to the duct are 90 m/s, 105 kPa, and 25°C, respectively. Heat is added to the air as it flows through the duct and it leaves the duct with a velocity of 200 m/s. Find the pressure and temperature at the exit and the total rate at which heat is being added to the air. Ignore the effects of friction.

**10.19.** Air is heated as it flows through a constant area duct by an electric heating coil wrapped uniformly around the duct. The air enters the duct at a velocity of 100 m/s, a temperature of 20°C, and a pressure of 101.3 kPa, and the heat transfer rate is 40 kJ/kg per unit length of duct. Plot the variations of exit Mach number, exit temperature, and exit pressure with the length of the duct. Neglect the effects of fluid friction.

**10.20.** Air enters a combustion chamber at a velocity of 100 m/s, a pressure of 90 kPa, and a temperature of 40°C. As a result of the combustion, heat is added to the air at a rate of 500 kJ/kg. Find the exit velocity and Mach number. Also find the heat addition that would be required to choke the flow. Neglect the effects of friction and assume that the properties of the gas in the combustion chamber are the same as those of air.

**10.21.** A fuel–air mixture which can be assumed to have the properties of air, enters a constant area combustion chamber at a velocity of 10 m/s and a temperature of 100°C. What amount of heat must be added per unit mass to cause the flow at the exit to be choked? Also find the exit Mach number and temperature if the actual heat addition due to combustion is 1000 kJ/kg.

**10.22.** Fuel and air are thoroughly mixed in the proportion of 1 : 40 by mass before entering a constant area combustion chamber. The pressure, temperature, and velocity at the inlet to the chamber are 50 kPa, 30°C, and 80 m/s respectively. The heating value of the fuel is 40 MJ/kg of fuel. Assuming steady flow and that the properties of the gas mixture are the same as those of air, determine the pressure, the stagnation temperature, and the Mach number at the exit of the combustion chamber. Neglect the effects of friction.

**10.23.** An air–fuel mixture enters a constant area combustion chamber at a velocity of 100 m/s, a pressure of 70 kPa, and a temperature of 150°C. Assuming that the fuel–air ratio is 0.04, that the heating value of the fuel is 30 MJ/kg and that the mixture has the properties of air, calculate the Mach number of the gases after combustion is completed and the change of stagnation temperature and stagnation pressure across the combustion chamber. Neglect the effects of friction and assume that the properties of the gas in the combustion chamber are the same as those of air.

**10.24.** An air–fuel mixture flows through a constant area combustion chamber. The velocity, pressure, and temperature at the entrance to the chamber are 130 m/s, 170 kPa, and 120°C respectively. If the enthalpy of reaction is 650 kJ/kg of mixture, find the Mach number and pressure at the exit of the chamber. Neglect the effects of friction and assume that the properties of the gas in the combustion chamber are the same as those of air.

**10.25.** An air–fuel mixture enters a constant area combustion chamber at a Mach number of 0.25, a pressure of 70 kPa, and a temperature of 35°C. If the heat transfer to the gases in the combustion chamber is 1.2 MJ/kg of mixture, determine the Mach number at the exit of the chamber and the change in stagnation temperature through the chamber. Neglect the effects of friction and assume that the properties of the gas in the combustion chamber are the same as those of air.

**10.26.** In a gas turbine plant, air from the compressor enters the combustion chamber at a pressure of 420 kPa, a temperature of 110°C, and a velocity of 80 m/s. Fuel having an effective heating value of 35 000 kJ/kg is sprayed into the air stream and burnt. Two types of injection systems are available. One gives a fuel-to-air mass ratio of 0.015, the other a ratio of 0.021. The temperature entering the turbine should not be less than 750°C but should not exceed the temperature determined by the metallurgical limit of the blade material. Which of the two injection systems should be used?

**10.27.** Air enters a constant area duct with a Mach number of 2.5, a stagnation temperature of 300°C, and a stagnation pressure of 1.2 MPa. If the flow is choked, determine the stagnation pressure and stagnation temperature at the exit to the duct and the heat transfer per unit mass if there is a normal shock at the inlet of the duct.

**10.28.** Air at a temperature of 300 K, flowing at a Mach number of 1.5, enters a constant area duct which feeds a convergent nozzle. At the exit of the nozzle the Mach number is 1.0 and the ratio of the nozzle exit area to the duct area is 0.98. If a normal shock occurs in the duct just upstream of the nozzle inlet, calculate the amount and direction of heat exchange with the air flow through the duct. Ignore the effect of friction on the flow in the duct and assume that the flow downstream of the shock is isentropic.

**10.29.** A jet engine is operating at an altitude of 7000 m. The mass of air passing through the engine is 45 kg/s and the heat addition in the combustion chamber is 500 kJ/kg. The cross-sectional area of the combustion chamber is 0.5 m², and the air enters the chamber at a pressure of 80 kPa and a temperature of 80°C. After the combustion chamber, the products of combustion, which can be assumed to have the properties of air, are expanded through a convergent nozzle to match the atmospheric pressure at the nozzle exit. Estimate the nozzle exit diameter and the nozzle exit velocity assuming the flow in the nozzle is isentropic. State the assumptions that have been made.

**10.30.** Air enters a 7.5 cm diameter pipe at a pressure of 1.3 kPa, a temperature of 200°C, and a Mach number of 1.8. Heat is added to the flow as a result of a chemical reaction taking place in the duct. Find the heat transfer rate necessary to choke the flow in the pipe. Assume that the air behaves as a perfect gas with constant specific heats and neglect changes in the composition of the gas stream due to the chemical reaction.

**10.31.** Air enters a 15 cm diameter pipe at a pressure of 1.3 MPa and a temperature of 20°C, and with a velocity of 60 m/s. Assuming that the friction factor is 0.004 and that the flow is effectively isothermal, find the Mach number at a point in the pipe where the pressure is 300 kPa and the length of the pipe to this point.

**10.32.** Consider subsonic air flow through a pipe with a diameter of 2.5 cm which is 3 m long. At the inlet to the pipe the pressure is 200 kPa and the Mach number is 0.35. If the mean value of the friction factor is assumed to be equal to 0.006, determine the exit Mach numbers assuming that the flow is isothermal.

**10.33.** Air at a pressure of 550 kPa and a temperature of 30°C flows through a pipe with a diameter of 0.3 m and a length of 140 m. If the mass flow rate is at its maximum value, find, assuming that the flow is isothermal and that the average friction factor is 0.0025, the Mach number at the inlet to the pipe, and the mass flow rate through the pipe.

**10.34.** In long pipelines, such as those used to convey natural gas, the temperature of the gas can usually be considered constant. In one such case, the gas leaves a pumping station at a pressure of 320 kPa and a temperature of 25°C with a Mach number of 0.10. At some other point in the flow, the pressure is measured and found to be 130 kPa. Calculate the Mach number of the flow at this section and determine how much heat has been added to or removed from the gas per unit mass between the pumping station and the point where the measurements are made. The gas can be assumed to have the properties of methane.

**10.35.** Natural gas is to be pumped through a 36 in diameter pipe connecting two compressor stations 40 miles apart. At the upstream station the pressure is 100 psig and the Mach number is 0.025. Find the pressure at the downstream station. Assume that there is sufficient heat transfer through the pipe to maintain the gas at a temperature of 70°F. The gas can be assumed to have a specific heat ratio of 1.3 and the friction factor can be assumed to be 0.004.

**10.36.** A 3 km long pipeline with a diameter of 0.10 m is used to transport methane at a rate of 1.0 kg/s. If the gas remains essentially at a constant temperature of 10°C and if the pressure at the exit to the pipe is 150 kPa, find the inlet velocity. Assume an average friction factor of 0.004. Methane has a molar mass of 16 and a specific heat ratio of 1.3.

**10.37.** Air enters a convergent duct at a Mach number of 0.75, a pressure of 500 kPa, and a temperature of 35°C. The exit area of the duct is half the inlet area. If, as a result of heat transfer, the Mach number at the duct exit is also 0.75, find the heat transfer rate per kilogram of air. Neglect the effects of friction.

**10.38.** A combustible gas mixture is contained in a long insulated tube at a pressure of 150 kPa and a temperature of 30°C. The gas is ignited at one end of the tube leading to the propagation of a detonation wave down the pipe. If the combustion causes a heat "release" of 1 MJ/kg of gas, find the pressure and temperature behind the detonation wave and the velocity at which the wave is moving down the pipe. Assume that the gas mixture has the properties of air.

# CHAPTER 11

# Generalized Quasi-One-Dimensional Flow

**11.1**
**INTRODUCTION**

The effects of area change, of friction and of heat transfer on compressible flow have been more or less separately considered in the preceding chapters. Brief attention was given to some situations in which at least two of these effects had to be simultaneously considered, e.g., variable area adiabatic flow with friction, variable area frictionless flow with heat transfer, and one-dimensional flow with both friction and heat transfer were considered. In general, however, the effects of area change, of friction, and of heat transfer are all simultaneously important in the flow. In addition, there may be mass addition or removal from the system, work may be done by or on the flow, and there may be objects in the flow and the drag force on these objects may influence the overall momentum change. Such a general flow situation is shown in Fig. 11.1.

The mass addition or removal may be associated with the use of a porous wall through which the gas passes, or it may be a way of modeling a combustion process, e.g., when considering the flow in the combustion chamber of a solid-fuelled rocket engine there is at the walls, i.e., at the surface of the propellant, an effective injection of gas into the flow as a result of the combustion. This is illustrated schematically in Fig. 11.2.

Attention will therefore be given in the present chapter to the analysis of flows in which all of the following effects can exist:

- The flow area changes
- Wall friction effects are important
- Heat exchange has an effect on the flow

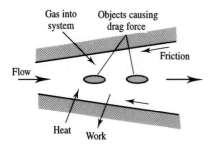

**FIGURE 11.1**
Generalized one-dimensional flow.

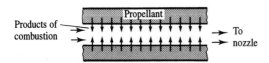

**FIGURE 11.2**
Simple model of flow in a solid-fuelled rocket motor combustion chamber.

- Gas is being injected into and removed from the flow
- Work is being done by or on the flow
- There is a drag force on the flow

In the analysis presented in this chapter it will be assumed that the flow can be adequately modeled as being steady and one-dimensional. It will also be assumed that the injected gas has the same properties as the gas in the main flow stream.

## 11.2
## GOVERNING EQUATIONS AND INFLUENCE COEFFICIENTS

As discussed in the preceding chapters, consider flow through a differentially short portion of the flow shown in Fig. 11.3.

As indicated in Fig. 11.3, gas is being injected into the system at a rate $d\dot{m}$ with velocity of $V_i$ at an angle $\phi_i$ to the main flow direction, the injected gas

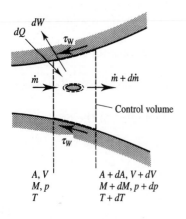

**FIGURE 11.3**
Flow situation considered.

having a temperature of $T_i$. The rate of heat addition and the rate work is being done on the system are $dQ$ and $dW$ respectively.

The application of the governing equations to the flow through the control volume shown in Fig. 11.3 will now be considered.

The perfect gas law applied at the inlet and at the outlet of the system gives:

$$p = \rho RT \quad \text{and} \quad p + dp = (\rho + d\rho)R(T + dT) \tag{11.1}$$

The second of these equations gives to first order accuracy, i.e., by assuming that the term $d\rho \times dT$ is negligible because $d\rho$, $dT$, etc. are very small:

$$p + dp = \rho RT + RT\, d\rho + \rho R\, dT$$

Dividing this by the first of the equations given in eq. (11.1) then gives as also shown in an earlier chapter:

$$\frac{dp}{p} = \frac{d\rho}{\rho} + \frac{dT}{T} \tag{11.2}$$

Next, consider the definition of the Mach number at the inlet and outlet of the system, i.e.,

$$M^2 = \frac{V^2}{a^2} = \frac{V^2}{\gamma RT}$$

and

$$(M + dM)^2 = \frac{(V + dV)^2}{(a + da)^2} = \frac{(V + dV)^2}{\gamma R(T + dT)}$$

i.e.,

$$M^2\left(1 + \frac{dM^2}{M^2}\right) = \frac{V^2(1 + dV/V)^2}{\gamma RT(1 + dT/T)} \tag{11.3}$$

i.e., to first order of accuracy:

$$M^2\left(1 + \frac{dM^2}{M^2}\right) = \frac{V^2}{\gamma RT}\left(1 + 2\frac{dV}{V}\right)\left(1 - \frac{dT}{T}\right)$$

i.e.,

$$M^2\left(1 + \frac{dM^2}{M^2}\right) = \frac{V^2}{\gamma RT}\left(1 + 2\frac{dV}{V} - \frac{dT}{T}\right)$$

Hence, it follows that to first order of accuracy:

$$\frac{dM^2}{M^2} = \frac{2\,dV}{V} - \frac{dT}{T} \tag{11.4}$$

Next, consider the conservation of mass principle applied to the control value shown in Fig. 11.3. This gives:

$$(\rho + d\rho)(V + dV)(A + dA) - \rho VA = d\dot{m} \tag{11.5}$$

From this it follows that to first order of accuracy:

$$\rho VA + VA\, d\rho + \rho A\, dV + \rho V\, dA - \rho VA = d\dot{m}$$

Dividing this equation through by $\dot{m} = \rho VA$ then gives:

$$\frac{dV}{V} + \frac{d\rho}{\rho} + \frac{dA}{A} = \frac{d\dot{m}}{\dot{m}} \tag{11.6}$$

Next, consider conservation of momentum. If $dF_D$ is the net drag force acting on any components in the flow, then this gives:

$$\begin{array}{l} \text{Net pressure} \\ \text{force on 1} \end{array} - \begin{array}{l} \text{Net pressure} \\ \text{force on 2} \end{array} + \begin{array}{l} \text{Net pressure force} \\ \text{on curved surface} \end{array}$$

$$- \begin{array}{l} \text{Component of net shear} \\ \text{force in flow direction} \end{array} - \text{Net drag force}$$

$$= \begin{array}{l} \text{Rate momentum leaves} \\ \text{system through 2} \end{array} - \begin{array}{l} \text{Rate momentum enters} \\ \text{system through 1} \end{array} - \begin{array}{l} \text{Rate momentum is} \\ \text{injected through wall} \end{array}$$

i.e., since the force due to the wall shear stress is equal to the product of the shear stress and the surface area of the portion of the duct being considered:

$$pA - (p + dp)(A + dA) + p\,dA - \tau_w\,dA_w\cos\phi_w - dF_D$$
$$= (\dot{m} + d\dot{m})(V + dV) - \dot{m}V - d\dot{m}V_i\cos\phi_i \tag{11.7}$$

where $\tau_w$ is again the wall shear stress.

In writing this equation, the expression for the pressure force on the curved surface that was derived earlier has been used. In this equation, $dA_w$ is the area of the wall surface. As before, this is written as:

$$dA_w = P\,dx \tag{11.8}$$

where $P$ is the perimeter of the duct through which the gas is flowing. $\phi_w$ is the angle the walls make to the $x$-direction. In order for the quasi-one-dimensional assumption to apply, $\phi_w$ must remain small and $\cos\phi_w$ will, therefore, be taken as 1. Equation (11.7) can, therefore, be written to first order of accuracy as:

$$-dpA - \tau_w P\,dx - dF_D = \dot{m}\,dV + V\,d\dot{m} - d\dot{m}V_i\cos\phi_i \tag{11.9}$$

Hence setting $\dot{m} = \rho V A$ and introducing the Fanning friction factor defined as before by:

$$\tau_w = f\frac{\rho V^2}{2} \tag{11.10}$$

eq. (11.9) can be written, after dividing through by $\dot{m}$, as:

$$-\frac{dp}{\rho V} - \frac{f}{2}\frac{P}{A}V\,dx - \frac{dF_D}{\rho V A} = dV + V\frac{d\dot{m}}{\dot{m}} - \frac{d\dot{m}V_i\cos\phi_i}{\dot{m}}$$

i.e.,
$$-\frac{dp}{p}\frac{p}{\rho V^2} - \frac{f}{2}\frac{P}{A}dx - \frac{dF_D}{\rho V^2 A} = \frac{dV}{V} + \frac{d\dot{m}}{\dot{m}} - \frac{d\dot{m}}{\dot{m}}\frac{V_i}{V}\cos\phi_i \tag{11.11}$$

Then defining:

$$r_i = \frac{V_i}{V}\cos\phi_i \tag{11.12}$$

and recalling that:

$$M^2 = \frac{V^2}{a^2} = \frac{V^2 \rho}{\gamma p}$$

eq. (11.11) can be written as:

$$-\frac{1}{\gamma M^2}\frac{dp}{p} - \frac{f}{2}\frac{P}{A}dx - \frac{dF_D}{\rho V^2 A} - \frac{dV}{V} - \frac{d\dot{m}}{\dot{m}} + r_i\frac{d\dot{m}}{\dot{m}} = 0 \qquad (11.13)$$

This can be written as:

$$\frac{dp}{p} + \frac{\gamma M^2}{2}\left(f\frac{P}{A}dx + 2\frac{dF_D}{\gamma p M^2 A} - 2r_i\frac{d\dot{m}}{\dot{m}}\right) + \gamma M^2\frac{dV}{V} + \gamma M^2\frac{d\dot{m}}{\dot{m}} = 0 \quad (11.14)$$

It has proved to be convenient to introduce, in conjunction with this equation, the so-called "impulse function," $I$, which is defined by:

$$I = pA + \dot{m}V \qquad (11.15)$$

i.e., since $\dot{m} = \rho V A$ and $M^2 = V^2\rho/\gamma p$:

$$I = pA(1 + \gamma M^2) \qquad (11.16)$$

From this it follows that:

$$I + dI = (p + dp)(A + dA)[1 + \gamma(M^2 + dM^2)]$$

Hence, to first order of accuracy:

$$dI = pA(1 + \gamma M^2) + pA\gamma M^2 + p\,dA(1 + \gamma M^2)$$
$$+ A\,dp(1 + \gamma M^2) - pA(1 + \gamma M^2)$$

from which it follows that:

$$\frac{dI}{I} = \frac{dp}{p} + \frac{dA}{A} + \frac{\gamma M^2}{1 + \gamma M^2}\frac{dM^2}{M^2} \qquad (11.17)$$

Next, consider the application of the conservation of energy principle to the flow through the control volume being considered. This gives:

$$(\dot{m} + d\dot{m})\left[c_p(T + dT) + \frac{(V + dV)^2}{2}\right] - \dot{m}\left[c_p T + \frac{V^2}{2}\right]$$
$$-d\dot{m}\left[c_p T_i + \frac{V_i^2}{2}\right] = dQ - dW \quad (11.18)$$

where $dW$ is the work done by the system.

Equation (11.18) can be written as:

$$\dot{m}\left[c_p(T + dT) + \frac{(V + dV)^2}{2} - c_p T - \frac{V^2}{2}\right]$$
$$-d\dot{m}\left[c_p(T - T_i) + \frac{V^2 - V_i^2}{2}\right] = dQ - dW \quad (11.19)$$

i.e., dividing through by $\dot{m}$ and defining $dq$ and $dw$ as the heat exchange and work done per unit mass of flowing fluid, this equation becomes:

$$\left[ c_p(T + dT) + \frac{(V + dV)^2}{2} - c_p T - \frac{V^2}{2} \right]$$

$$- \frac{d\dot{m}}{\dot{m}} \left[ c_p(T - T_i) + \frac{(V^2 - V_i^2)}{2} \right] = dq - dw \quad (11.20)$$

Now if $T_0$ and $T_0 + dT_0$ are the stagnation temperatures at the inlet and exit of the control volume, then:

$$c_p T_0 = c_p T - \frac{V^2}{2}$$

and: $\qquad\qquad c_p(T_0 + dT_0) = c_p(T + dT) + \frac{(V + dV)^2}{2}$

The first term in eq. (11.20) is, therefore, equal to $c_p(T_0 + dT_0) - c_p T_0 = c_p \, dT_0$. Therefore, eq. (11.20) can be written as:

$$\frac{dT_0}{T_0} = \left\{ dq - dw - \frac{d\dot{m}}{\dot{m}} \left[ c_p(T - T_i) + \frac{(V^2 - V_i^2)}{2} \right] \right\} \Big/ c_p T_0 \qquad (11.21)$$

This equation indicates that the increase in stagnation temperature is a measure of the net effect of heat exchange, work, and injected enthalpy. Thus, instead of specifying these quantities separately, only the increase in the stagnation temperature need be specified.

It is convenient to note that:

$$T_0 = T\left( 1 + \frac{\gamma - 1}{2} M^2 \right)$$

and consequently that:

$$T_0 + dT_0 = (T + dT)\left[ 1 + \frac{\gamma - 1}{2}(M^2 + dM^2) \right]$$

Subtracting these two equations then gives to first order of accuracy:

$$dT_0 = \left( 1 + \frac{\gamma - 1}{2} M^2 \right) dT + \frac{\gamma - 1}{2} T \, dM^2$$

Dividing this by the first of the above equations then gives:

$$\frac{dT_0}{T_0} = \frac{dT}{T} + \left[ \frac{(\gamma - 1)M^2/2}{1 + (\gamma - 1)M^2/2} \right] \frac{dM^2}{M^2} \qquad (11.22)$$

It is also noted that the change in entropy through the control volume shown in Fig. 11.3 is given by:

$$\frac{ds}{c_p} = \frac{dT}{T} - \frac{(\gamma - 1)}{\gamma} \frac{dp}{p} \qquad (11.23)$$

Lastly it is recalled that the stagnation pressure is given by:

$$p_0 = p\left[1 + \frac{\gamma - 1}{2}M^2\right]^{\frac{\gamma}{\gamma - 1}}$$

so:

$$p_0 + dp_0 = (p + dp)\left[1 + \frac{\gamma - 1}{2}(M^2 + dM^2)\right]^{\frac{\gamma}{\gamma - 1}}$$

hence:

$$\frac{p_0 + dp_0}{p_0} = \frac{(p + dp)}{p}\left[1 + \frac{(\gamma - 1)\,dM^2/2}{1 + (\gamma - 1)M^2/2}\right]^{\frac{\gamma}{\gamma - 1}}$$

which is, to first order of accuracy:

$$1 + \frac{dp_0}{p_0} = \left(1 + \frac{dp}{p}\right)\left[1 + \left(\frac{\gamma}{\gamma - 1}\right)\left(\frac{\gamma - 1}{2}\right)\frac{dM^2}{1 + (\gamma - 1)M^2/2}\right]$$

i.e.,

$$\frac{dp_0}{p_0} = \frac{dp}{p} + \frac{\gamma M^2/2}{1 + (\gamma - 1)M^2/2}\frac{dM^2}{M^2} \tag{11.24}$$

Equations (11.2), (11.4), (11.6), (11.14), (11.17), (11.22), (11.23), and (11.24) describe the general quasi-one-dimensional flow being considered. They represent a set of eight equations that involve the following quantities:

$$\frac{dp}{p}, \quad \frac{dT}{T}, \quad \frac{d\rho}{\rho}, \quad \frac{dM^2}{M^2}, \quad \frac{dV}{V}, \quad \frac{dA}{A}, \quad \frac{d\dot{m}}{\dot{m}}, \quad \frac{dI}{I}, \quad \frac{dT_0}{T_0}, \quad \frac{ds}{c_p}, \quad \frac{dp_0}{p_0}$$

as well as the quantity:

$$\left(f\frac{P}{A}dx + 2\frac{dF_D}{\gamma p M^2 A} - 2r_i\frac{d\dot{m}}{\dot{m}}\right)$$

Of these quantities, the following will, in most cases, be specified:

$$\frac{dA}{A}, \quad \frac{d\dot{m}}{\dot{m}}, \quad \frac{dT_0}{T_0}$$

$$\left(f\frac{P}{A}dx + 2\frac{dF_D}{\gamma p M^2 A} - 2r_i\frac{d\dot{m}}{\dot{m}}\right)$$

This leaves the following to be determined:

$$\frac{dp}{p}, \quad \frac{dT}{T}, \quad \frac{d\rho}{\rho}, \quad \frac{dM^2}{M^2}, \quad \frac{dV}{V}, \quad \frac{dI}{I}, \quad \frac{ds}{c_p}, \quad \frac{dp_0}{p_0}$$

This represents a set of eight variables that are described by the eight equations derived above, these eight equations being given in somewhat modified form below for completeness:

$$\frac{dp}{p} - \frac{dT}{T} - \frac{d\rho}{\rho} = 0$$

$$\frac{dM^2}{M^2} - \frac{2\,dV}{V} + \frac{dT}{T} = 0$$

$$\frac{dV}{V} + \frac{d\rho}{\rho} + \frac{dA}{A} - \frac{d\dot{m}}{\dot{m}} = 0$$

$$\frac{dp}{p} + \gamma M^2 \frac{dV}{V} + \gamma M^2 \frac{d\dot{m}}{\dot{m}} + \frac{\gamma M^2}{2}\left(f\frac{P}{A}dx + 2\frac{dF_D}{\gamma p M^2 A} - 2r_i\frac{d\dot{m}}{\dot{m}}\right) = 0$$

$$\frac{dI}{I} - \frac{dp}{p} - \frac{dA}{A} - \frac{\gamma M^2}{1 + \gamma M^2}\frac{dM^2}{M^2} = 0$$

$$\frac{dT_0}{T_0} - \frac{dT}{T} - \left[\frac{(\gamma - 1)M^2/2}{1 + (\gamma - 1)M^2/2}\right]\frac{dM^2}{M^2} = 0$$

$$\frac{ds}{c_p} - \frac{dT}{T} + \frac{(\gamma - 1)}{\gamma}\frac{dp}{p} = 0$$

$$\frac{dp_0}{p_0} - \frac{dp}{p} - \left[\frac{\gamma M^2/2}{1 + (\gamma - 1)M^2/2}\right]\frac{dM^2}{M^2} = 0$$

This set of equations give the following form:

$$\begin{bmatrix} 1 & -1 & -1 & 0 & 0 & 0 & 0 & 0 \\ 0 & 1 & 0 & 1 & -2 & 0 & 0 & 0 \\ 0 & 0 & 1 & 0 & 1 & 0 & 0 & 0 \\ 1 & 0 & 0 & 0 & \gamma M^2 & 0 & 0 & 0 \\ -1 & 0 & 0 & -X & 0 & 1 & 0 & 0 \\ 0 & 1 & 0 & -Y & 0 & 0 & 0 & 0 \\ \gamma-1/\gamma & -1 & 0 & 0 & 0 & 0 & 1 & 0 \\ -1 & 0 & 0 & -Z & 0 & 0 & 0 & 1 \end{bmatrix} \begin{bmatrix} dp/p \\ dT/T \\ d\rho/\rho \\ dM^2/M^2 \\ dV/V \\ dI/I \\ ds/c_p \\ dp_0/p_0 \end{bmatrix} = \begin{bmatrix} 0 \\ 0 \\ d\dot{m}/\dot{m} - dA/A \\ S + W \\ dA/A \\ dT_0/T_0 \\ 0 \\ 0 \end{bmatrix}$$

$$(11.25)$$

where:

$$S = -\frac{\gamma M^2}{2}\left(f\frac{P}{A}dx + 2\frac{dF_D}{\gamma p M^2 A}\right)$$

$$W = -\gamma M^2(1 - r_i)\frac{d\dot{m}}{\dot{m}}$$

$$X = \frac{\gamma M^2}{1 + \gamma M^2}$$

$$(11.26)$$

$$Y = \frac{(\gamma - 1)M^2/2}{1 + (\gamma - 1)M^2/2}$$

$$Z = \frac{\gamma M^2/2}{1 + (\gamma - 1)M^2/2}$$

The matrix in eq. (11.25) can be inverted to give expressions for each of the variables. This procedure gives a result of the form:

$$
\begin{bmatrix}
dp/p \\
dT/T \\
d\rho/\rho \\
dM^2/M^2 \\
dV/V \\
dI/I \\
ds/c_p \\
dp_0/p_0
\end{bmatrix}
=
\begin{bmatrix}
F_{p1} & F_{p2} & F_{p3} & F_{p4} & 0 & F_{p6} & 0 & 0 \\
F_{T1} & F_{T2} & F_{T3} & F_{T4} & 0 & F_{T6} & 0 & 0 \\
F_{\rho1} & F_{\rho2} & F_{\rho3} & F_{\rho4} & 0 & F_{\rho6} & 0 & 0 \\
F_{M1} & F_{M2} & F_{M3} & F_{M4} & 0 & F_{M6} & 0 & 0 \\
F_{V1} & F_{V2} & F_{V3} & F_{V4} & 0 & F_{V6} & 0 & 0 \\
F_{I1} & F_{I2} & F_{I3} & F_{I4} & 1 & F_0 & 0 & 0 \\
0 & F_{s2} & 0 & F_{s4} & 0 & F_{s6} & 0 & 1 \\
0 & F_{02} & 0 & 1 & 0 & F_{06} & 0 & 0
\end{bmatrix}
\begin{bmatrix}
0 \\
0 \\
d\dot{m}/\dot{m}-dA/A \\
V + W \\
dA/A \\
dT_0/T_0 \\
0 \\
0
\end{bmatrix}
$$

$$(11.27)$$

The $F$ values are basically what are termed the "influence coefficients." These coefficients are given by:

$$F_{p1} = \frac{\gamma M^2}{M^2 - 1}, \qquad F_{p2} = -\frac{\gamma(\gamma - 1)M^4}{2(M^2 - 1)}$$

$$F_{p3} = \frac{\gamma M^2}{M^2 - 1}, \qquad F_{p4} = \frac{1 + (\gamma - 1)M^2}{M^2 - 1}$$

$$F_{p6} = -\frac{\gamma M^2[1 + (\gamma - 1)M^2/2]}{M^2 - 1}, \qquad F_{T1} = \frac{(\gamma - 1)M^2}{M^2 - 1}$$

$$F_{T2} = -\frac{(\gamma - 1)M^2(\gamma M^2 - 1)}{2(M^2 - 1)}, \qquad F_{T3} = \frac{(\gamma - 1)M^2}{M^2 - 1}$$

$$F_{T4} = -\frac{(\gamma - 1)M^2}{M^2 - 1}, \qquad F_{T6} = -\frac{(\gamma M^2 - 1)[1 + (\gamma - 1)M^2/2]}{M^2 - 1}$$

$$F_{\rho1} = \frac{1}{M^2 - 1}, \qquad F_{\rho2} = -\frac{(\gamma - 1)M^2}{2(M^2 - 1)}$$

$$F_{\rho3} = \frac{M^2}{M^2 - 1}, \qquad F_{\rho4} = -\frac{1}{M^2 - 1}$$

$$F_{\rho6} = -\frac{1 + (\gamma - 1)M^2/2}{M^2 - 1}, \qquad F_{M1} = -\frac{2[1 + (\gamma - 1)M^2/2]}{M^2 - 1}$$

$$F_{M2} = \frac{(\gamma M^2 - 1)[1 + (\gamma - 1)M^2/2]}{M^2 - 1}, \qquad F_{M3} = -\frac{2[1 + (\gamma - 1)M^2/2]}{M^2 - 1}$$

$$F_{M4} = \frac{2[1 + (\gamma - 1)M^2/2]}{M^2 - 1}, \qquad F_{M6} = -\frac{(\gamma M^2 - 1)[1 + (\gamma - 1)M^2/2]}{M^2 - 1}$$

$$F_{V1} = -\frac{1}{M^2 - 1}, \qquad F_{V2} = \frac{(\gamma - 1)M^2}{2(M^2 - 1)}$$

$$F_{V3} = -\frac{1}{M^2 - 1}, \qquad F_{V4} = \frac{1}{M^2 - 1}$$

$$F_{V6} = -\frac{1 + (\gamma - 1)M^2/2}{M^2 - 1}, \qquad F_{I1} = \frac{\gamma M^2}{\gamma M^2 + 1}$$

$$F_{I2} = \frac{\gamma M^2}{\gamma M^2 + 1}, \qquad F_{I3} = \frac{\gamma M^2}{\gamma M^2 + 1}$$

$$F_{I4} = \frac{1}{\gamma M^2 + 1}, \qquad F_{s2} = -\frac{(\gamma - 1)M^2}{2}$$

$$F_{s4} = -\frac{\gamma - 1}{\gamma}, \qquad F_{s6} = 1 + \frac{\gamma - 1}{2} M^2$$

$$F_{02} = \frac{\gamma M^2}{2}, \qquad F_{06} = -\frac{\gamma M^2}{2}$$

Equation (11.27) represents a set of eight differential equations in the eight variables. Consideration will, for the moment, be focused on the equation for $dM^2/M^2$. Using the values of the coefficients listed above, this equation is:

$$\frac{dM^2}{M^2} = -\frac{2[1 + (\gamma - 1)M^2/2]}{(M^2 - 1)} \left( \frac{d\dot{m}}{\dot{m}} - \frac{dA}{A} \right)$$

$$- \frac{2[1 + (\gamma - 1)M^2/2]}{(M^2 - 1)}(-\gamma M^2)\left( \frac{f}{2}\frac{P}{A}dx + \frac{dF_D}{\gamma p M^2 A}\frac{d\dot{m}}{\dot{m}} - r_i\frac{d\dot{m}}{\dot{m}} \right)$$

$$- \frac{(\gamma M^2 + 1)[1 + (\gamma - 1)M^2/2]}{(M^2 - 1)}\left( \frac{dT_0}{T_0} \right)$$

i.e.,

$$\frac{dM^2}{M^2} = -\frac{2[1 + (\gamma - 1)M^2/2]}{(1 - M^2)}\left( \frac{dA}{A} \right) + \frac{(\gamma M^2 + 1)[1 + (\gamma - 1)M^2/2]}{(1 - M^2)}\left( \frac{dT_0}{T_0} \right)$$

$$+ \frac{2[1 + (\gamma - 1)M^2/2](1 + \gamma M^2)}{(1 - M^2)}\left( \frac{d\dot{m}}{\dot{m}} \right)$$

$$+ \frac{\gamma M^2[1 + (\gamma - 1)M^2/2]}{(1 - M^2)}\left( f\frac{P}{A}dx + \frac{2}{\gamma p M^2 A}dF_D - 2r_i\frac{d\dot{m}}{\dot{m}} \right) \qquad (11.28)$$

Similar equations could be derived from the other variables. However, in most cases it is possible to use eq. (11.28) to solve for the variation of $M^2$ and then use algebraic equations to solve for the variations of the other quantities. This is discussed in the next section. Before entering into that discussion, however, some general points regarding eq. (11.28) need to be discussed. It will be noted that every term on the right hand side of eq. (11.28) has $1 + (\gamma - 1)M^2/2$ in the numerator and has $1 - M^2$ as the denominator. Equation (11.28) can, therefore, be written as:

$$\frac{dM^2}{M^2} = \left[\frac{1+(\gamma-1)M^2/2}{(1-M^2)}\right]H \tag{11.29}$$

where:     $H = -2\left(\dfrac{dA}{A}\right) + (\gamma M^2 + 1)\left(\dfrac{dT_0}{T_0}\right) + 2(1+\gamma M^2)\left(\dfrac{d\dot{m}}{\dot{m}}\right)$

$$+ \gamma M^2\left(f\frac{P}{A}dx + \frac{2}{\gamma p M^2 A}dF_D - 2r_i\frac{d\dot{m}}{\dot{m}}\right) \tag{11.30}$$

The function $H$ will depend on the values of the input qualities and on $M$. It will be seen that since $1 + (\gamma-1)M^2/2$ and $M^2$ are always positive, the sign of $dM^2$ depends on the sign of $H/(1-M^2)$. Hence, if $H$ is positive, $M$ will increase in subsonic flow and decrease in supersonic flow, whereas if $H$ is negative, $M$ will decrease in subsonic flow and increase in supersonic flow. Consider, as a particular case, flow in a nozzle when both friction and heat transfer are important but when there is no mass transfer and no drag causing bodies in the flow. In this case, eqs. (11.29) and (11.30) give:

$$\frac{dM^2}{M^2} = \left[\frac{1+(\gamma-1)M^2/2}{(1-M^2)}\right]\left[-2\left(\frac{dA}{A}\right) + (\gamma M^2+1)\left(\frac{dT_0}{T_0}\right) + \gamma M^2 f\frac{P}{A}dx\right]$$

If heat is being added to the flow, $dT_0$ is positive. The friction term is, of course, always positive. The term in the brackets on the right hand side of this equation can then be either positive or negative depending on the sign of $dA$ and on the relative magnitude of the terms. If $M$ is less than 1 and the area is decreasing, i.e., $dA$ is negative, $dM^2$ will always be positive, i.e., the Mach number will always increase in the subsonic flow in the divergent portion of the nozzle. However, if the flow is to go from subsonic to supersonic, the sonic point, i.e., the point when $M = 1$, must occur where:

$$-2\left(\frac{dA}{A}\right) + (\gamma M^2+1)\left(\frac{dT_0}{T_0}\right) + \gamma M^2 f\frac{P}{A}dx = 0$$

Because the case where the second two terms are positive is being considered, this is only possible if $dA$ is positive. Hence, when friction is important and there is heat addition, the sonic point occurs in the divergent position of the nozzle as indicated in Fig. 11.4.

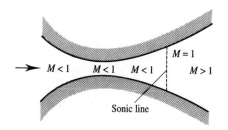

**FIGURE 11.4**
Flow in a nozzle with friction and heat addition.

Thus in generalized flows in a nozzle, the Mach number will not usually be equal to 1 at the throat. This result has previously been reached in Chapters 9 and 10 for the particular cases being dealt with there.

## 11.3
## SOLUTION PROCEDURE FOR GENERALIZED FLOW WITH NO SONIC POINT

Attention for the moment will be given to situations in which the Mach number does not pass through a value of 1 and in which, therefore, the term $1 - M^2$ in eq. (11.29) remains nonzero. Equation (11.29) can then be solved numerically to give the variation of $M$ along the duct, the equation of this purpose being written by dividing through by $dx$ as:

$$\frac{dM^2}{dx} = \left\{ \frac{M^2[1 + (\gamma - 1)M^2/2]}{(1 - M^2)} \right\} G \tag{11.31}$$

where:
$$G = -\frac{2}{A}\frac{dA}{dx} + \frac{(\gamma M^2 + 1)}{T_0}\frac{dT_0}{dx} + \frac{2(1 + \gamma M^2)}{\dot{m}}\frac{d\dot{m}}{dx}$$

$$+ \gamma M^2 \left( f\frac{P}{A} + \frac{2}{\gamma p M^2 A}\frac{dF_D}{dx} - \frac{2r_i}{\dot{m}}\frac{d\dot{m}}{dx} \right)$$

The values of the remaining flow variables could then be found by integrating the differential equations for their variations which are listed in eq. (11.27). However, once eq. (11.31) has been used to solve for $M$, it is usually more convenient to use algebraic equations, from which the equations listed in (11.27) were derived, to obtain the values of the other variables. Thus since, as discussed before:

$$\frac{T_0}{T} = 1 + \frac{\gamma - 1}{2}M^2$$

it follows that:

$$\frac{T_2}{T_1} = \left[ \frac{1 + (\gamma - 1)M_1^2/2}{1 + (\gamma - 1)M_2^2/2} \right] \frac{T_{02}}{T_{01}} \tag{11.32}$$

the subscripts 1 and 2 referring to conditions at any two points in the flow as indicated in Fig. 11.5. Since $T_{02}$ and $T_{01}$ are specified inputs and since $M_1$ and $M_2$ are related by solving eq. (11.31), eq. (11.32) allows $T_2/T_1$ to be found.
Now:

$$\dot{m} = \rho V A = \left( \frac{p}{RT} \right) MaA = \left( \frac{p}{RT} \right) M\sqrt{\gamma RT}A$$

hence:
$$\dot{m} = \sqrt{\frac{\gamma}{RT}}pMA \tag{11.33}$$

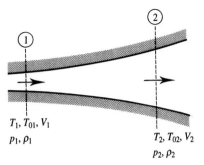

**FIGURE 11.5**
Conditions at two points considered.

$T_1, T_{01}, V_1$
$p_1, \rho_1$

$T_2, T_{02}, V_2$
$p_2, \rho_2$

Applying this between the two points indicated in Fig. 11.5 then gives, on rearrangement:

$$\frac{p_2}{p_1} = \frac{\dot{m}_2 \, M_1 \, A_1}{\dot{m}_1 \, M_2 \, A_2} \sqrt{\frac{T_2}{T_1}} \tag{11.34}$$

Since $\dot{m}_1$, $\dot{m}_2$, $A_1$, and $A_2$ are given inputs and since $M_1$ and $M_2$ are related by solving eq. (11.31) and since eq. (11.32) gives $T_2/T_1$, eq. (11.34) allows $p_2/p_1$ to be determined.

Next it is recalled that since, as already noted:

$$V = Ma = M\sqrt{\gamma RT}$$

it follows that:

$$\frac{V_2}{V_1} = \frac{M_2}{M_1} \sqrt{\frac{T_2}{T_1}} \tag{11.35}$$

Also, since the equation of state gives:

$$\frac{p}{\rho} = RT$$

it follows that:

$$\frac{\rho_2}{\rho_1} = \frac{p_2}{p_1} \frac{T_1}{T_2} \tag{11.36}$$

Next, it is recalled that because of the way the stagnation pressure and temperature are defined, isentropic relations give:

$$\frac{p_0}{p} = \left(\frac{T_0}{T}\right)^{\gamma/\gamma-1}$$

from which it follows that:

$$\frac{p_{02}}{p_{01}} = \frac{p_2}{p_1} \left(\frac{T_2}{T_1}\right)^{\gamma/\gamma-1} \left(\frac{T_{02}}{T_{01}}\right)^{\gamma/\gamma-1} \tag{11.37}$$

Lastly, it is noted that the entropy equation can be directly integrated to give:

$$\frac{s_2 - s_1}{c_p} = \ln\left(\frac{T_2}{T_1}\right) - \frac{\gamma - 1}{\gamma}\ln\left(\frac{p_2}{p_1}\right) \tag{11.38}$$

To summarize, then, generalized quasi-one-dimensional flow is described by the following set of equations:

$$\frac{dM^2}{dx} = \left\{\frac{M^2[1 + (\gamma - 1)M^2/2]}{(1 - M^2)}\right\}G$$

$$G = -\frac{2}{A}\frac{dA}{dx} + \frac{(\gamma M^2 + 1)}{T_0}\frac{dT_0}{dx} + \frac{2(1 + \gamma M^2)}{\dot{m}}\frac{d\dot{m}}{dx}$$

$$+ \gamma M^2\left(f\frac{P}{A} + \frac{2}{\gamma p M^2 A}\frac{dF_D}{dx} - \frac{2r_i}{\dot{m}}\frac{d\dot{m}}{dx}\right)$$

$$\frac{T_2}{T_1} = \left[\frac{1 + (\gamma - 1)M_1^2/2}{1 + (\gamma - 1)M_2^2/2}\right]\frac{T_{02}}{T_{01}}$$

$$\frac{p_2}{p_1} = \frac{\dot{m}_2}{\dot{m}_1}\frac{M_1}{M_2}\frac{A_1}{A_2}\sqrt{\frac{T_2}{T_1}}$$

$$\frac{V_2}{V_1} = \frac{M_2}{M_1}\sqrt{\frac{T_2}{T_1}}$$

$$\frac{\rho_2}{\rho_1} = \frac{p_2}{p_1}\frac{T_1}{T_2}$$

$$\frac{p_{02}}{p_{01}} = \frac{p_2}{p_1}\left(\frac{T_2}{T_1}\right)^{\gamma/\gamma-1}\left(\frac{T_{02}}{T_{01}}\right)^{\gamma/\gamma-1}$$

$$\frac{s_2 - s_1}{c_p} = \ln\left(\frac{T_2}{T_1}\right) - \frac{\gamma - 1}{\gamma}\ln\left(\frac{p_2}{p_1}\right)$$

In obtaining a solution, the first of these equations is, as discussed before, numerically solved and the remaining equations are then used to get the changes in the other variables. While there are many excellent and relatively simple methods available that can be used to obtain the numerical solution to eq. (11.31), the very simplest explicit procedure will be used here since the purpose of the discussion is to illustrate the ideas involved in the analysis of generalized one-dimensional flow and not to discuss numerical techniques. For the present purposes, the duct will be broken down into a series of sections of length $dx$ shown in Fig. 11.6.

For any section such as that shown in Fig. 11.6, eq. (11.31) gives approximately:

$$dM^2 = \left\{\frac{M_1^2[1 + (\gamma - 1)M_1^2/2]}{(1 - M_1^2)}\right\}G_1\,dx \tag{11.39}$$

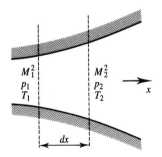

**FIGURE 11.6**
Section considered in numerical solution.

where $G_1$ is the value of $G$ evaluated using conditions existing at section 1. The Mach number at the second section is then given by:

$$M_2^2 = M_1^2 + dM^2, \qquad M_2 = \sqrt{M_2^2} \qquad (11.40)$$

With $M_2$ found, the values of the other variables $T_2$, $p_2$, ... can be found.

To illustrate the procedure, attention will be focused on the simplified case where there is no mass addition or removal, i.e., $\dot{m} = 0$, and in which $F_D = 0$. In this case:

$$G = -\frac{2}{A}\frac{dA}{dx} + \frac{(\gamma M^2 + 1)}{T_0}\frac{dT_0}{dx} + \gamma M^2 f \frac{P}{A} \qquad (11.41)$$

It will also be assumed that a circular duct is being considered and that the following equation is adequate to represent the variation of duct diameter $D$ with distance along the duct $x$:

$$D = D_i + A_D x + B_D x^2 \qquad (11.42)$$

where $A_D$ and $B_D$ are specified coefficients and the subscript $i$ refers to initial conditions at $x = 0$ as indicated in Fig. 11.7. From the above equation it follows that since:

$$A = \pi \frac{D^2}{4}$$

then:        $$\frac{dA}{dx} = \frac{\pi D}{2}\frac{dD}{dx} = \frac{\pi D}{2}(A_D + B_D x) \qquad (11.43)$$

It will be assumed that a similar equation is adequate to represent the variation of $T_0$, i.e., it will be assumed that:

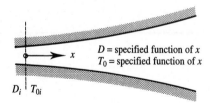

**FIGURE 11.7**
Situation considered in numerical solution.

$D$ = specified function of $x$
$T_0$ = specified function of $x$

$$T_0 = T_{0i} + A_T x + B_T x^2 \qquad (11.44)$$

where $A_T$ and $B_T$ are again specified coefficients and the subscript $T_{0i}$ refers to the initial value of $T_0$ at $x = 0$ as indicated in Fig. 11.7.

The above equation for the stagnation point variation gives:

$$\frac{dT_0}{dx} = A_T + B_T x \qquad (11.45)$$

The friction factor, $f$, is assumed to be specified and $fP/A$ is, since a circular duct is being considered, set equal to $4f/D$.

A simple computer program written in BASIC based on this procedure is listed in Program 11.1 and its use is illustrated in Examples 11.1 and 11.2.

**Program 11.1**

```
 10  CLS
 20  PRINT" "
 30  PRINT"VARIABLE AREA FLOW WITH FRICTION AND HEAT ADDITION"
 40  PRINT" ":PRINT" "
 50  INPUT" GAMMA ";GA
 60  GM=(GA-1)/2
 70  GI=GA/(GA-1)
 80  INPUT" INITIAL MACH NUMBER ";MI
 90  INPUT" INITIAL PRESSURE ";PI
100  INPUT" INITIAL TEMPERATURE ";TI
110  INPUT" INITIAL DIAMETER ";DI
120  INPUT" DUCT LENGTH ";L
130  INPUT" FANNING FRICTION FACTOR ";F
140  SI=TI*(1+GM*MI*MI)
150  POI=PI*((1+GM*MI*MI)^GI)
160  AI=3.1416*DI*DI/4
170  INPUT" FIRST DIAMETER VARIATION COEFF. ";AD
180  INPUT" SECOND DIAMETER VARIATION COEFF. ";BD
190  INPUT" FIRST STAGNATION TEMP. VARIATION COEFF. ";AE
200  INPUT" SECOND STAGNATION TEMP. VARIATION COEFF. ";BE
210  INPUT" NUMBER OF STEPS ALONG DUCT ";N
220  DX=L/(N-1)
230  PRINT "X","A","T0","M","T","P"
240  PRINT" "
250  X=0.0
260  M=MI
270  P=PI
280  T=TI
290  S=SI
300  A=AI
310  D=DI
320  AX=1.571*D*(AD+2*BD*X)
330  SX=AE+2*E*X
340  G=A-2*AX/A+(GA*M*M+1)*SX/S+GA*M*M*F*4/D
350  MX=M*M*(1+GM*M*M)*G/(1-M*M)
360  M2=M*M+MX*DX
370  MN=M2^0.5
380  X=X+DX
390  SN=SI+AE*X+BE*X*X
400  DN=DI+AD*X+BD*X*X
410  AN=3.1416*DN*DN/4
420  TN=TI*(1+GM*MI*MI)*SN/((1+GM*MN*MN)*SI)
430  PN=(PI*MI*AI/(MN*AN))*((TN/TI)^0.5)
440  PRINT USING "#########.####";X,AN,SN,MN,TN,PN
450  IF M>0.95 AND M<1.05 THEN GOTO 520
460  IF X>=L THEN GOTO 520
470  M=MN
480  P=PN
```

```
490   D = DN
500   A = AN
510   GOTO 320
520   STOP
530   END
```

## EXAMPLE 11.1

Air enters a conical combustion chamber at a Mach number of 0.22, a pressure of 2200 kPa, and a temperature of 475°C. The combustion chamber is 0.8 m long and has an inlet diameter of 0.3 m and an exit diameter of 0.25 m. As a result of the combustion, the stagnation temperature increases by 450°C across the chamber. It can be assumed that the rate of increase of stagnation temperature is constant. The friction factor can be assumed to be 0.004. Find the exit Mach number, pressure, and temperature. Assume the gas has the properties of air and that the flow is steady and one-dimensional.

### Solution

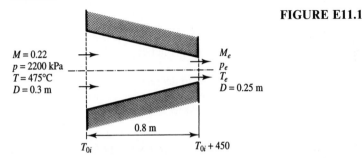

**FIGURE E11.1**

The flow situation is shown in Fig. E11.1. Since the chamber is conical, the diameter decreases linearly. Hence:

$$D = 0.3 - (0.3 - 0.25)\frac{x}{0.8} = 0.3 - 0.0625x$$

Similarly, since the stagnation temperature can be assumed to increase linearly:

$$T_0 = T_i\left(1 + \frac{\gamma - 1}{2}M_i^2\right) + 450\frac{x}{0.8}$$

$$= (475 + 273)(1 + 0.2 \times 0.22^2) + 450\frac{x}{0.8} = 755.2 + 562.5x$$

The program listed above has, therefore, been run with the following inputs:

$\gamma = 1.4$
Initial Mach number = 0.22
Initial pressure = 2 200 000 (Pa)
Initial temperature = 748 (K)
Initial diameter = 0.3 (m)
Duct length = 0.8 (m)
Fanning friction factor = 0.004
First diameter variation coefficient = −0.0625 (m/m)
Second diameter variation coefficient = 0
First stagnation temperature variation coefficient = 562.5 (K/m)
Second stagnation temperature variation coefficient = 0

Number of steps along duct = 100
(Calculations with various numbers of steps should be undertaken to ensure that the results do not depend on the chosen number of steps.)

The computed output conditions are:

Exit Mach number = 0.521
Exit pressure = 1 653 951 Pa = 1654 kPa
Exit temperature = 1143.2 K = 870.2°C

### EXAMPLE 11.2

Air enters an axisymmetric nozzle at a Mach number of 1.2 and with a pressure and temperature of 30 kPa and 400°C respectively. The inlet diameter of the nozzle is 3 cm and the exit diameter is 6 cm, the nozzle being 9 cm long. The wall of the nozzle is parallel to its axis at the exit. Find the Mach number, the pressure, the temperature, and the velocity of the air at the exit of the nozzle for the following three cases:

1. The flow in the nozzle is assumed to be adiabatic and frictionless
2. The flow in the nozzle is assumed to be adiabatic and the friction factor is assumed to be equal to 0.005
3. The stagnation temperature is assumed to decrease linearly by 150 K as the air flows through the nozzle as a result of heat transfer to the wall. The friction factor can again be assumed to be equal to 0.005

*Solution*

**FIGURE E11.2**

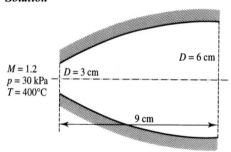

$M = 1.2$
$p = 30$ kPa
$T = 400$°C

$D = 3$ cm

$D = 6$ cm

9 cm

The flow situation is shown in Fig. E11.2. It will be assumed that the nozzle shape can be adequately represented by:

$$D = D_i + A_D x + B_D x^2 = 0.03 + A_D x + B_D x^2$$

But the specified nozzle shape is such that:

At $x = 0.09$:   $D = 0.06$

At $x = 0.09$:   $dD/dx = 0$

Substituting these into the nozzle shape equation then gives:

$$0.09 A_D + 0.0081 B_D = 0.03$$

and:
$$A_D + 0.18 B_D = 0$$

Solving between these two equations then gives:

$$A_D = 0.6667, \qquad B_D = -3.704$$

The inlet stagnation temperature is given by:

$$T_{0I} = T_i\left(1 + \frac{\gamma - 1}{2}M_i^2\right)$$

$$= (400 + 273)(1 + 0.2 \times 1.2^2) = 866.8 \text{ K}$$

The three cases will now be separately considered using the program listed above.

**Case 1.** Here $f$ can be assumed to be zero and the change in stagnation temperature can be assumed to be zero. The program listed above has, therefore, been run with the following inputs for this case:

$\gamma = 1.4$
Initial Mach number $= 1.2$
Initial pressure $= 30\,000$ (Pa)
Initial temperature $= 673$ (K)
Initial diameter $= 0.03$ (m)
Duct length $= 0.09$ (m)
Fanning friction factor $= 0$
First diameter variation coefficient $= 0.6667$ (m/m)
Second diameter variation coefficient $= -3.704$ (m/m$^2$)
First stagnation temperature variation coefficient $= 0$
Second stagnation temperature variation coefficient $= 0$
Number of steps along duct $= 100$

The computed output conditions are:

Exit Mach number $= 3.005$
Exit pressure $= 308.9$ Pa
Exit temperature $= 2029.1$ K $= 1756.1°$C

From which it follows that:

$$\text{Exit velocity} = M_{\text{exit}}a_{\text{exit}} = M_{\text{exit}}\sqrt{\gamma R T_{\text{exit}}} = 2714 \text{ m/s}$$

**Case 2.** The only difference here is that:

Fanning friction factor $= 0.005$

The computed output conditions are:

Exit Mach number $= 2.842$
Exit pressure $= 331.4$ Pa
Exit temperature $= 2221.8$ K $= 1948.8°$C

From which it follows that:

$$\text{Exit velocity} = M_{\text{exit}}a_{\text{exit}} = M_{\text{exit}}\sqrt{\gamma R T_{\text{exit}}} = 2685 \text{ m/s}$$

**Case 3.** Since the stagnation temperature decreases linearly:

$$T_0 = 866.8 - 1667.7x$$

Hence the only difference in the input to the program as compared to that in Case 2 is:

First stagnation temperature variation coefficient $= -1667.7$ (K/m)
Second stagnation temperature variation coefficient $= 0$

The computed output conditions are:

Exit Mach number $= 4.053$
Exit pressure $= 167.3$ Pa
Exit temperature $= 1107.2$ K $= 834.2°$C

From which it follows that:

$$\text{Exit velocity} = M_{\text{exit}}a_{\text{exit}} = M_{\text{exit}}\sqrt{\gamma RT_{\text{exit}}} = 2703 \text{ m/s}$$

The result of the calculations for the three cases are summarized in Table E11.2.

**TABLE E11.2**

| Case | $M$ | $p$ (Pa) | $T$ (°C) | $V$ (m/s) |
|------|-----|----------|----------|-----------|
| 1 | 3.005 | 308.9 | 1756 | 2714 |
| 2 | 2.842 | 331.4 | 1948 | 2685 |
| 3 | 4.053 | 167.3 | 834 | 2703 |

## 11.4
## FLOW WITH A SONIC POINT

As mentioned before, difficulties develop in the solution of eq. (11.31) if there is a point in the flow at which $M = 1$, i.e., if there is a sonic point in the flow. A sonic point can occur in the flow if, for example, $M$ goes from a subsonic value to a supersonic value or from a supersonic value to a subsonic value. To illustrate how such situations can be dealt with, attention will continue to be restricted to the case where there is no mass transfer and where $F_D$ can be assumed to be zero. In this case, as discussed before, eq. (11.31) reduces to:

$$\frac{dM^2}{dx} = \frac{M^2[1 + (\gamma - 1)M^2/2]}{(1 - M^2)}\left[-\frac{2}{A}\frac{dA}{dx} + \frac{(\gamma M^2 + 1)}{T_0}\frac{dT_0}{dx} + \gamma M^2 f\frac{P}{A}\right] \quad (11.46)$$

The difficulty at the sonic point arises because the term $1 - M^2$ in the denominator tends to 0 as $M$ tends to 1. The solution must, therefore, as discussed before, be such that as $M$ tends to 1:

$$-\frac{2}{A}\frac{dA}{dx} + \frac{\gamma M^2 + 1}{T_0} + \gamma M^2 f\frac{p}{A} \to 0$$

i.e., since $M = 1$ at this point, the sonic point will occur when:

$$-\frac{2}{A}\frac{dA}{dx}+\frac{1+\gamma}{T_0}\frac{dT_0}{dx}+\gamma f\frac{P}{A}=0 \tag{11.47}$$

Since the variation of $A$ and $T_0$ with $x$ and the value of $f$ are specified, this equation can be used to find the value of $x$ at which $M=1$. The value of $dM^2/dx$ at this point is then undefined ($=0/0$) according to eq. (11.46). However, by using l'Hospital's rule, its value at the sonic point can be found as follows:

$$\frac{dM^2}{dx}\bigg|_{M=1}=\frac{\gamma+1}{2}\lim_{M\to1}\left[-\frac{2}{A}\frac{dA}{dx}+\frac{\gamma M^2+1}{T_0}\frac{dT_0}{dx}+\gamma M^2 f\frac{P}{A}\right]\bigg/(1-M^2)$$

$$=\frac{\gamma+1}{2}\lim_{M\to1}\left[-2\frac{d}{dx}\left(\frac{1}{A}\frac{dA}{dx}\right)+\frac{2\gamma M}{T_0}\frac{dT_0}{dx}\frac{dM}{dx}\right.$$

$$\left.+(\gamma M^2+1)\frac{d}{dx}\left(\frac{1}{T_0}\frac{dT_0}{dx}\right)+2\gamma Mf\frac{P}{A}\frac{dM}{dx}\right]\bigg/\left(-2M\frac{dM}{dx}\right)$$

i.e., since the point where $M=1$ is being considered, this gives:

$$-\frac{8}{(\gamma+1)}\left(\frac{dM}{dx}\right)^2-2\gamma f\frac{P}{A}\left(\frac{dM}{dx}\right)-\left(\frac{2\gamma}{T_0}\frac{dT_0}{dx}\right)\left(\frac{dM}{dx}\right)$$

$$=-2\frac{d}{dx}\left(\frac{1}{A}\frac{dA}{dx}\right)+(1+\gamma)\frac{d}{dx}\left(\frac{1}{T_0}\frac{dT_0}{dx}\right) \tag{11.48}$$

Because eq. (11.47) gives the value of $x$ at which the sonic point occurs, this equation can be used to solve for $dM/dx$ at the sonic point. Equation (11.48) gives two values of $dM/dx$ at the sonic point, one value corresponding to the flow on the subsonic side of the sonic line and the other to the flow on the supersonic side of the sonic line.

The calculation of flows with a sonic point therefore involves the following steps:

1. Using eq. (11.47) find the position of the sonic point
2. Use eq. (11.48) to find the values of $dM/dx$ at the sonic point
3. Use these values of $dM/dx$ to advance one step upstream and downstream of the sonic point
4. Apply the same procedure as before to advance in the upstream and downstream directions away from the sonic point

The procedure is illustrated in Fig. 11.8.

## 11.5
## CONCLUDING REMARKS

A procedure has been outlined for solving steady quasi-one-dimensional flows in which there can be area changes, friction effects, stagnation temperature changes and drag forces. The solution for such flows where there is no point in

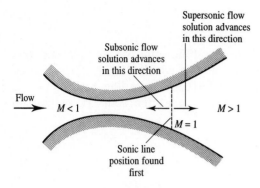

Supersonic flow
solution advances
in this direction

Subsonic flow
solution advances
in this direction

Flow

$M < 1$          $M > 1$

$M = 1$

Sonic line
position found
first

**FIGURE 11.8**
Solution for flows with a sonic
point.

the flow at which $M = 1$, i.e., no sonic point, has been considered in some
detail. Brief consideration has been given to the solution of flows in which
there exists a sonic point.

## PROBLEMS

**11.1.** Consider steady air flow through a duct that has a circular cross-sectional
shape. The inlet diameter of the duct is 6 cm and the duct has a length of
1.5 m. The air enters the duct with a Mach number of 0.35 and a temperature
of 40°C. Heat is added to the flow in the duct at a uniform rate which is such
that the stagnation temperature increases by 246 K in the duct. If the friction
factor is assumed to be 0.003, determine the Mach number and temperature
variations along the duct if (1) its diameter increases linearly by 20 percent,
(2) its diameter decreases linearly by 2 percent and (3) its diameter remains
constant. Assume that the air behaves as a perfect gas and that the flow is
steady.

**11.2.** Air enters a conical convergent duct at a Mach number of 0.2 with a pressure of
600 kPa and a temperature of 50°C. The inlet diameter of the duct is 0.05 m
and its length is 1 m. If the exit area of the duct is 70 percent of that at the inlet
and if heat is added to the air such that the stagnation temperature at the outlet
is twice that at the inlet, find the Mach number, temperature, and pressure at
the outlet of the duct assuming a friction factor of 0.003.

**11.3.** Consider steady air flow through a constant area circular duct which has a
diameter of 10 cm and a length of 1 m. The Mach number, pressure, and
temperature at the inlet to the duct are 0.3, 200 kPa, and 80°C. Heat is
added at a uniform rate to the air as it flows through the duct causing the
stagnation temperature to increase by 300 K. If the friction factor can be
assumed to be 0.003, find the Mach number, pressure, and temperature at
the outlet of the duct.

# CHAPTER 12

# Numerical Analysis of One-Dimensional Flow

## 12.1
## INTRODUCTION

Developments that have occurred in the field of digital computers over the past thirty to forty years have made possible the introduction of a number of procedures for numerical modeling of complex fluid flows. To ignore these developments in computational fluid dynamics (CFD) in a book of the present type would be unwarranted. However, to try to give a detailed discussion of the various procedures presently being used in CFD would be equally unwarranted. Therefore, only a very brief introduction to the numerical analysis of one-dimensional compressible fluid flows will be presented.

The numerical analysis of any flow problem involves the following steps:

1. *Creation of a model of the flow:* e.g., it must be decided whether the flow can be assumed to be steady, whether the flow can be assumed to be one- or two-dimensional or whether it has to be treated as three-dimensional, whether viscous effects must be included, whether compressibility effects are important, etc. Once the flow model has been selected, the equations governing the flow can be obtained.
2. *Defining the boundary conditions:* i.e., the values of the flow variables on the boundaries of the flow regime being considered. For example, in the case of flow through a duct as shown in Fig. 12.1, the values of velocity or the derivative of velocity on the inlet plane AB, on the walls AD and BC, and on the exit plane CD may be specified.

   Instead of velocity values, the values of other flow variables such as pressure may be specified on the boundaries.

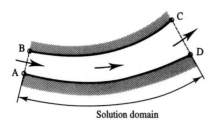

**FIGURE 12.1**
Boundary surfaces of flow through a duct.

Great care has to be exercised in specifying the boundary conditions to ensure that the problem is not under- or over-specified.

If a problem in which the values of the variables are time dependent is being considered it is necessary to specify the initial conditions, i.e., the values of all the flow variables at some initial instant of time, in addition to the boundary conditions.

3. *Creation of numerical approximations to the governing equations:* Many methods of obtaining these numerical approximations are available, e.g., finite-difference, finite-volume, finite-element, boundary element, etc.

4. *Solution of the numerical equations for the flow variables:* Because the equations governing fluid flows are often non-linear, i.e., some of the terms in the governing equations involve products of the flow variables, some form of iterative technique often has to be incorporated into the solution procedure.

As mentioned above, attention will be restricted to one-dimensional compressible flows in the present chapter. Now it was shown in the earlier chapters that relatively simple analytical solutions can be obtained for such one-dimensional flows. It is really not necessary, therefore, to resort to numerical methods for the predictions of one-dimensional flows. However, a consideration of the numerical analysis of one-dimensional flows serves as a good introduction to the numerical analysis of more complex compressible flows. Further, solutions for some one-dimensional flow situations, particularly for those involving unsteady flow, are sometimes more rapidly obtained by using numerical methods than by using analytical methods.

The governing equations and some general aspects of numerical solutions will first be discussed in this chapter. Simple numerical solutions of steady flows will then be considered, and attention will lastly be given to the numerical solution of unsteady flows.

Numerical solutions involving finite-difference approximations have, of course, been introduced in earlier chapters. An attempt will be made in the present chapter to present a more rigorous discussion of the numerical methods and to apply these methods directly to the basic governing equations rather than to derived forms of these equations as was done in earlier chapters. Several computer programs written in FORTRAN are listed in this chapter.

## 12.2
## GOVERNING EQUATIONS

The equations for steady one-dimensional compressible flow have been extensively discussed in earlier chapters. Because both steady and unsteady flows will be discussed in this chapter, the governing equations for unsteady flow will be derived here using procedures that are extensions of those used earlier for steady flows. Attention will again be restricted to the flow of a calorically and thermally perfect continuum gas. Adiabatic, frictionless flow is considered.

The governing equations are, of course, derived by applying conservation of mass, momentum, and energy to flow through a control volume. Because one-dimensional flow is being considered, a control volume of the type shown in Fig. 12.2 will be used.

Conservation of mass will first be considered. For the flow through the control volume, this requires that:

$$\begin{array}{ccc} \text{Rate mass enters} \\ \text{control volume} \end{array} - \begin{array}{c} \text{Rate mass leaves} \\ \text{control volume} \end{array} = \begin{array}{c} \text{Rate of increase of mass} \\ \text{in control volume} \end{array} \quad (12.1)$$

Because $dx$ is very small, the mass of gas in the control volume is $\rho A\, dx$. The above equation therefore gives since the rate at which mass enters the control volume is $\rho VA$:

$$\rho VA - \left[\rho VA + \frac{\partial}{\partial x}(\rho AV)\, dx\right] = \frac{\partial}{\partial t}(\rho A\, dx)$$

i.e., since $A$ is not a function of time:

$$A\frac{\partial \rho}{\partial t} + \frac{\partial}{\partial x}(\rho AV) = 0 \quad (12.2)$$

**FIGURE 12.2**
Control volume considered.

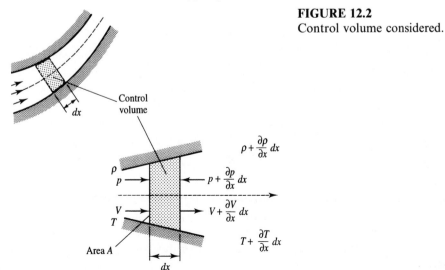

It will be noted that since the flow variables vary with both $x$ and $t$, partial derivatives are used.

Conservation of momentum is next considered. It requires for the flow through the control volume:

$$\begin{array}{c} \text{Rate momentum leaves} \\ \text{control volume} \end{array} - \begin{array}{c} \text{Rate momentum enters} \\ \text{control volume} \end{array}$$

(12.3)

$$+ \begin{array}{c} \text{Rate of increase of momentum} \\ \text{in control volume} \end{array} = \begin{array}{c} \text{Net force acting on control} \\ \text{volume in flow direction} \end{array}$$

Now it was shown earlier that the net pressure force on the control volume in the $x$-direction is:

$$F_x = -A \frac{\partial p}{\partial x} dx \qquad (12.4)$$

Also, since $dx$ is small, the momentum of the gas in the control volume is $\rho V A \, dx$. Hence, eq. (12.3) gives:

$$\left[ \rho V^2 A + \frac{\partial}{\partial x}(\rho V^2 A)\, dx - \rho V^2 A \right] + \frac{\partial}{\partial t}(\rho V A)\, dx = -A \frac{\partial p}{\partial x} dx$$

i.e.,
$$A \frac{\partial}{\partial t}(\rho V) + \frac{\partial}{\partial x}(\rho V^2 A) + A \frac{\partial p}{\partial x} = 0 \qquad (12.5)$$

The first term in this equation is:

$$A \rho \frac{\partial V}{\partial t} + A V \frac{\partial \rho}{\partial t}$$

which, using the continuity equation (12.2), can be written as:

$$A \rho \frac{\partial V}{\partial t} - \frac{\partial}{\partial x}(\rho A V)$$

i.e., as:
$$A \rho \frac{\partial V}{\partial t} - \frac{\partial}{\partial x}(\rho V^2 A) + \rho A V \frac{\partial V}{\partial x}$$

Substituting this back into eq. (12.2) gives:

$$A \rho \frac{\partial V}{\partial t} + \rho A V \frac{\partial V}{\partial x} + A \frac{\partial p}{\partial x} = 0$$

i.e.,
$$\rho \frac{\partial V}{\partial t} + \rho V \frac{\partial V}{\partial x} + \frac{\partial p}{\partial x} = 0 \qquad (12.6)$$

Lastly, conservation of energy is applied to the flow through the control volume. This gives:

Rate enthalpy + kinetic energy    Rate enthalpy + kinetic energy
   enter control volume    $-$       leave control volume

$$= \begin{array}{l} \text{Rate of increase of internal energy} \\ +\text{kinetic energy in control volume} \end{array} \quad (12.7)$$

Now, again noting that the rate at which mass enters the control volume is $\rho VA$ and that, as mentioned above, because $dx$ is small, the mass of the gas in the control volume is $\rho A\, dx$, the above equation can be written as:

$$\rho VA\left(h+\frac{V^2}{2}\right) - \left\{\rho VA\left(h+\frac{V^2}{2}\right) + \frac{\partial}{\partial x}\left[\rho VA\left(h+\frac{V^2}{2}\right)\right] dx\right\}$$

$$= \frac{\partial}{\partial t}\left[\rho A\left(e+\frac{V^2}{2}\right)\right] dx$$

i.e.,
$$A\frac{\partial}{\partial t}\left[\rho\left(e+\frac{V^2}{2}\right)\right] + \frac{\partial}{\partial x}\left[\rho VA\left(h+\frac{V^2}{2}\right)\right] = 0 \quad (12.8)$$

This equation can be written as:

$$\rho A\frac{\partial}{\partial t}\left(e+\frac{V^2}{2}\right) + \rho VA\frac{\partial}{\partial x}\left(h+\frac{V^2}{2}\right)$$

$$+\left(e+\frac{V^2}{2}\right)A\frac{\partial\rho}{\partial t} + \left(h+\frac{V^2}{2}\right)\frac{\partial}{\partial x}(\rho VA) = 0 \quad (12.9)$$

But the continuity equation gives:

$$A\frac{\partial\rho}{\partial t} = -\frac{\partial}{\partial x}(\rho AV)$$

so eq. (12.9) gives:

$$\rho A\frac{\partial}{\partial t}\left(e+\frac{V^2}{2}\right) + \rho VA\frac{\partial}{\partial x}\left(h+\frac{V^2}{2}\right) + (h-e)\frac{\partial}{\partial x}(\rho VA) = 0$$

i.e.,
$$\rho A\frac{\partial e}{\partial t} + \rho AV\frac{\partial h}{\partial x} + (h-e)\frac{\partial}{\partial x}(\rho VA) + \rho AV\frac{\partial V}{\partial t} + \rho AV^2\frac{\partial V}{\partial x} = 0$$

The last two terms in this equation are then combined using the momentum equation, i.e., eq. (12.6), giving:

$$\rho A\frac{\partial e}{\partial t} + \rho AV\frac{\partial h}{\partial x} + (h-e)\frac{\partial}{\partial x}(\rho VA) - AV\frac{\partial p}{\partial x} = 0 \quad (12.10)$$

But:
$$h = e + \frac{p}{\rho}$$

so:
$$\frac{\partial h}{\partial x} = \frac{\partial e}{\partial x} + \frac{1}{\rho}\frac{\partial p}{\partial x} - \frac{p}{\rho^2}\frac{\partial\rho}{\partial x} \quad (12.11)$$

Substituting this result into eq. (12.10) then gives:

$$\rho A \frac{\partial e}{\partial t} + \rho A V \frac{\partial e}{\partial x} + \frac{p V A}{\rho} \frac{\partial \rho}{\partial x}(h - e)\frac{\partial}{\partial x}(\rho V A) = 0 \qquad (12.12)$$

But:     $e = c_v T,$     $h - e = (\gamma - 1)c_v T,$     $\dfrac{p}{\rho} = c_v(\gamma - 1)T$

Hence, eq. (12.12) becomes:

$$c_v \rho A \frac{\partial T}{\partial t} + c_v \rho A V \frac{\partial T}{\partial x} - (\gamma - 1)c_v T V A \frac{\partial \rho}{\partial x}$$

$$+ (\gamma - 1)c_v T\left[ V A \frac{\partial \rho}{\partial x} + \rho A \frac{\partial V}{\partial x} + \rho V \frac{\partial A}{\partial x} \right] = 0$$

i.e.,     $$\frac{\partial T}{\partial t} + V \frac{\partial T}{\partial x} + (\gamma - 1)T\left[ \frac{\partial V}{\partial x} + \frac{V}{A} \frac{\partial A}{\partial x} \right] = 0 \qquad (12.13)$$

Lastly, the equation of state gives:

$$p = \rho R T \qquad (12.14)$$

Equations (12.2), (12.6), (12.13), and (12.14), which are repeated below, are one form of the governing equations for one-dimensional, unsteady, compressible flow:

$$A \frac{\partial \rho}{\partial t} + \frac{\partial}{\partial x}(\rho A V) = 0$$

$$\rho \frac{\partial V}{\partial t} + \rho V \frac{\partial V}{\partial x} + \frac{\partial p}{\partial x} = 0$$

$$\frac{\partial T}{\partial t} + V \frac{\partial T}{\partial x} + (\gamma - 1)T\left[ \frac{\partial V}{\partial x} + \frac{V}{A} \frac{\partial A}{\partial x} \right] = 0$$

$$p = \rho R T \qquad (12.15)$$

These equations, together, represent a set of four equations in the four variables $p$, $V$, $T$, and $\rho$. The variation of $A$ with $x$ is assumed to be specified. If the flow is isentropic, the entropy equation gives:

$$\frac{p}{\rho^\gamma} = \text{constant} \qquad (12.16)$$

At first sight, then, it may appear that in isentropic flow, the flow is overspecified because there appear to be five equations involving only four unspecified variables. However, eq. (12.16) in conjunction with the equation of state gives, as discussed before, for isentropic flow:

$$\rho = C T^{1/\gamma - 1} \qquad (12.17)$$

where $C$ is a constant. Differentiating eq. (12.17) and then dividing by the original equation gives:

$$\frac{1}{\rho}\frac{\partial \rho}{\partial t} = \frac{1}{(\gamma - 1)T}\frac{\partial T}{\partial t}$$

and:

$$\frac{1}{\rho}\frac{\partial \rho}{\partial x} = \frac{1}{(\gamma - 1)T}\frac{\partial T}{\partial x}$$

Substituting these equations into the energy equation (12.13) then gives:

$$(\gamma - 1)\frac{T}{\rho}\frac{\partial \rho}{\partial t} + (\gamma - 1)\frac{VT}{\rho}\frac{\partial \rho}{\partial x} + (\gamma - 1)T\frac{\partial V}{\partial x} + (\gamma - 1)\frac{VT}{A}\frac{\partial A}{\partial x} = 0$$

Dividing this equation through by:

$$(\gamma - 1)\frac{T}{\rho A}$$

then gives:

$$A\frac{\partial \rho}{\partial t} + VA\frac{\partial \rho}{\partial x} + \rho A\frac{\partial V}{\partial x} + \rho V\frac{\partial A}{\partial x} = 0 \qquad (12.18)$$

This equation can be written as:

$$A\frac{\partial \rho}{\partial t} + \frac{\partial}{\partial x}(\rho V A) = 0$$

This is, of course, the continuity equation, i.e., in isentropic flow the energy equation adds no additional information to the solution and does not have to be independently solved. This is because, in isentropic flow, one state variable, i.e., the entropy, is constant.

The equations for one-dimensional flow given in this section will be used in the numerical solutions discussed in this chapter. It is possible to develop more efficient numerical solution procedures by writing these equations in other forms. This will not, however, be done here because the purpose is to illustrate the basic ideas involved in the numerical solution of compressible flow problems, not to discuss the most efficient and accurate solution procedures.

## 12.3
## DISCRETIZATION

The numerical solution of the differential equations that describe fluid flows usually involves the introduction of a series of distinct nodal points over the solution domain and then the solution of the numerical form of the governing equations to give the values of the flow variables at these nodal points rather than for their values continuously over the whole flow field. To do this, some form of numerical approximation for the derivatives in the governing equations is introduced. These numerical approximations for the derivatives, which are expressed in terms of the values of the variables at the nodal points, can be

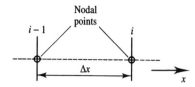

**FIGURE 12.3**
Nodal points considered.

obtained in a number of different ways. In this chapter, because only a simple introduction to the topic is being given, attention will be restricted to the use of finite-difference methods for the numerical approximation of the derivatives.

Formally, finite-difference approximations are obtained by using Taylor series and then truncating the series because the distance between the nodal points is chosen to be small. For example, consider two adjacent nodal points, $i$ and $i - 1$, in a spatially one-dimensional flow situation as shown in Fig. 12.3. If the value of the first derivative of, say, velocity, $V$, at nodal point $i$ is required, it is noted that:

$$V_{i-1} = V_i + \frac{dV}{dx}\bigg|_i (-\Delta x) + \frac{d^2V}{dx^2}\bigg|_i \frac{(-\Delta x)^2}{2!} + \cdots \qquad (12.19)$$

Here $\Delta x$ is the distance between the nodal points and, because a backward difference is being considered, i.e., because the $x$-value at the point $i - 1$ is smaller than the $x$-value at point $i$, the step from point $i$ to point $i - 1$ is $-\Delta x$. Equation (12.19) can be rearranged to give:

$$\frac{dV}{dx}\bigg|_i = \frac{V_i - V_{i-1}}{\Delta x} + \frac{d^2V}{dx^2}\bigg|_i \frac{\Delta x}{2!} + \cdots \qquad (12.20)$$

If $\Delta x$ is small, the higher order terms, i.e., the terms in eq. (12.19) that involve $\Delta x^2, \Delta x^3, \Delta x^4, \ldots$ will be negligible and eq. (12.20) then gives the following first order accurate finite-difference approximation for the spatial derivative (it is first order accurate because terms involving $\Delta x^2$ and higher powers of $\Delta x$ have been neglected:

$$\frac{dV}{dx}\bigg|_i = \frac{V_i - V_{i-1}}{\Delta x} \qquad (12.21)$$

This is, therefore, a first order, backward-difference approximation for the spatial derivative at nodal point $i$. The smaller the value of $\Delta x$ used, the better will be the accuracy of this approximation. This will be discussed in the next section

If unsteady flow is involved, a forward difference in time can be obtained by using a Taylor series in time to give:

$$V_i^1 = V_i^0 + \frac{dV}{dt}\bigg|_i^0 \Delta t + \frac{d^2V}{dx^2}\bigg|_i^0 \frac{\Delta t^2}{2!} + \cdots \qquad (12.22)$$

Superscripts 0 and 1 refer to conditions at the beginning and end of the time step $\Delta t$ respectively. If $\Delta t$ is chosen to be small, the higher order terms, i.e., the terms that involve $\Delta t^2$, $\Delta t^3$, $\Delta t^4$, ... will be negligible and eq. (12.22) then gives the following first order accurate finite-difference approximation:

$$\left. \frac{dV}{dt} \right|_i^0 = \frac{V_i^1 - V_i^0}{\Delta t} \tag{12.23}$$

Finite-difference approximations for the other derivatives that occur in the governing equations can be obtained by applying the same procedure as that outlined above.

## 12.4
## STABILITY AND ACCURACY

Consider a flow situation in which there is a sudden change in the boundary conditions. As a result of the change, the flow variables will all change. However, if after the change, the boundary conditions are kept constant, the flow variables will move to new steady state values. If a numerical analysis of the flow transient is undertaken, a result similar to that shown as curve A in Fig. 12.4 would be expected.

Because numerical approximations to the space and time derivations are being used, a difference between the exact and the numerical solution can be anticipated. This is the numerical error. Now, using a Taylor expansion, as discussed in the previous section:

$$G_i^{j+1} = G_i^j + \left. \frac{\partial G}{\partial t} \right|_i^j \Delta t + \left. \frac{\partial^2 G}{\partial t^2} \right|_i^j \frac{\Delta t^2}{2!} + \cdots$$

$$G_{i+1}^j = G_i^j + \left. \frac{\partial G}{\partial x} \right|_i^j \Delta x + \left. \frac{\partial^2 G}{\partial x^2} \right|_i^j \frac{\Delta x^2}{2!} + \cdots \tag{12.24}$$

where $G$ is some flow variable and the subscripts and superscripts have the same meaning as in the previous section. Equation (12.24) can be used to give:

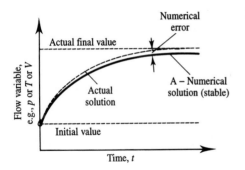

**FIGURE 12.4**
Effect of numerical error on solution.

$$\frac{\partial G}{\partial t}\bigg|_{i}^{j} = \left(\frac{G_{i}^{j+1} - G_{i}^{j}}{\Delta t}\right) - \frac{\partial^2 G}{\partial t^2}\bigg|_{i}^{j} \frac{\Delta t}{2} - \cdots$$

$$\frac{\partial G}{\partial x}\bigg|_{i}^{j} = \left(\frac{G_{i+1}^{j} - G_{i}^{j}}{\Delta x}\right) - \frac{\partial^2 G}{\partial x^2}\bigg|_{i}^{j} \frac{\Delta x}{2} - \cdots$$

(12.25)

If first order finite-difference approximations are used in the numerical analysis, i.e., if the following approximations are used:

$$\frac{\partial G}{\partial t}\bigg|_{i}^{j} = \frac{G_{i}^{j+1} - G_{i}^{j}}{\Delta t}$$

$$\frac{\partial G}{\partial x}\bigg|_{i}^{j} = \frac{G_{i+1}^{j} - G_{i}^{j}}{\Delta x}$$

(12.26)

then it will be seen as mentioned in the previous section that errors depending on the values of $\Delta t$ and $\Delta x$ and on the values of the second and higher order derivatives will be incurred when using these finite-difference approximations. This is the source of the numerical error shown in Fig. 12.4. These errors are reduced by using small values of $\Delta t$ and $\Delta x$, particularly if the flow variables are changing sharply in time or space.

With some solution schemes, a different type of problem can arise—the solution can become unstable. This is illustrated in Fig. 12.5. In this case, the numerical solution moves away from the actual solution by an amount that grows in some way with time, the numerical value eventually growing outside the limit allowed by the computational device being used.

To study, in a very simple way, how such instability can develop, it is noted that the governing equations, such as eq. (12.22), have the following simplified generic form:

$$\frac{\partial G}{\partial t} + V \frac{\partial G}{\partial x} = 0$$

(12.27)

where $G$ is a flow variable. The actual equations are more complex than eq. (12.27) but an analysis of the stability of eq. (12.27) should give an indication of the stability problems involved in the solution of the actual flow equations. It should be noted that eq. (12.27) basically states that the rate of change of a

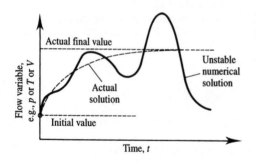

**FIGURE 12.5**
Development of numerical instability.

quantity in a control volume is the net rate at which the quantity is being "convected" into the flow by the flow.

Consider the explicit finite difference form of eq. (12.27). It is:

$$\frac{G_i^{j+1} - G_i^j}{\Delta t} = -V_i^j \left( \frac{G_i^j - G_{i-1}^j}{\Delta x} \right)$$

i.e.,

$$G_i^{j+1} = G_i^j - \left( V_i^j \frac{\partial t}{\partial x} \right)(G_i^j - G_{i-1}^j) \tag{12.28}$$

Assume that $G_i^j = G_{i-1}^j - \Delta G$, i.e., that the value of $G$ at point $i$ is less by an amount $\Delta G$ than the value of $G$ at point $(i-1)$. Equation (12.23) then gives:

$$G_i^{j+1} = G_i^j + \left( \frac{V_i^j \Delta t}{\Delta x} \right) \Delta G \tag{12.29}$$

Hence, if the quantity:

$$S = \frac{V_i^j \Delta t}{\Delta x} \tag{12.30}$$

is greater than 1, eq. (12.29) indicates that $G_i^{j+1} > G_{i-1}^j$. But this is physically unrealistic because $G$ is increasing at point $i$ because $G_{i-1} > G_i$ so it is physically impossible for $G_{i-1}^{j+1}$ to be greater than $G_{i-1}^j$. Hence, if $S > 1$ oscillations will develop in the solution that will eventually lead to instability. Hence, for an explicit finite-difference scheme to be stable it is necessary that:

$$S = \frac{V_i \Delta t}{\Delta x} < 1 \tag{12.31}$$

This is only a guide to the criteria that will ensure the stability of an explicit finite-difference solution to the actual flow equations.

An explicit finite-difference scheme is only stable if the relative values of $\Delta t$ and $\Delta x$ are correctly chosen, i.e., it is conditionally stable. Other finite-difference schemes exist that are unconditionally stable, i.e., stable no matter what values of $\Delta t$ and $\Delta x$ are used.

**Two-Step Procedure**

A procedure that has a higher order of accuracy than that discussed above is a two-step procedure using so-called predictor and corrector steps. To understand the basic procedure, consider eq. (12.27). Basically the attempt is to use the following approximation:

$$\left. \frac{\partial G}{\partial t} \right|_i = \frac{1}{2} \left( \left. \frac{\partial G}{\partial t} \right|_i^{j+1} + \left. \frac{\partial G}{\partial t} \right|_i^j \right) \tag{12.32}$$

However, since conditions at time $j + 1$ are not known, a two-step procedure is used.

The first step gives, using a backward spatial derivative:

$$\left.\frac{\partial G}{\partial t}\right|_i^j = -V_i^j\left(\frac{\partial G_i^j - G_{i-1}^j}{\Delta x}\right) = F_i^j$$

so:
$$\bar{G}_i^{j+1} = G_i^j + H_i^j \Delta t \tag{12.33}$$

The overbar on $\bar{G}_i^{j+1}$ indicating that it is a predicted or approximate value of $G_i^{j+1}$. This is the predictor step.

Using the values of $\bar{G}_i^{j+1}$ so obtained, an approximation to the timewise gradient of $G$ at time $j + 1$ is:

$$\left.\frac{\partial G}{\partial t}\right|_i^{j+1} = -\bar{V}_i^{j+1}\left(\frac{\bar{G}_{i+1}^{j+1} - \bar{G}_i^{j+1}}{\Delta x}\right) = \bar{F}_i^{j+1} \tag{12.34}$$

This is a corrector step, a forward spatial derivative being used in the correction step.

The actual values at time $j + 1$ are then calculated using:

$$G_i^{j+1} = G_i^j + \frac{\Delta t}{2}\left(\left.\frac{\partial G}{\partial t}\right|_i^{j+1} + \left.\frac{\partial G}{\partial t}\right|_i^j\right)$$

$$= G_i^j + \frac{\Delta t}{2}(\bar{F}_i^{j+1} + \bar{F}_i^j) \tag{12.35}$$

Many other numerical schemes that limit the numerical errors and that provide unconditional stability are available.

## 12.5
## UNDER-RELAXATION

An iterative approach often has to be used in obtaining the numerical solution to a fluid flow problem. For example, if the equation governing the flow has the form:

$$\frac{dV}{dx} = F(V, x) \tag{12.36}$$

one way to obtain a numerical solution to such an equation is to initially guess the values of $V$ at all the nodal points. Using these guessed values then allows the values of the function $F$ to be found at all the nodal points. Using a finite-difference approximation to the derivative then gives:

$$\frac{V_i - V_{i-1}}{\Delta x} = F(V_i^0, x)$$

i.e.,
$$V_i = V_{i-1} + \Delta x F(V_i^0, x) \tag{12.37}$$

In this equation, $V_i^0$ is the guessed value of $V_i$ and $V_i^1$ is the updated value of $V_i$. The values of $V_i^1$ so found can then be used to determine updated values of $F$ and further values of $V_i$ at the nodes can then be determined, the process being continued until convergence is reached, i.e., until the values of $V_i$ cease changing to within some prescribed tolerance. In this discussion it has

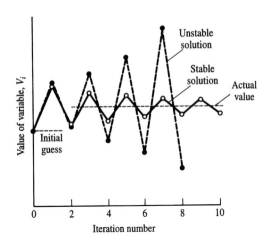

**FIGURE 12.6**
Oscillations and development of numerical instability during iterative solution.

been assumed that the boundary conditions determine some initial value of $V$, i.e., determine $V_1$.

The problem with the above procedure is that oscillatory values can be obtained and this can mean that a very large number of iterations is required before convergence is obtained. In some cases, the oscillations can grow and instability can develop. This is illustrated in Fig. 12.6.

This problem can be overcome in many cases by using damping, i.e., by noting that the change in the value of $V_i$ from one iteration to the next is, according to eq. (12.37), given by:

$$V_i^1 - V_i^0 = V_{i-1} + \Delta x F(V_i^0, x) \tag{12.38}$$

Then, instead of applying this full change in $V_i$, a fraction $r$ of the change is actually used, i.e., the new value of $V_i$ is taken as:

$$V_i^1 = V_i^0 + r(V_{i-1} + \Delta x F - V_i^0) \tag{12.39}$$

The factor $r$, which is here taken to be less than 1, is called the "damping" or "under-relaxation" factor. Its effect on the solution can resemble that shown in Fig. 12.7. If $r = 1$, eq. (12.39) becomes:

$$V_i^1 = V_{i-1} + \Delta x F$$

This is the same result as given by eq. (12.38), i.e., there is no damping when $r = 1$. The smaller the value of $r$, the more the damping. However, if a very small value of $r$ is used, a very slow convergence to the final value can result. The value of $r$ used should be so chosen to be small enough to ensure that the oscillations in the solution are damped and stability is achieved but large enough to ensure that relatively rapid convergence is achieved.

The use of an under-relaxation factor is illustrated in the following simplified example.

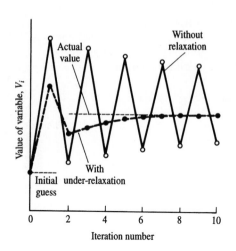

**FIGURE 12.7**
Effect of the use of an under-relaxation factor on an iterative solution.

### EXAMPLE 12.1

Consider incompressible flow through a 0.5 m long channel whose area variation is such that a constant negative pressure gradient exists along the channel. The fluid enters the channel at a velocity of 1 m/s.

Recalling that the momentum equation gives for steady incompressible flow:

$$V\frac{dV}{dx} = -\frac{1}{\rho}\frac{dp}{dx}$$

the magnitude of the applied pressure gradient is such that this equation becomes:

$$V\frac{dV}{dx} = 1600$$

Solve this equation iteratively by dividing the flow into a series of twenty steps as shown in Fig. E12.1a and by setting $V_i = V_1 = 1$ m/s at all nodal points to start the iterative process. Carry out calculations for relaxation factors, $r$, of 1, 0.8, 0.6, and 0.4 and compare the variations of the velocities at the center point, i.e., at $i = 11$, and at the exit, i.e., at $i = 21$, with iteration number for these values of $r$. Consider a maximum of 25 iterations.

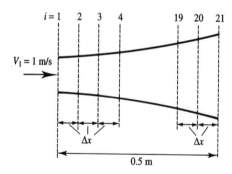

**FIGURE E12.1a**

**Solution**

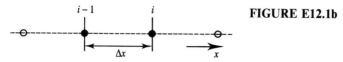

**FIGURE E12.1b**

Consider again two adjacent nodal points as shown in Fig. E12.1b. The finite-difference form of the given governing equation is then:

$$V_i \frac{V_i - V_{i-1}}{\Delta x} = 1600$$

i.e.,

$$V_i = V_{i-1} + \frac{1600\Delta x}{V_i}$$

In the situation under consideration here:

$$\Delta x = \frac{0.5}{20} = \frac{1}{40}$$

so the finite-difference equation given above becomes:

$$V_i = V_{i-1} + \frac{40}{V_i}$$

In applying this equation, the values of $V_i$ are all guessed and these values are used to determine the values of the term $40/V_i$ in the equation given above. This allows new values of $V_i$ to be found starting at point $i = 2$ and using $V_1 = 1.0$ and then marching forward to $i = 3, \ldots, 21$. A relaxation factor is used so the above equation for $V_i$ is actually written as:

$$V_i = V_i^0 + r\left(V_{i-1} + \frac{40}{V_i} - V_i^0\right)$$

As discussed above, $V_i^0$ is set equal to the inlet velocity, i.e., 1 m/s, at all points to start the procedure.

While the process is easily carried out using a spreadsheet program, in order to illustrate the procedure, its implementation using a simple BASIC program is given in Program E12.1.

**Program E12.1**

```
 10   DIM V(21)
 20   CLS
 30   PRINT" "
 40   PRINT"EFFECT OF RELAXATION FACTOR ON AN INCOMPRESSIBLE FLOW
      SOLUTION
 50   PRINT" ":PRINT" "
 60   INPUT " RELAXATION FACTOR ";R
 70   FOR I=1 TO 21
 80   V(I)=1.0
 90   NEXT I
100   PRINT "ITER.","V MID","VEND"
110   FOR J=1 TO 25
120   FOR I=2 TO 21
130   V(I)=V(I)+R*(V(I-1)+40.0/V(I)-V(I))
140   NEXT I
150   PRINT J,V(11),V(21)
160   NEXT J
```

**TABLE E12.1**

| Iter. Num. | $V_{11}$ $r = 1.0$ | $V_{21}$ $r = 1.0$ | $V_{11}$ $r = 0.8$ | $V_{21}$ $r = 0.8$ | $V_{11}$ $r = 0.6$ | $V_{21}$ $r = 0.6$ | $V_{11}$ $r = 0.4$ | $V_{21}$ $r = 0.4$ |
|---|---|---|---|---|---|---|---|---|
| 0 | 1.00 | 1.00 | 1.00 | 1.00 | 1.00 | 1.00 | 1.00 | 1.00 |
| 1 | 401.00 | 801.00 | 143.82 | 159.16 | 60.64 | 61.00 | 27.66 | 27.67 |
| 2 | 3.89 | 4.56 | 110.75 | 152.83 | 60.19 | 61.96 | 28.61 | 28.63 |
| 3 | 131.84 | 225.15 | 75.43 | 138.65 | 58.09 | 62.86 | 29.49 | 29.56 |
| 4 | 7.70 | 9.92 | 49.57 | 117.94 | 54.27 | 63.61 | 30.27 | 30.46 |
| 5 | 76.35 | 120.93 | 35.44 | 95.00 | 49.28 | 64.09 | 30.92 | 31.34 |
| 6 | 11.40 | 15.39 | 29.53 | 74.39 | 43.92 | 64.15 | 31.40 | 32.19 |
| 7 | 55.38 | 84.69 | 27.54 | 58.92 | 38.97 | 63.66 | 31.69 | 33.02 |
| 8 | 14.64 | 20.25 | 26.98 | 49.01 | 34.91 | 62.53 | 31.79 | 33.82 |
| 9 | 44.90 | 67.41 | 26.84 | 43.50 | 31.90 | 60.79 | 31.70 | 34.60 |
| 10 | 17.35 | 24.35 | 26.81 | 40.79 | 29.84 | 58.52 | 31.45 | 35.35 |
| 11 | 38.85 | 57.68 | 26.80 | 39.58 | 28.54 | 55.89 | 31.08 | 36.08 |
| 12 | 19.57 | 27.71 | 26.80 | 39.08 | 27.76 | 53.11 | 30.63 | 36.77 |
| 13 | 35.07 | 51.68 | 26.80 | 38.89 | 27.31 | 50.38 | 30.15 | 37.43 |
| 14 | 21.33 | 30.38 | 26.80 | 38.81 | 27.06 | 47.87 | 29.65 | 38.05 |
| 15 | 32.58 | 47.77 | 26.80 | 38.79 | 26.93 | 45.68 | 29.18 | 38.61 |
| 16 | 22.69 | 32.47 | 26.80 | 38.78 | 26.87 | 43.86 | 28.75 | 39.13 |
| 17 | 30.89 | 45.13 | 26.80 | 38.78 | 26.83 | 42.42 | 28.37 | 39.58 |
| 18 | 23.74 | 34.08 | 26.80 | 38.78 | 26.81 | 41.32 | 28.04 | 39.96 |
| 19 | 29.72 | 43.31 | 26.80 | 38.78 | 26.81 | 40.51 | 27.76 | 40.28 |
| 20 | 24.53 | 35.29 | 26.80 | 38.78 | 26.80 | 39.93 | 27.54 | 40.53 |
| 21 | 28.90 | 42.03 | 26.80 | 38.78 | 26.80 | 39.53 | 27.36 | 40.71 |
| 22 | 25.13 | 36.20 | 26.80 | 38.78 | 26.80 | 39.26 | 27.22 | 40.83 |
| 23 | 28.31 | 41.12 | 26.80 | 38.78 | 26.80 | 39.08 | 27.11 | 40.88 |
| 24 | 25.57 | 36.88 | 26.80 | 38.78 | 26.80 | 38.97 | 27.03 | 40.89 |
| 25 | 27.89 | 40.47 | 26.80 | 38.78 | 26.80 | 38.89 | 26.96 | 40.85 |

All velocities in m/s.

```
170  STOP
180  END
```

Results obtained using this program are given in Table E12.1.

It will be seen that with $r = 1$ and with $r = 0.4$, convergence has not been reached after 25 iterations, whereas with $r = 0.8$, convergence has been reached after approximately 15 iterations. With $r = 0.6$ convergence has very nearly been reached after 25 iterations.

It should be noted that the governing equation:

$$V \frac{dV}{dx} = 1600$$

can be directly integrated to give, since $V = 1$ when $x = 0$:

$$V^2 - 1 = 3200x$$

This gives the velocities at the center point and at the exit as 28.30 m/s and 40.01 m/s respectively, which may be compared with the converged values given by the numerical solution, i.e., 26.80 m/s and 38.78 m/s. The difference between the analytical and numerical values is due to numerical error resulting from the relatively coarse grid used. If 101 nodal points had been used instead of 21, the numerically predicted velocities would have been 27.92 m/s and 39.71 m/s, which is within approximately 1 percent of the analytical values.

## 12.6
## STEADY ISENTROPIC SUBSONIC FLOW

Attention will first be given to steady isentropic flow. In this case the governing equations are:

$$\frac{d}{dx}(\rho A V) = 0 \qquad (12.40)$$

$$V\frac{dV}{dx} + \frac{1}{\rho}\frac{dp}{dx} = 0 \qquad (12.41)$$

$$\frac{p}{\rho^\gamma} = \text{constant} \qquad (12.42)$$

It will be noted that because the flow variables depend only on $x$ in steady flow, these equations do not involve partial derivatives.

Of course, as shown in earlier chapters, analytical solutions to eqs. (12.40)–(12.43) are easily obtained. The purpose here is not, therefore, to imply that it is necessary to solve this set of equations numerically. It is simply to illustrate some of the considerations that are involved in obtaining numerical solutions to the equations governing compressible fluid flows.

In order to illustrate how a numerical solution to the above set of equations can be obtained, consideration will be given to a flow situation in which the values of all the variables at some initial section are known. This situation is shown in Fig. 12.8. In this case, eqs. (12.40) and (12.42) give:

$$\rho A V = \rho_1 A_1 V_1 \qquad (12.43)$$

$$\frac{p}{\rho^\gamma} = \frac{p_1}{\rho_1^\gamma} \qquad (12.44)$$

Because eq. (12.2) is nonlinear, i.e., there are terms in this equation that involve the products of functions of the variables, an iterative form of solution

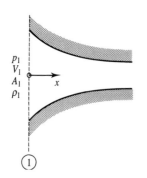

**FIGURE 12.8**
Flow situation being considered.

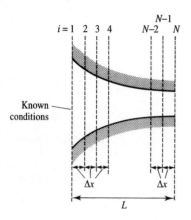

**FIGURE 12.9**
Grid points used.

will be adopted here. Now attention is being restricted to subsonic flows in this section and in such flows the changes in density are relatively small (as compared to those that can be anticipated in supersonic flows). The iterative finite-difference procedure used here therefore involves the following steps:

1. Divide the flow domain into a series of equally spaced segments each of length $\Delta x$ which is such that:

$$\Delta x = \frac{L}{N-1} \qquad (12.45)$$

   $N$ being the number of grid points and $L$ the length of the solution domain. These are shown in Fig. 12.9. (As will be shown in the example given below, the actual value of $\Delta x$ does not have to be specified because equally spaced nodal points are being used.)

2. Specify the conditions on the inlet plane, i.e., specify $\rho_1$, $p_1$, and $V_1$. Also specify the cross-sectional area at each nodal point, i.e., specify $A_1, A_2, \ldots, A_{N-1}, A_N$.

3. As a first guess, assume that the effects of the density changes are negligible, i.e., assume that $\rho_i = \rho_1$ at all sections. This allows first approximations to the velocities at all points $2, \ldots, N$ to be found using the continuity equation which in this case gives:

$$V_i = \frac{V_1 A_1}{A_i} \qquad (12.46)$$

4. The pressure at points $2, \ldots, N$ are then found using a first order finite-difference approximation to eq. (12.41). Using:

$$\frac{dV}{dx} = \frac{V_i - V_{i-1}}{\Delta x}$$

and:

$$\frac{dp}{dx} = \frac{p_i - p_{i-1}}{\Delta x}$$

eq. (12.41) gives:

$$V_i \left( \frac{V_i - V_{i-1}}{\Delta x} \right) + \frac{1}{\rho_i} \left( \frac{p_i - p_{i-1}}{\Delta x} \right) = 0$$

i.e., 

$$p_i = p_{i-1} - \rho_i V_i (V_i - V_{i-1}) \qquad (12.47)$$

Applying this equation sequentially from points $i = 2, \ldots, N$ allows the values of $p_i$ at each of these points to be found.

5. Using the values of $p_i$ found in step 4 above, find updated values of the density at the nodal points by using eq. (12.44), which gives:

$$\rho_i = \rho_1 \left( \frac{p_i}{p_1} \right)^{1/\gamma} \qquad (12.48)$$

However, because an iterative solution procedure is being used, under-relaxation, discussed in the previous section, will be adopted and eq. (12.48) is therefore actually written as:

$$\rho_i = \rho_i^0 + r \left[ \rho_1 \left( \frac{p_i}{p_1} \right)^{1/\gamma} - \rho_i^0 \right] \qquad (12.49)$$

the superscript 0 again indicating the density at the previous iteration. The under-relaxation factor, $r$, is chosen to be less than 1.

6. Using the continuity equation, written as:

$$V_i = \frac{\rho_1 V_1 A_1}{\rho_i A_i} \qquad (12.50)$$

find updated values of the velocity at each nodal point.

7. Repeat steps 4–6 until a converged solution is obtained. This could, for example, be assumed to have been reached when:

$$\frac{|\rho_i - \rho_i^0|}{\rho_i^0} < 0.001 \qquad (12.51)$$

at all points from $i = 2, \ldots, N$. However, in implementing this procedure, it is important to limit the total number of iterations because if $r$ is chosen to be too large, oscillations in the calculated values may occur and convergence, if it occurs, may only occur after an unacceptably large number of iterations. Alternatively, if $r$ is chosen to be too small convergence may again only occur after an unacceptably large number of iterations.

The implementation of the above procedure is illustrated in the following example.

**EXAMPLE 12.2**

Consider subsonic air flow through the convergent nozzle shown in Fig. E12.2a. The nozzle shape is such that the area variation with distance along the nozzle is given by:

$$A(x) = a + bx + cx^2$$

Determine numerically how the exit Mach number and exit density vary with the inlet velocity, $V_1$. The inlet pressure and temperature remain constant and

**FIGURE E12.2a**

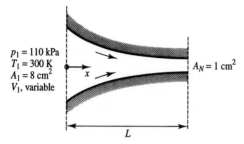

$p_1 = 110$ kPa
$T_1 = 300$ K
$A_1 = 8$ cm$^2$
$V_1$, variable

$x$

$A_N = 1$ cm$^2$

$L$

equal to 110 kPa and 300 K respectively. Also show some typical variations of density along the nozzle.

### Solution

The values of the coefficients in the area equation in terms of the nozzle inlet and exit areas will first be determined. Assuming that the walls of the nozzle are parallel at the exit of the nozzle, the following are known about the area variation:

When $x = 0$,    $A = A_1$
When $x = L$,    $A = A_N$
When $x = L$,    $dA/dx = 0$

where $L$ is the length of the nozzle as indicated in Fig. E12.2a.

Hence, using the given equation for the variation of $A$, the above conditions give:

$$a = A_1$$

$$a + bL + cL^2 = A_N$$

$$b + 2cL = 0$$

These are three equations in the three unknowns $a$, $b$, and $c$. Solving between them gives:

$$a = A_1$$

$$b = -\frac{2(A_1 - A_N)}{L}$$

$$c = \frac{(A_1 - A_N)}{L^2}$$

Therefore, the variation of the area with distance along the nozzle is given by:

$$A = A_1 - 2(A_1 - A_N)\frac{x}{L} + (A_1 - A_N)\frac{x^2}{L^2}$$

The area variation with $x$ is therefore determined by specifying the values of $A_1$ and $A_N$. The value of $L$ does not have to be specified because:

$$\frac{x}{L} = \frac{i-1}{N-1}$$

A computer program, written in FORTRAN, that implements the procedure described above and uses the form of the area variation just discussed is given in Program E12.2. The program writes the output to the screen and to a file called CONVNOZ.DAT.

**Program E12.2**

```
C
*******************************************************************************
C *
C *                       CONVNOZ
C *                    ***********
C *
C * THIS PROGRAM CALCULATES STEADY SUBSONIC FLOW THROUGH A
C * CONVERGENT NOZZLE
C *
C
*******************************************************************************
C
      REAL*8 V(300),P(300),A(300),RH(300),T(300),M(300)
C
C     N = NUMBER OF GRID POINTS
C     P = PRESSURE (pascals)
C     V = VELOCITY (m/s)
C     T = DENSITY (K)
C     RH = DENSITY (kg/m^3)
C     A = AREA (m^2)
C     VS = ACOUSTIC SPEED (m/s)
C     X = DISTANCE ALONG NOZZLE (m)
C     M = MACH NUMBER
C
      OPEN (1,FILE = 'CONVNOZ.DAT')
C
      WRITE(*,750)
      WRITE(1,750)
C
C     NUMBER OF GRID POINTS
C
      N = 200
      N1 = N-1
C
      WRITE(1,3000) N
      WRITE(*,3000) N
C
C     RELAXATION FACTOR
C
      REX = 0.2
C
C     PROPERTIES FOR AIR
C
      GAM = 1.4
      RG = 287.0
C
C     INITIAL CONDITIONS
C
      P(1) = 110000.0
      T(1) = 300.0
      V(1) = 10.0
      RH(1) = P(1)/(RG*T(1))
C
C     INLET AND EXIT AREA
C
      A(1) = 0.0008
      A(N) = 0.0001
C
C     NOZZLE GEOMETRIC PARAMETERS
C
      AS = A(1)
      CS = -(A(N)-A(1))
      BS = -2.0*CS
C
C     SET AREAS AND INITIALIZE DENSITY
C
```

```
      DO 151 I = 2,N
        X = (I-1.0)/(N-1.0)
        A(I) = AS + BS*X + CS*X*X
        RH(I) = RH(1)
  151 CONTINUE
C
C   ITERATE UNTIL P CONVERGES
C
      J = 0
      ORH = 1.0
C
  220 J = J + 1
C
C   CHECK FOR NON-CONVERGENCE
C
      IF (J.LT.500) THEN
C
      DO 210 I = 2,N
        V(I) = RH(1)*V(1)*A(1)/(RH(I)*A(I))
  210 CONTINUE
C
      DO 260 I = 2,N
        P(I) = P(I-1)-RH(I)*V(I)*(V(I)-V(I-1))
  260 CONTINUE
C
      DO 310 I = 2,N
        RH(I) = RH(I) + REX*(RH(1)*(P(I)/P(1))**(1.0/GAM)-RH(I))
        VS = SQRT(GAM*P(I)/RH(I))
        M(I) = V(I)/VS
        IF(M(I).GT.0.99) STOP ' M > 0.99 '
  310 CONTINUE
C
C   TEST FOR CONVERGENCE OF RH
C
      IF (ABS((RH(N/2)-ORH)/RH(N/2)).LT.0.000001) GO TO 2500
C
      ORH = RH(N/2)
      GOTO 220
C
      ELSE
        STOP 'NON-CONVERGENCE'
      END IF
C
 2500 CONTINUE
      WRITE(*,5000)
      WRITE(1,5000)
C
      DO 410 I = 1,N
        WRITE(*,2000) I,P(I),V(I),RH(I),A(I),M(I)
        WRITE(1,2000) I,P(I),V(I),RH(I),A(I),M(I)
  410 CONTINUE
      CLOSE(1)
      STOP
  750 FORMAT(' FLOW IN A CONVERGENT NOZZLE',/
     $,' --------------------------------',//)
 2000 FORMAT(I4,2F10.2,3F10.4)
 3000 FORMAT(' NO. OF POINTS =',I4,//)
 5000 FORMAT(' I      P       V       D       A       M',
     $/' ───────  ───────  ───────  ───────  ───────')
      END
```

This program has been run with various values of the inlet velocity $V(1)$ between 5 m/s and 25.5 m/s and the variations of the variables on the exit plane with $V(1)$ so obtained are shown in Fig. E12.2b while typical variations of density along the nozzle are shown in Fig. E12.2c.

As the inlet velocity increases, the exit Mach number, of course, also increases until it reaches a value of 1. When this occurs, the solution basically becomes

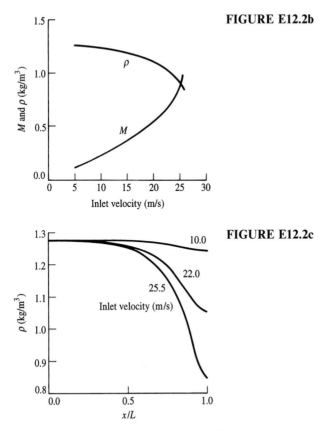

**FIGURE E12.2b**

**FIGURE E12.2c**

unstable for the reasons discussed before, i.e., because for a fixed inlet pressure there is a maximum possible inlet velocity. For this reason, the calculation is stopped if the exit Mach number gets above a value of 0.99.

## 12.7
## STEADY ISENTROPIC SUPERSONIC FLOW

The subsonic flow solution procedure discussed in the previous section started with the assumption that the density was constant. This assumption allowed first approximations for the density and the velocity at all nodal points to be determined. This assumption is often not a good starting point for the solution when the flow is supersonic because in this case the density changes are dominant. However, it will be recalled that in supersonic flow at high Mach numbers for a given stagnation temperature there is a fixed maximum velocity that can be reached. This means that in supersonic flows at high Mach numbers, the velocity changes will be relatively small compared to the density changes. Hence, in order to apply the solution procedure outlined in the previous section to supersonic flow, steps 3–6 given in that section must be modified. For supersonic flow these steps become, steps 1 and 2 being unchanged:

3. As a first guess, assume that the velocity changes are small, i.e., as a first guess set $V_i = V_1$ at all sections. This allows first approximations to the densities at all points 2, ..., $N$ to be found using the continuity equation which in this case gives:

$$\rho_i = \frac{\rho_1 A_1}{A_i} \tag{12.52}$$

4. Using the values of $\rho_i$ so found, the pressures at points 2, ..., $N$ are then found using:

$$p_i = p_1 \left( \frac{\rho_i}{\rho_1} \right)^{\gamma} \tag{12.53}$$

5. Updated values of the velocities at points 2, ..., $N$ are then found using a first-order finite-difference approximation to eq. (12.42) which gives:

$$V_i = V_{i-1} - \frac{(p_i - p_{i-1})}{\rho_i V_i} \tag{12.54}$$

6. Updated values of the density are then found using the continuity equation which gives:

$$\rho_i = \frac{\rho_1 V_1 A_1}{V_i A_i}$$

An under-relaxation procedure is actually again used so this equation is written in the following way:

$$\rho_i = \rho_i^0 + r \left( \frac{\rho_1 V_1 A_1}{V_i A_i} - \rho_i^0 \right) \tag{12.55}$$

Superscript 0 again indicates the value at the beginning of the iteration step.

The implementation of the above procedure is illustrated in the following example. In this example, the procedure is used in conjunction with that outlined in the previous section to compute a flow that involves both subsonic and supersonic Mach numbers.

**EXAMPLE 12.3**
Consider air flow through the convergent–divergent nozzle shown in Fig. E12.3a. Assuming that the flow goes from subsonic to supersonic in the nozzle, write a computer program that will give the distribution of the variables along the nozzle in terms of the values existing at the throat, i.e., at the sonic point. If the pressure on the nozzle inlet section is 200 kPa use the results given by the program to find the throat and exit section pressures.

*Solution*
The values of the variables are all expressed in terms of those existing at the sonic point, i.e., the following are defined:

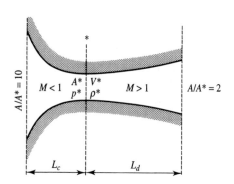

**FIGURE E12.3a**

$$P = \frac{p}{p^*}, \qquad D = \frac{\rho}{\rho^*}, \qquad E = \frac{A}{A^*}$$

$$U = \frac{V}{a^*} = \frac{V}{\sqrt{\gamma p^* / \rho^*}} \tag{i}$$

The $x$ coordinate is expressed in terms of the length of the nozzle section being considered, i.e., the following is defined:

$$X = \frac{x}{L} \tag{ii}$$

When calculating the flow in the divergent section of the nozzle, $L$ is the length of the divergent section while when calculating the flow in the convergent section of the nozzle, $L$ is the length of the convergent section.

In terms of these variables, the governing equations, i.e., eqs. (12.40), (12.41), and (12.42), give:

$$DUE = \frac{V^*}{\sqrt{\gamma p^* / \rho^*}} = 1 \tag{iii}$$

$$\gamma U \frac{dU}{dX} + \frac{1}{D} \frac{dP}{dX} = 0 \tag{iv}$$

$$\frac{P}{D^\gamma} = 1 \tag{v}$$

At the throat, i.e., at the sonic point where $V = a^*$, $p = p^*$, $A = A^*$, and $\rho = \rho^*$, the following conditions apply:

$$P = 1, \qquad D = 1, \qquad U = 1, \qquad E = 1 \tag{vi}$$

The solution is obtained by marching upstream and downstream from the throat as shown in Fig. E12.3b.

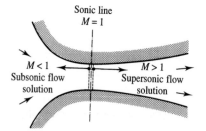

**FIGURE E12.3b**

The supersonic solution starts with a value of $U$, which at this point is effectively the Mach number, that is slightly above 1, say 1.0001 while the subsonic flow solution is started with a value of $U$ that is slightly below 1, say 0.9999.

The shape of the nozzle is assumed to be defined in the same basic way as in Example 12.2, i.e., in the convergent subsonic flow portion of the nozzle the area variation is given by:

$$A = a + bx + cx^2$$

with the requirements that:

When $x = 0$,     $A = A^*$
When $x = L_c$,     $A = A_I$
When $x = 0$,     $dA/dx = 0$

where $L_c$ is the length of the convergent portion of the nozzle as indicated in Fig. E12.3a.

Applying these conditions to the flow area variation equation gives:

$$a = A^*$$

$$a + bL_c + cL_c^2 = A_I$$

$$b = 0$$

Solving between these equations to obtain expressions for $a$, $b$, and $c$ gives the variation of the area with distance along the nozzle in the supersonic convergent portion of the nozzle as:

$$A = A^* + (A_I - A^*)\frac{x^2}{L_c^2}$$

Dividing this equation through by $A^*$ then gives:

$$E = 1 + (E_I - 1)X^2$$

In this equation, $X = x/L_c$. For the nozzle being considered, $E_I = 10$.

In the divergent portion of the nozzle the same form of variation is assumed, i.e., it is again assumed that:

$$A = a + bx + cx^2$$

with the requirements that:

When $x = 0$,     $A = A^*$
When $x = L_d$,     $A = A_E$
When $x = 0$,     $dA/dx = 0$

where $L_d$ is the length of the divergent portion of the nozzle as indicated in Fig. E12.2b.

Applying these conditions to the flow area variation equation gives:

$$a = A^*$$

$$a + bL_d + cL_d^2 = A_E$$

$$b = 0$$

Solving between these equations to obtain expressions for $a$, $b$, and $c$ gives the variation of the area with distance along the nozzle in the subsonic divergent portion of the nozzle. Dividing the equation so obtained through by $A^*$ then gives:

$$E = 1 + (E_E - 1)X^2$$

In this equation, $X = x/L_d$. For the nozzle being considered, $E_E = 2$.

The finite-difference form of the dimensionless momentum equation, i.e., eq. (iv), is:

$$P_i = P_{i-1} - \gamma D_i U_i(U_i - U_{i-1})$$

for subsonic flow and:

$$U_i = U_{i-1} - \frac{(P_i - P_{i-1})}{\gamma D_i U_i}$$

for supersonic flow.

The initial guesses for the dimensionless density and velocity values are, as discussed before:

Supersonic flow:   $U_i = U_1,\ D_i = \dfrac{1}{U_1 E_i}$

Subsonic flow:    $D_i = D_1,\ U_i = \dfrac{1}{R_1 E_i}$

A computer program, written in FORTRAN, that implements the procedure described above is listed in Program E12.3. The program writes the output to the screen and to a file called CONDINOZ.DAT.

**Program E12.3**

```
C
*****************************************************************************
C *
C *                      CONDINOZ
C *                    ***********
C *
C * THIS PROGRAM CALCULATES STEADY FLOW THROUGH A
C * CONVERGENT-DIVERGENT NOZZLE. THE SOLUTION MARCHES OUTWARDS
C * FROM THE THROAT WHERE THE LOCAL MACH NUMBER IS 1.0
C * THE SUPERSONIC PORTION IS CALCULATED FIRST.
C *
C
*****************************************************************************
C
      REAL*8 U(500),P(500),E(500),D(500)
C
C     N=NUMBER OF GRID POINTS
C     P=PRESSURE (p/p*)
C     U=VELOCITY (V/V*)
C     D=DENSITY (rho/rho*)
C     E=AREA (A/A*)
C     VS=ACOUSTIC SPEED (a/a*)
C     X=DISTANCE ALONG NOZZLE (x/L)
C
      OPEN (1,FILE='CONDINOZ.DAT')
C
      WRITE(*,750)
      WRITE(1,750)
  750 FORMAT(' FLOW IN A CONVERGENT/DIVERGENT NOZZLE',/
     $,' ----------------------------------',/,/)
C
C     NUMBER OF GRID POINTS
C
      N=250
      N1=N-1
C
C     PROPERTIES FOR AIR
C
      GAM=1.4
```

```
C
C    INITIAL CONDITIONS FOR SUPERSONIC PORTION OF NOZZLE
C
     P(1) = 1.0
     D(1) = 1.0
     U(1) = 1.0001
C
     E(1) = 1.0
     E(N) = 2.0
C
C    NOZZLE GEOMETRIC PARAMETERS
C
     AS = E(1)
     CS = -(E(N)-E(1))
     BS = -2.0*CS
C
C    INITIALIZE E, U & D - ASSUME U = CONSTANT THROUGH NOZZLE
C
     DO 200 I = 2,N
       X = (I-1.0)/(N-1.0)
       E(I) = AS + BS*X + CS*X*X
       U(I) = U(1)
       D(I) = D(1)*E(1)/E(I)
 200 CONTINUE
C
C    ITERATE UNTIL D CONVERGES
C
     J = 0
     ORH = 1.0
C
 225 J = J + 1
C
C    CHECK FOR NON-CONVERGENCE
C
     IF (J.LT.500) THEN
C
     DO 300 I = 2,N
       P(I) = P(1)*(D(I)/D(1))**(GAM)
 300 CONTINUE
C
C    D IS UPDATED USING AN UNDER-RELAXATION TECHNIQUE
C    D(NEW) = D(OLD) + 20% OF (D(NEW) - D(OLD))
C
     DO 250 I = 2,N
       U(I) = U(I-1)-(P(I)-P(I-1))/(GAM*D(I)*U(I))
       D(I) = D(I) + 0.2*(D(1)*E(1)*U(1)/(E(I)*U(I))-D(I))
 250 CONTINUE
C
C    TEST FOR CONVERGENCE OF RH
C
     IF (ABS((D(N/2)-ORH)/D(N/2)).LT.0.000001) GOTO 1000
       ORH = D(N/2)
     GOTO 225
C
1000 CONTINUE
     ELSE
       STOP 'NON-CONVERGENCE IN SUPERSONIC PORTION'
     END IF
C
     WRITE(*,3000) N
     WRITE(1,3000) N
3000 FORMAT(' NO. OF POINTS =',I4)
     WRITE(*,5100)
     WRITE(1,5100)
5100 FORMAT(/,10X,' SUPERSONIC PORTION OF NOZZLE',/)
     WRITE(*,5000)
     WRITE(1,5000)
5000 FORMAT(' I   P   U   RH   A   M',
    $/' ---  ---  ---  ---  ---  ---')
```

```
      DO 400 I=1,N
        VS=SQRT(P(I)/D(I))
        WRITE(*,2000) I,P(I),U(I),D(I),E(I),U(I)/VS
        WRITE(1,2000) I,P(I),U(I),D(I),E(I),U(I)/VS
  400 CONTINUE
 2000 FORMAT(I4,5F10.4)
C
C     INITIAL CONDITIONS FOR SUBSONIC PORTION OF NOZZLE
C
C     INLET AREA OF NOZZLE
C
      E(N)=10.0
C
C     NOZZLE GEOMETRIC PARAMETERS
C
      AS=E(1)
      CS=-(E(N)-E(1))
      BS=-2.0*CS
C
C     INITIALIZE E AND D
C
      DO 151 I=2,N
        X=(I-1.0)/(N-1.0)
        E(I)=AS+BS*X+CS*X*X
        D(I)=D(1)
  151 CONTINUE
C
      U(1)=0.9999
C
C     ITERATE UNTIL D CONVERGES
C
      J=0
      ORH=1.0
C
  220 J=J+1
C
C     CHECK FOR NON-CONVERGENCE
C
      IF (J.LT.500) THEN
C
        DO 210 I=1,N
          U(I)=D(1)*U(1)*E(1)/(D(I)*E(I))
  210   CONTINUE
C
        DO 260 I=2,N
          P(I)=P(I-1)-GAM*D(I)*U(I)*(U(I)-U(I-1))
  260   CONTINUE
C
        DO 310 I=2,N
          D(I)=D(I)+0.2*(D(1)*(P(I)/P(1))**(1.0/GAM)-D(I))
  310   CONTINUE
C
C     TEST FOR CONVERGENCE OF D
C
        IF (ABS((D(N/2)-ORH)/D(N/2)).LT.0.000001) GOTO 1100
          ORH=D(N/2)
        GOTO 220
C
 1100   CONTINUE
      ELSE
          STOP 'NON-CONVERGENCE IN SUBSONIC PORTION'
      END IF
C
      WRITE(*,5200)
      WRITE(1,5200)
 5200 FORMAT(/,10X,' SUBSONIC PORTION OF NOZZLE',/)
      WRITE(*,5000)
      WRITE(1,5000)
C
```

```
      DO 410 I=1,N
        VS=SQRT(P(I)/D(I))
        WRITE(*,2000) I,P(I),U(I),D(I),E(I),U(I)/VS
        WRITE(1,2000) I,P(I),U(I),D(I),E(I),U(I)/VS
  410 CONTINUE
        CLOSE(1)
        STOP
        END
```

This program gives:

$P$ at inlet $= P_I = 1.853$
$P$ at exit $= P_E = 0.178$

Hence:
$$p^* = \frac{p^*}{p_I} p_I = \frac{p_I}{P_I} = \frac{200}{1.853} = 107.9 \text{ kPa}$$

$$p_E = \frac{p_E \, p^*}{p^* \, p_I} p_i = \frac{P_E}{P_I} \times 200 = \frac{0.178}{1.853} \times 200 = 19.21 \text{ kPa}$$

## 12.8
## UNSTEADY ONE-DIMENSIONAL FLOW

In order to illustrate how numerical solutions for unsteady compressible gas flows can be obtained, attention will be given to the flow that occurs in a long constant area channel following some sudden change. Because the possibility that shock waves can occur must be allowed for, the flow cannot, in general, be assumed to be isentropic. The governing equations are, therefore, since $A$ is assumed to be constant:

$$\frac{\partial \rho}{\partial t} + \frac{\partial}{\partial x}(\rho V) = 0 \tag{12.56}$$

$$\frac{\partial V}{\partial t} + V\frac{\partial V}{\partial x} + \frac{1}{\rho}\frac{\partial p}{\partial x} = 0 \tag{12.57}$$

$$\frac{\partial T}{\partial t} + V\frac{\partial T}{\partial x} + (\gamma - 1)T\frac{\partial V}{\partial x} = 0 \tag{12.58}$$

$$p = \rho RT \tag{12.59}$$

A finite-difference solution of these equations will be considered. At time zero the values of all the variables are specified. The following finite-difference forms of the above equations are used, it being assumed that a set of $N$ uniformly spaced points are used:

$$\left.\frac{\partial \rho}{\partial t}\right|_i = -\left[\frac{\rho_{i+1}V_{i+1} - \rho_{i-1}V_{i-1}}{2\Delta x}\right] \tag{12.60}$$

$$\left.\frac{\partial V}{\partial t}\right|_i = -\left[\frac{p_{i+1} - p_{i-1}}{2\rho_i\Delta x}\right] - V_i\left[\frac{V_i - V_{i-1}}{\Delta x}\right] \tag{12.61}$$

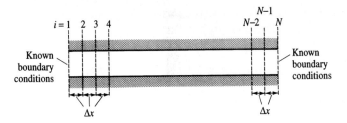

**FIGURE 12.10**
Application of boundary conditions.

$$\left.\frac{\partial T}{\partial t}\right|_i = -(\gamma - 1)T_i\left[\frac{V_{i+1} - V_{i-1}}{2\Delta x}\right] - V_i\left[\frac{T_i - T_{i-1}}{\Delta x}\right] \qquad (12.62)$$

In writing the above equations, it has been assumed that the direction of the x-axis can be chosen so that the values of $V_i$ are all positive. It will be seen that central difference approximations have been used except that "upwind" type approximations, i.e., in the present situation backward differences, for the convective terms in eqs. (12.57) and (12.58), i.e., the terms $V\,\partial V/\partial x$ and $V\,\partial T/\partial x$, have been used. This improves the stability of the solution procedure.

In addition to eqs. (12.60)–(12.62), the solution procedure uses the perfect gas equation written as:

$$p_i = \rho_i RT_i \qquad (12.63)$$

A two-step, predictor–corrector procedure is used. This involves the following steps, it being assumed that the boundary conditions allow the values of the variables at $i = 1$ and $i = N$ to be found (see Fig. 12.10).

1. The initial conditions define the values of the variables $\rho_i$, $V_i$, $T_i$, and $p_i$ at all points at time $t = 0$.
2. Using these initial values, use eqs. (12.60), (12.61), and (12.62) to get first approximations to the derivatives at points $i = 2, \ldots, N - 1$, i.e., to get:

$$\left.\frac{\partial \rho}{\partial t}\right|_i^0, \qquad \left.\frac{\partial V}{\partial t}\right|_i^0, \qquad \left.\frac{\partial T}{\partial t}\right|_i^0$$

3. Using these, predict the values of the variables at time $t = \Delta t$ using:

$$\rho_i^1 = \rho_i^0 + \left.\frac{\partial \rho}{\partial t}\right|_i^0 \Delta t$$

$$V_i^1 = V_i^0 + \left.\frac{\partial V}{\partial t}\right|_i^0 \Delta t$$

$$T_i^1 = T_i^0 + \left.\frac{\partial T}{\partial t}\right|_i^0 \Delta t$$

where $\Delta t$ is a selected time step and superscripts 0 and 1 refer to conditions at the beginning and end of the time step.

　　Use these values in eq. (12.63) to get predicted values of the pressure, i.e., set:

$$p_i^1 = \rho_i^1 RT_i^1$$

This is basically the predictor step.

4. Use these predicted values in eqs. (12.60), (12.61), and (12.62) to get second approximations to the derivatives at points $i = 2, \ldots, N-1$, i.e., to get:

$$\left.\frac{\partial \rho}{\partial t}\right|_i^1, \qquad \left.\frac{\partial V}{\partial t}\right|_i^1, \qquad \left.\frac{\partial T}{\partial t}\right|_i^1$$

5. Use these values to get corrected values of the variables at time $t = \Delta t$ using:

$$\rho_i^1 = \rho_i^0 + \frac{1}{2}\left[\left.\frac{\partial \rho}{\partial t}\right|_i^0 + \left.\frac{\partial \rho}{\partial t}\right|_i^1\right]\Delta t$$

$$V_i^1 = V_i^0 + \frac{1}{2}\left[\left.\frac{\partial V}{\partial t}\right|_i^0 + \left.\frac{\partial V}{\partial t}\right|_i^1\right]\Delta t$$

$$T_i^1 = T_i^0 + \frac{1}{2}\left[\left.\frac{\partial T}{\partial t}\right|_i^0 + \left.\frac{\partial T}{\partial t}\right|_i^1\right]\Delta t$$

$$p_i^1 = \rho_i^1 RT_i^1$$

This is basically the corrector step.

6. Having in this way established the values of the variables at time $\Delta t$, the same procedure can be used to advance the solution to time $2\Delta t$ and so on.

The use of the above procedure is illustrated in the following examples.

### EXAMPLE 12.4

Consider a long tube filled with air and divided into two sections by a diaphragm. The pressure on one side of the diaphragm is 150 kPa while that on the other side is 100 kPa. The air temperature throughout the tube is 300 K.

　　Write a computer program that will calculate the flow that develops if the diaphragm suddenly ruptures. Restrict attention to a distance of 1 m upstream and downstream of the position of the diaphragm. Give representative distributions of pressure, velocity, and temperature. Show the effect of the number of nodal points used on the form of the solution.

### Solution

The situation considered is basically that in a shock tube with a relatively small pressure difference applied across the diaphragm. The initial conditions are shown in Fig. E12.4a.

　　$x$ is chosen in the direction shown in Fig. E12.4a because the air flow that develops after the diaphragm ruptures will be from the high pressure section towards the low-pressure section, i.e., $x$ is chosen to be in the direction of the induced flow.

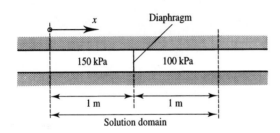

**FIGURE E12.4a**

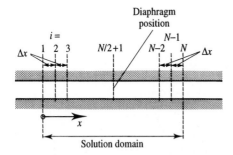

**FIGURE E12.4b**

Attention will be restricted to a maximum time that is short enough to ensure that the flow created when the diaphragm ruptures has not reached a distance of 1 m either upstream or downstream of the diaphragm. Hence, it will be assumed in the solution that at points 1 and $N$ shown in Fig. E12.4b the variables retain their initial undisturbed values, i.e., it will be assumed that:

$$p_1 = 150 \text{ kPa}, \qquad T_1 = 300 \text{ K}, \qquad V_1 = 0$$

and: $\qquad p_N = 100 \text{ kPa}, \qquad T_N = 300 \text{ K}, \qquad V_N = 0$

at all times considered.

Because the solution domain is chosen to have a length of 2 m, the spatial grid size is given by:

$$\Delta x = \frac{2}{N-1} m$$

Since the diaphragm is at point $(N/2 + 1)$, initially, i.e., at time $t = 0$, the following conditions apply:

For $i < N/2 + 1$:  $\qquad p_i = 150 \text{ kPa}, \qquad T_i = 300 \text{ K}, \qquad V_i = 0$
For $i > N/2 + 1$:  $\qquad p_i = 100 \text{ kPa}, \qquad T_i = 300 \text{ K}, \qquad V_i = 0$

A computer program (Program E12.4) based on the procedure outlined above has then been used. This program is again written in FORTRAN. The time step and the maximum time to which the solution is continued are inputs. It writes the output to the screen and to a file called CONACHAN.DAT.

**Program E12.4**

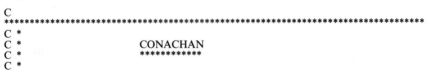

```
C
**********************************************************************************
C *
C *                              CONACHAN
C *                           ***********
C *
```

```
C * THIS PROGRAM CALCULATES UNSTEADY FLOW IN A CONSTANT AREA
C * CHANNEL - IT IS SETUP TO CALCULATE THE FLOW WITHIN A SHOCK TUBE
C * JUST AFTER THE RUPTURE DISK HAS BURST AND PRIOR TO EITHER THE
C * INDUCED SHOCK OR EXPANSION WAVES REFLECTING FROM THE ENDS OF
C * THE SHOCK TUBE.
C *
C
*******************************************************************************
C
      DIMENSION U(500),P(500),T(500),R(500)
     $ ,UI(500),PI(500),TI(500),RI(500)
     $ ,DT(2,500),DU(2,500),DR(2,500)
C
C    N = NUMBER OF GRID POINTS
C    P = PRESSURE (Pascals)
C    T = TEMPERATURE (Kelvin)
C    U = VELOCITY (m/s)
C    R = DENSITY (kg/m^3)
C    DL = DUCT LENGTH (m)
C    TIMX = MAXIMUM TIME (s)
C    DTIM = TIME STEP (s)
C    RG = GAS CONSTANT (J/kg K)
C
      OPEN (1,FILE = 'CONACHAN.DAT')
C
      WRITE(*,750)
      WRITE(1,750)
  750 FORMAT(' UNSTEADY COMPRESSIBLE FLOW IN A CONSTANT AREA DUCT',/
     $,' --------------------------------------------------',//)
C
C                        INPUT DATA
C
      N = 200
      DL = 2.0
      DTIM = 0.0000004
      TIMX = 0.00175
C
C    THE FOLLOWING CONSTANTS APPLY TO AIR
C
      RG = 287.0
      GAMMA = 1.4
C
C    THE FOLLOWING DETERMINE THE UPSTREAM AND DOWNSTREAM
C    PRESSURE AND TEMPERATURE
C
      PI(1) = 150000.0
      TI(1) = 300.0
      PI(N) = 100000.0
      TI(N) = 300.0
C
C    INITIALIZE GRID BOUNDARY DATA VALUES AT 1 AND N
C
      RI(1) = PI(1)/(RG*TI(1))
      UI(1) = 0.0
      RI(N) = PI(N)/(RG*TI(N))
      UI(N) = 0.0
C
C    N2 IS JUNCTION BETWEEN HIGH AND LOW PRESSURE ZONES
C
      N2 = N/2 + 1
      N1 = N-1
C
C    INITIALIZE GRID DATA VALUES FROM 2 TO N1
C
      DO 500 I = 2,N1
        IF(I.LT.N2) THEN
        PI(I) = PI(1)
        TI(I) = TI(1)
        RI(I) = RI(1)
```

```
          UI(I) = UI(1)
       ELSE
          PI(I) = PI(N)
          TI(I) = TI(N)
          RI(I) = RI(N)
          UI(I) = UI(N)
       END IF
  500 CONTINUE
C
       WRITE(*,3000) N
       WRITE(1,3000) N
 3000 FORMAT(' NO. OF POINTS  =',I4)
C
       TIM = 0.0
       DX = DL/N1
       N2 = N/2
C
  900 CONTINUE
C
C    UPDATE THE FLOW FIELD
C
       DO 150 I = 1,N
          P(I) = PI(I)
          T(I) = TI(I)
          U(I) = UI(I)
          R(I) = RI(I)
  150 CONTINUE
C
C    CALCULATE THE TIME DERIVATIVES
C
       DO 100 I = 2,N1
       DR(1,I) = -(U(I + 1)*R(I + 1)-U(I-1)*R(I-1))/(2.0*DX)
       DU(1,I) = -(P(I + 1)-P(I-1))/(2.0*DX*R(I))
      $    -U(I)*(U(I)-U(I-1))/DX
       DT(1,I) = -(GAMMA-1)*T(I)*(U(I + 1)-U(I-1))/(2.0*DX)
      $    -U(I)*(T(I)-T(I-1))/DX
  100 CONTINUE
C
C    PREDICT NEW VALUES OF THE FLOW FIELD
C
       DO 200 I = 2,N1
       T(I) = T(I) + DT(1,I)*DTIM
       R(I) = R(I) + DR(1,I)*DTIM
       U(I) = U(I) + DU(1,I)*DTIM
       P(I) = R(I)*RG*T(I)
  200 CONTINUE
C
C    CORRECT THE TIME DERIVATIVES
C
       DO 250 I = 2,N1
       DR(2,I) = -(U(I + 1)*R(I + 1)-U(I-1)*R(I-1))/(2.0*DX)
       DU(2,I) = -(P(I + 1)-P(I-1))/(2.0*DX*R(I))
      $    -U(I)*(U(I)-U(I-1))/DX
       DT(2,I) = -(GAMMA-1)*T(I)*(U(I + 1)-U(I-1))/(2.0*DX)
      $    -U(I)*(T(I)-T(I-1))/DX
  250 CONTINUE
C
C    CORRECT THE NEW VALUES OF THE FLOW FIELD
C
       DO 275 I = 2,N1
       TI(I) = TI(I) + (DT(1,I) + DT(2,I))*DTIM/2.0
       RI(I) = RI(I) + (DR(1,I) + DR(2,I))*DTIM/2.0
       UI(I) = UI(I) + (DU(1,I) + DU(2,I))*DTIM/2.0
       PI(I) = RI(I)*RG*TI(I)
  275 CONTINUE
C
       TIM = TIM + DTIM
       IF(TIM.GT.TIMX) GO TO 300
       GO TO 900
```

```
C
 300 WRITE(*,6000) TIM
     WRITE(1,6000) TIM
6000 FORMAT(' TIME  =',F12.6,/)
     WRITE(*,5000)
     WRITE(1,5000)
5000 FORMAT(' I   P   T   U   RHO   X'
    $,/,' ---   ---   ---   ---   ---   ---',/)
     DO 400 I=1,N
     X=(I-1)*DX
     WRITE(*,1000) I,PI(I),TI(I),UI(I),RI(I),X
     WRITE(1,1000) I,PI(I),TI(I),UI(I),RI(I),X
 400 CONTINUE
1000 FORMAT(I4,F12.0,F10.0,F10.3,F12.3,4X,F6.3,3X,F6.3)
C
     CLOSE(1)
     STOP
     END
```

A typical set of results obtained with $N = 100$ are shown in Fig. E12.4c. The shock wave, the expansion wave, and the division between the flow traversed by the shock wave and by the expansion wave will be noted. Also shown in Fig. E12.4c are the values of the pressure, velocities, and temperature in the various zones obtained using the analytical procedure outlined in Chapter 7.

It will be seen from Fig. E12.4c that, because of the relatively small number of grid points used, the shock wave is "smeared out." The effect of increasing the number of grid points on the solution is shown in Fig. E12.4d from which it will be seen that the details of the shock wave are well captured when a larger number of grid points are used.

The above results were for a relatively small pressure difference across the diaphragm. A typical set of results for a high pressure difference ($p_1 = 1000$ kPa, $p_N = 100$ kPa with 1000 grid points) is shown in Fig. E12.4e.

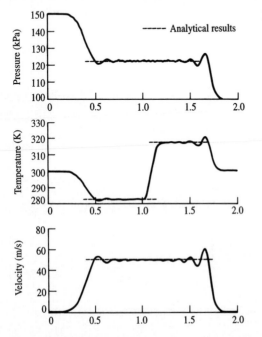

**FIGURE E12.4c**

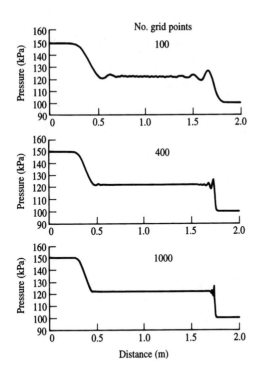

**FIGURE E12.4d**

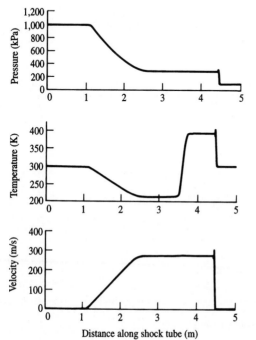

**FIGURE E12.4e**

**EXAMPLE 12.5**

Air is contained in a long closed pipe at a pressure of 150 kPa and a temperature of 300 K. The end of this pipe is suddenly opened to the ambient in which the pressure is 100 kPa. Numerically determine the velocity at which the air leaves the tube. Also show typical pressure and velocity distributions in the tube.

***Solution***

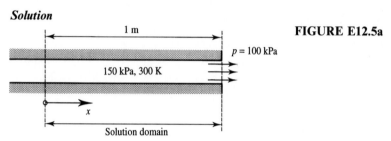

**FIGURE E12.5a**

The flow situation is shown in Fig. E12.5a. The solution over a length of 1 m of the tube will be considered and the $x$-coordinate will be measured in the direction shown because the air will flow out of the tube when the end is suddenly opened.

The initial conditions at time $t = 0$ are then:

For $i = 1, \ldots, N - 1$:     $p_i = 150$ kPa,     $T_i = 300$ K,     $V_i = 0$
For $i = N$:   $p_i = 100$ kPa

The boundary conditions are assumed to be:

For $i = 1$:     $p_i = 150$ kPa,     $T_i = 300$ K,     $V_i = 0$

For $i = N$:     $p_i = 100$ kPa,     $\dfrac{\partial V_i}{\partial x} = 0$

The last of these boundary conditions, i.e., $\partial V / \partial x = 0$ at $i = N$, follows from the condition that the pressure is not changing in the $x$-direction at the exit so the velocity must also not be changing in the $x$-direction at the exit.

The computer program for the solution of this flow is basically the same as that given in Example 12.4. A partial listing of this program that attempts to indicate the differences between this program and the program used in Example 12.4 is given in Program E12.5.

**Program E12.5**

```
C
************************************************************************
C *
C *                        CONACHAO
C *                     ***********
C *
C * THIS PROGRAM CALCULATES UNSTEADY FLOW IN A CONSTANT AREA
C * CHANNEL - IT IS SETUP TO CALCULATE THE FLOW THAT DEVELOPS
C * FOLLOWING THE SUDDEN LOWERING OF THE PRESSURE AT ONE END OF
C * THECHANNEL DUE, FOR EXAMPLE, TO THE SUDDEN OPENING OF THE END
C * OF THE CHANNEL TO THE AMBIENT.
C *
C
************************************************************************
C
      DIMENSION U(500),P(500),T(500),R(500)
C        :
```

```
      OPEN (1,FILE='CONACHAO.DAT')
C
      WRITE(*,750)
       ⋮
C
      N=300
      N1=N-1
      DL=1.0
      DTIM=0.0000004
      TIMX=0.0015000
C
C     THE FOLLOWING CONSTANTS APPLY TO AIR
C
      RG=287.0
      GAMMA=1.4
C
C     THE FOLLOWING DETERMINE THE INITIAL PRESSURE AND TEMPERATURE
C
      PI(1)=150000.0
      TI(1)=300.0
      PI(N)=100000.0
      TI(N)=300.0
C
C     INITIALIZE GRID BOUNDARY DATA VALUES AT 1 AND N
C
      RI(1)=PI(1)/(RG*TI(1))
      UI(1)=0.0
      RI(N)=PI(N)/(RG*TI(N))
      UI(N)=0.0
C
C     INITIALIZE GRID DATA VALUES FROM 2 TO N1
C
      DO 500 I=2,N1
        PI(I)=PI(1)
        TI(I)=TI(1)
        RI(I)=RI(1)
        UI(I)=UI(1)
  500 CONTINUE
C
      WRITE(*,3000) N
      WRITE(1,3000) N
 3000 FORMAT(' NO. OF POINTS =',I4)
C
      TIM=0.0
      DX=DL/N1
C
  900 CONTINUE
C
C     UPDATE THE FLOW FIELD
C
      DO 150 I=1,N
       ⋮
      P(I)=R(I)*RG*T(I)
  200 CONTINUE
C
C     APPLY THE ZERO DERIVATIVE END CONDITION
C
      U(N)=U(N1)
      R(N)=R(N1)
      T(N)=T(N1)
C
C     CORRECT THE TIME DERIVATIVES
C
      DO 250 I=2,N1
       ⋮
      PI(I)=RI(I)*RG*TI(I)
  275 CONTINUE
C
C     APPLY THE ZERO DERIVATIVE END CONDITION
```

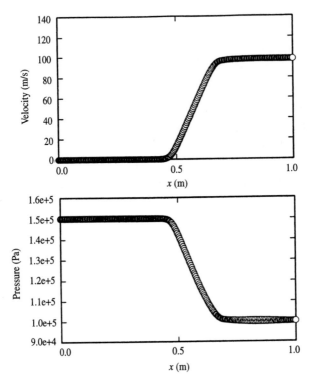

**FIGURE E12.5b**

```
C
      UI(N) = UI(N1)
      RI(N) = RI(N1)
      TI(N) = TI(N1)
C
      TIM = TIM + DTIM
      IF(TIM.GT.TIMX) GO TO 300
        ⋮
      CLOSE(1)
      STOP
      END
```

Typical velocity and pressure distributions along the pipe given by this program are shown in Fig. E12.5b. The expansion wave generated by the opening of the end of the tube can be clearly seen. The results given by the program indicate that the air is discharged from the pipe at a velocity of 97.5 m/s.

### EXAMPLE 12.6

Air is contained in a long closed pipe at a pressure of 100 kPa and a temperature of 300 K. The pressure at one end of this pipe is suddenly increased to 150 kPa. Numerically determine the velocity at which the air enters the tube. Also show typical pressure and velocity distributions in the tube.

### Solution

The flow situation is shown in Fig. E12.6a. The solution over a length of 1 m of the tube will be considered and the x coordinate will be measured in the direction shown because the air will flow into the tube when the pressure is increased.

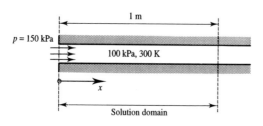

**FIGURE E12.6a**

The initial conditions at time $t = 0$ are then:

For $i = 2, \ldots, N$:    $p_i = 100$ kPa,    $T_i = 300$ K,    $V_i = 0$
For $i = 1$:    $p_i = 150$ kPa

The boundary conditions are assumed to be:

For $i = N$:    $p_i = 100$ kPa,    $T_i = 300$ K,    $V_i = 0$

For $i = 1$:    $p_i = 150$ kPa,    $\dfrac{\partial V_i}{\partial x} = 0$

The last of these boundary conditions, i.e., $\partial V / \partial x = 0$ at $i = 1$, follows from the condition that the pressure is not changing in the $x$ direction at the inlet so the velocity must also not be changing in the $x$ direction at the inlet.

The computer program for the solution of this flow is basically the same as that given in Example 12.4. A partial listing that attempts to indicate the differences between this program and the program used in Example 12.4 is given in Program E12.6.

**Program E12.6**

```
C
**************************************************************************
C *
C *                       CONACHAP
C *                    ************
C *
C * THIS PROGRAM CALCULATES UNSTEADY FLOW IN A CONSTANT AREA
C * CHANNEL - IT IS SETUP TO CALCULATE THE FLOW THAT DEVELOPS
C * FOLLOWING THE SUDDEN INCREASE OF THE PRESSURE AT ONE END OF
C * THE CHANNEL DUE, FOR EXAMPLE, TO THE SUDDEN OPENING OF A VALVE
C * AT THE END OF THE CHANNEL.
C *
C
**************************************************************************
C
      DIMENSION U(500),P(500),T(500),R(500)
   $ ,UI(500),PI(500),TI(500),RI(500)
       :
C
      OPEN (1,FILE='CONACHAP.DAT')
C
C
      WRITE(*,750)
       :
C
      N = 300
      N1 = N-1
      DL = 1.0
      DTIM = 0.0000004
      TIMX = 0.0015000
C
```

```
C     THE FOLLOWING CONSTANTS APPLY TO AIR
C
      RG = 287.0
      GAMMA = 1.4
C
C     THE FOLLOWING DETERMINE THE INITIAL PRESSURE AND TEMPERATURE
C
      PI(1) = 150000.0
      TI(1) = 300.0
      PI(N) = 100000.0
      TI(N) = 300.0
      UI(N) = 0.0
C
C     INITIALIZE GRID BOUNDARY DATA VALUES AT 1 AND N
C
      RI(1) = PI(1)/(RG*TI(1))
      RI(N) = PI(N)/(RG*TI(N))
      UI(1) = 0.0
C
C     INITIALIZE GRID DATA VALUES FROM 2 TO N1
C
      DO 500 I = 2,N1
        PI(I) = PI(N)
        TI(I) = TI(N)
        RI(I) = RI(N)
        UI(I) = UI(N)
  500 CONTINUE
C
      WRITE(*,3000) N
      WRITE(1,3000) N
 3000 FORMAT(' NO. OF POINTS  = ',I4)
C
      TIM = 0.0
      DX = DL/N1
C
  900 CONTINUE
C
C     UPDATE THE FLOW FIELD
C
      DO 150 I = 1,N
        :
      P(I) = R(I)*RG*T(I)
  200 CONTINUE
C
C     APPLY THE ZERO DERIVATIVE END CONDITION
C
      U(1) = U(2)
      R(1) = R(2)
      T(1) = T(2)
C
C     CORRECT THE TIME DERIVATIVES
C
      DO 250 I = 2,N1
        :
      PI(I) = RI(I)*RG*TI(I)
  275 CONTINUE
C
C     APPLY THE ZERO DERIVATIVE END CONDITION
C
      UI(1) = UI(2)
      RI(1) = RI(2)
      TI(1) = TI(2)
C
      TIM = TIM + DTIM
      IF(TIM.GT.TIMX) GO TO 300
        :
      CLOSE(1)
      STOP
      END
```

Typical velocity and pressure distributions along the pipe given by this program are shown in Fig. E12.6b. The shock wave generated by the pressure rise at the inlet to the tube can be clearly seen. The "ringing" near the shock wave can also be noted. The results given by the program indicate that the air flows into the pipe at a velocity of 100 m/s.

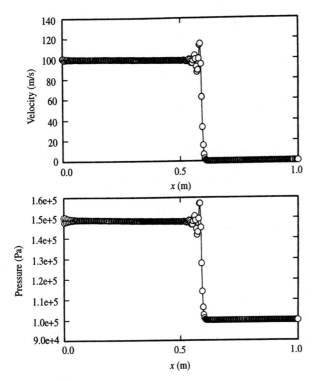

**FIGURE E12.6b**

## 12.9
## CONCLUDING REMARKS

The discussion given in this chapter together with the examples should indicate that it is relatively easy to obtain numerical solutions for all the one-dimensional flows discussed in earlier chapters. The real power of the numerical approach is, however, felt when problems for which it is difficult to obtain analytical solutions are considered. For example, it is easy to obtain a numerical solution for unsteady flow in a variable area channel whereas analytical solutions to such flows can be difficult to obtain.

## PROBLEMS

**12.1.** A computer program (CONDINOZ) that calculates steady flow in a convergent–divergent nozzle was given in this chapter. Use this program to find the pressure distribution in a nozzle through which air is flowing if the nozzle has inlet and outlet areas that are three times the throat area and if the inlet pressure is 200 kPa.

**12.2.** Modify the program (CONDINOZ) for steady flow in a convergent–divergent nozzle given in this chapter to include the effects of a normal shock wave in the divergent portion of the nozzle. Assume that the position at which the shock wave occurs is specified. Use the program as given above to calculate the flow upstream of the shock. This will give the Mach number ahead of the shock since $V/a^* = (v/a)(a/a^*) = M(T/T^*)^{0.5}$. Using the value of $M$ so found, apply the normal shock relations to get the values of the variables downstream of the shock wave. Then use the subsonic numerical procedure to calculate the flow downstream of the shock wave. Apply this program to the nozzle described in Problem 12.1, the shock wave occurring halfway down the divergent portion of the nozzle. Find the pressure distribution along the nozzle.

**12.3.** Air enters a convergent conical duct at a steady rate at a Mach number of 1.5 and a pressure of 80 kPa. The diameter of the nozzle at the exit is 60 percent of the value at the inlet. Use a computer program based on the supersonic flow procedure outlined in the chapter to determine the Mach number and pressure variation along the duct.

**12.4.** Numerically find the Mach number in the flow generated behind the shock wave in a constant area air filled shock tube when the high pressure section is initially at a pressure of 200 kPa and the low pressure section is initially at a pressure of 100 kPa and the temperature is initially 25°C throughout the tube. (Use CONACHAN.)

**12.5.** Consider a long tube initially containing air at rest at a pressure of 60 kPa and a temperature of 30°C. If the pressure at one end of the tube is suddenly increased to 120 kPa, numerically calculate the flow conditions that will exist behind the shock wave that propagates down the tube. The solution for a 2 m long portion of the tube should be considered.

**12.6.** Air is flowing out of a long tube at a velocity of 50 m/s at a pressure and temperature of 100 kPa and 30°C respectively. If a valve at the end of this tube is suddenly closed, numerically determine the pressure acting on this valve.

**12.7.** Consider the general situation described in Problem 12.5 but assume that the initial pressure is 100 kPa and that the pressure at the inlet only increases to 110 kPa. Numerically study the effect of the initial temperature on the velocity of the wave produced. Initial temperatures between 0°C and 100°C should be considered. Compare the wave velocities obtained with the speed of sound.

# CHAPTER 13

# An Introduction to Two-Dimensional Compressible Flow

## 13.1
## INTRODUCTION

The discussion given in the preceding chapters has mainly been concerned with one-dimensional flows. This is because a large number of flows that occur in engineering practice can be adequately modeled by assuming that the flow is one-dimensional and because an understanding of many features of compressible flows can be gained by considering one-dimensional flow. However, this assumption is not always adequate. There are a number of flows in which it is necessary to account for the two- or three-dimensional nature of the flow. For example, the flows shown in Fig. 13.1 cannot be treated as one-dimensional.

In this chapter, attention will mainly be given to the analysis of steady two-dimensional isentropic flows, i.e., the effects of viscosity and heat transfer

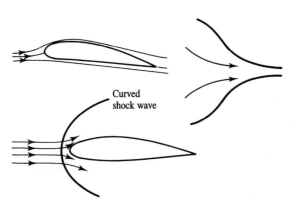

Curved
shock wave

**FIGURE 13.1**
Two-dimensional flows.

will be neglected in most of the present chapter. The extension of the methods discussed in this chapter to three-dimensional flows is relatively straightforward and will not be discussed here. Also, while the equations for two-dimensional flow can be written in compact form using vector notation, this notation will not be used here in an effort to concentrate on developing an understanding of the meaning and implications of the equations. The governing equations will, therefore, be expressed in terms of cartesian coordinates.

The governing equations will first be developed and methods of solving these equations will then be discussed.

## 13.2
## GOVERNING EQUATIONS

The equations governing two-dimensional compressible flow will be derived in this section.

### Continuity Equation

Consider flow through the control volume shown in Fig. 13.2. Conservation of mass requires that since steady flow is being considered:

$$\begin{matrix} \text{Rate mass} \\ \text{enters AB} \end{matrix} - \begin{matrix} \text{Rate mass} \\ \text{leaves CD} \end{matrix} + \begin{matrix} \text{Rate mass} \\ \text{enters AC} \end{matrix} - \begin{matrix} \text{Rate mass} \\ \text{leaves BC} \end{matrix} = 0 \qquad (13.1)$$

Because two-dimensional flow is being considered, attention can be restricted to unit width of the control volume. Since the mass flow rate through AB is $pu \times$ area of AB $= pu\,dy$ and since, similarly, the mass flow rate through AD is $pv\,dx$, eq. (13.1) gives, since $dx$ and $dy$ are, by assumption, small:

$$pu\,dy - \left[(pu\,dy) + \frac{\partial}{\partial x}(pu\,dy)\,dx\right] + pv\,dx - \left[(pv\,dx) + \frac{\partial}{\partial y}(pv\,dx)\,dy\right] = 0$$

i.e.,
$$\frac{\partial}{\partial x}(pu) + \frac{\partial}{\partial y}(pv) = 0 \qquad (13.2)$$

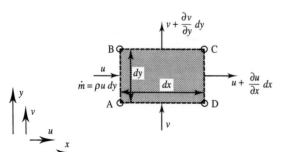

**FIGURE 13.2**
Control volume considered.

This is the continuity or conservation of mass equation for two-dimensional compressible flow. It basically states that the rate of change of $\rho u$ in the $x$ direction is associated with an equal but negative rate of change of $\rho v$ in the $y$ direction.

## Momentum Equation

Since viscous effects are being neglected, the conservation of momentum equation gives in the $x$ direction for the control volume shown in Fig. 13.2:

$$\begin{matrix} \text{Rate } x \text{ momentum} \\ \text{leaves CD} \end{matrix} - \begin{matrix} \text{Rate } x \text{ momentum} \\ \text{enters AB} \end{matrix} + \begin{matrix} \text{Rate } x \text{ momentum} \\ \text{leaves BC} \end{matrix}$$

$$\begin{matrix} \text{Rate } x \text{ momentum} \\ \text{enters AB} \end{matrix} = \begin{matrix} \text{Pressure force} \\ \text{on AB} \end{matrix} - \begin{matrix} \text{Pressure force} \\ \text{on CD} \end{matrix} \qquad (13.3)$$

In writing this equation, it has been recalled that steady flow is being considered and it has been noted that the pressure acting on faces AD and BC of the control volume does not contribute any force in the $x$ direction. All body forces such as those that could arise due to electrical and magnetic fields have also been neglected.

Since a control volume of unit width is being considered and since the rate $x$ momentum enters through a face of the control volume is (mass flow rate through face $\times u$), eq. (13.3) becomes:

$$\left[ (\rho u^2 \, dy) + \frac{\partial}{\partial x} (\rho u^2 \, dy) \, dx - (\rho u^2 \, dy) \right] + \left[ (\rho v u \, dx) + \frac{\partial}{\partial y} (\rho v u \, dx) \, dy - (\rho v u \, dx) \right]$$

$$= p \, dy - \left[ p \, dy + \frac{\partial}{\partial x} (p \, dy) \, dx \right]$$

i.e., dividing through by $dx \, dy$:

$$\frac{\partial}{\partial x} (\rho u^2) + \frac{\partial}{\partial x} (\rho v u) = -\frac{\partial p}{\partial x} \qquad (13.4)$$

The left hand side of this equation can be expressed as:

$$u \frac{\partial (\rho u)}{\partial x} + \rho u \frac{\partial u}{\partial x} + u \frac{\partial}{\partial y} (\rho v) + \rho v \frac{\partial u}{\partial y}$$

i.e., as:

$$u \left[ \frac{\partial}{\partial x} (\rho u) + \frac{\partial}{\partial y} (\rho v) \right] + \rho u \frac{\partial u}{\partial x} + \rho v \frac{\partial u}{\partial y}$$

The bracketed term is zero by virtue of the continuity equation (13.2). Equation (13.4) can therefore be written as:

$$\rho u \frac{\partial u}{\partial x} + \rho v \frac{\partial u}{\partial y} = -\frac{\partial p}{\partial x} \qquad (13.5)$$

Similarly, conservation of momentum in the $y$ direction gives:

$$\begin{array}{c} \text{Rate } y \text{ momentum} \\ \text{leaves BC} \end{array} - \begin{array}{c} \text{Rate } y \text{ momentum} \\ \text{enters AD} \end{array} + \begin{array}{c} \text{Rate } y \text{ momentum} \\ \text{leaves CD} \end{array}$$

$$- \begin{array}{c} \text{Rate } y \text{ momentum} \\ \text{enters AB} \end{array} = \begin{array}{c} \text{Pressure force} \\ \text{on AD} \end{array} - \begin{array}{c} \text{Pressure force} \\ \text{on BC} \end{array}$$

i.e., $\quad \rho v^2\, dx + \dfrac{\partial}{\partial y}(\rho v^2\, dx)\, dy - \rho v^2\, dx + \rho uv\, dy + \dfrac{\partial}{\partial x}(\rho uv\, dy)\, dx - \rho uv\, dy$

$$= p\, dx - \left[ p\, dx + \frac{\partial}{\partial y}(p\, dx)\, dy \right]$$

i.e., 
$$\frac{\partial}{\partial x}(\rho uv) + \frac{\partial}{\partial y}(\rho v^2) = -\frac{\partial p}{\partial y}$$

By again using the continuity equation, this equation can be expressed as:

$$\rho u \frac{\partial v}{\partial x} + \rho v \frac{\partial v}{\partial y} = -\frac{\partial p}{\partial y} \qquad (13.6)$$

**Energy Equation**

Because adiabatic flow is being considered, conservation of energy gives for flow through the control volume shown in Fig. 13.2:

$$\begin{array}{c} \text{Rate (enthalpy} + \text{kinetic energy)} \\ \text{leave CD} \end{array} - \begin{array}{c} \text{Rate (enthalpy} + \text{kinetic energy)} \\ \text{enter AB} \end{array}$$

$$\begin{array}{c} \text{Rate (enthalpy} + \text{kinetic energy)} \\ \text{leave BC} \end{array} - \begin{array}{c} \text{Rate (enthalpy} + \text{kinetic energy)} \\ \text{enter AD} \end{array} = 0$$

$$(13.7)$$

But the rate at which (enthalpy + kinetic energy) enters AB is equal to the mass flow rate through AB $\times [c_p T + (u^2 + v^2)/2]$, i.e.,

$$\rho u\, dy[c_p T + (u^2 + v^2)/2]$$

Similarly the rate at which (enthalpy + kinetic energy) enters AD is:

$$\rho v\, dx[c_p T + (u^2 + v^2)/2]$$

Hence eq. (13.7) gives:

$$\rho u \, dy \left[ c_p T + \frac{(u^2 + v^2)}{2} \right] + \frac{\partial}{\partial x} \left\{ \rho u \, dy \left[ c_p T + \frac{(u^2 + v^2)}{2} \right] \right\} dx$$

$$- \rho u \, dy \left[ c_p T + \frac{(u^2 + v^2)}{2} \right] + \rho v \, dx \left[ c_p T + \frac{(u^2 + v^2)}{2} \right]$$

$$+ \frac{\partial}{\partial y} \left\{ \rho v \, dx \left[ c_p T + \frac{(u^2 + v^2)}{2} \right] \right\} dy$$

$$- \rho v \, dx \left[ c_p T + \frac{(u^2 + v^2)}{2} \right] = 0$$

i.e.,
$$\frac{\partial}{\partial x} \left\{ \rho u \left[ c_p T + \frac{(u^2 + v^2)}{2} \right] \right\} + \frac{\partial}{\partial y} \left\{ \rho v \left[ c_p T + \frac{(u^2 + v^2)}{2} \right] \right\} = 0 \quad (13.8)$$

This equation can be written as:

$$\left[ c_p T + \frac{(u^2 + v^2)}{2} \right] \left[ \frac{\partial}{\partial x} (\rho u) + \frac{\partial}{\partial y} (\rho v) \right] + u \frac{\partial}{\partial x} \left[ c_p T + \frac{(u^2 + v^2)}{2} \right]$$

$$\times v \frac{\partial}{\partial y} \left[ c_p T + \frac{(u^2 + v^2)}{2} \right] = 0$$

But, by virtue of the continuity equation (13.2), the first term in this equation is zero. Hence the conservation of energy equation can be written as:

$$u \frac{\partial}{\partial x} \left[ c_p T + \frac{(u^2 + v^2)}{2} \right] + v \frac{\partial}{\partial y} \left[ c_p T + \frac{(u^2 + v^2)}{2} \right] = 0 \quad (13.9)$$

Consider the change in any quantity $Z$ over a short length $ds$ of a streamline as shown in Fig. 13.3. The change will be given by:

$$dZ = \frac{\partial Z}{\partial x} dx + \frac{\partial Z}{\partial y} dy \quad (13.10)$$

The rate of change in $Z$ as it moves along the streamline is $dZ/dt$ where $dt$ is the time taken for the gas particles to move from A to B. Hence the rate of change in $Z$ is given by:

$$\frac{\partial Z}{\partial t} = \frac{\partial x}{\partial t} \frac{\partial Z}{\partial x} + \frac{\partial y}{\partial t} \frac{\partial Z}{\partial y} \quad (13.11)$$

i.e., since $dx/dt = u$ and $dy/dt = v$:

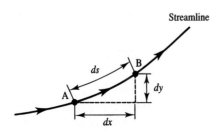

Streamline    **FIGURE 13.3**
Changes over length of streamline considered.

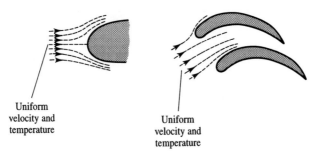

Uniform
velocity and
temperature

Uniform
velocity and
temperature

**FIGURE 13.4**
Uniform upstream flow.

$$\frac{\partial Z}{\partial t} = u\frac{\partial Z}{\partial x} + v\frac{\partial Z}{\partial y} \qquad (13.12)$$

If the quantity $Z$ is not changing along the streamline, i.e., if $dZ/dt = 0$, this equation shows that:

$$u\frac{\partial Z}{\partial x} + v\frac{\partial Z}{\partial y} = 0 \qquad (13.13)$$

Comparing this with eq. (13.9) then shows that the conservation of energy equation indicates that the quantity:

$$c_p T + \frac{(u^2 + v^2)}{2}$$

is constant along a streamline. Now, in most flows, the flow can be assumed to originate from a region of uniform flow as indicated in Fig. 13.4.

In such cases, the value of $[c_p T + (u^2 + v^2)/2]$ will initially be the same on all streamlines. The conservation of energy equation then shows that this quantity will remain the same everywhere in the flow. Hence, if the velocity components are determined from the continuity and momentum equations, the energy gives the temperature at any point in the flow as:

$$c_p T + \left(\frac{u^2 + v^2}{2}\right) = c_p T_1 + \left(\frac{u_1^2 + v_1^2}{2}\right) = c_p T_0 \qquad (13.14)$$

the subscript 1 referring to conditions in the initial uniform flow.

## 13.3
## VORTICITY CONSIDERATIONS

As previously discussed, attention is here being directed to flows in which the effects of viscosity are negligible. Consider the fluid particles in the flow. Only if there are tangential forces acting on the surface of these particles can there be a change in the net rate at which the particles rotate. This is illustrated in

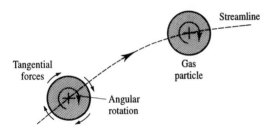

**FIGURE 13.5**
Changes in rotational motion of particles produced by viscous stresses. Tangential forces on particle are required if rotational motion is changing along streamline.

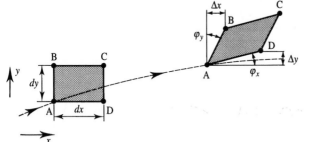

**FIGURE 13.6**
Distortion of fluid particle.

Fig. 13.5. But the only source of a tangential force is viscosity. Hence, if viscous effects are neglected, there can be no change in the rate at which the fluid particles rotate. If the flow originates in a uniform freestream (see Fig. 13.4) the fluid particles will have no initial rotation and so they will have no rotation anywhere in the flow.

Consider a fluid particle that is initially rectangular in shape with side lengths $dx$ and $dy$. As it moves through the flow, this particle will distort as indicated in Fig. 13.6. The net amount by which the particle has rotated in the counter clockwise direction is $(\phi_x - \phi_y)/2$. The net rate at which the fluid particles are rotating (i.e., the vorticity) is then given by:

$$\omega = \frac{1}{2}\left(\frac{\partial \phi_x}{\partial t} - \frac{\partial \phi_y}{dt}\right) \tag{13.15}$$

Now, consider the velocities of corner points A, B, and D as indicated in Fig. 13.7.

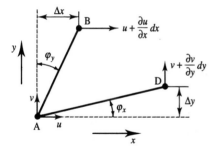

**FIGURE 13.7**
Velocities of corner points of fluid particle considered.

From this it follows that:

$$\frac{\partial \phi_x}{\partial t} = \frac{\partial(\Delta y)/\partial t}{\partial x} = \left(\frac{v + (\partial v/\partial x) - v}{dx}\right) = \frac{\partial v}{\partial x} \qquad (13.16)$$

and:

$$\frac{\partial \phi_y}{\partial t} = \frac{\partial(\Delta x)/\partial t}{\partial y} = \left(\frac{v + (\partial v/\partial y) - u}{dy}\right) = \frac{\partial u}{\partial y} \qquad (13.17)$$

Substituting these into eq. (13.15) then gives the rate of fluid particle rotation as:

$$\omega = \frac{1}{2}\left(\frac{\partial v}{\partial x} - \frac{\partial u}{\partial y}\right) \qquad (13.18)$$

If viscous forces are negligible and if the flow initially has no rotation, it follows that everywhere in the flow:

$$\frac{\partial u}{\partial x} \cancel{/} \frac{\partial u}{\partial y} = 0 \qquad (13.19)$$

Using this, the $x$ momentum equation, eq. (13.5) can be written as:

$$\rho u \frac{\partial u}{\partial x} + \rho v \frac{\partial v}{\partial x} = -\frac{\partial p}{\partial x}$$

i.e.,

$$\frac{\partial}{\partial x}\left(\frac{u^2 + v^2}{2}\right) = -\frac{1}{\rho}\frac{\partial p}{\partial x} \qquad (13.20)$$

Similarly, the $y$ momentum equation, eq. (13.6), can be written using eq. (13.19) as:

$$\rho u \frac{\partial u}{\partial y} + \rho v \frac{\partial v}{\partial y} = -\frac{\partial p}{\partial y}$$

i.e.,

$$\frac{\partial}{\partial y}\left(\frac{u^2 + v^2}{2}\right) = -\frac{1}{\rho}\frac{\partial p}{\partial y} \qquad (13.21)$$

But, since isentropic flow is being considered:

$$T = Cp^{\frac{\gamma-1}{\gamma}}$$

Hence:

$$\frac{\partial T}{\partial x} = \frac{Cp^{\frac{\gamma-1}{\gamma}}}{p}\frac{\gamma-1}{\gamma}\frac{\partial p}{\partial x}$$

$$= \frac{\gamma-1}{\gamma}\frac{T}{p}\frac{\partial p}{\partial x}$$

$$= \frac{\gamma-1}{\gamma}\frac{1}{R\rho}\frac{\partial p}{\partial x}$$

i.e.,

$$\frac{\partial T}{\partial x} = \frac{1}{c_p \rho} \frac{\partial p}{\partial x} \tag{13.22}$$

Similarly:

$$\frac{\partial T}{\partial y} = \frac{1}{c_p \rho} \frac{\partial p}{\partial y} \tag{13.23}$$

Substituting eqs. (13.22) and (13.23) into eqs. (13.20) and (13.21) respectively gives:

$$\frac{\partial}{\partial x}\left(\frac{u^2 + v^2}{2}\right) + c_p \frac{\partial T}{\partial x} = 0$$

i.e., since $c_p$ is assumed constant:

$$\frac{\partial}{\partial x}\left[c_p T + \left(\frac{u^2 + v^2}{2}\right)\right] = 0 \tag{13.24}$$

and:

$$\frac{\partial}{\partial y}\left[c_p T + \left(\frac{u^2 + v^2}{2}\right)\right] = 0 \tag{13.25}$$

These two equations together indicate that the quantity:

$$c_p T + \left(\frac{u^2 + v^2}{2}\right)$$

remains constant in an irrotational isentropic flow. This is the same as the result deduced from energy considerations. Thus, in irrotational isentropic flow, momentum and energy conservation considerations give the same result.

## 13.4
## THE VELOCITY POTENTIAL

The equations governing the velocity field in two-dimensional irrotational flow are the continuity equation, eq. (13.2), and the irrotationality equation, eq. (13.19), i.e.,

$$\frac{\partial}{\partial x}(\rho u) + \frac{\partial}{\partial y}(\rho v) = 0 \tag{13.2}$$

$$\frac{\partial v}{\partial x} - \frac{\partial u}{\partial y} = 0 \tag{13.19}$$

The boundary conditions on these equations are that $u$ and $v$ are prescribed in the initial flow, e.g., $u = u_\infty$, $v = 0$ well upstream of the body considered, and that the velocity component normal to any solid surface is zero.

If a quantity, $\Phi$, termed the velocity potential, is introduced, $\Phi$ being such that:

$$u = \frac{\partial \Phi}{\partial x}, \qquad v = \frac{\partial \Phi}{\partial y} \tag{13.26}$$

then it will be seen that the left hand side of eq. (13.19) becomes:

$$\frac{\partial^2 \Phi}{\partial x\,\partial y} - \frac{\partial^2 \Phi}{\partial y\,\partial x}$$

This will always be zero so it follows that the velocity potential function, as defined by eq. (13.26), satisfies the irrotationality equation (13.19). The continuity equation (13.2), must then be used to solve for $\Phi$. This equation gives:

$$\frac{\partial}{\partial x}\left(\rho\frac{\partial \Phi}{\partial x}\right) + \frac{\partial}{\partial y}\left(\rho\frac{\partial \Phi}{\partial y}\right) = 0$$

i.e.,
$$\rho\left(\frac{\partial^2 \Phi}{\partial x^2} + \frac{\partial^2 \Phi}{\partial y^2}\right) + \left(\frac{\partial \Phi}{\partial x}\frac{\partial \rho}{\partial x} + \frac{\partial \Phi}{\partial y}\frac{\partial \rho}{\partial y}\right) = 0 \tag{13.27}$$

But, since isentropic flow is being considered:

$$\rho = Cp^{\frac{1}{\gamma}} \tag{13.28}$$

Hence:
$$\frac{\partial \rho}{\partial x} = \frac{1}{\gamma}\frac{Cp^{\frac{1}{\gamma}}}{p}\frac{\partial p}{\partial x} = \frac{\rho}{\gamma p}\frac{\partial p}{\partial x} = \frac{1}{a^2}\frac{\partial p}{\partial x}$$

Similarly, it can be shown that:

$$\frac{\partial \rho}{\partial y} = \frac{1}{a^2}\frac{\partial p}{\partial y} \tag{13.29}$$

Using these in the momentum equations (13.5) and (13.6) and using eq. (13.26) gives:

$$\frac{\partial \rho}{\partial x} = -\frac{\rho}{a^2}\left(\frac{\partial \Phi}{\partial x}\frac{\partial^2 \Phi}{\partial x^2} + \frac{\partial \Phi}{\partial y}\frac{\partial^2 \Phi}{\partial x\,\partial y}\right) \tag{13.30}$$

and:
$$\frac{\partial \rho}{\partial y} = -\frac{\rho}{a^2}\left(\frac{\partial \Phi}{\partial x}\frac{\partial^2 \Phi}{\partial x\,\partial y} + \frac{\partial \Phi}{\partial y}\frac{\partial^2 \Phi}{\partial y^2}\right) \tag{13.31}$$

Substituting these two equations into eq. (13.27) then gives:

$$\frac{\partial^2 \Phi}{\partial x^2} + \frac{\partial^2 \Phi}{\partial y^2} - \frac{1}{a^2}\left[\left(\frac{\partial \Phi}{\partial x}\right)^2\frac{\partial^2 \Phi}{\partial x^2} + \frac{\partial \Phi}{\partial x}\frac{\partial \Phi}{\partial y}\frac{\partial^2 \Phi}{\partial x\,\partial y}\right]$$
$$-\frac{1}{a^2}\left[\frac{\partial \Phi}{\partial y}\frac{\partial \Phi}{\partial x}\frac{\partial^2 \Phi}{\partial x\,\partial y} + \left(\frac{\partial \Phi}{\partial y}\right)^2\frac{\partial^2 \Phi}{\partial y^2}\right] = 0$$

i.e.,

$$\frac{\partial^2 \Phi}{\partial x^2} + \frac{\partial^2 \Phi}{\partial y^2} - \frac{1}{a^2}\left[\left(\frac{\partial \Phi}{\partial x}\right)^2\frac{\partial^2 \Phi}{\partial x^2} + 2\frac{\partial \Phi}{\partial x}\frac{\partial \Phi}{\partial y}\frac{\partial^2 \Phi}{\partial x\,\partial y} + \left(\frac{\partial \Phi}{\partial y}\right)^2\frac{\partial^2 \Phi}{\partial y^2}\right] = 0 \tag{13.32}$$

This can be written as:

$$\frac{\partial^2 \Phi}{\partial x^2} + \frac{\partial^2 \Phi}{\partial y^2} - \frac{1}{a^2}\left[ u^2 \frac{\partial^2 \Phi}{\partial x^2} + 2uv \frac{\partial^2 \Phi}{\partial x\, \partial y} + v^2 \frac{\partial^2 \Phi}{\partial y^2}\right] = 0 \qquad (13.33)$$

Beside $\Phi$, this equation contains the speed of sound $a$. An expression relating $a$ to $\Phi$ is, therefore, required. This is supplied by the energy equation which, as discussed above, gives (see eq. (13.14)):

$$c_p T + \left( \frac{u^2 + v^2}{2} \right) = c_p T_0 \qquad (13.34)$$

where $T_0$ is the stagnation temperature which is a constant throughout the flow. Hence, since:

$$a^2 = \gamma RT \quad \text{and} \quad c_p = \gamma R/(\gamma - 1)$$

eq. (13.34) gives:

$$a^2 + \frac{\gamma - 1}{2}(u^2 + v^2) = a_0^2$$

i.e., using eq. (13.26):

$$a^2 = a_0^2 - \left( \frac{\gamma - 1}{2} \right)\left[ \left( \frac{\partial \Phi}{\partial x} \right)^2 + \left( \frac{\partial \Phi}{\partial y} \right)^2 \right] \qquad (13.35)$$

Equations (13.32) and (13.35) together describe the variation of $\Phi$ in irrotational, isentropic, flow. In low speed flow (i.e., $M \ll 1$), the density variation is negligible and eq. (13.27) gives:

$$\frac{\partial^2 \Phi}{\partial x^2} + \frac{\partial^2 \Phi}{\partial y^2} = 0 \qquad (13.36)$$

Hence, in low speed flow, the variation of $\Phi$ is governed by Laplace's equations. In incompressible flow, then, it is relatively easy to determine $\Phi$, eq. (13.36) being a linear equation. In compressible flow, however, the compressibility effects give rise to the nonlinear terms in eq. (13.32), i.e., terms such as $(\partial \Phi/\partial x)^2 (\partial^2 \Phi/\partial x^2)$ which involve the product of functions of $\Phi$. This makes the determination of $\Phi$ in compressible flows significantly more difficult than in incompressible flows. To solve for $\Phi$ in compressible flows, the following methods can be used:

- Full numerical solutions
- Transformation of variables to give a linear governing equation
- Linearized solutions

The second method is only applicable in a few situations and will not be discussed here. Linearized solutions will be discussed in the next section and a very brief discussion of numerical methods will be given in a later section.

## 13.5
## LINEARIZED SOLUTIONS

In order to keep the drag low on objects in high-speed flows, the objects are usually kept relatively slender in order to minimize the disturbance they produce in the flow. With such slender objects, the differences between the values of the flow variables near the object and the values of these variables in the freestream flow ahead of the object are small, e.g., consider the situation shown in Fig. 13.8. If the components of $V$ are $u_\infty + u_p$ and $v_p$ then, for a slender object, the perturbation velocities $u_p$ and $v_p$ will be very small compared to $u_\infty$. This assumption is the basis for the analysis given in the present section.

Now, in the undisturbed flow ahead of the object, $u = u_\infty$ and $v = 0$ so the velocity potential is, by virtue of eq. (13.26), given by:

$$\frac{\partial \Phi}{\partial x} = u_\infty \quad \text{and} \quad \frac{\partial \Phi}{\partial y} = 0$$

i.e.,

$$\Phi = u_\infty x$$

The velocity potential in the flow will, therefore, be written as:

$$\Phi = u_\infty x + \Phi_p$$

Where $\Phi_p$ is the perturbation velocity potential which must be such that:

$$u_p = \frac{\partial \Phi_p}{\partial x}, \qquad v_p = \frac{\partial \Phi_p}{\partial y} \tag{13.37}$$

Substituting the above relations into the potential function equation in the form given in eq. (13.33) leads to:

$$\frac{\partial^2 \Phi_p}{\partial x^2} + \frac{\partial^2 \Phi_p}{\partial y^2} - \frac{1}{a^2}\left[(u_\infty + u_p)^2 \frac{\partial^2 \Phi_p}{\partial x^2} + 2(u_\infty + u_p)v_p \frac{\partial^2 \Phi_p}{\partial x\,\partial y} + v_p^2 \frac{\partial^2 \Phi_p}{\partial y^2}\right] = 0$$

i.e.,

$$\frac{\partial^2 \Phi_p}{\partial x^2} + \frac{\partial^2 \Phi_p}{\partial y^2} - \left(\frac{u_\infty}{a}\right)^2 \left[\left(1 + \frac{u_p}{u_\infty}\right)^2 \frac{\partial^2 \Phi_p}{\partial x^2}\right.$$

$$\left. + 2\left(1 + \frac{u_p}{u_\infty}\right)\left(\frac{v_p}{u_\infty}\right)\frac{\partial^2 \Phi_p}{\partial x\,\partial y} + \left(\frac{v_p}{u_\infty}\right)^2 \frac{\partial^2 \Phi_p}{\partial y^2}\right] = 0 \quad (13.38)$$

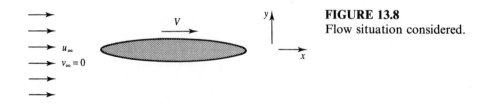

**FIGURE 13.8**
Flow situation considered.

If the perturbation velocities are small, i.e., if:

$$\frac{u_p}{u_\infty} \ll 1 \quad \text{and} \quad \frac{v_p}{u_\infty} \ll 1 \tag{13.39}$$

eq. (13.38) can be approximated by:

$$\frac{\partial^2 \Phi_p}{\partial x^2} + \frac{\partial^2 \Phi_p}{\partial y^2} - \left(\frac{u_\infty}{a}\right)^2 \frac{\partial^2 \Phi_p}{\partial x^2} = 0$$

i.e.,
$$\left[1 - M_\infty^2 \left(\frac{a_\infty}{a}\right)^2\right] \frac{\partial^2 \Phi_p}{\partial x^2} + \frac{\partial^2 \Phi_p}{\partial y^2} = 0 \tag{13.40}$$

But it was shown above that $c_p T + V^2/2$ is the same everywhere in the flow, so:

$$c_p T_\infty = \frac{u_\infty^2}{2} = c_p T + \frac{(u_\infty + u_p)^2 + v_p^2}{2}$$

i.e.,
$$c_p (T_\infty - T) = \frac{2 u_p u_\infty + u_p^2 + v_p^2}{2} \tag{13.41}$$

But, since:

$$a^2 = \gamma R T = \gamma c_p \left(1 - \frac{1}{\gamma}\right) T = (\gamma - 1) c_p T$$

eq. (13.41) gives:

$$\left(\frac{a}{a_\infty}\right)^2 = 1 - (\gamma - 1)\left(\frac{u_p u_\infty}{a_\infty^2} + \frac{1}{2}\frac{u_p^2}{a_\infty^2} + \frac{1}{2}\frac{v_p^2}{a_\infty^2}\right)$$

$$= 1 - (\gamma - 1)M_\infty^2 \left[\frac{u_p}{u_\infty} + \frac{1}{2}\left(\frac{u_p}{u_\infty}\right)^2 + \frac{1}{2}\left(\frac{v_p}{u_\infty}\right)^2\right]$$

For small $u_p/u_\infty$ and $v_p/u_\infty$, this gives:

$$\left(\frac{a}{a_\infty}\right)^2 = 1 - (\gamma - 1)M_\infty^2 \frac{u_p}{u_\infty} \tag{13.42}$$

Substituting this into eq. (13.40) then gives approximately:

$$\left[1 - M_\infty^2 + (\gamma - 1)M_\infty^4 \frac{u_p}{u_\infty}\right] \frac{\partial^2 \phi_p}{\partial x^2} + \frac{\partial^2 \phi_p}{\partial y^2} = 0 \tag{13.43}$$

Provided that $M_\infty$ is not very large and provided that $M_\infty$ is not near 1 then:

$$(\gamma - 1)M_\infty^4 \frac{u_p}{a_\infty} \ll 1 - M_\infty^2$$

and eq. (13.43) then gives:

$$(1 - M_\infty^2)\frac{\partial^2 \Phi_p}{\partial x^2} + \frac{\partial^2 \Phi_p}{\partial y^2} = 0 \tag{13.44}$$

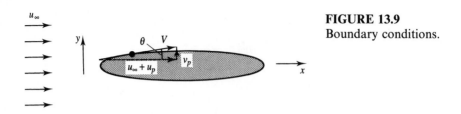

**FIGURE 13.9**
Boundary conditions.

This is the linearized velocity potential equation. It applies provided $M_\infty$ is not very near 1 or very large. Equation (13.44) is a linear equation and is, therefore, much easier to solve than the full, nonlinear velocity potential equation.

In order to solve eq. (13.44) it is necessary to specify the boundary conditions on the solution. Well upstream of the body, as indicated in Fig. 13.9, the flow is undisturbed and $\Phi_p$ is zero. On the surface of the body, there is no velocity component normal to the surface, i.e., the velocity vector is tangent to the surface. Hence, if $\theta$ is the angle the surface makes to the $x$ direction as indicated in Fig. 13.9:

$$\tan\theta = \frac{dy}{dx}\bigg|_s = \frac{v_p}{u_\infty + u_p}\bigg|_s \tag{13.45}$$

The subscript $s$ indicates conditions on the surface.

Now:

$$\frac{v_p}{u_\infty + u_p} = \frac{v_p/u_\infty}{1 + u_p/u_\infty} = \frac{v_p}{u_\infty} - \left(\frac{v_p}{u_\infty}\right)\left(\frac{u_p}{u_\infty}\right) \tag{13.46}$$

Hence, since $u_p/u_\infty$ and $v_p/u_\infty$ are $\ll 1$, these two equations give:

$$\frac{v_p|_s}{u_\infty} = \frac{dy}{dx}\bigg|_s \tag{13.47}$$

$v_p$, of course, being equal to $\partial\Phi_p/\partial y$.   $\partial\Phi_p/\partial y$

The solution of the velocity potential equation allows the velocity components at all points in the flow to be determined. The pressure variation through the flow can then be deduced from the velocity distribution. To do this, it is noted that the energy equation gives (see eq. (13.41)):

$$T_\infty - T = \frac{2u_p u_\infty + u_p^2 + v_p^2}{2c_p}$$

i.e.,   $$\frac{T}{T_\infty} = 1 - \left(\frac{\gamma - 1}{2}\right)M_\infty^2\left[2\left(\frac{u_p}{u_\infty}\right) + \left(\frac{u_p}{u_\infty}\right)^2 + \left(\frac{v_p}{u_\infty}\right)^2\right] \tag{13.48}$$

But, since isentropic flow is being considered:

$$\frac{p}{p_\infty} = \left(\frac{T}{T_\infty}\right)^{\frac{\gamma}{\gamma-1}}$$

so eq. (13.48) gives:

$$\frac{p}{p_\infty} = \left\{1 - \left(\frac{\gamma - 1}{2}\right)M_\infty^2\left[2\left(\frac{u_p}{u_\infty}\right) + \left(\frac{u_p}{u_\infty}\right)^2 + \left(\frac{v_p}{u_\infty}\right)^2\right]\right\}^{\frac{\gamma}{\gamma-1}} \quad (13.49)$$

The second term in this equation involves only terms like $u_p/u_\infty$ and is, therefore, small. Hence, since $(1 + \epsilon)^n \approx 1 + n\epsilon$ when $\epsilon \ll 1$, eq. (13.49) gives approximately:

$$\frac{p}{p_\infty} = 1 - \frac{\gamma M_\infty^2}{2}\left[2\frac{u_p}{u_\infty} + \left(\frac{u_p}{u_\infty}\right)^2 + \left(\frac{v_p}{u_\infty}\right)^2\right] \quad (13.50)$$

Further, since $v_p/u_\infty$ and $u_p/u_\infty$ are small, the second two terms in the bracket are much smaller than the first so approximately:

$$\frac{p}{p_\infty} = 1 - \gamma\frac{M_\infty^2}{2}\left(2\frac{u_p}{u_\infty}\right) \quad (13.51)$$

It is usual to express the pressure distribution in terms of a pressure coefficient defined by:

$$C_p = \frac{p - p_\infty}{\frac{1}{2}\rho_\infty u_\infty^2}$$

$$= \frac{(p/p_\infty) - 1}{\frac{1}{2}(\rho_\infty/p_\infty)u_\infty^2}$$

i.e., $$C_p = \frac{(p/p_\infty) - 1}{\gamma M_\infty^2/2} \quad (13.52)$$

because $a_\infty^2 = \gamma p_\infty/\rho_\infty$.

Combining this with eq. (13.51) then gives for linearized flows:

$$C_p = -\frac{2u_p}{u_\infty} \quad (13.53)$$

Using eq. (13.37), this can be written as:

$$C_p = -\frac{2}{u_\infty}\frac{\partial \Phi_p}{\partial x} \quad (13.54)$$

Thus, once the distribution of $\Phi_p$ has been determined, the variation of $C_p$ through the flow field can be found, i.e., the variation of $C_p$ about the surface of any body in the flow can be determined.

## 13.6
## LINEARIZED SUBSONIC FLOW

It is to be expected that in subsonic flow over a body the solution for $\phi_p$ in the actual flow can be related to the solution for the flow that would exist over the same body if the flow was incompressible. To show that this is, indeed, the case, eq. (13.44) which governs the actual flow is written as:

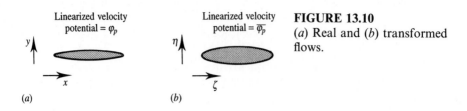

Linearized velocity potential = $\varphi_p$

Linearized velocity potential = $\overline{\Phi_p}$

**FIGURE 13.10**
(a) Real and (b) transformed flows.

(a)

(b)

$$\beta^2 \frac{\partial^2 \Phi_p}{\partial x^2} + \frac{\partial^2 \Phi_p}{\partial y^2} = 0 \qquad (13.55)$$

where:

$$\beta^2 = 1 - M_\infty^2 \qquad (13.56)$$

The coordinates in the real space, $x$ and $y$, are now transformed into coordinates $\zeta$ and $\eta$ in a transformed space using:

$$\zeta = x, \qquad \eta = \beta y \qquad (13.57)$$

Moreover, let the linearized velocity potential in the transformed space, $\overline{\Phi_p}$, be related to the linearized velocity potential in the real space, $\Phi_p$ by:

$$\overline{\Phi_p} = \beta \Phi_p \qquad (13.58)$$

This is illustrated in Fig. 13.10. Substituting eqs. (13.57) and (13.58) into eq. (13.55) and noting that $\zeta$ depends only on $x$ and $\eta$ only on $y$ gives:

$$\beta^2 \frac{\partial^2 (\overline{\Phi_p}/\beta)}{\partial \zeta^2} \left(\frac{d\zeta}{dx}\right)^2 + \frac{\partial^2 (\overline{\Phi_p}/\beta)}{\partial \eta^2} \left(\frac{d\eta}{dy}\right)^2 = 0$$

i.e., since $d\eta/dy = \beta$:

$$\beta \frac{\partial^2 \overline{\Phi_p}}{\partial \zeta^2} + \beta \frac{\partial^2 \overline{\Phi_p}}{\partial \eta^2} = 0$$

i.e.,

$$\frac{\partial^2 \overline{\Phi_p}}{\partial \zeta^2} + \frac{\partial^2 \overline{\Phi_p}}{\partial \eta^2} = 0 \qquad (13.59)$$

Thus the variation of $\overline{\Phi_p}$ in the transformed plane is governed by Laplace's equation.

Now consider the boundary conditions at the surface as discussed in the derivation of eq. (13.47). Let:

$$y = f(x)$$

describe the shape of the actual body and let:

$$\eta = F(\zeta)$$

describe the shape of the body in the transformed plane.

The boundary condition in the actual flow is:

$$\left.\frac{\partial \Phi_p}{\partial y}\right|_s = u_\infty \frac{df}{dx} \qquad (13.60)$$

whereas in the transformed plane the boundary condition is:

$$\left.\frac{\partial\overline{\Phi}_p}{\partial\eta}\right|_s = u_\infty \frac{dF}{d\zeta}$$  (13.61)

But:

$$\left.\frac{\partial\overline{\Phi}_p}{\partial\eta}\right|_s = \left[\frac{\partial(\Phi_p\beta)}{\partial y}\frac{\partial y}{\partial\eta}\right]\Bigg|_s = \left[\frac{\partial(\Phi_p\beta)}{\partial y}\frac{1}{\beta}\right]\Bigg|_s$$

i.e.,

$$\left.\frac{\partial\overline{\Phi}_p}{\partial\eta}\right|_s = \left.\frac{\partial\Phi_p}{\partial y}\right|_s$$  (13.62)

Equation (13.62) shows that the left hand side of eqs. (13.60) and (13.61) are equal, so:

$$\frac{df}{dx} = \frac{dF}{d\zeta}$$  (13.63)

This means that the shape of the body is the same in real and the transformed planes, i.e., the function that relates $y$ to $x$ on the surface of the body in the real plane is identical to the function that relates $\eta$ to $\zeta$ on the surface of the body in the transformed plane. Since $\Phi_p$ is determined by eq. (13.59) which is identical to the equation that applies in incompressible flow and since the shape of the body is the same in the real and transformed planes, it follows that $\overline{\Phi}_p$ is the same as the linearized velocity potential function that would exist in incompressible flow over the body being considered. Hence, if $\overline{\Phi}_p$ is determined from the solution from incompressible flow over the body, the actual linearized velocity potential is given by:

$$\Phi_p = \overline{\Phi}_p/\beta = \overline{\Phi}_p/\sqrt{1-M^2}$$

$$x = \zeta$$

$$y = \eta/\beta = \eta/\sqrt{1-M^2}$$

Now, eq. (13.54) gives:

$$C_p = -\frac{2}{u_\infty}\frac{\partial\Phi_p}{\partial x}$$

$$= -\frac{2}{u_\infty}\frac{\partial(\overline{\Phi}_p/\beta)}{\partial\zeta}$$

i.e.,

$$C_p = \frac{1}{\beta}\left[-\frac{2}{u_\infty}\frac{\partial\overline{\Phi}_p}{\partial\zeta}\right]$$  (13.64)

But the variation of $\overline{\Phi}_p$ with $\zeta$ is the same as in incompressible flow over the body shape being considered, i.e., by virtue of eq. (13.54), eq. (13.64) gives:

$$C_p = \frac{1}{\beta}C_{p0} = \frac{C_{p0}}{\sqrt{1-M_\infty^2}}$$  (13.65)

where $C_{p0}$ is the value of the pressure coefficient that would exist in incompressible flow over the body being considered. This means that if the pressure coefficient distribution is determined in incompressible irrotational flow, the pressure coefficient distribution is compressible flow at Mach number $M_\infty$ can

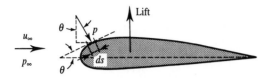

**FIGURE 13.11**
Calculation of lift from pressure distribution.

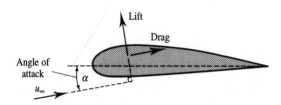

**FIGURE 13.12**
Angle of attack of airfoil.

be found by applying a 'compressibility correction factor' equal to $1/\sqrt{1 - M_\infty^2}$. This only applies in subsonic flows, i.e., to flows in which $M_\infty < 1$.

Consider the lift force on a body, i.e., the force normal to the upstream flow, resulting from the variation in the pressure over the surface of the body. If $L$ is the lift per unit span, it will be seen from Fig. 13.11 that:

$$L = \int (p - p_\infty) \cos \theta \, ds$$

the integral being carried out over the surface of the body. This equation can be written as:

$$C_L\left( = \frac{L}{\frac{1}{2} \rho_\infty u_\infty^2 c} \right) = \int C_p \cos \theta \, d\left( \frac{s}{c} \right) \tag{13.66}$$

where $C_L$ is the lift coefficient and $c$ is the wing chord. From eq. (13.65) it follows that:

$$C_L = \frac{C_{L0}}{\sqrt{1 - M_\infty^2}} \tag{13.67}$$

$C_{L0}$ being the lift coefficient that would exist in incompressible flow. Since $1 - M_\infty^2 < 1$, the above equations indicate that compressibility increases the coefficient of lift.

Now for smaller angles of attack, $\alpha$, as defined in Fig. 13.12, $C_L = a\alpha$ in incompressible flow. In compressible flow then $C_L = a\alpha/\sqrt{1 - M_\infty^2}$. This is shown in Fig. 13.13.

## 13.7
## LINEARIZED SUPERSONIC FLOW

In supersonic flow disturbances are, as discussed in Chapter 2, propagated along Mach lines as indicated in Fig. 13.14. The flow upstream of the Mach line is undisturbed by the presence of the wave. It is to be expected, therefore,

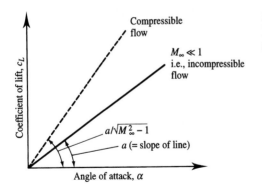

**FIGURE 13.13**
Effect of compressibility on lift coefficient variation.

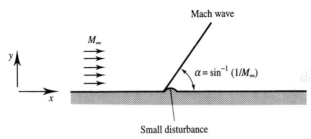

Mach wave

$M_\infty$

$\alpha = \sin^{-1}(1/M_\infty)$

Small disturbance

**FIGURE 13.14**
Mach wave.

that the perturbation velocity potential, $\Phi_p$, will, in supersonic flow, be constant along a Mach line. Now, along a Mach line:

$$\frac{y}{x} = \frac{1}{\sqrt{M_\infty^2 - 1}}$$

i.e.,

$$x = \sqrt{M_\infty^2 - 1}\, y$$

where it has again been noted that because $\sin \alpha = 1/M_\infty$, $\tan \alpha = 1/\sqrt{M_\infty^2 - 1}$.

Hence, it is to be expected that:

$$\phi_p = f(x - \lambda y) = f(\eta) \tag{13.68}$$

where:

$$\lambda = \sqrt{M_\infty^2 - 1} \tag{13.69}$$

and:

$$\eta = x - \lambda y \tag{13.70}$$

If this is, indeed, a solution it must satisfy eq. (13.44) which can be written as:

$$\lambda^2 \frac{\partial^2 \Phi_p}{dx^2} - \frac{\partial^2 \Phi_p}{dy^2} = 0 \tag{13.71}$$

Now eq. (13.68) gives:

$$\frac{\partial \Phi_p}{dx} = \frac{df}{d\eta}\frac{\partial \eta}{dx} = \frac{df}{d\eta}$$

from which it follows that:

$$\frac{\partial^2 \Phi_p}{dx^2} = \left(\frac{d^2 f}{d\eta^2}\right)\left(\frac{\partial \eta}{dx}\right)^2 = \frac{d^2 f}{d\eta^2} \tag{13.72}$$

Similarly, eq. (13.68) gives:

$$\frac{\partial^2 \Phi_p}{dy^2} = \left(\frac{d^2 f}{d\eta^2}\right)\left(\frac{\partial \eta}{dy}\right)^2 = \lambda^2 \frac{d^2 f}{d\eta^2} \tag{13.73}$$

Substituting these last two equations into the left hand side of eq. (13.71) gives this left hand side as:

$$\lambda^2 \frac{d^2 f}{d\eta^2} - \lambda^2 \frac{d^2 f}{d\eta^2}$$

which is always zero. This proves that eq. (13.68) is, indeed, a solution to the linearized velocity potential equation for supersonic flow.

Next consider the pressure coefficient at the surface of a body in a supersonic flow. It will be recalled that the linearized boundary condition at the surface as given in eq. (13.47) requires that:

$$\frac{v_p|_s}{u_\infty} = \tan \theta \tag{13.74}$$

But since linearized flow is being considered, $\theta$ must remain small and $\tan \theta$ can be approximated by $\theta$. Equation (13.74) therefore, in linearized flow, can be written as:

$$\frac{v_p|_s}{u_\infty} = \theta \tag{13.75}$$

It is then noted that using eq. (13.68) gives:

$$u_p = \frac{\partial \Phi_p}{\partial x} = \frac{df}{d\eta}\frac{\partial \eta}{\partial x} = \frac{df}{\partial \eta}$$

$$v_p = \frac{\partial \Phi_p}{\partial y} = \frac{df}{d\eta}\frac{\partial \eta}{\partial y} = -\lambda \frac{df}{d\eta}$$

Dividing these two equations gives:

$$\frac{u_p}{v_p} = -\frac{1}{\lambda}$$

i.e.,

$$u_p = -\frac{v_p}{\lambda} \tag{13.76}$$

Combining this with eq. (13.75) then gives:

$$u_p|_s = -\frac{u_\infty}{\lambda}\theta \tag{13.77}$$

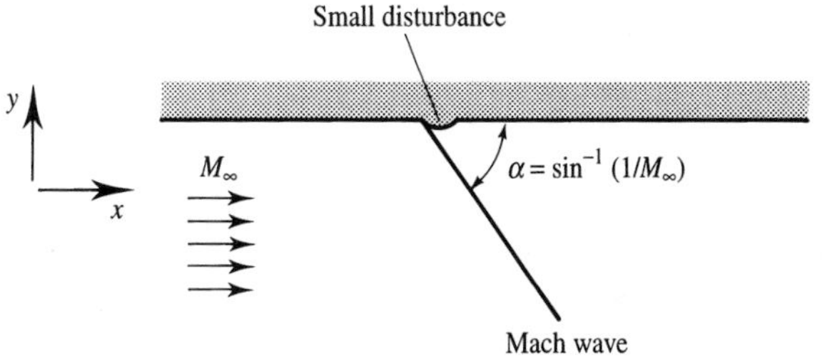

Small disturbance

**FIGURE 13.15**
Mach waves on lower
surface.

$\alpha = \sin^{-1}(1/M_\infty)$

$M_\infty$

$y$

$x$

Mach wave

But eq. (13.53) gives $C_p = -2u_p/u_\infty$ so at the surface of a body in supersonic flow:

$$C_p = \frac{2\theta}{\lambda} = \frac{2\theta}{\sqrt{M_\infty^2 - 1}} \tag{13.78}$$

This equation shows that, provided the assumption of linearized flow is applicable, the pressure coefficient on the surface of a body in supersonic flow is proportional to the angle the surface makes to the oncoming flow.

It should be noted that the above analysis only applies to flow in which the Mach waves run upward as shown in Fig. 13.14. Consider the waves generated on the underside of a surface as shown schematically in Fig. 13.15.

On the lower surface of a body it is to be expected then that:

$$\Phi_p = h(x + \lambda y) = h\zeta \tag{13.79}$$

where:

$$\zeta = x + \lambda y \tag{13.80}$$

Applying the same type of analysis as that presented above for the upper surface then gives for the lower surface:

$$C_p = \frac{-2\theta}{\sqrt{M_\infty^2 - 1}} \tag{13.81}$$

Equations (13.78) and (13.81) define the distribution of the pressure coefficient over a slender body in supersonic flow. Consider a body of the type shown in Fig. 13.16. On the forward part of the body between A and B, $\theta$ is positive and $C_p$ is, by virtue of eq. (13.78) positive. Also on the forward part of the body between A and D, $\theta$ is negative and $C_p$ is thus also, by virtue of eq. (13.81), positive. Hence, $C_p$ is positive everywhere along BAD. On the rear portion of the body, $\theta$ is negative between B and C and positive between D and C. Hence, by virtue of eqs. (13.78) and (13.81), $C_p$ is negative everywhere along BCD.

$M_\infty$

**FIGURE 13.16**
Type of body considered.

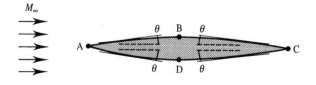

$\theta$   B   $\theta$

A

C

$\theta$   D   $\theta$

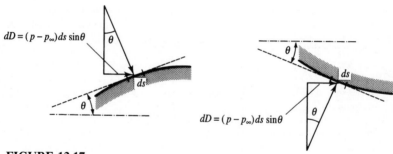

**FIGURE 13.17**
Generation of drag force. $p - p_\infty$ is the pressure relative to freestream pressure.

Next, consider the drag force, $D$, on a body such as that shown in Fig. 13.16. The drag force arises because of the pressure variation about the surface and, considering the forces acting on a small portion of the surface indicated in Fig. 13.17. It will be seen that:

$$D = \int (p - p_\infty) \sin \theta \, ds \qquad (13.82)$$

the integration being carried out over the surface of the body, the upper and lower surfaces being separately considered. Since linearized theory is being used, which applies only to slender bodies, $\sin \theta \approx \theta$ and $ds \approx dx$. Hence, eq. (13.82) can be written as:

$$D = \int (p - p_\infty) \theta \, dx \qquad (13.83)$$

The drag coefficient $C_D$ is defined as usual in two-dimensional flow by:

$$C_D = \frac{D}{\frac{1}{2} \rho u_\infty^2 c} \qquad (13.84)$$

$c$ being the chord of the body. Unit span, i.e., unit distance at right angles to the flow, has been considered.

Equation (13.83) thus gives:

$$C_D = \int_{\text{upper surface}} C_p \theta \, d(x/c) + \int_{\text{lower surface}} C_p \theta \, d(x/c) \qquad (13.85)$$

Consider the body shown in Fig. 13.16. Since $C_p$ is positive between A and B and between A and C, this portion BAC of the surface will give a positive contribution to the drag. But $C_p$ is negative between B and D and between C and D and, therefore, because of the way the surface slopes, this also gives a positive contribution to the drag. Thus there will be the net drag force on the body. Now, viscosity is being neglected in the present analysis. In incompressible flow, i.e., flow at $M_\infty \approx 0$, the drag is always predicted to be zero if viscosity is neglected. In supersonic flow, however, a drag force arises even when viscosity is neglected. This type of pressure drag, associated with supersonic flow, is termed "wave drag."

### EXAMPLE 13.1

A thin wing can be modeled as a 1 m wide flat plate set at an angle of 3° to the upstream flow. If this wing is placed in a flow with a Mach number of 3 and a static pressure of 50 kPa, find using linearized theory the pressure on the upper and lower surfaces of the airfoil and the lift and drag per meter span.

*Solution*

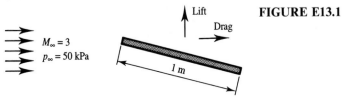

**FIGURE E13.1**

The angle that the upper surface makes to the flow is $-3° = -0.0524$ radians. Hence, eq. (13.78) gives:

$$C_{p \text{ upper}} = \frac{-2 \times 0.0524}{\sqrt{9-1}} = -0.0371$$

i.e.,

$$\frac{p_{\text{upper}} - p_\infty}{\frac{1}{2}\rho_\infty u_\infty^2} = -0.0371$$

But:

$$\tfrac{1}{2}\rho_\infty u_\infty^2 = \frac{p_\infty}{2}\frac{\rho_\infty}{p_\infty}u_\infty^2 = \frac{\gamma p_\infty M_\infty^2}{2}$$

so:

$$\frac{p_{\text{upper}} - 50}{1.4 \times 50 \times 9/2} = -0.0371$$

Therefore:

$$p_{\text{upper}} = 38.31 \text{ kPa}$$

Similarly, since $\theta$ for the lower surface is also $-3°$, eq. (13.81) gives:

$$\frac{p_{\text{lower}} - 50}{1.4 \times 50 \times 9/2} = +0.0371$$

Therefore:

$$p_{\text{lower}} = 61.69 \text{ kPa}$$

The lift will be given by:

$$L = (p_{\text{lower}} A - p_{\text{upper}} A) \cos \theta$$

where $A$ is the platform area of the wing, i.e., $1 \text{ m} \times 1 \text{ m}$. Since $\theta$ is small, it is consistent with the previous assumptions to set $\cos \theta = 1$. Hence:

$$L = (61.69 - 38.31) \times 1 \times 1 = 23.38 \text{ kN}$$

Similarly, the drag in the airfoil is given by:

$$D = (p_{\text{lower}} A - p_{\text{upper}} A) \sin \theta$$

$$= (p_{\text{lower}} - p_{\text{upper}}) A\theta$$

$$= (61.69 - 38.31) \times 0.0524$$

$$= 1.23 \text{ kN}$$

Hence, the pressures on the upper and lower surfaces are 38.31 kPa and 61.69 kPa respectively, and the lift and drag are 23.38 kN and 1.23 kN respectively.

If the oblique shock and expansion wave results given in Chapters 6 and 7 are used, the values of the pressures and the lift and drag are very close to the values given by the approximate linearized theory.

**EXAMPLE 13.2**

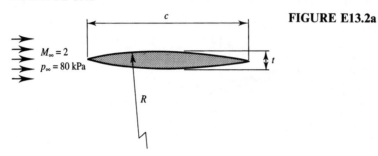

**FIGURE E13.2a**

An airfoil of the shape shown in Fig. E13.2a is placed in a flow at a Mach number of 2 and a pressure of 80 kPa. Derive expressions for the pressure variations along the upper and lower surfaces.

*Solution*

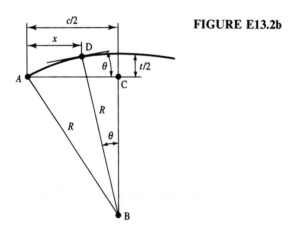

**FIGURE E13.2b**

Consider the geometrical parameters shown in Fig. E13.2b. Considering the triangle ABC shown in the figure, it will be seen that:

$$R^2 = \left(\frac{c}{2}\right)^2 + \left(R - \frac{t}{2}\right)^2$$

$$= \frac{c^2}{4} + R^2\left(1 - \frac{t}{2R}\right)^2$$

But, by assumption, $t/R \ll 1$ so this equation gives, approximately:

$$R^2 = \frac{c^2}{4} + R^2 - \frac{2tR}{2}$$

i.e.,
$$R = \frac{c^2}{4t} \tag{i}$$

Next consider point D which lies distance $x$ from the leading edge. It will be seen that:

$$x = \frac{c}{2} - R \sin \theta$$

which since $\sin \theta \approx \theta$ gives:

$$x = \frac{c}{2} - R\theta$$

i.e.,
$$\theta = \frac{c/2 - x}{R}$$

Hence, using eq. (i) gives:

$$\theta = 2\left(\frac{t}{c}\right) - \left(\frac{4xt}{c^2}\right) \tag{ii}$$

Equation (13.78) then gives:

$$C_p = \frac{2\theta}{\sqrt{4-1}} = \frac{2}{\sqrt{3}} \times 2 \times \frac{t}{c} \times \left[1 - 2\left(\frac{x}{c}\right)\right]$$

i.e., since:
$$C_p = \frac{p - p_\infty}{\gamma p_\infty M_\infty^2 / 2}$$

it follows that:

$$p = p_\infty + \frac{\gamma p_\infty M_\infty^2}{2} \times \frac{4}{\sqrt{3}} \times \left(\frac{t}{c}\right) \times \left[1 - 2\left(\frac{x}{c}\right)\right]$$

$$= 80 + \frac{1.4 \times 80 \times 4}{2} \times \frac{4}{\sqrt{3}} \times \left(\frac{t}{c}\right) \times \left[1 - 2\left(\frac{x}{c}\right)\right]$$

$$= 80 + 517\left(\frac{t}{c}\right)\left[1 - 2\left(\frac{x}{c}\right)\right]$$

The pressure on the surface therefore, varies linearly with $x/c$. The variation is shown in Fig. E13.2c. The pressure variations along the upper and lower surfaces are of course the same.

**FIGURE E13.2c**

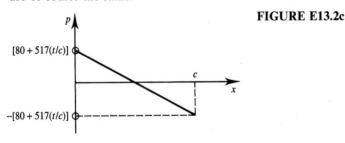

## 13.8
## METHOD OF CHARACTERISTICS

The method of characteristics has been used for many years to compute supersonic irrotational flows. Although the method has a strong analytical basis, its practical implementation is, essentially, always numerical and it is then used to compute the values of the flow variables at a series of distinct points in the flow rather than continuously throughout the flow field.

To understand the basic idea on which the method of characteristics is based, consider supersonic flow over a wall which suddenly changes direction by a small amount as illustrated in Fig. 13.18.

A Mach wave is generated as indicated in Fig. 13.18, the flow being uniform before and after the wave, the flow variables changing discontinuously across the wave. Along the wave itself, the derivatives are indeterminate, the wave sharply dividing one region of uniform flow from another region of uniform flow. Thus, in supersonic flow there can exist lines along which the derivatives are indeterminate and across which the flow variables can change discontinuously. Such lines are called characteristic lines.

To investigate the existence of these lines in a general steady two-dimensional irrotational flow the velocity potential $\Phi$ will again be used. The variation of $\Phi$ is governed by eq. (13.33) which can be written as:

$$\left(1 - \frac{u^2}{a^2}\right)\frac{\partial^2 \Phi}{\partial x^2} + \left(1 - \frac{v^2}{a^2}\right)\frac{\partial^2 \Phi}{\partial y^2} - \frac{2uv}{a^2}\frac{\partial^2 \Phi}{\partial x\,\partial y} = 0 \qquad (13.86)$$

Now, consider the change in any flow variable, $f$, in a direction, determined by small changes in the coordinates $dx$ and $dy$ as illustrated in Fig. 13.19. The change in the variable, $df$, is given by:

$$df = \frac{\partial f}{\partial x}dx + \frac{\partial f}{\partial y}dy \qquad (13.87)$$

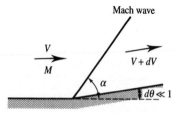

**FIGURE 13.18**
Effect of a small change in wall direction.
$\sin \alpha = 1/M$.

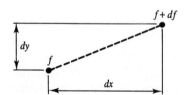

**FIGURE 13.19**
Change in variable considered.

If $f = \partial\Phi/\partial x$ then this will give:

$$d\left(\frac{\partial\Phi}{\partial x}\right) = \frac{\partial^2\Phi}{\partial x^2}dx + \frac{\partial^2\Phi}{\partial y\,\partial x}dy$$

But $u = \partial\Phi/\partial x$ so this equation gives:

$$du = \frac{\partial^2\Phi}{\partial x^2}dx + \frac{\partial^2\Phi}{\partial x\,\partial y}dy \tag{13.88}$$

This gives the change in the velocity component $u$ associated with small specified changes in coordinates $x$ and $y$ in terms of the derivatives of $\Phi$.
If $f = \partial\Phi/\partial y$, eq. (13.87) gives:

$$d\left(\frac{\partial\Phi}{\partial y}\right) = \frac{\partial^2\Phi}{\partial x\,\partial y}dx + \frac{\partial^2\Phi}{\partial y^2}dy$$

i.e., since $v = \partial\Phi/\partial y$:

$$dv = \frac{\partial^2\Phi}{\partial x\,\partial y}dx + \frac{\partial^2\Phi}{\partial y^2}dy \tag{13.89}$$

Consider eqs. (13.86), (13.88), and (13.89), i.e.,

$$\left(1 - \frac{u^2}{a^2}\right)\frac{\partial^2\Phi}{\partial x^2} + \left(1 - \frac{v^2}{a^2}\right)\frac{\partial^2\Phi}{\partial y^2} - \left(\frac{2uv}{a^2}\right)\frac{\partial^2\Phi}{\partial x\,\partial y} = 0$$

$$(dx)\frac{\partial^2\Phi}{\partial x^2} + (dy)\frac{\partial^2\Phi}{\partial x\,\partial y} = du$$

$$(dy)\frac{\partial^2\Phi}{\partial y^2} + (dx)\frac{\partial^2\Phi}{\partial x\,\partial y} = dv$$

If $u$, $v$, $a$, $du$, $dv$, $dx$, and $dy$ are specified, these three equations are the linear equations in $\partial^2\Phi/\partial x^2$, $\partial^2\Phi/\partial y^2$, $\partial^2\Phi/\partial x\,\partial y$.
Solving, for example, for $\partial^2\Phi/\partial x\,\partial y$ gives:

$$\frac{\partial^2\Phi}{\partial x\,\partial y} = \frac{(1 - u^2/a^2)\,du\,dy + (1 - v^2/a^2)\,dv\,dx}{(1 - u^2/a^2)(dy)^2 + (2uv/a^2)\,dx\,dy + (1 - v^2/a^2)(dx)^2} \tag{13.90}$$

In general, this equation can be solved for any chosen values of $dx$ and $dy$, i.e., for any chosen direction, to give $\partial^2\Phi/\partial x\,\partial y$ at a selected point in the flow. This is illustrated in Fig. 13.20. However, for the reasons discussed above, in supersonic flow it is possible to have $\partial^2\Phi/\partial x\,\partial y$ indeterminate in certain directions. In these directions, eq. (13.82) must give an indeterminate value. For this to be the case, eq. (13.82) must give:

$$\frac{\partial^2\Phi}{\partial x\,\partial y} = \frac{0}{0}$$

i.e., when the direction along which the changes in $du$ and $dv$ give an indeterminate value of $\partial^2\Phi/\partial x\,\partial y$ is chosen, both the numerator and the denominator of eq. (13.82) must be zero. Consider the denominator. Along

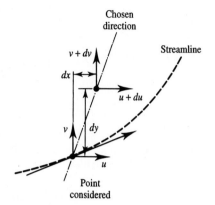

Chosen direction

Streamline

$v + dv$

$dx$

$u + du$

$v$

$dy$

$u$

Point considered

**FIGURE 13.20**
Changes in variables.

characteristic lines, i.e., the lines on which $\partial^2\Phi/\partial x\,\partial y$ are indeterminate, this denominator is zero so:

$$\left(1 - \frac{u^2}{a^2}\right)(dy)^2 + \left(\frac{2uv}{a^2}\right)dx\,dy + \left(1 - \frac{v^2}{a^2}\right)(dx)^2 = 0$$

Dividing by $(dx)^2$ gives:

$$\left(1 - \frac{u^2}{a^2}\right)\left(\frac{dy}{dx}\bigg|_{ch}\right)^2 + \left(\frac{2uv}{a^2}\right)\left(\frac{dy}{dx}\bigg|_{ch}\right) + \left(1 - \frac{v^2}{a^2}\right) = 0 \qquad (13.91)$$

The subscript $ch$ on $dy/dx$ indicates that the slope of the characteristic line is being considered. Equation (13.91) is a quadratic equation for $dy/dx|_{ch}$. Solving gives:

$$\frac{dy}{dx}\bigg|_{ch} = \frac{-(2uv/a^2) \pm \sqrt{(2uv/a^2)^2 - 4(1 - u^2/a^2)(1 - v^2/a^2)}}{2(1 - u^2/a^2)} \qquad (13.92)$$

There are thus two possible values for $dy/dx|_{ch}$, i.e., there are two characteristic directions associated with each point in the flow.

Equation (13.92) can be rearranged to give:

$$\frac{dy}{dx}\bigg|_{ch} = \frac{-(uv/a^2) \pm \sqrt{(u^2 + v^2)/a^2 - 1}}{(1 - u^2/a^2)} \qquad (13.93)$$

If $\theta$ is the direction that the velocity vector makes to the $x$-coordinate direction as shown in Fig. 13.21 then:

$$u^2 + v^2 = V^2, \qquad u = V\cos\theta, \qquad v = V\sin\theta \qquad (13.94)$$

$V$ being the magnitude of the velocity vector at the point considered as shown in Fig. 13.21. Substituting these values into eq. (13.93) then gives:

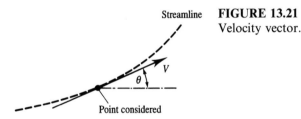

**FIGURE 13.21**
Velocity vector.

$$\frac{dy}{dx}\bigg|_{ch} = \frac{-M^2 \cos\theta \sin\theta \pm \sqrt{M^2 - 1}}{1 - M^2 \cos^2\theta} \tag{13.95}$$

where:
$$M = \frac{V}{a} \tag{13.96}$$

Now, the discussion given earlier indicates that the characteristic lines are in some way related to the Mach waves. Equation (13.95) is, therefore, rewritten in terms of the local Mach angle, $\alpha$, which, as discussed before, is given by:

$$M = \frac{1}{\sin\alpha}, \qquad \sqrt{M^2 - 1} = \frac{1}{\tan\alpha} \tag{13.97}$$

Using this, eq. (13.95) can be written as:

$$\frac{dy}{dx}\bigg|_{ch} = \frac{-\cos\theta \sin\theta \pm \cos\alpha \sin\alpha}{\sin^2\alpha - \cos^2\alpha} \tag{13.98}$$

After much manipulation and rearrangement, it can be shown that this equation gives:

$$\frac{dy}{dx}\bigg|_{ch} = \tan(\theta \pm \alpha) \tag{13.99}$$

[As a check, it can be noted that if $\theta$ is zero, i.e., the flow is in the $x$-direction, eq. (13.98) gives:

$$\frac{dy}{dx}\bigg|_{ch} = \pm \frac{\cos\alpha \sin\alpha}{\sin^2\alpha - 1} = \pm \frac{\cos\alpha \sin\alpha}{\cos^2\alpha} = \pm \tan\alpha$$

which is the same result as given by eq. (13.99).]

According to eq. (13.99), characteristic lines are thus at angles of $+\alpha$ and $-\alpha$ to the local flow direction as indicated in Fig. 13.22. This is, of course, exactly what would be anticipated.

Equation (13.99) determines the direction of the characteristic lines. It was obtained by using the fact that the denominator of eq. (13.90) must be zero along these lines. But the numerator of this equation must also be zero along these lines. Therefore, along the characteristic lines:

$$\left(1 - \frac{u^2}{a^2}\right) du\, dy + \left(1 - \frac{v^2}{a^2}\right) dv\, dx = 0 \tag{13.100}$$

i.e., along these lines:

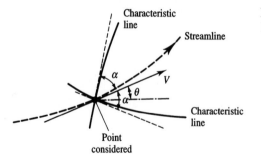

**FIGURE 13.22**
Directions of characteristic lines.

$$\frac{dv}{du} = -\frac{(1 - u^2/a^2)}{(1 - v^2/a^2)(dy/dx)} \tag{13.101}$$

Substituting for $dy/dx$ along the characteristic lines from eq. (13.93) then gives:

$$\frac{dv}{du} = \frac{uv/a^2 \pm \sqrt{(u^2 + v^2)/a^2 - 1}}{(1 - v^2/a^2)} \tag{13.102}$$

Using eq. (13.94) then gives:

$$\frac{dv}{du} = \frac{M^2 \sin\theta \cos\theta \pm \sqrt{M^2 - 1}}{1 - M^2 \sin^2\theta} \tag{13.103}$$

But since:

$$v = V \sin\theta$$

and since:

$$dv = \frac{\partial v}{\partial V} dV + \frac{\partial v}{\partial \theta} d\theta$$

it follows that:

$$dv = \sin\theta \, dV + V \cos\theta \, d\theta$$

Similarly since:

$$u = V \cos\theta$$

it follows that:

$$du = \frac{\partial u}{\partial V} dV + \frac{\partial u}{\partial \theta} d\theta$$

Therefore:

$$du = \cos\theta \, dV - V \sin\theta \, d\theta$$

Dividing these two equations gives:

$$\frac{dv}{du} = \frac{\sin\theta \, dV + V \cos\theta \, d\theta}{\cos\theta \, dV - V \sin\theta \, d\theta}$$

i.e.,

$$\frac{dv}{du} = \frac{\sin\theta(dV/V) + \cos\theta \, d\theta}{\cos\theta(dV/V) - \sin\theta \, d\theta} \tag{13.104}$$

Substituting this into eq. (13.103) then gives:

$$\frac{\sin\theta(dV/V) + \cos\theta \, d\theta}{\cos\theta(dV/V) - \sin\theta \, d\theta} = \frac{M^2 \sin\theta \cos\theta \pm \sqrt{M^2 - 1}}{1 - M^2 \sin^2\theta}$$

i.e.,   $\sin\theta\left(\dfrac{dV}{V}\right) + \cos\theta\, d\theta - M^2\sin^3\theta\left(\dfrac{dV}{V}\right) - M^2\sin^2\theta\cos\theta\, d\theta$

$$= M^2\sin\theta\cos^2\theta\left(\dfrac{dV}{V}\right) \pm \sqrt{M^2-1}\cos\theta\left(\dfrac{dV}{V}\right)$$

$$-M^2\sin^2\theta\cos\theta\, d\theta \pm \sqrt{M^2-1}\sin\theta\, d\theta$$

i.e.,   $[\cos\theta \pm \sqrt{M^2-1}\sin\theta]\, d\theta$

$$= [-\sin\theta + M^2\sin\theta(\sin^2 + \cos^2\theta) \pm \sqrt{M^2-1}\cos\theta]\left(\dfrac{dV}{V}\right)$$

i.e.,   $d\theta = -\left[\dfrac{\sin\theta(M^2-1) \pm \sqrt{M^2-1}\cos\theta}{\pm\sqrt{M^2-1}\sin\theta - \cos\theta}\right]\left(\dfrac{dV}{V}\right)$

i.e.,   $d\theta = \left[\dfrac{\sqrt{M^2-1}\sin\theta \pm \cos\theta}{\pm\sqrt{M^2-1}\sin\theta - \cos\theta}\right]\sqrt{M^2-1}\left(\dfrac{dV}{V}\right)$

i.e.,   $$d\theta = \pm\sqrt{M^2-1}\,\dfrac{dV}{V} \qquad (13.105)$$

This is the equation governing the changes in the variables along the characteristic lines. It will be noted that it is an ordinary differential equation whereas the original equation considered, i.e., eq. (13.86), was a partial differential equation. Furthermore, eq. (13.105), can be integrated along the characteristic lines to give:

$$\int d\theta \pm \int \sqrt{M^2-1}\,\dfrac{dV}{V} = \text{constant} \qquad (13.106)$$

The integral:

$$\int \sqrt{M^2-1}\,\dfrac{dV}{V}$$

has been encountered before in Chapter 7 where it was shown that it could be written as:

$$\nu(M) = \int_0^M \dfrac{\sqrt{M^2-1}}{[1+(\gamma-1)M^2/2]}\dfrac{dM}{M} \qquad (13.107)$$

where $\nu$ is the Prandtl–Meyer angle (given the symbol $\theta$ in Chapter 7) which is arbitrarily taken as zero when $M = 1$. The form of the relation between $\nu$ and $M$ is given in Chapter 7 where it is noted that values of $\nu$ are also listed in isentropic tables. Using this and noting that $\int d\theta = \theta$, the initial value being arbitrary, eq. (13.106) gives:

$$\theta \pm \nu = \text{constant} \qquad (13.108)$$

the constant having a different value on each of the two characteristics passing through any point.

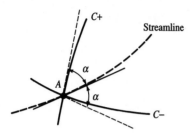

**FIGURE 13.23**
Characteristic lines considered.

Consider the two characteristic lines $C+$ and $C-$ passing through point A as shown in Fig. 13.23. On $C+$:

$$\left.\frac{dy}{dx}\right|_{ch} = \tan(\theta + \alpha), \qquad \theta - \nu = K+ \tag{13.109}$$

On $C-$:

$$\left.\frac{dy}{dx}\right|_{ch} = \tan(\theta - \alpha), \qquad \theta + \nu = K-$$

$K+$ and $K-$ being constants.

Thus for the type of flow being considered here, the governing equation, i.e., eq. (13.108), is an algebraic equation and thus much easier to solve than the partial differential equation with which the analysis started.

To see how the above equations can be used to solve for the variation of the flow variables, consider a point 3 which lies at the point of intersection of the characteristic lines passing through two points 1 and 2 as shown in Fig. 13.24. For the characteristic passing through point 1:

$$\theta_1 + \nu_1 = (K-)_1 = \theta_3 + \nu_3[= (K-)_3] \tag{13.110}$$

Similarly for the characteristic passing through point 2:

$$\theta_2 - \nu_2 = (K+)_2 = \theta_3 - \nu_3[= (K+)_3] \tag{13.111}$$

i.e.,

$$\theta_3 + \nu_3 = (K-)_1$$

$$\theta_3 - \nu_3 = (K+)_2$$

First adding these two equations and then subtracting them gives:

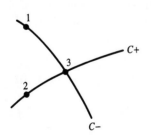

**FIGURE 13.24**
Points considered.

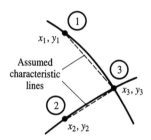

$x_1, y_1$

Assumed
characteristic
lines

$x_3, y_3$

$x_2, y_2$

**FIGURE 13.25**
Straight line approximations used.

$$\theta_3 = [(K-)_1 + (K+)_2]/2 \tag{13.112}$$

$$\nu_3 = [(K-)_1 - (K+)_2]/2 \tag{13.113}$$

which can also be written as:

$$\theta_3 = (\theta_1 + \theta_2)/2 + (\nu_1 + \nu_2)/2 \tag{13.114}$$

$$\nu_3 = (\theta_1 - \theta_2)/2 + (\nu_1 + \nu_2)/2 \tag{13.115}$$

This allows $\theta_3$ and $\nu_3$ to be found. Since $\nu_3$ depends only on $M$, this allows $M_3$ and hence $\alpha_3$ to be found. Since the stagnation pressure and temperature are constant throughout the flow field and are determined by the initial conditions, once $M_3$ is found, $p_3$, $T_3$, $a_3$, and $\rho_3$ can all be found and then $V_3(= M_3 a_3)$ can be found. The velocity components can then be found using $u_3 = V_3 \cos \theta_3$ and $v_3 = V_3 \sin \theta_3$. Therefore, all the flow variables at point 3 can be found. However, the location of point 3 has not yet been established. The characteristic lines are, in general, curved, their local slope depending on the local values of $\nu$ and $\theta$. However, if points 1 and 3 and 2 and 3 are close together, the characteristic lines can be assumed to be straight with a slope equal to the average of the values at the end points as indicated in Fig. 13.25. Considering the situation shown in Fig. 13.25, it is assumed that:

$$[(\theta_1 - \alpha_1) + (\theta_3 - \alpha_3)]/2 = \tan^{-1}\left[\frac{y_3 - y_1}{x_3 - x_1}\right] \tag{13.116}$$

$$[(\theta_2 + \alpha_2) + (\theta_3 + \alpha_3)]/2 = \tan^{-1}\left[\frac{y_3 - y_2}{x_3 - x_2}\right] \tag{13.117}$$

Since $\theta_3$ and $\alpha_3$ are determined by solving eqs. (13.114) and (13.115), these two equations determine $x_3$ and $y_3$.

The procedure discussed above was for an "internal" point, i.e., a point 3 in the flow field that did not lie on a boundary. If a point lies on the boundary, the flow direction at this point will be determined by the slope of the boundary, e.g., consider the point 5 shown in Fig. 13.26 which lies on a solid wall. The flow direction at this point $\theta_5$ is equal to the slope of the wall as indicated. Consider the characteristic line between points 4 and 5 as shown in Fig. 13.26. Since $(K-)_4 = (K-)_5$ it follows that:

$$\theta_4 + \nu_4 = \theta_5 + \nu_5$$

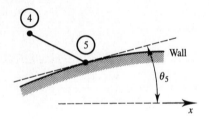

**FIGURE 13.26**
Conditions at wall.

Hence, since $\theta_5$ is known, $\nu_5$ is given by:

$$\nu_5 = \theta_4 - \theta_5 + \nu_4 \tag{13.118}$$

With $\nu_5$ determined, the values of all the flow properties at 5 can be determined as discussed before. The characteristic line between 4 and 5 is, of course, assumed to be straight which determines the position of the point 5.

The procedure for using the method of characteristic lines to numerically calculate the flow in a duct is as follows:

1. The conditions on some initial line must be specified, e.g., conditions on the line AB in Fig. 13.27 must be specified
2. The shape of the walls, e.g., AD and BC in Fig. 13.27, must be specified
3. Using the initial values of the variables on line A, determine the stagnation pressure, temperature, etc.
4. Starting with a series of chosen points on line AB, march the solution forward to the points defined by the intersection of characteristics with each other or with the wall as indicated
5. At each point, use the calculated values of $\nu$ and $\theta$ to get $p$, $T$, $V$, $\rho$, $u$, and $v$.

A computer program based on this procedure can fairly easily be developed, but this will not be done here. Instead, a simple example of the application of the method in which hand calculations are used is given below. In any real application, far more points would be used.

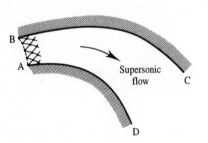

**FIGURE 13.27**
Initial conditions.

**EXAMPLE 13.3**

**FIGURE E13.3a**

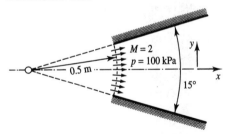

Consider flow through a channel whose walls diverge at an included angle of 15° as shown in Fig. E13.3a. The flow is uniform and radial with a Mach number of 2 and a pressure of 100 kPa at the inlet to the channel. Assuming steady, irrotational, two-dimensional flow, find the variation of pressure along the center line.

*Solution*

**FIGURE E13.3b**

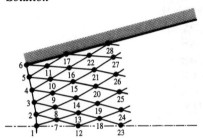

The flow will be symmetrical about the center line so attention will be restricted to flow in the upper half of the channel, the flow direction being known on the center line. Six equally spaced points on the initial line will be used as shown in Fig. E13.3b. The values of $\theta$ at these six points are 0°, 1.5°, 3°, 4.5°, 6°, and 7.5°, and the values of $x$ and $y$ at these points are given by:

$$x = -(0.5 - 0.5 \cos \theta), \qquad y = 0.5 \sin \theta$$

Some calculated values of $M$, $\alpha$, $\theta$, $\nu$, and $p$ at various points are listed in Table E13.3. The values of $\nu$ and $\alpha$ at points 1–6 are obtained by noting that the Mach number at these points is 2.

To illustrate how the values are obtained, consider the following points:

*Point 9:* Equations (13.114) and (13.115) give:

$$\theta_9 = (\theta_4 + \theta_3)/2 + (\nu_4 - \nu_3)/2 = 3.75°$$

$$\nu_9 = (\theta_4 - \theta_3)/2 + (\nu_4 + \nu_3)/2 = 27.13°$$

The Mach number at point 9 is then determined from the value of $\nu_9$. This could, for example, be done using isentropic tables or relations or the software. Once $M_9$ is found, $p_9$ can be found using:

$$p_9 = \frac{p_9/p_0}{p_1/p_0} p_1$$

$p_1/p_0$ and $p_1$ being known and $p_9/p_0$ also being known since $M_9$ is known.

**TABLE E13.3**

| Point | $\alpha$ | $\theta$ | $v$ | $M$ | $p$ |
|-------|-------|-------|-------|-------|-------|
| 1 | 30 | 0 | 26.38 | 2 | 100 |
| 2 | 30 | 1.5 | 26.38 | 2 | 100 |
| 3 | 30 | 3 | 26.38 | 2 | 100 |
| 4 | 30 | 4.5 | 26.38 | 2 | 100 |
| 5 | 30 | 6 | 26.38 | 2 | 100 |
| 6 | 30 | 7.5 | 26.38 | 2 | 100 |
| 7 | 24.53 | 0.75 | 27.13 | 2.029 | 95.46 |
| 8 | 29.53 | 2.25 | 27.13 | 2.029 | 95.46 |
| 9 | 29.53 | 3.75 | 27.13 | 2.029 | 95.46 |
| 10 | 29.53 | 5.25 | 27.13 | 2.029 | 95.46 |
| 11 | 29.53 | 6.75 | 27.13 | 2.029 | 95.46 |
| 12 | 29.12 | 0 | 27.88 | 2.055 | 91.78 |
| 13 | 29.12 | 1.5 | 27.88 | 2.055 | 91.78 |
| 14 | 29.12 | 3 | 27.88 | 2.055 | 91.78 |
| 15 | 29.12 | 4.5 | 27.88 | 2.055 | 91.78 |
| 16 | 29.12 | 6 | 27.88 | 2.055 | 91.78 |
| 17 | 29.12 | 7.5 | 27.88 | 2.055 | 91.78 |
| 18 | 28.66 | 0.075 | 28.63 | 2.085 | 87.64 |
| 19 | 28.66 | 2.25 | 28.63 | 2.085 | 87.64 |
| 20 | 28.66 | 3.75 | 28.63 | 2.085 | 87.64 |
| 21 | 28.66 | 5.25 | 28.63 | 2.085 | 87.64 |
| 22 | 28.66 | 6.75 | 28.63 | 2.085 | 87.64 |
| 23 | 28.28 | 0 | 29.38 | 2.111 | 84.12 |
| 24 | 28.28 | 1.5 | 29.38 | 2.111 | 84.12 |
| 25 | 28.28 | 3 | 29.38 | 2.111 | 84.12 |
| 26 | 28.28 | 4.5 | 29.38 | 2.111 | 84.12 |
| 27 | 28.28 | 6 | 29.38 | 2.111 | 84.12 |
| 28 | 28.28 | 7.5 | 29.38 | 2.111 | 84.12 |

*Point 12:* Equation (13.118) gives:

$$v_{12} = (\theta_7 - \theta_{12}) + v_7 = 27.88°$$

it having been noted that since point 12 lies on the center line, $\theta_{12} = 0°$.
    The value of $M_{12}$ can then be found using this value of $v_{12}$ and $p_{12}$ can then be found.

    The position of the points on the center line can be found by assuming the characteristic lines to be straight lines. Consider the points shown in Fig. E13.3c. Using eq. (13.116) gives:

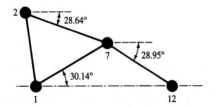

**FIGURE E13.3c**

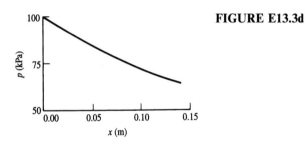

**FIGURE E13.3d**

$$\tan\left[\frac{(\theta_1 + \alpha_1)}{2} + \frac{(\theta_7 + \alpha_7)}{2}\right] = \frac{y_7 - y_1}{x_7 - x_1}$$

$$\tan\left[\frac{(\theta_2 - \alpha_2)}{2} + \frac{(\theta_7 - \alpha_7)}{2}\right] = \frac{y_7 - y_2}{x_7 - x_2}$$

i.e.,
$$y_7 - y_1 = 0.581(x_7 - x_1)$$
$$y_7 - y_2 = -0.546(x_7 - x_2)$$

But $x_1 = 0$, $y_1 = 0$, $x_2 = -0.0002$, $y_2 = 0.0131$. Therefore:
$$y_7 = 0.581x_7, \qquad y_7 = -0.546x_7 + 0.0130$$

Solving gives:
$$x_7 = 0.0115, \qquad y_7 = 0.0067$$

Also:
$$\tan\left[\left(\frac{\theta_7 - \alpha_7}{2}\right) + \left(\frac{\theta_{12} - \alpha_{12}}{2}\right)\right] = \frac{y_{12} - y_7}{x_{12} - x_7}$$

Hence, since $y_{12} = 0$,
$$x_{12} = x_7 - y_7/(-0.5057)$$

i.e., $x_{12} = 0.0248$.

Using this procedure, the $x$ coordinates of the points on the center line can be found. Alternatively, they could be determined by using a scale drawing using the values of the angles shown in Fig. E13.3c. With the $x$ values of the points on the center line found in this way, a graph showing the variation of $p$ with $x$ along the center line can be drawn, such a graph being given in Fig. E13.3d.

The method of characteristics thus reduces the computation of supersonic flow to a set of simple, repetitive steps. As mentioned before, it is relatively easy to develop a computer program based on the method. The disadvantage of the method is that it applies only in supersonic flows and that other methods must be used to find the initial conditions in a mixed subsonic–supersonic flow.

## 13.9
## NUMERICAL SOLUTIONS

The calculation of two- and three-dimensional compressible flows is today usually undertaken by using numerical methods, i.e., by using computational fluid dynamics. A wide range of software based on many different numerical

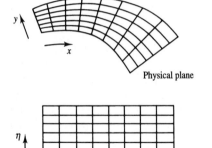

Physical plane

Transformed plane

**FIGURE 13.28**
Coordinate transformation.

methods such as the finite-element method, the finite-difference method, and the finite-volume method is available.

Two main difficulties arise in developing software for the numerical calculation of multidimensional compressible fluid flows:

- Because the bodies that are usually of interest are complex in shape, with some types of numerical method a coordinate transformation that converts the nonorthogonal mesh in the real coordinate system into an orthogonal mesh in the transformed coordinate system has to be adopted. This is schematically illustrated in Fig. 13.28. With some numerical methods, such as the finite-element method, this is not necessary.
- If shock waves occur in the flow, some form of iterative procedure has to be used because the shock wave shape is not known. One approach involves guessing a shock shape, carrying out the solutions upstream and downstream of the shock, and then altering the shock shape to match the two solutions correctly. This is shown very schematically in Fig. 13.29. A number of other approaches are also used to deal with this difficulty, most of these relying on some form of adaptive modification of the mesh.

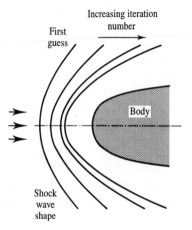

Increasing iteration number

First guess

Body

Shock wave shape

**FIGURE 13.29**
Shock adaptation during iterative form of solution.

## 13.10
## CONCLUDING REMARKS

The equations for two-dimensional compressible flow have been derived in this chapter and some methods of solving these equations have been presented. Approximate linearized methods which assume that the disturbances produced in the flow are small have been discussed in some detail and a brief discussion of the method of characteristics has also been given. Some brief remarks about numerical methods of solving the equations have also been presented.

## PROBLEMS

**13.1.**  Air flows with a Mach number of 2.8 over a flat plate which is set at an angle of 7° to the upstream flow. The pressure in the upstream flow is 100 kPa. Find the lift and drag coefficients using linearized theory.

**13.2.**  A thin symmetrical supersonic airfoil has parabolic upper and lower surfaces with a maximum thickness occurring at midchord. Using linearized theory, compute the drag coefficient on this airfoil when it is set at an angle of attack of 0°.

**13.3.**  The pressure coefficient at a certain point on a two-dimensional airfoil in a very low Mach number air flow is found to be −0.5. Using linearized theory, estimate the pressure coefficients that would exist at the same point on this airfoil in flows at Mach numbers of 0.5 and 0.8.

**13.4.**  A thin airfoil can be approximated as a flat plate. The airfoil is set at an angle of 10° to an air flow with a Mach number of 2, a temperature of −50°C, and a pressure of 50 kPa. Using linearized theory, find the pressures on the upper and lower surfaces of this wing.

**13.5.**  An airfoil has a triangular cross-sectional shape. The lower surface of the airfoil is flat and the ratio of the maximum thickness to the chord is 0.1. The maximum thickness occurs at a distance of 0.3 times the chord downstream of the leading edge. If this airfoil is placed with its lower surface at an angle of attack of 2° to an airflow in which the Mach number is 3, use linearized theory to determine the distribution of the pressure coefficient over the surface of the airfoil.

**13.6.**  A symmetrical double-wedge airfoil has a maximum thickness equal to 0.05 times the chord. This airfoil is placed at an angle of attack of 5° to an airstream with a Mach number of 2, a pressure of 50 kPa and a temperature of −50°C. Find the lift and drag acting on the airfoil using linearized theory and using shock wave and expansion wave results.

# CHAPTER 14

# Hypersonic Flow

## 14.1
## INTRODUCTION

Hypersonic flow was loosely defined in Chapter 1 as flow in which the Mach number is greater than about 5. No real reasons were given at that point as to why supersonic flows at high Mach numbers were different from those at lower Mach numbers. However, it is the very existence of these differences that really defines hypersonic flow. The nature of these hypersonic flow phenomena and, therefore, the real definition of what is meant by hypersonic flow, will be presented in the next section.

Hypersonic flows have, up to the present, mainly been associated with the reentry of orbiting and other high altitude bodies into the atmosphere. For example, a typical Mach number with altitude variation for a reentering satellite is shown in Fig. 14.1. It will be seen from this figure that because of the high velocity that the craft had to possess to keep it in orbit, very high

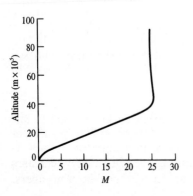

**FIGURE 14.1**
Typical variation of Mach number with altitude during reentry.

**FIGURE 14.2**
Proposed hypersonic passenger aircraft. (*Courtesy of McDonnell Douglas.*)

Mach numbers—values that are well into the hypersonic range—exist during reentry.

Discussions and studies of passenger aircraft that can fly at hypersonic speeds at high altitudes are also continuing. A typical proposed such vehicle is shown in Fig. 14.2.

The present chapter, which presents a brief introduction to hypersonic flow, is the first of three interrelated chapters. One of the characteristics of hypersonic flow is the presence of so-called "high temperature gas effects," and these effects will be discussed more fully in Chapter 15. Hypersonic flow is also conventionally associated with high altitudes where the air density is very low, and "low-density flows" will be discussed in Chapter 16.

## 14.2
## CHARACTERISTICS OF HYPERSONIC FLOW

As mentioned above, hypersonic flows are usually loosely described as flows at very high Mach numbers, say greater than about 5. However, the real definition of hypersonic flows is that they are flows at such high Mach numbers that phenomena occur that do not exist at low supersonic Mach numbers. These phenomena are discussed in this section.

One of the characteristics of hypersonic flow is the presence of an interaction between the oblique shock wave generated at the leading edge of the body and the boundary layer on the surface of the body. Consider the oblique shock wave formed at the leading edge of a wedge in a supersonic flow as shown in Fig. 14.3. As the Mach number increases, the shock angle decreases and the shock therefore lies very close to the surface at high Mach numbers.

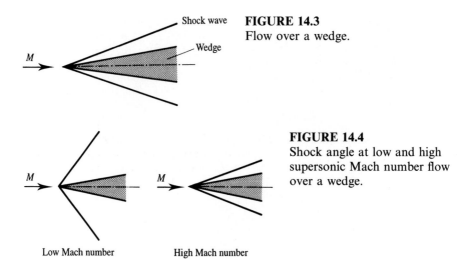

**FIGURE 14.3**
Flow over a wedge.

**FIGURE 14.4**
Shock angle at low and high supersonic Mach number flow over a wedge.

This is illustrated in Fig. 14.4. Because the shock wave lies close to the surface at high Mach numbers, there is an interaction between the shock wave and the boundary layer on the wedge surface. In order to illustrate this shock wave–boundary layer interaction, consider the flow of air over a wedge having a half-angle of 5° at various Mach numbers. The shock angle for any selected value of $M$ can be obtained from the oblique shock relations or charts or using the software (see Chapter 6). The angle between the shock wave and the wedge surface is then given by the difference between the shock angle and the wedge half-angle. The variation of this angle with Mach number is shown in Fig. 14.5. It will be seen from Fig. 14.5 that, as the Mach number increases, the shock wave lies closer and closer to the surface. Further, hypersonic flow normally only exists at relatively low ambient pressures which means that the Reynolds numbers tend to be low and the boundary layer thickness, therefore, tends to be relatively large. The boundary layer thickness also tends to increase with increasing Mach number. In hypersonic flow, then, the shock wave tends to lie close to the surface and the boundary layer tends to be thick.

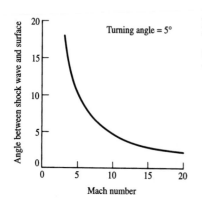

**FIGURE 14.5**
Variation of angle between shock wave and surface with Mach number for flow over a wedge.

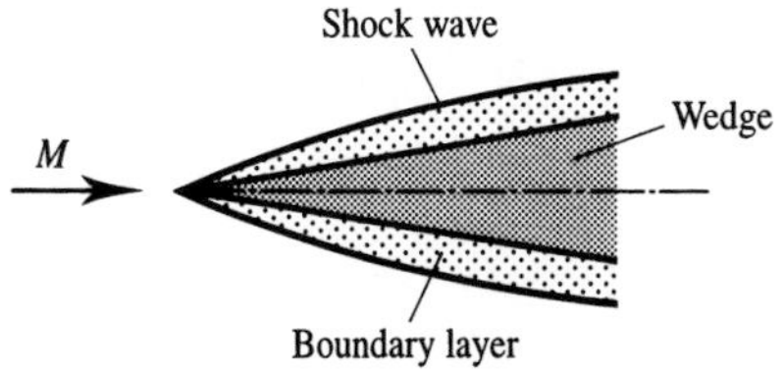

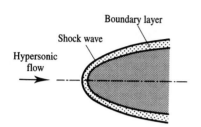

**FIGURE 14.6**
Interaction between shock wave and boundary layer in hypersonic flow over a wedge.

**FIGURE 14.7**
Interaction between shock wave and boundary layer in hypersonic flow over a curved body.

As a consequence, interaction between the shock wave and the boundary layer flow usually occurs, the shock being curved as a result and the flow resembling that shown in Fig. 14.6.

The above discussion used the flow over a wedge to illustrate interaction between the shock wave and the boundary layer flow in hypersonic flow. This interaction occurs, in general, for all body shapes as illustrated in Fig. 14.7.

Another characteristic of hypersonic flows is the high temperatures that are generated behind the shock waves in such flows. In order to illustrate this, consider the flow through a normal shock wave occurring ahead of a blunt body at a Mach number of 36 at an altitude of 59 km in the atmosphere. The flow situation is shown in Fig. 14.8. These were approximately the conditions that occurred during the reentry of some of the earlier manned spacecraft, the flow over such a craft being illustrated in Fig. 14.9. The flow situation shown in Fig. 14.8 is therefore an approximate model of the situation shown in Fig. 14.9.

Now, conventional relationships for a normal shock wave at a Mach number of 36 give:

$$\frac{T_1}{T_2} = 253$$

But at 59 km in atmosphere:

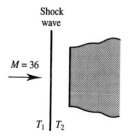

**FIGURE 14.8**
Normal shock wave in situation considered.

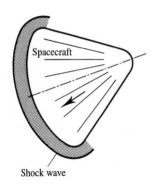

**FIGURE 14.9**
Flow over reentering spacecraft.

$$T = 258 \text{ K} \quad (\text{i.e.,} \; -15°\text{C})$$

Hence, the conventional normal shock wave relations give the temperature behind the shock wave as:

$$T_2 = 258 \times 253 = 65\,200 \text{ K}$$

At temperatures as high as these a number of so-called high temperature gas effects will become important. For example, the values of the specific heats, $c_p$ and $c_v$, and their ratio, $\gamma$, change at higher temperatures, their values depending on temperature. For example, the variation of $\gamma$ of nitrogen with temperature is shown in Fig. 14.10. It will be seen from this figure that changes in $\gamma$ may have to be considered at temperatures above about 500°C.

Another high-temperature effect arises from the fact that, at ambient conditions, air is made up mainly of nitrogen and oxygen in their diatomic form. At high temperatures, these diatomic gases tend to dissociate into their monoatomic form and at still higher temperatures, ionization of these monoatomic atoms tends to occur. Dissociation occurs under the following circumstances:

*For* 2000 K $< T <$ 4000 K: $O_2 \rightarrow 2O$, i.e., the oxygen molecules break down to O molecules

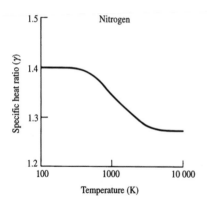

**FIGURE 14.10**
Variation of specific heat ratio of nitrogen with temperature.

*For* 4000 K < *T* < 9000 K: $N_2 \to 2N$, i.e., the nitrogen molecules break down to N molecules

When such dissociation occurs, energy is "absorbed." It should also be clearly understood that the range of temperatures given indicates that not all of the air is immediately dissociated once a certain temperature is reached. Over the temperature ranges indicated above the air will, in fact, consist of a mixture of diatomic and monoatomic molecules, the fraction of monatomic molecules increasing as the temperature increases.

Similarly, ionization occurs under the following circumstances:

*For T* > 9000 K:

$$O \to O^+ + e^-$$

$$N \to N^+ + e^-$$

When ionization occurs, energy is again "absorbed." There will again be a range of temperatures over which air will consist of a mixture of ionized and unionized atoms, the fraction of ionized atoms increasing as the temperature increases.

Other chemical changes can also occur at high temperatures, e.g., there can be a reaction between the nitrogen and the oxygen to form nitrous oxides at high temperatures. This and the other effects mentioned above are illustrated by the results given in Fig. 14.11. This figure shows the variation of the composition of air with temperature. It will be seen, therefore, that at high Mach numbers, the temperature rise across a normal shock may be high enough to cause specific heat changes, dissociation, and, at very high Mach numbers, ionization. As a result of these processes, conventional shock relations do not apply. For example, for the case discussed above, of a normal shock wave at a Mach number of 36 at an altitude of 59 km in the atmosphere,

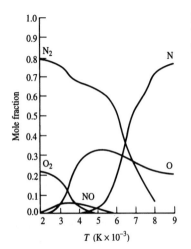

**FIGURE 14.11**
Variation of equilibrium composition of air with temperature.

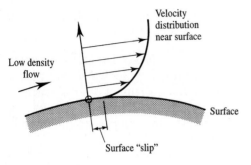

**FIGURE 14.12**
Surface slip in low-density flow.

as a result of these high temperature gas effects the actual temperature behind the shock wave is approximately:

$$T_2 \approx 11\,000 \text{ K}$$

rather than the value of 65 200 K indicated by the normal shock relations for a perfect gas.

There are several other phenomena that are often associated with high Mach number flow and whose existence helps define what is meant by a hypersonic flow. For example, as mentioned above, since most hypersonic flows occur at high altitudes, the presence of low density effects such as the existence of "slip" at the surface, i.e., of a velocity jump at the surface (see Fig. 14.12) is often taken as an indication that hypersonic flow exists.

The details of these high temperature and low density effects will be discussed in the next two chapters. The remainder of this chapter will be devoted to a discussion of an approximate method of calculating the pressures and forces acting on a surface placed in a hypersonic flow.

## 14.3
## NEWTONIAN THEORY

Although the details of the flow about a surface in hypersonic flow are difficult to calculate because of the complexity of the phenomena involved, the pressure distribution about a surface placed in a hypersonic flow can be estimated quite accurately using the approximate approach discussed below. Because the flow model used is essentially the same as one that was incorrectly suggested by Newton for the calculation of forces on bodies in incompressible flow, it is referred to as the "Newtonian model."

First, consider the flow over a flat surface inclined at an angle to a hypersonic flow. This flow situation is shown in Fig. 14.13. Only the flow over the upstream face of the surface will, for the moment, be considered. Because the shock wave lies so close to the surface in hypersonic flow, the flow will essentially be unaffected by the surface until the flow reaches the surface, i.e., until it "strikes" the surface, at which point it will immediately become

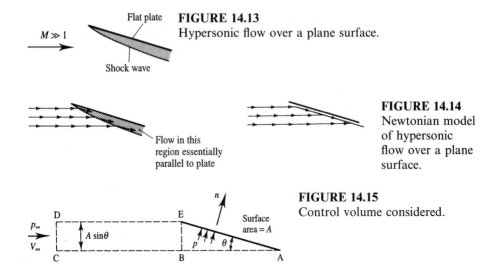

Flat plate

$M \gg 1$

Shock wave

**FIGURE 14.13**
Hypersonic flow over a plane surface.

Flow in this
region essentially
parallel to plate

**FIGURE 14.14**
Newtonian model
of hypersonic
flow over a plane
surface.

**FIGURE 14.15**
Control volume considered.

D                    E

$p_\infty$            Surface
                      area $= A$
$A \sin\theta$

$V_\infty$        $p$     $\theta$

C            B              A

$n$

parallel to the surface. Hence, the flow over the upstream face of a plane surface at hypersonic speeds resembles that shown in Fig. 14.14.

In order to find the pressure on the surface, consider the momentum equation applied to the control volume shown in Fig. 14.15. Because the flow at the surface is all assumed to be turned parallel to the surface, no momentum leaves the control volume in the $n$ direction, so the force on the control volume in this direction is equal to the product of the rate mass enters the control volume and the initial velocity component in the $n$ direction, i.e., is given by:

$$\text{Mass flow rate} \times \text{Velocity in } n \text{ direction} = (\rho_\infty V_\infty A \sin\theta)(V_\infty \sin\theta)$$

$$= \rho_\infty V_\infty^2 \sin^2\theta A$$

where $A$ is the area of the surface.

Now if $p$ is the pressure acting on the upstream face of the surface, the net force acting on the control volume in the $n$ direction is given by:

$$pA - p_\infty A$$

In deriving this result, it has been noted that since the flow is not affected by the presence of the surface until it effectively reaches the surface, the pressure on ABCDE (see Fig. 14.15) is everywhere equal to $p_\infty$ and that the forces on BC and DE are therefore equal and opposite and cancel. The force on AB is $p_\infty A \cos\theta$ which has a component in the direction normal to the surface of $p_\infty A \cos^2\theta$. Similarly the force on CD is $p_\infty A \sin\theta$, which has a component in the direction normal to the surface of $p_\infty A \sin^2\theta$. Hence, the sum of the forces on AB and CD in the direction normal to the surface is $p_\infty A \cos^2\theta + p_\infty A \sin^2\theta$, i.e., equal to $p_\infty A$.

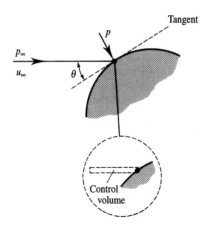

**FIGURE 14.16**
Control volume considered in dealing with flow over a curved surface.

Combining the above two results then gives:

$$p - p_\infty = \rho_\infty V_\infty^2 \sin^2 \theta \qquad (14.1)$$

This result can be expressed in terms of a dimensionless pressure coefficient, defined as before by:

$$C_p = \frac{p - p_\infty}{\frac{1}{2}\rho_\infty V_\infty^2} \qquad (14.2)$$

to give:

$$C_p = 2 \sin^2 \theta \qquad (14.3)$$

From the above analysis it therefore follows that the pressure coefficient is determined only by the angle of the surface to the flow.

The above analysis was for flow over a flat surface. However, it will also apply to a small portion of a curved surface such as that shown in Fig. 14.16.

Therefore, the local pressure acting at any point on the surface will be given as before by:

$$p - p_\infty = \rho_\infty V_\infty^2 \sin^2 \theta$$

from which it can be deduced that:

$$C_p = 2 \sin^2 \theta \qquad (14.4)$$

where $\theta$ is the local angle of inclination of the surface and $C_p$ is the local pressure coefficient.

Equation (14.4) can be written as:

$$\frac{p - p_\infty}{p_\infty} = \frac{\rho_\infty}{p_\infty} V_\infty^2 \sin^2 \theta$$

Hence, since $a_\infty^2 = \gamma p_\infty / \rho_\infty$, this equation gives:

$$\frac{p - p_\infty}{p_\infty} = \gamma M_\infty^2 \sin^2 \theta$$

i.e.,

$$\frac{p}{p_\infty} = 1 + \gamma M_\infty^2 \sin^2 \theta \qquad (14.5)$$

**EXAMPLE 14.1**

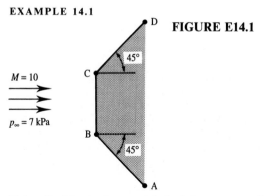

**FIGURE E14.1**

Using the Newtonian model, find the pressure distribution on the body shown in Fig. E14.1. The flow is two-dimensional and the Mach number and the pressure in the freestream ahead of the body are 10 and 7 Pa respectively.

**Solution**

Because each surface of the body is flat, the pressure is constant along each surface. Now, as shown above, for any surface:

$$p = p_\infty [1 + \gamma M_\infty^2 \sin^2 \theta]$$

For surface BC, $\theta = 90°$ while for surfaces AB and CD, $\theta = 45°$. Hence:

$$p_{AB} = p_{CD} = 7[1 + 1.4 \times 10^2 \times \sin^2 45°] = 497 \text{ Pa}$$

and:

$$p_{BC} = 7[1 + 1.4 \times 10^2 \times \sin^2 90°] = 987 \text{ Pa}$$

## 14.4
## MODIFIED NEWTONIAN THEORY

Consider hypersonic flow over a symmetrical body of arbitrary shape such as is shown in Fig. 14.17. At any point on the surface, as shown above, the pressure is given by:

$$p - p_\infty = \rho_\infty V_\infty^2 \sin^2 \theta$$

Hence at the "stagnation" point where $\theta = 90°$ and where, therefore, $\sin \theta = 1$ the pressure, $p_S$, is given by:

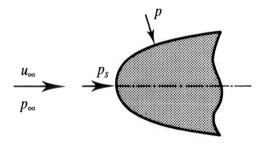

**FIGURE 14.17**
Form of body being considered.

$$p_S - p_\infty = \rho_\infty V_\infty^2$$

Hence:
$$\left(\frac{p_S - p_\infty}{\frac{1}{2}\rho_\infty V_\infty^2}\right) = 2$$

i.e., the pressure coefficient at the stagnation point is given by:

$$C_{ps} = 2 \tag{14.6}$$

From these relations it follows that the pressure distribution about the surface can be written as:

$$\frac{C_p}{C_{ps}} = \sin^2 \theta \tag{14.7}$$

or as:

$$\frac{p - p_\infty}{p_S - p_\infty} = \sin^2 \theta \tag{14.8}$$

Now the Newtonian theory does not really apply near the stagnation point. However, the shock wave in this region is, as previously discussed, effectively a normal shock wave and, therefore, the pressure on the surface at the stagnation point can be found using normal shock relations and then the Newtonian relation can be used to determine the pressure distribution around the rest of the body. This means, for example, that eq. (14.7) can be written as:

$$\frac{C_p}{C_{ps_N}} = \sin^2 \theta \tag{14.9}$$

where $C_{ps_N}$ is the pressure coefficient at the stagnation point as given by the normal shock relations. This is, basically, the modified Newtonian equation.

Now it will be recalled from Chapter 5 that the normal shock relations give:

$$\frac{p_S}{p_\infty} = \frac{[(\gamma + 1)M_\infty^2/2]^{\gamma/(\gamma-1)}}{[2\gamma M_\infty^2/(\gamma + 1) - (\gamma - 1/\gamma + 1)]^{1/(\gamma-1)}} \tag{14.10}$$

i.e., since:

$$C_p = \frac{p - p_\infty}{\frac{1}{2}\rho_\infty V_\infty^2}$$

$$= \frac{(p/p_\infty) - 1}{\gamma M_\infty^2/2} \tag{14.11}$$

Therefore, the normal shock relation (14.10) gives:

$$C_{ps_N} = \frac{[(\gamma + 1)M_\infty^2/2]^{\gamma/(\gamma-1)}}{[2\gamma M_\infty^2/(\gamma + 1) - (\gamma - 1)/(\gamma + 1)]^{1/(\gamma-1)}[\gamma M_\infty^2/2]} \tag{14.12}$$

If $M_\infty$ is very large the above equation tends to:

$$C_{ps_N} = \frac{[(\gamma + 1)/2]^{\gamma/(\gamma-1)}}{[2\gamma/(\gamma + 1)]^{1/(\gamma-1)}[\gamma/2]} \tag{14.13}$$

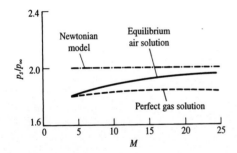

**FIGURE 14.18**
Typical variation of stagnation point pressure coefficient with Mach number.

For $\gamma = 1.4$ this gives the limiting value of $C_{ps_N}$ for large values of $M_\infty$ as 1.839. Hence, assuming a perfect gas and a large freestream Mach number, the modified Newtonian theory gives:

$$C_p = 1.839 \sin^2 \theta \qquad (14.14)$$

As discussed in the first section of this chapter, when the Mach number is very large, the temperature behind the normal shock wave in the stagnation point region becomes so large that high temperature gas effects become important and these affect the value of $C_{ps_N}$. The way in which the flow behind a normal shock wave is calculated when these high temperature gas effects become important will be discussed in the next chapter. The relation between the perfect gas normal shock results, the normal shock results with high temperature effects accounted for, and the Newtonian result is illustrated by the typical results shown in Fig. 14.18.

The results shown in Fig. 14.18 and similar results for other situations indicate that the stagnation pressure coefficient given by the high Mach number form of the normal shock relations for a perfect gas, i.e., eq. (14.13), applies for Mach numbers above about 5 and that it gives results that are within 5 percent of the actual values for Mach numbers up to more than 10. Therefore, the modified Newtonian equation, as given in eq. (14.9), using the high Mach number limit of the perfect gas normal shock to give the stagnation point pressure coefficient, will give results that are of adequate accuracy for values of $M_\infty$ up to more than 10. At higher values of $M_\infty$, the unmodified Newtonian equation gives more accurate results than this form of the modified equation. Of course, the modified Newtonian equation with the stagnation pressure coefficient determined using high temperature normal shock results will apply at all hypersonic Mach numbers.

It should be noted that eq. (14.7) gives:

$$\frac{p - p_\infty}{p_\infty} = C_{ps} \frac{1}{2} \frac{\rho_\infty}{p_\infty} V_\infty^2 \sin^2 \theta$$

i.e., again using $a_\infty^2 = \gamma p_\infty / \rho_\infty$ the above equation gives:

$$\frac{p - p_\infty}{p_\infty} = C_{ps} \frac{\gamma}{2} M_\infty^2 \sin^2 \theta$$

i.e.,
$$\frac{p}{p_\infty} = 1 + C_{ps} \frac{\gamma}{2} M_\infty^2 \sin^2 \theta \qquad (14.15)$$

**EXAMPLE 14.2**
Consider two-dimensional flow over a 30 cm thick body with a circular leading
edge that is moving through air at a Mach number of 14, the ambient air pressure
being 2 Pa. Using the modified Newtonian model in conjunction with the strong
shock limit of the perfect gas relation for a normal shock, find the pressure
variation around the leading edge in terms of the distance from the stagnation
point.

**Solution**

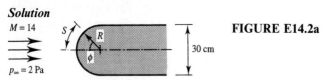

FIGURE E14.2a

It was shown above that the strong shock solution for air gives $C_{ps} = 1.839$.
Hence, the modified Newtonian model is assumed to give:

$$\frac{p}{p_\infty} = 1 + \frac{1.839 \times 1.4}{2} M_\infty^2 \sin^2 \theta$$

$$= 1 + 1.287 M_\infty^2 \sin^2 \theta$$

In the present situation, it will be seen from Fig. E14.2a that if $S$ is the
distance around the leading edge from the stagnation point, then $\theta = \pi/2 - \phi$
so $\sin \theta = \cos \phi$. Hence, since $\phi = S/R$, the pressure variation is given by:

$$p = p_\infty \left[ 1 + 1.287 M_\infty^2 \cos^2 \left( \frac{S}{0.15} \right) \right]$$

$$= 2 \left[ 1 + 1.287 \times 14 \times 14 \cos^2 \left( \frac{S}{0.15} \right) \right]$$

$$= 2 + 504.6 \cos^2 \left( \frac{S}{0.15} \right)$$

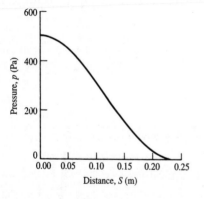

FIGURE E14.2b

It has been noted that the radius of the leading edge, $R$, is equal to half the thickness of the body, i.e., 30/2 cm.

Now $S$ varies from 0 to $\pi R/2$, i.e., from 0 to 0.2356 m around the leading edge. The pressure variation is therefore as shown in Table E14.2b and plotted in Fig. E14.2b.

**TABLE E14.2**

| $S$ (m) | $p$ (Pa) |
| --- | --- |
| 0.00 | 506.6 |
| 0.02 | 497.7 |
| 0.04 | 471.6 |
| 0.06 | 430.1 |
| 0.08 | 376.2 |
| 0.10 | 313.7 |
| 0.12 | 246.9 |
| 0.14 | 180.7 |
| 0.16 | 119.7 |
| 0.18 | 68.3 |
| 0.20 | 29.9 |
| 0.22 | 7.5 |
| 0.2356 | 2.0 |

## 14.5
## FORCES ON A BODY

The Newtonian or modified Newtonian model gives the pressure distribution on the upstream faces (e.g., faces AB and BC of the two-dimensional wedge-shaped body shown in Fig. 14.19) of a body in a hypersonic flow to an accuracy that is acceptable for many purposes. To find the net force acting on a body it is also necessary to know the pressures acting on the downstream faces of the body (e.g., face AC of the body shown in Fig. 14.19).

Now, as discussed above, in hypersonic flow, it is effectively only when the flow reaches the surface that it is influenced by the presence of the surface. The flow that does not reach the surface is therefore unaffected by the body. The flow leaving the upstream faces of the body therefore turns parallel to the original flow. Since the flow is then all parallel to the original flow direction and since the pressure in the outer part of the flow that was not affected by the presence of the body is $p_\infty$, the pressure throughout this downstream flow will be $p_\infty$. From this it follows that the pressure acting on the downstream faces of body in Newtonian hypersonic flow is $p_\infty$. This is illustrated in Fig. 14.20. The downstream faces on which the pressure is $p_\infty$ are often said to lie in the "shadow of the upstream flow."

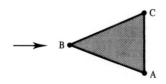

**FIGURE 14.19**
Two-dimensional flow over a wedge-shaped body in hypersonic flow.

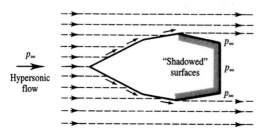

**FIGURE 14.20**
"Shadowed" areas of a body in hypersonic flow.

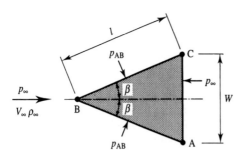

**FIGURE 14.21**
Pressures acting on faces of wedged-shaped body.

In calculating the forces on a body in hypersonic flow using the Newtonian or modified Newtonian model, the pressure will, therefore, be assumed to be $p_\infty$ on the downstream or "shadowed" portions of the body surface. There are more rigorous and elegant methods of arriving at this assumption, but the above discussion gives the basis of the argument.

To illustrate how the pressure drag force on a body is calculated using the Newtonian or modified Newtonian approach, consider again flow over a two-dimensional wedge-shaped body shown in Fig. 14.21. The force on face AB of the body per unit width is equal to $p_{AB}l$. This contributes $p_{AB}l \sin \beta$ to the drag. But $l \sin \beta$ is equal to $W/2$, i.e., equal to the projected area of face AB. Hence the pressure force on AB contributes $p_{AB} W/2$ to the drag. Because the wedge is symmetrically placed with respect to the freestream flow, the pressure on BC will be equal to that on AB so the pressure force on BC will also contribute $p_{AB} W/2$ to the drag. Therefore, since AC is a shadowed surface on which the pressure is assumed to be $p_\infty$, the drag on the wedge per unit width is given by:

$$D = 2\frac{p_{AB} W}{2} - p_\infty W = (p_{AB} - p_\infty)W \tag{14.16}$$

Now the drag coefficient for the type of body being considered is defined by:

$$C_D = \frac{D}{\frac{1}{2}\rho_\infty V_\infty^2 \text{ projected area}}$$

but since unit width is being considered, the projected area normal to the freestream flow direction is equal to $W$ hence:

$$C_D = \frac{D}{\frac{1}{2}\rho_\infty V_\infty^2 W} \tag{14.17}$$

Combining eqs. (14.16) and (14.17) then gives:

$$C_D = \frac{p_{AB} - p_\infty}{\frac{1}{2}\rho_\infty V_\infty^2}$$

Using eq. (14.8) this equation then gives:

$$C_D = \frac{(p_S - p_\infty)\sin^2\beta}{\frac{1}{2}\rho_\infty V_\infty^2} = C_{ps}\sin^2\beta \tag{14.18}$$

where $\beta$ is the half-angle of the wedge.

The Newtonian model gives $C_{ps} = 2$ so the Newtonian model gives for flow over a wedge:

$$C_D = 2\sin^2\beta \tag{14.19}$$

It must be stressed that the above analysis only gives the pressure drag on the surface. In general, there will also be a viscous drag on the body. However, if the body is relatively blunt, i.e., if the wedge angle is not very small, the pressure drag will be much greater than the viscous drag.

To illustrate how the drag on an axisymmetric body is calculated, consider flow over a conical body symmetrically placed with respect to the freestream. Consider a small portion of the surface of this body on the forward face of this body and the equivalent small portion of the surface on the shadowed face as shown in Fig. 14.22. The force on the foward-facing section is equal to $p_{AB}2\pi r\, dS$. This contributes an amount equal to $p_{AB}2\pi r\, dS \sin\theta$ to the drag. But $2\pi r\, dS \sin\beta$ is equal to $2\pi r\, dr$, i.e., equal to the projected area of the section. Hence, since the pressure on the shadowed wall section is assumed to be $p_\infty$, the net contribution to the drag force by the two wall sections is given by:

$$dD = (p_{AB} - p_\infty)2\pi r\, dr$$

This can be integrated to get the total drag force acting on the cone. This gives:

$$D = \int_0^R (p_{AB} - p_\infty)2\pi r\, dr$$

i.e., since $p_{AB}$ is a constant:

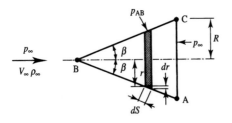

**FIGURE 14.22**
Surface sections used in determining drag force on a conical body.

$$D = (p_{AB} - p_\infty)\pi R^2 \tag{14.20}$$

Using eq. (14.8), this gives:

$$D = (p_S - p_\infty) \sin^2 \beta \pi R^2 \tag{14.21}$$

The drag coefficient based on the projected frontal area $\pi R^2$ is therefore given by:

$$C_D = \frac{(p_S - p_\infty) \sin^2 \beta}{\frac{1}{2} \rho_\infty V_\infty^2} = C_{ps} \sin^2 \beta \tag{14.22}$$

Again, since the Newtonian model gives $C_{ps} = 2$, the Newtonian model gives for flow over a cone:

$$C_D = 2 \sin^2 \beta \tag{14.23}$$

### EXAMPLE 14.3

Using the modified Newtonian model and using the limiting perfect gas normal shock result to get the stagnation point pressure, derive an expression for the drag on a cylinder whose axis is normal to a hypersonic air flow. Use this result to find the drag force per m span on a 0.1 m diameter cylinder moving at a Mach number of 8 through air at an ambient pressure of 10 Pa.

*Solution*

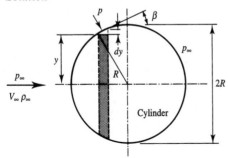

**FIGURE E14.3**

Considering the surface sections shown in Fig. E14.3, the net contribution to the drag force on the cylinder resulting from the pressures acting on the two surface sections is equal to $(p - p_\infty) \times$ projected area of surface sections, i.e., equal to $(p - p_\infty) \, dy$. The total drag on the cylinder is therefore given by:

$$D = 2 \int_0^R (p - p_\infty) \, dy$$

Now the modified Newtonian model gives using the normal shock relations to obtain the stagnation point pressure:

$$\frac{p - p_\infty}{p_{SN} - p_\infty} = \sin^2 \beta$$

i.e., since $\sin \beta = \sqrt{R^2 - y^2}/R$, it follows that:

$$\frac{p - p_\infty}{p_{SN} - p_\infty} = \frac{R^2 - y^2}{R^2}$$

Hence:
$$D = 2 \int_0^R (p_{SN} - p_\infty)\left(\frac{R^2 - y^2}{R^2}\right) dy$$

$$= 2(p_{SN} - p_\infty)\frac{2R}{3}$$

Since the frontal area of the cylinder per unit span is $2R$, the above result can be written in terms of the drag coefficient as:

$$C_D = \tfrac{2}{3} C_{ps_N} = \frac{2 \times 1.839}{3} = 1.226$$

it having been noted that the strong shock solution for air gives $C_{ps_N} = 1.839$ for air.

Now:

$$D = C_D \tfrac{1}{2} \rho_\infty V_\infty^2 2R$$

$$= C_D \frac{p_\infty}{2} \frac{\rho_\infty V_\infty^2}{p_\infty} 2R$$

$$= C_D \frac{p_\infty}{2} \gamma M_\infty^2 2R$$

Hence since $C_D$ was shown to be 1.226 according to the form of the modified Newtonian model adopted, using the specified conditions, the drag force on the cylinder is given by:

$$D = 1.226 \times 10 \times 1.4 \times 8^2 \times 0.1 = 109.9 \text{ N}$$

**EXAMPLE 14.4**
Using the Newtonian model, derive an expression for the drag on a sphere. Use this result to find the drag force on 30 cm diameter sphere moving at a Mach number of 10 through air at an ambient pressure of 0.03 kPa.

**Solution**

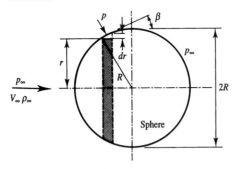

FIGURE E14.4

Considering the surface sections shown in Fig. E14.4, the net contribution to the drag force on the sphere resulting from the pressures acting on the two surface sections is equal to $(p - p_\infty) \times$ projected area of surface sections, i.e., equal to $(p - p_\infty)2\pi r \, dr$. The total drag on the sphere is therefore given by:

$$D = \int_0^R (p - p_\infty)2\pi r \, dr$$

Now the Newtonian model gives:

$$\frac{p - p_\infty}{p_S - p_\infty} = \sin^2 \beta$$

i.e., since $\sin \beta = \sqrt{R^2 - r^2}/R$, it follows that:

$$\frac{p - p_\infty}{p_S - p_\infty} = \frac{R^2 - r^2}{R^2}$$

Hence:
$$D = \int_0^R (p_S - p_\infty) 2\pi r \left( \frac{R^2 - r^2}{R^2} \right) dr$$

$$= (p_S - p_\infty) \pi \frac{R^2}{2}$$

Since the frontal area of the sphere is $\pi R^2$, the above result can be written in terms of the drag coefficient as:

$$C_D = \frac{C_{ps}}{2} = 1$$

it having again been noted that the Newtonian model gives $C_{ps} = 2$.

Now:
$$D = C_D \tfrac{1}{2} \rho_\infty V_\infty^2 \pi R^2$$

$$= C_D \frac{p_\infty}{2} \frac{\rho_\infty V_\infty^2}{p_\infty} \pi R^2$$

$$= C_D \frac{p_\infty}{2} \gamma M_\infty^2 \pi R^2$$

Hence since $C_D$ was shown to be 1 according to the Newtonian model, using the specified conditions, the drag force on the sphere is given by:

$$D = 1 \times 30 \times 1.4 \times 10^2 \times \pi 0.3^2 = 1187.5 \text{ N}$$

## 14.6
## CONCLUDING REMARKS

In hypersonic flow, because the temperatures are very high and because the shock waves lie close to the surface, the flow field is difficult to compute. However, because the flow behind the shock waves is all essentially parallel to the surface, the pressure variation along a surface can be easily estimated using the Newtonian model. Modifications to this model have been discussed and the calculation of drag forces on bodies in hypersonic flow using this method has been considered.

## PROBLEMS

**14.1.** A flat plate is set at an angle of 3° to an air flow at a Mach number of 8 in which the pressure is 1 kPa. Estimate the pressure acting on the lower surface of this plate.

**14.2.** Air, at a pressure of 10 Pa, flowing at a Mach number of 8 passes over a body which has a semicircular leading edge with a radius of 0.15 m. Assuming the flow to be two-dimensional, find the pressure acting on this nose portion of the body at a distance of 0.1 m around the surface measured from the leading edge of the body.

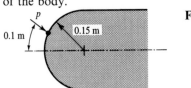

**FIGURE P14.2**

**14.3.** Using the Newtonian model, estimate the pressures acting on surfaces 1, 2, and 3 of the body shown in Fig. P14.3.

**FIGURE P14.3**

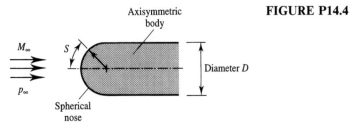

**14.4.** Consider hypersonic flow over the body shape indicated in Fig. P14.4. Using Newtonian theory, derive an expression for the pressure distribution around the surface of this body in terms of the distance from the stagnation point, $S$.

Axisymmetric body                    **FIGURE P14.4**

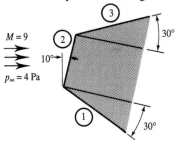

**14.5.** Consider two-dimensional air flow at a Mach number of 7 over the body shown in Fig. P14.5. The pressure in the flow ahead of the body is 12 Pa. Using the Newtonian method, find the pressures acting on the surfaces 1, 2, and 3 indicated in the figure.

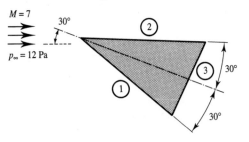

**FIGURE P14.5**

**14.6.** A wedge-shaped body has an included angle of 40° and a base width of 1.5 m. Using the Newtonian model and assuming two-dimensional flow, find the drag on the wedge per m width when it is moving through air in which the ambient pressure is 10 Pa at a Mach number of 7.

**14.7.** The axisymmetric body shown in Fig. P14.7 is an approximate model of some earlier spacecraft. Using the Newtonian model, derive an expression for the drag coefficient for this body in hypersonic flow.

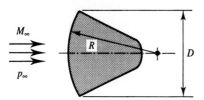

**FIGURE P14.7**

## CHAPTER 15

# High Temperature Flows

## 15.1
## INTRODUCTION

Chapter 1 contains a discussion of some of the assumptions that are commonly adopted in the analysis of compressible gas flows, i.e., of the assumptions that are commonly used in modeling such flows. It was explained in Chapter 1 that most such analyses are based on the assumptions that:

- The specific heats of the gas are constant
- The perfect gas law, $p/\rho = RT$, applies
- There are no changes in the physical nature of the gas in the flow
- The gas is in thermodynamic equilibrium

However, as discussed in Chapter 14, if the temperature in the flow becomes very high, it is possible that some of these assumptions may cease to be valid. To investigate again (a discussion of this was also given in Chapter 14) whether it is possible to get such high temperatures in a flow, consider the flow of air through a normal shock wave. The air will be assumed to have a temperature of 216.7 K (i.e., $-56.3°C$) ahead of the shock. This is the temperature in the so-called standard atmosphere between an altitude of approximately 11 000 m and approximately 25 000 m. The situation considered is, therefore, as shown in Fig. 15.1($a$).

If the assumptions discussed above apply, it was shown in Chapter 5 that:

$$\frac{T_2}{216.7} = \frac{[2\gamma M_1^2 - (\gamma - 1)][2 + (\gamma - 1)M_1^2]}{(\gamma + 1)^2 M_1^2}$$

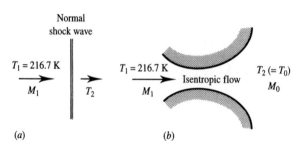

**FIGURE 15.1**
Flow situations considered:
(a) normal shock;
(b) isentropic deceleration.

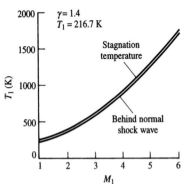

**FIGURE 15.2**
Mach number dependence of temperature behind normal shock wave and in isentropic flow considered.

Hence, since $\gamma = 1.4$ for air, the temperature downstream of the shock will be given by:

$$\frac{T_2}{216.7} = \frac{(2.8M_1^2 - 0.4)(2 + 0.4M_1^2)}{5.76M_1^2} \tag{15.1}$$

The variation of $T_2$ with $M_1$ given by this equation is shown in Fig. 15.2.

If instead of passing through a normal shock, the flow is brought to rest isentropically (see Fig. 15.1(b)), the temperature attained, i.e., the stagnation temperature, is given by:

$$\frac{T_2}{216.7} = 1 + \frac{\gamma - 1}{2}M_1^2$$

i.e., since the case of $\gamma = 1.4$ is being considered:

$$\frac{T_2}{216.7} = 1 + 0.2M_1^2 \tag{15.2}$$

The variation of $T_2$ with $M_1$ given by the equation is also shown in Fig. 15.2.

It will be seen from Fig. 15.2 that at Mach numbers of roughly 5 or greater, $T_2$ exceeds 1000°C in both types of flow considered. When temperatures as high as this exist in a flow, it seems prudent to investigate the applicability of the assumptions on which the analysis of the flow is based, i.e., in particular, to consider whether at such temperatures the specific heats can still be assumed to be independent of temperature, whether the perfect gas law is still applicable, and whether dissociation of the gas molecules and, perhaps, even ionization of the atoms is likely to occur. These effects will be examined in this chapter. However, only a brief introduction to the very important topic of high temperature gas flows can be given in this chapter despite the fact that such flows occur in a number of situations of great practical importance.

## 15.2
## EFFECT OF TEMPERATURE ON SPECIFIC HEATS

The first high temperature gas effect considered here is the possibility of changes in the specific heats of the gas at high temperatures. The increase of internal energy that results when the temperature of a gas is increased is associated with an increase in the energy possessed by the gas molecule. Now the increase in the energy of a molecule can be associated with an increase in the translational kinetic energy or with an increase in the rotational kinetic energy or with an increase in the vibrational kinetic energy of the molecule. This is illustrated in Fig. 15.3.

In addition, at high temperatures, changes in the energy associated with the electron motion can occur. Therefore:

$$\Delta e = \Delta e_{\text{trans}} + \Delta e_{\text{rot}} + \Delta e_{\text{vib}} + \Delta e_{\text{el}} \tag{15.3}$$

where $\Delta e$ is the change in internal energy, $\Delta e_{\text{trans}}$ is the change in translational energy, $\Delta e_{\text{rot}}$ is the change in rotational energy, $\Delta e_{\text{vib}}$ the change in vibrational energy, and $\Delta e_{\text{el}}$ is the change in electron energy. Measuring $e$ from 0 at absolute zero temperature this gives:

$$e = e_{\text{trans}} + e_{\text{rot}} + e_{\text{vib}} + e_{\text{el}} \tag{15.4}$$

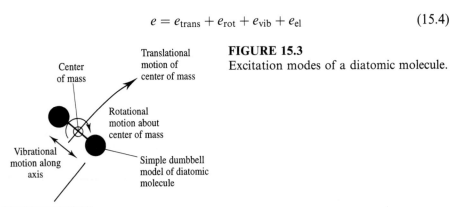

**FIGURE 15.3**
Excitation modes of a diatomic molecule.

The specific heat at constant volume, $c_v$, is given by:

$$c_v = \frac{\partial e}{\partial T} \tag{15.5}$$

So it follows that:

$$c_v = \frac{\partial e_{\text{trans}}}{\partial T} + \frac{\partial e_{\text{rot}}}{\partial T} + \frac{\partial e_{\text{vib}}}{\partial T} + \frac{\partial e_{\text{el}}}{\partial T} \tag{15.6}$$

The effect of $e_{\text{el}}$ can be neglected for most high temperature gas flow applications. This assumption will be adopted in the following analysis.

First consider a monatomic gas such as helium (He), argon (Ar), or neon (Ne). The molecules (atoms) of a monatomic gas have no rotational or vibrational energy so for such a gas:

$$c_v = \frac{\partial e_{\text{trans}}}{\partial T} \tag{15.7}$$

Statistical thermodynamics gives:

$$e_{\text{trans}} = \tfrac{3}{2} RT \tag{15.8}$$

So for a monatomic gas, eq. (15.7) gives:

$$c_v = \tfrac{3}{2} R \tag{15.9}$$

Since $c_p - c_v = R$ this gives:

$$c_p = \tfrac{5}{2} R \tag{15.10}$$

Dividing eq. (15.10) by eq. (15.9) then gives for a monatomic gas:

$$\gamma = \tfrac{5}{3} \tag{15.11}$$

Hence, the specific heats, and, therefore, the specific heat ratio, $\gamma$, of a monatomic gas, do not change with temperature unless the temperature is so high that the change in electron energy becomes important.

Gases with diatomic molecules, such as nitrogen ($N_2$) and oxygen ($O_2$) will next be considered. At very low temperatures, the rotational and vibrational modes are not excited and $c_v$ and $c_p$ have the same value as a monatomic gas. However, at temperatures above 100 K, the rotational mode becomes excited and $c_v$ increases. This is illustrated in Fig. 15.4. At higher temperatures, the vibrational mode becomes excited and $c_v$ then again increases. This is also shown in Fig. 15.4.

Approximate temperatures at which the rotational energy mode become fully excited for various diatomic gases are listed in Table 15.1. These results indicate that under all conditions conventionally encountered in gas flows, the rotational mode will be fully excited.

Now, statistical thermodynamics indicates that for a diatomic gas:

$$e_{\text{rot}} = RT \tag{15.12}$$

So for a diatomic gas:

$$e = \tfrac{3}{2} RT + RT + e_{\text{vib}} \tag{15.13}$$

Therefore, for a diatomic gas:

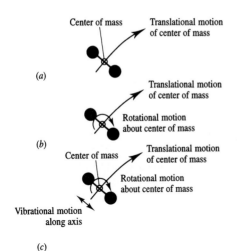

(a)

(b)

(c)

**FIGURE 15.4**
Effect of temperature on excitation modes of a diatomic molecule. (a) "low" temperatures; (b) "intermediate" temperatures; (c) "high" temperatures.

$$c_v = \tfrac{3}{2}R + R + \frac{\partial e_{\text{vib}}}{\partial T} \tag{15.14}$$

At temperatures below that at which the vibrational mode is excited, say, very roughly below 2000 K for air, this gives:

$$c_v = \tfrac{5}{2}R \tag{15.15}$$

Therefore, again using $c_p - c_v = R$, gives:

$$c_p = \tfrac{7}{2}R \tag{15.16}$$

These two equations then give:

$$\gamma = \tfrac{7}{5} = 1.4 \tag{15.17}$$

the value that has been used extensively for air in this book.

At higher gas temperatures, the vibrational mode becomes excited. A simple model of a diatomic molecule gives:

$$\frac{\partial e_{\text{vib}}}{\partial T} = R \left[ \frac{\theta_{\text{vib}}}{2} \right]^2 \frac{e^{\theta_{\text{vib}}/T}}{[e^{\theta_{\text{vib}}/T} - 1]^2} \tag{15.18}$$

**TABLE 15.1**
**Rotational excitational temperatures for various gases**

| Gas | Rotational excitation temperature (K) |
|---|---|
| Hydrogen ($H_2$) | 86 |
| Nitrogen ($N_2$) | 3 |
| Oxygen ($O_2$) | 2 |
| Carbon monoxide (CO) | 3 |

**TABLE 15.2**
$\theta_{vib}$ **values for various gases**

| Gas | $\theta_{vib}$ (K) |
|---|---|
| Hydrogen (H₂) | 6140 |
| Oxygen (O₂) | 2260 |
| Nitrogen (N₂) | 3340 |
| Carbon monoxide (CO) | 3120 |

This is an approximate equation and may not be adequate in some situations. It only applies to a diatomic gas. More complex equations are available for gases with more complex molecules. $\theta_{vib}$ depends on the type of gas involved and some approximate values are given in Table 15.2.

Substituting eq. (15.18) into eq. (15.14) then gives, if electronic energy is neglected:

$$c_v = \tfrac{5}{2}R + R\left[\frac{\theta_{vib}}{T}\right]^2 \frac{e^{\theta_{vib}/T}}{[e^{\theta_{vib}/T} - 1]^2} \tag{15.19}$$

Again using $c_p - c_v = R$, this equation indicates that if the vibration mode is excited:

$$c_p = \tfrac{7}{2}R + R\left[\frac{\theta_{vib}}{T}\right]^2 \frac{e^{\theta_{vib}/T}}{[e^{\theta_{vib}/T} - 1]^2} \tag{15.20}$$

Hence, using eqs. (15.19) and (15.20):

$$\gamma = 1.4\left\{\frac{1 + \frac{2}{7}\left[\frac{\theta_{vib}}{T}\right]^2 e^{\theta_{vib}/T}/[e^{\theta_{vib}/T} - 1]^2}{1 + \frac{2}{5}\left[\frac{\theta_{vib}}{T}\right]^2 e^{\theta_{vib}/T}/[e^{\theta_{vib}/T} - 1]^2}\right\} \tag{15.21}$$

A simple computer program, based on the above equations, that can be used to find the specific heats of a diatomic gas at any temperature is listed in Program 15.1.

**Program 15.1**

```
 10  CLS
 20  INPUT" THETA - VIB OF GAS IN K ";THV
 30  INPUT" MOL. MASS ";M
 40  R=8314.3/M
 50  PRINT" ":PRINT" ":PRINT" "
 60  INPUT" TEMPERATURE IN K (NEG. TO END) ";T
 70  IF T<0.0 THEN GOTO 280
 80  PRINT" "
 90  TR=THV/T
100  ETR=EXP(TR)
110  FAC=TR*TR*ETR/((ETR-1)^2)
120  CVR=5/2+FAC
130  CPR=7/2+FAC
140  GAM=CPR/CVR
150  CV=CVR*R
```

```
160  CP = CPR*R
170  PRINT" gamma = ";:PRINT USING "##.####";GAM
180  PRINT" CP/R = ";:PRINT USING "##.####"; CPR
190  PRINT" CV/R = ";:PRINT USING "##.####";CVR
200  PRINT" CP = ";:PRINT USING "#####.##";CP
210  PRINT" CV = ";:PRINT USING "#####.##";CV
220  PRINT" ":PRINT" ":PRINT" "
230  PRINT" ANY KEY TO CONTINUE" 240 A$ = ""
250  A$ = INKEY$
260  IF A$ = "" THEN GOTO 250
270  GOTO 50
280  STOP
290  END
```

It will be noted that if the temperature, $T$, is large, i.e., if $(\theta_{vib}/T)$ is small, then since $e^x$ tends to $1 + x$ for small $x$:

$$\left[\frac{\theta_{vib}}{T}\right]^2 \frac{e^{\theta_{vib}/T}}{[e^{\theta_{vib}/T} - 1]^2} \approx \frac{[e^{\theta_{vib}/T} - 1]^2}{e^{\theta_{vib}/T}} \approx 1 \tag{15.22}$$

Hence the above equations indicate that the following limiting values will be reached by a diatomic gas at high temperatures:

$$c_v = \tfrac{7}{2} R, \qquad c_p = \tfrac{9}{2} R, \qquad \gamma = \tfrac{9}{7} \tag{15.23}$$

Thus for a diatomic gas the specific heats are as follows:

*T very low:* $\qquad\qquad\qquad c_v = \tfrac{3}{2} R, \qquad c_p = \tfrac{5}{2} R, \qquad \gamma = \tfrac{5}{3}$

*T intermediate values:* $\qquad c_v = \tfrac{5}{2} R, \qquad c_p = \tfrac{7}{2} R, \qquad \gamma = \tfrac{7}{5}$ $\qquad$ (15.24)

*T high:* $\qquad\qquad\qquad\quad c_v = \tfrac{7}{2} R, \qquad c_p = \tfrac{9}{2} R, \qquad \gamma = \tfrac{9}{7}$

This is illustrated by the results for nitrogen shown in Fig. 15.5.

To illustrate the effect of vibrational energy on high temperature flows, consider the variation of the stagnation temperature with $M$. Since in gen-

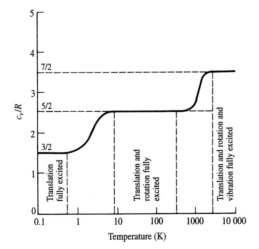

**FIGURE 15.5**
Specific heat variation with temperature for nitrogen.

erating the stagnation conditions the flow is brought to rest adiabatically, the energy equation gives:

$$h + \frac{V^2}{2} = h_0 \tag{15.25}$$

which gives:

$$\frac{V^2}{2} = h_0 - h$$

$$= \int_T^{T_0} c_p \, dT$$

$$= R \int_T^{T_0} \left[ \frac{7}{2} + \left( \frac{\theta_{vib}}{T} \right)^2 \frac{e^{\theta_{vib}/T}}{(e^{\theta_{vib}/T} - 1)^2} \right] dT$$

$$= R \left[ \frac{7}{2} + \frac{\theta_{vib}}{e^{\theta_{vib}/T} - 1} \right]\Bigg|_T^{T_0} \tag{15.26}$$

i.e., 
$$\frac{V^2}{2} = R \left\{ \tfrac{7}{2}(T_0 - T) + \theta_{vib} \left[ \frac{1}{e^{\theta_{vib}/T} - 1} - \frac{1}{e^{\theta_{vib}/T} - 1} \right] \right\} \tag{15.27}$$

The speed of sound is, as before, given by:

$$a^2 = \gamma R T$$

so the above equation can be written as:

$$M^2 = \left( \frac{2}{\gamma} \right) \left\{ \frac{7}{2} \left[ \frac{T_0}{T} - 1 \right] + \left( \frac{\theta_{vib}}{T} \right) \left[ \frac{1}{e^{\theta_{vib}/T_0} - 1} - \frac{1}{e^{\theta_{vib}/T} - 1} \right] \right\} \tag{15.28}$$

The specific heat ratio, $\gamma$, is given by eq. (15.21).

Equation (15.28) gives a relation between $T_0$, $M$, and $T$. For the situation discussed above, i.e., for flow with $T = 216.7$ K, it gives the variation of $T_0$ with $T$ that is shown in Fig. 15.6. Also shown in this figure is the variation given by assuming that $c_p$ is constant and $\gamma = 1.4$.

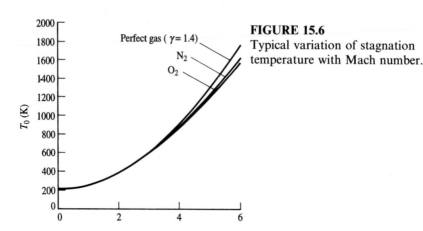

**FIGURE 15.6**
Typical variation of stagnation temperature with Mach number.

The vibrational excitation will be seen to have a significant effect on $T_0$ when $M$ is high.

The stagnation pressure can be found by noting that the entropy change between any two points in the flow is given by:

$$s_2 - s_1 = \int_{T_1}^{T_2} c_p \frac{dT}{T} - \int_{p_1}^{p_2} R \frac{dp}{p} \tag{15.29}$$

Since the flow being considered is isentropic, i.e., $s_2 - s_1 = 0$, and because $R$ is constant, this equation gives:

$$R \ln \left( \frac{p_0}{p} \right) = \int_{T}^{T_0} c_p \frac{dT}{T} \tag{15.30}$$

Using eq. (15.20) then gives:

$$\ln \left( \frac{p_0}{p} \right) = \int_{T}^{T_0} \left[ \frac{7}{2} + \left( \frac{\theta_{\text{vib}}}{T} \right)^2 \frac{e^{\theta_{\text{vib}}/T}}{(e^{\theta_{\text{vib}}/T} - 1)^2} \right] \frac{dT}{T}$$

$$= \frac{7}{2} \ln \left( \frac{T_0}{T} \right) + \left[ \left( \frac{\theta_{\text{vib}}}{T} \right) \frac{e^{\theta_{\text{vib}}/T}}{(e^{\theta_{\text{vib}}/T} - 1)} - \ln(e^{\theta_{\text{vib}}/T} - 1) \right] \Bigg|_{T}^{T_0}$$

i.e.,

$$\frac{p_0}{p} = \left[ \frac{T_0}{T} \right]^{7/2} \left[ \frac{e^{\theta_{\text{vib}}/T_0} - 1}{e^{\theta_{\text{vib}}/T} - 1} \right] \exp \left\{ \left( \frac{\theta_{\text{vib}}}{T_0} \right) \left[ \frac{e^{\theta_{\text{vib}}/T_0}}{e^{\theta_{\text{vib}}/T_0} - 1} \right] - \left( \frac{\theta_{\text{vib}}}{T} \right) \left[ \frac{e^{\theta_{\text{vib}}/T}}{e^{\theta_{\text{vib}}/T} - 1} \right] \right\} \tag{15.31}$$

Since $(T_0/T)$ is given by eq. (15.28), this equation allows $p_0$ to be found for any values of $M$, $T$, and $p$. If the density is required, it can then be found, using $p/\rho = RT$, i.e., using $\rho_0/\rho = (p_0/p)(T/T_0)$.

As another example of high temperature effects on gas flows consider the flow through a normal shock wave under such conditions that the effects of vibrational excitation are significant. The variables are defined in Fig. 15.7.

Explicit expressions for the changes across the shock wave will not be derived. Instead, a procedure that allows the downstream conditions to be found for any prescribed values of the upstream conditions will be discussed.

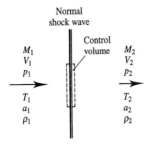

**FIGURE 15.7**
Flow through a normal shock wave.

Considering the flow through the control volume shown in Fig. 15.7, conservation of mass, momentum, and energy give, as before:

$$\rho_1 V_1 = \rho_2 V_2 \tag{15.32}$$

$$p_1 + \rho_1 V_1^2 = p_2 + \rho_2 V_2^2 \tag{15.33}$$

$$h_1 + \frac{V_1^2}{2} = h_2 + \frac{V_2^2}{2} \tag{15.34}$$

The continuity equation, eq. (15.32), gives:

$$V_2 = \rho_1 V_1 / \rho_2 \tag{15.35}$$

Substituting this into the momentum equation, eq. (15.33), then gives:

$$p_1 + \rho_1 V_1^2 = p_2 + \rho_1^2 V_1^2 / \rho_2$$

i.e.,

$$p_2 = p_1 + \rho_1 V_1^2 \left[ 1 - \frac{\rho_1}{\rho_2} \right] \tag{15.36}$$

Substituting eq. (15.35) into the energy equation, eq. (15.34), gives:

$$\frac{V_1^2}{2} \left[ 1 - \left( \frac{\rho_1}{\rho_2} \right)^2 \right] = h_2 - h_1 \tag{15.37}$$

But:

$$h_2 - h_1 = \int_{T_1}^{T_2} c_p \, dT \tag{15.38}$$

So using eq. (15.20) and the integral discussed in the derivation of equation (15.27), the following is obtained:

$$h_2 - h_1 = R \left\{ \tfrac{7}{2}(T_2 - T_1) + \theta_{\text{vib}} \left[ \left( \frac{1}{e^{\theta_{\text{vib}}/T_2} - 1} \right) - \left( \frac{1}{e^{\theta_{\text{vib}}/T_1} - 1} \right) \right] \right\} \tag{15.39}$$

Lastly it is noted that the perfect gas equation gives $p/\rho = RT$, i.e., gives:

$$\left( \frac{p_2}{p_1} \right) \left( \frac{\rho_1}{\rho_2} \right) = \left( \frac{T_2}{T_1} \right) \tag{15.40}$$

Equations (15.36), (15.37), (15.39), and (15.40) constitute a set of four simultaneous equations in the four unknowns $p_2$, $\rho_2$, $(h_2 - h_1)$, and $T_2$. These can be solved to give the values of the variables using, for example, an iterative technique. For example, $T_2$ can be guessed, e.g., it could be set equal to $2T_1$, and eq. (15.39) can then be used to solve for $h_2 - h_1$. Substituting this value into eq. (15.37) then allows the corresponding value of $\rho_2$ to be found. Substituting this into eq. (15.36) then gives $p_2$. Substituting these values of $p_2$ and $\rho_2$ into eq. (15.40) then gives a new value of $T_2$. The whole procedure can then be repeated until the values of $T_2$ ceases to change. Much more efficient and elegant solution procedures are available but that given here should indicate the basic ideas involved.

A simple computer program written in BASIC that implements this procedure is listed in Program 15.2. The inputs are $M_1$ and $T_1$. The program uses an under-relaxation procedure for stability, i.e., if $T_{2i}$ is the value of $T_2$ at any stage of the iteration and if $T_{2n}$ is the calculated value of $T_2$ at the end of the iteration, then the value of $T_2$ at the beginning of the next iteration is taken as:

$$T_2 = T_{2i} + r(T_{2n} - T_{2i})$$

where $r$ is a chosen factor that is less than 1. A very small value of $r$ has been used here. The program is intended for use for values of $M_1$ that are above roughly 3.

**Program 15.2**

```
10   CLS
20   INPUT" THETA - VIB OF GAS IN K ";THV
30   INPUT" MOL. MASS ";M
40   R=8314.3/M
50   PRINT" ":PRINT" ":PRINT" "
60   INPUT" UPSTREAM TEMPERATURE IN K (NEG. TO END) ";T1
70   IF T1<0.0 THEN GOTO 390
80   INPUT" UPSTREAM MACH NUMBER ";M1
90   PRINT" "
100  TR=THV/T1
110  ETR1=EXP(TR)
120  FAC=TR*TR*ETR1/((ETR1-1)^2)
130  CVR=5/2+FAC
140  CPR=7/2+FAC
150  GAM=CPR/CVR
160  CV=CVR*R
170  CP=CPR*R
180  TRP=((2.8*M1*M1-0.4)*(2+0.4*M1*M1))/(2.4*2.4*M1*M1)
190  T2P=T1*TRP
200  T2=2*T1
210  TR=THV/T2
220  ETR2=EXP(TR)
230  H21=R*(3.5*(T2-T1)+THV*((1/(ETR2-1))-(1/(ETR1-1))))
240  R12=SQR(1-2.0*H21/(GAM*R*T1*M1*M1))
250  P21=1+GAM*M1*M1*(1-R12)
260  TN=P21*R12*T1
270  TDIF=ABS(TN-T2)
280  IF TDIF<0.2 THEN GOTO 310
290  T2=T2+0.05*(TN-T2)
300  GOTO 210
310  PRINT" T2 = ";:PRINT USING "#####.####";T2;:PRINT" K"
320  PRINT" T2 PERFECT = ";:PRINT USING "#####.####";T2P;:PRINT" K"
330  PRINT" ":PRINT" ":PRINT" "
340  PRINT" ANY KEY TO CONTINUE"
350  A$=""
360  A$=INKEY$
370  IF A$="" THEN GOTO 360
380  GOTO 50
390  STOP
400  END
```

The program starts by finding the value of $\gamma_1$ using eq. (15.21). Equation (15.37) is then used in the form:

$$\left(\frac{M_1^2 \gamma R T_1}{2}\right)\left[1 - \left(\frac{\rho_1}{\rho_2}\right)^2\right] = h_2 - h_1$$

and $(\rho_1/\rho_2)$ is found.

Equation (15.36) is then used in the form:

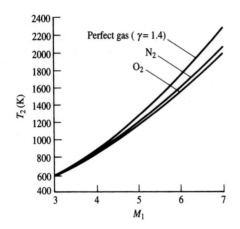

**FIGURE 15.8**
Results for a normal shock wave in air with $T_1 = 216.7$ K.

$$\frac{p_2}{p_1} = 1 + \gamma_1 M_1^2 \left(1 - \frac{\rho_1}{\rho_2}\right)$$

and eq. (15.40) is then used to determine a new value of $T_2$.

The program has been used to determine the variation of $T_2$ with $M_1$ for the case of $T_1 = 216.7$ K discussed before. The results are compared in Fig. 15.8 with those obtained before assuming constant specific heats.

All the other flows discussed earlier, such as isentropic flow through a nozzle, can be analyzed for the case where the vibrational excitation of a diatomic gas becomes important using the approach outlined above for this case. These analyses, and those discussed above, indicate that the effects of specific heat changes, caused by the excitation of the vibrational modes in diatomic gases and in gases with more complex molecules, are likely to be important in high Mach number flows.

## 15.3
## PERFECT GAS LAW

The perfect gas law gives:

$$\frac{p}{\rho R T} = 1 \tag{15.41}$$

Therefore the quantity:

$$Z = \frac{p}{\rho R T} \tag{15.42}$$

which is termed the "compressibility factor," can be used as a measure of the deviation of the behavior of the gas from that of a perfect gas. For a gas that obeys the perfect gas law $Z = 1$.

One of the most widely discussed gas equation that accounts for the deviations from eq. (15.41) is that due to van der Waals. It was derived by trying to account for the modification of the wall pressure in a gas that results from the net attraction exerted on a molecule near the wall by the rest of the molecules in the gas and by trying to account for the volume taken up by the gas molecules in a dense gas. The former effect is accounted for by altering the pressure by $a/v^2$, where $a$ is a constant that depends on the type of gas and $v$ is the specific volume, whereas the latter effect is accounted for by replacing the specific volume in the perfect gas equation by $(v - b)$, $b$ also being a constant that depends on the type of gas. The van der Waals modified gas equation is then:

$$\left(p + \frac{a}{v^2}\right)(v - b) = RT \tag{15.43}$$

Values of $a$ and $b$ for various gases have been experimentally determined by fitting eq. (15.43) to the measured variation of $v$ with $p$ and $T$.

The van der Waals equation can be written in terms of the density $\rho(= 1/v)$ as:

$$p = \frac{\rho RT}{1 - b\rho} + a\rho^2$$

Now, the van der Waals equation, eq. (15.43) defines the variation of $p$ with $v$ for a given value of $T$. The form of the variation for various values of $T$ is as shown in Fig. 15.9. For temperatures below a certain value, the curves display a minimum whereas for temperatures above this value, the curves have no minimum. The temperature that divides the two types of behavior is indicated in Fig. 15.9 as $T_c$. Now, on the $T_c$ curve, there will be a point, marked C in Fig. 15.9, at which the slope is zero, i.e., at which $dp/dv = 0$. Because the slope of the curve, $dp/dv$, is never positive on the $T_c$ curve, the slope has a maximum at point C, i.e., $d^2p/dv^2$ is 0 at point C. Thus point C is that at which $dp/dv = 0$ and $d^2p/dv^2 = 0$. Using the van der Waals equation, eq. (15.43), thus indicates that $p_c$ and $T_c$ are such that:

$$0 = \frac{RT_c}{(v_c - b)^2} + \frac{2a}{v_c^3} \tag{15.44}$$

$$0 = \frac{2RT_c}{(v_c - b)^3} - \frac{6a}{v_c^4}$$

Solving between these then gives:

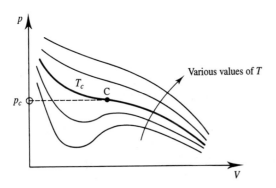

**FIGURE 15.9**
Isotherms given by the van der Waals equation.

$$a = \frac{27}{64}\frac{R^2 T_c^2}{p_c}, \qquad b = \frac{R T_c}{8 p_c} \qquad (15.45)$$

Substituting these values back into eq. (15.43) then gives:

$$\left(p + \frac{27}{64}\frac{R^2 T_c^2}{p_c v^2}\right)\left(v - \frac{R T_c}{8 p_c}\right) = RT$$

i.e.,

$$\left[\frac{pv}{RT} + \frac{27}{64}\left(\frac{RT}{pv}\right)\left(\frac{T_c}{T}\right)^2\left(\frac{p}{p_c}\right)\right]\left[1 - \frac{1}{8}\left(\frac{RT}{pv}\right)\left(\frac{T_c}{T}\right)\left(\frac{p}{p_c}\right)\right] = 1 \quad (15.46)$$

Therefore, defining:

$$p_r = \frac{p}{p_c}, \qquad T_r = \frac{T}{T_c} \qquad (15.47)$$

and recalling the definition of the factor $Z$ given in eq. (15.42), it will be seen that the van der Waals equation gives:

$$\left[Z + \frac{27}{64}\left(\frac{p_r}{Z T_r^2}\right)\right]\left[1 - \frac{1}{8Z}\frac{p_r}{T_r}\right] = 1 \qquad (15.48)$$

This equation indicates that:

$$Z = \text{function}(p_r, T_r) \qquad (15.49)$$

Although the van der Waals equation has not been found to give a very good description of the behavior of real gases over a wide range of temperatures, the form of behavior derived from this equation and given in eq. (15.49) does correlate the behavior of real gases very well. Values of $p_c$ and $T_c$ for various common gases are listed in Table 15.3.

The form of the variation of $Z$ with $p_r$ and $T_r$ for all gases obtained using such values of $p_c$ and $T_c$ is then as shown in Fig. 15.10. This graph, it must be stressed, applies quite accurately for almost all gases provided the correct values of $p_c$ and $T_c$ are used. It will be noted from Table 15.3 that for air, $p_c$ and $T_c$ are approximately 3910 kPa and 132 K respectively.

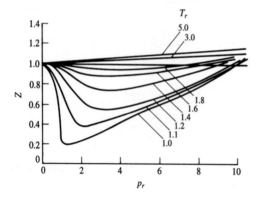

**FIGURE 15.10**
Variation of compressibility factor with reduced pressure, $p_r$.

**TABLE 15.3**
**Critical temperature and pressure values**

| Gas | $T_c$ (K) | $p_c$ (atm) |
|---|---|---|
| Air | 132.41 | 37.25 |
| Helium (He) | 5.19 | 2.26 |
| Hydrogen (H$_2$) | 33.24 | 12.80 |
| Nitrogen (N$_2$) | 126.2 | 33.54 |
| Oxygen (O$_2$) | 154.78 | 50.14 |
| Carbon monoxide (CO) | 132.91 | 34.26 |
| Carbon dioxide (CO$_2$) | 304.20 | 72.90 |

1 atmosphere (atm) = 101.3 kPa

For a perfect gas, $Z = 1$. It will be seen from Fig. 15.10, therefore, that the greatest deviations from perfect gas behavior occur when $T_r$ is near 1 and $p_r$ near 2. In most compressible flows encountered in practice, if $p_r$ is low $T_r$ also tends to be low and when $p_r$ is high, say of the order 2 to 3, $T_r$ tends also to be high. As a result, it will be seen that in such flows, $Z$ tends to be always near 1. This means that the use of the perfect gas equation will give good results in such flows.

If the circumstances are such that $Z$ is likely to deviate significantly from 1, a more complex equation has to be used to describe the gas behavior. The van der Waals equation discussed above is quite adequate for this purpose in many cases. However, other equations that provide a better description of the gas behavior over wider ranges of pressure and temperature have been developed. The Beattie–Bridgeman equation is typical of these equations. It gives:

$$p = \frac{RT(1-\epsilon)}{v^2}(v+B) - \frac{A}{v^2} \qquad (15.50)$$

where:

$$A = A_0(1 - a/v)$$

$$B = B_0(1 - b/v) \qquad (15.51)$$

$$\epsilon = c/vT^3$$

where $A_0$, $B_0$, $a$, $b$, and $c$ are constants that depend on the type of gas involved.

The general conclusion of this section is that the use of $p/\rho = RT$ will provide a quite adequate description of most compressible flows encountered in practice. As a check on the adequacy of the results obtained using this equation, the variations of pressure and temperature in such flows can be calculated using the perfect gas equation and then the results can be used to deduce the variations of $p_r$ and $T_r$ in the flow. Figure 15.10 can then be used to decide whether a more complex gas behavior equation should have been utilized in the analysis of the flow.

## 15.4
## DISSOCIATION AND IONIZATION

At high temperatures, the molecules of gases that consist of two or more atoms can start to break down into simpler molecules and into the atoms of which these molecules consist, i.e., dissociation of the molecules can occur. Examples of dissociation are:

$$N_2 \rightarrow N + N$$

$$O_2 \rightarrow O + O \tag{15.52}$$

$$2H_2O \rightarrow 2H_2 + O_2$$

As a very rough guide, if the temperature of a gas goes above 2000 K, the possibility that dissociation is occurring has to be considered.

Now, as illustrated in Fig. 15.11, dissociation occurs over a wide range of temperatures. In Fig. 15.11, the dissociation of $O_2$ into $O + O$ is illustrated. At low temperatures the gas essentially consists entirely of $O_2$ molecules. At high temperatures, O atoms start to exist and as the temperature further increases, the number of $O_2$ molecules decreases and the number of O atoms increases until at high temperatures the gas essentially consists entirely of O atoms. If the gas is in thermodynamic equilibrium, the amount of each constituent, i.e., $O_2$ and O in the above example, depends on the pressure and the temperature of the gas. The equilibrium constant, $K_p$, is used to determine how much of each constituent is present.

In order to define the equilibrium constant it is recalled that the pressure in a mixture of gases is made up of the sum of the partial pressures of all the constituent gases, i.e., that:

$$p = \sum_{i=1}^{N} p_i \tag{15.53}$$

$p_i$ being the partial pressure of gas $i$. Each gas separately satisfies the perfect gas equation, i.e., $p_i/\rho_i = R_i T_i$. From this it follows that:

$$\frac{p_i}{p} = \frac{n_i}{n_T} \tag{15.54}$$

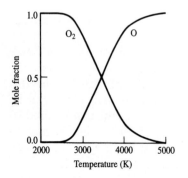

**FIGURE 15.11**
Effect of temperature on composition of oxygen at 1 atmosphere.

where $n_i$ is the number of moles of constituent $i$ and $n_T$ is the total number of moles.

If the dissociation equation is of the form:

$$i_A A = i_B B + i_C C \tag{15.55}$$

then the equilibrium constant, $K_p$, is defined as:

$$K_p = \frac{p_B^{i_B} p_C^{i_C}}{p_A^{i_A}} \tag{15.56}$$

where $p_A$, $p_B$, and $p_C$ are the partial pressures of constituents $A$, $B$, and $C$ respectively.

For example, for the reaction:

$$O_2 \rightarrow O + O$$

i.e.,

$$1O_2 \rightarrow 1O + 1O$$

it will be seen that $i_A = 1$, $i_B = 1$, and $i_C = 1$. It follows, therefore, that:

$$K_p = p_O^1 p_O^1 / p_{O_2}^1 = p_O^2 / p_{O_2}$$

Returning to the discussion of eq. (15.56), in view of eq. (15.54), eq. (15.56) gives:

$$
\begin{aligned}
K_p &= \left[\frac{n_B}{n_T}\right]^{i_B} p^{i_B} \left[\frac{n_C}{n_T}\right]^{i_C} p^{i_C} \Big/ \left[\frac{n_A}{n_T}\right]^{i_A} p^{i_A} \\
&= \left[\frac{n_B^{i_B} n_C^{i_C}}{n_A^{i_A}}\right] \frac{p^{i_B + i_C - i_A}}{n_T^{i_B + i_C - i_A}}
\end{aligned}
\tag{15.57}
$$

The equilibrium constant, $K_p$, is a function of temperature. Values for $O_2 = O + O$ and for $N_2 = N + N$ are listed in Table 15.4. The pressure used in expressing these values is in atmospheres. It should be noted that Table 15.4 gives values of $\log_{10} K_p$. It will be seen that $K_p$ is therefore very small for both $O_2$ and $N_2$ below 2000 K and that it remains small for $N_2$ up to 4000 K.

The way in which these values of $K_p$ are used to find the amount of dissociated gas present is illustrated in the following example.

**TABLE 15.4**
**Variation of $K_p$ for dissociation with temperature**

| $T$ (K) | $\log_{10} K_p$ for $O_2 \rightarrow 2O$ | $\log_{10} K_p$ for $N_2 \rightarrow 2N$ |
|---|---|---|
| 1000 | −44.71 | −99.08 |
| 2000 | −14.41 | −41.61 |
| 3000 | −4.19 | −22.32 |
| 4000 | 0.95 | −12.62 |
| 5000 | 4.05 | −6.75 |
| 10 000 | 10.36 | 5.37 |

**EXAMPLE 15.1**
Oxygen kept at a pressure of 25 kPa is heated to temperatures of 3000 K, 4000 K, and 5000 K. Find the amounts of $O_2$ and O present at each temperature.

**Solution**
The values of $K_p$ at each temperature are as follows, the values in Table 15.4 being used:

$$T = 3000 \text{ K:} \qquad K_p = 10^{-4.19} = 6.46 \times 10^{-5}$$
$$T = 4000 \text{ K:} \qquad K_p = 10^{0.95} = 8.91$$
$$T = 5000 \text{ K:} \qquad K_p = 10^{4.05} = 1.12 \times 10^4$$

The reaction being considered is:

$$O_2 \rightarrow aO_2 + bO$$

Mass balance requires:

$$2 = 2a + b \tag{i}$$

while the definition of the equilibrium constant $K_p$ gives, as discussed above:

$$K_p = p_0^2 / p_{02}$$

$$= \frac{(b/a+b)^2 p^2}{(a/a+b)p} = \frac{b^2}{a(a+b)}p$$

But $p = 25/101 = 0.248$ atm, so:

$$K_p = \frac{0.248 b^2}{a(a+b)} \tag{ii}$$

Since eq. (i) gives $b = 2(1 - a)$, eq. (ii) gives:

$$K_p = \frac{0.992(1-a)^2}{2a - a^2}$$

For each value of $K_p$, this allows $a$ to be found. $b$ can then be found using eq. (i). The results so obtained are given in Table E15.1.

At this pressure, then, there is essentially no dissociation below a temperature of 3000 K and the oxygen is fully dissociated above a temperature at 5000 K.

**TABLE E15.1**

| T (K) | a | b |
|-------|-------|-------|
| 3000 | 1 | 0 |
| 4000 | 0.053 | 1.895 |
| 5000 | 0 | 2 |

In many situations the flow involves a mixture of gases, the most common example of this being the flow of air. To illustrate how such situations are dealt with, consider the case of the flow of air. It will be assumed that air consists of a mixture of $N_2$ and $O_2$ and, therefore, that the reactions of interest are:

$$O_2 \rightarrow O + O$$
$$N_2 \rightarrow N + N$$

In fact, another reaction of great practical consequence can occur in high temperature air flows. This is the reaction $N + O \rightarrow NO$, a reaction that has extremely important environmental consequences. This reaction will, however, for simplicity, not be considered here.

The air pressure is made up of the partial pressures of the constituent gases, i.e.,

$$p = p_{O_2} + p_O + p_{N_2} + p_N \tag{15.58}$$

The partial pressures of the O and N are given by the definition of the equilibrium constant by:

$$\frac{p_O^2}{p_{O_2}} = K_{pO_2} \tag{15.59}$$

$$\frac{p_N^2}{p_{N_2}} = K_{pN_2} \tag{15.60}$$

the equilibrium constant $K_p$ being known functions of temperature.

In addition, an overall mass balance requires, since:

$$aN_2 + bO_2 = cN_2 + dN + eO_2 + fO$$

that:
$$2a = 2c + d$$

$$2b = 2e + f \tag{15.61}$$

The composition of the air in the undissociated state is assumed to be known, i.e., $a$ and $b$ are known. For example, under standard conditions, air consists of approximately 80% $N_2$ and approximately 20% $O_2$.

Equation (15.61) gives:

$$\frac{2c + d}{2e + f} = \frac{a}{b} \tag{15.62}$$

the right hand side of this equation being the ratio of the number of moles of nitrogen involved to the number of moles of oxygen involved. As mentioned above, it is a known quantity, equal to approximately 4.

Now, as discussed before, see eq. (15.54):

$$\frac{p_i}{p} = \frac{n_i}{n_T}$$

Hence:
$$\frac{p_{N_2}}{p} = \frac{c}{c + d + e + f}$$

i.e.,
$$c = \left[\frac{c + d + e + f}{p}\right] p_{N_2} \tag{15.63}$$

Similarly:
$$d = \left[\frac{c + d + e + f}{p}\right] p_N \tag{15.64}$$

$$e = \left[\frac{c + d + e + f}{p}\right] p_{O_2} \tag{15.65}$$

$$f = \left[\frac{c+d+e+f}{p}\right]p_O \tag{15.66}$$

Substituting these equations into eq. (15.62) then gives:

$$\frac{2p_{N_2}+p_N}{2p_{O_2}+p_O} = \frac{a}{b}(\approx 4) \tag{15.67}$$

Equations (15.58), (15.59), (15.60), and (15.67) constitute a set of four equations in the four unknowns $p_{O_2}$, $p_O$, $p_{N_2}$, and $p_N$. The values of these partial pressures are, therefore, easily solved for using the known values of the equilibrium constants. Once the values of $p_{O_2}$, $p_O$, $p_{N_2}$, and $p_N$ are found, the mole fraction of each constituent can be calculated. Typical results for air were given in Chapter 14. The results given there included the effect of NO production which was not considered in the above discussion.

**EXAMPLE 15.2**
Find the composition of air at a temperature of 5000 K and a pressure of 1 atmosphere.

*Solution*
The equilibrium constants for $O_2$ and $N_2$ at 5000 K are:

$$K_{p_{O_2}} = 10^{4.05} = 11\,200$$

$$K_{p_{N_2}} = 10^{-6.75} = 1.778 \times 10^{-7}$$

Therefore, equations (15.59) and (15.60) give:

$$p_{O_2} = p_O^2/11\,200.0 \tag{i}$$

$$p_{N_2} = p_N^2/0.000\,000\,1778 \tag{ii}$$

Substituting for $p_{O_2}$ and $p_{N_2}$ in eqs. (15.67) and (15.58) then gives:

$$\frac{p_N^2}{0.000\,000\,0889} + p_N = 4\left[\frac{p_O^2}{5600.0} + p_O\right]$$

and:

$$\frac{p_O^2}{11\,200.0} + p_O + \frac{p_N^2}{0.000\,000\,1778} + p_N = p$$

Solving between these two equations for $p_O/p$ and $p_N/p$ then gives:

$$\frac{p_O}{p} = 0.333, \qquad \frac{p_N}{p} = 3.44 \times 10^{-4}$$

Substituting these into eqs. (i) and (ii) above then gives:

$$\frac{p_{O_2}}{p} \approx 0, \qquad \frac{p_{N_2}}{p} = 0.666$$

From these results, it will be seen that, at this temperature, the oxygen is essentially all dissociated while hardly any of the nitrogen is dissociated, the air then consisting essentially of $0.8N_2 + 0.2O + 0.2O$.

Computer programs that implement the above procedure for finding the composition of air and other gases are easily developed.

**TABLE 15.5**
**Variation of $K_p$ for ionization with temperature**

| $T$ (K) | $\log_{10} K_p$ for ionization of N and O |
|---|---|
| 1000 | $-163.16$ |
| 2000 | $-78.09$ |
| 3000 | $-49.30$ |
| 4000 | $-34.69$ |
| 5000 | $-25.80$ |
| 10 000 | $-7.48$ |
| 15 000 | $-1.02$ |

If very high temperatures are involved in the flow, ionization of the atoms can occur. For example, in the case of air flow, the following ionization reactions become important:

$$N \rightarrow N^+ + e^-$$

$$O \rightarrow O^+ + e^- \tag{15.68}$$

The degree of ionization that has occurred at a given temperature is found using an equilibrium constant defined in the same way as in the above discussion of dissociation. Values of $K_p$ that apply to both of the reactions given in eq. (15.68) are listed in Table 15.5.

A consideration of the values given in Table 15.5 shows that ionization of N and O is not likely to be important at temperatures below roughly 10 000 K.

**EXAMPLE 15.3**
Find the amounts of $O^+$ in oxygen which has been heated to a temperature of 15 000 K.

*Solution*
The reaction being considered is:

$$O \rightarrow O^+ + e^-$$

there being a negligible amount of $O_2$ present at this temperature. Hence writing:

$$O \rightarrow aO + bO^+ + be^{-1}$$

it follows that:

$$K_p = \frac{p_O^+}{p_O} = \frac{\left[\dfrac{b}{a+2b}\right]p}{\left[\dfrac{a}{a+2b}\right]p} = \frac{b}{a}$$

But a mass balance requires:

$$b + a = 1$$

therefore:

$$K_p = \frac{b}{1-b}$$

Since, as will be seen from Table 15.5, $K_p = 0.0955$ at a temperature of 15 000 K, it follows that $b = 0.087$ and hence that $a = 0.913$. Therefore, approximately 9 percent of the oxygen atoms have ionized at this temperature.

The discussion to this point has been concerned with the composition of a gas undergoing dissociation and ionization. In calculating compressible gas flows at high temperatures it is the changes in the thermodynamic properties of the flow that are actually required. However, if the composition of the gas is known under any conditions, the enthalpy of the gas under these conditions can be obtained. To do this, the enthalpy of the gas is written in terms of a fixed energy, i.e., the energy of formation at a chosen temperature, plus the change in sensible enthalpy between the temperature at which the energy of formation is specified and the actual gas temperature, i.e., $h$ is written as:

$$h = h_f + \Delta h_{sen}$$

$\Delta h_{sen}$ being the change in sensible enthalpy and $h_f$ being the heat of formation. $\Delta h_{sen}$ is evaluated using the procedures discussed in the previous section. Values of $h_f$ for various gases are available.

The enthalpy of a mixture of gases is then found by adding the enthalpy of all the constituent gases. Computer programs for doing this for air are available. The results are also available in the form of a "Mollier chart." This gives the variation of $h$ with $s$ for various temperatures and pressures. The form of such a chart is illustrated in Fig. 15.12 which applies to air.

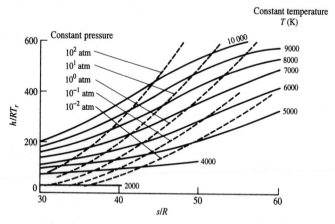

**FIGURE 15.12**
Form of Mollier chart for air ($T_f = 273$ K).

**EXAMPLE 15.4**
Air flowing at a velocity of 4000 m/s has a temperature of 1750°C and a pressure of 1 kPa. If it is isentropically brought to rest, find the temperature that then exists (i.e., find the stagnation temperature).

**Solution**
On the Mollier chart, the process is as shown in Fig. E15.4.

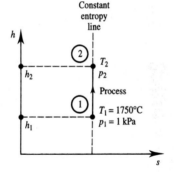

**FIGURE E15.4**

The energy equation gives:

$$h_2 = h_1 + \frac{V^2}{2}$$

From the diagram for $T = 2023$ K and $p = 0.0099$ atm:

$$\frac{h_1}{RT_r} = 36$$

so:

$$h_2 = 36 \times 287 \times 273 + \frac{4000^2}{2} = 10.82 \times 10^6 \text{ J/kg}$$

Hence:

$$\frac{h_2}{RT_r} = \frac{10.82 \times 10^6}{287 \times 273} = 138$$

Therefore, from the chart:

$$T_2 = 6000 \text{ K}$$

For comparison, if the variations in air properties had been ignored:

$$T_0 = T_1 \left[1 + \frac{\gamma - 1}{2} M_1^2\right]$$

$$= T_1 \left[1 + \frac{\gamma - 1}{2} \frac{V_1^2}{\gamma RT}\right]$$

$$= 2023 \times \left[1 + \frac{1.4 - 1}{2} \times \frac{4000^2}{1.4 \times 287 \times 2023}\right]$$

$$= 9987 \text{ K}$$

From these results, it will be seen that the dissociation of the air has a significant influence on the value of the stagnation temperature.

The flow through a normal shock wave can also be calculated using the Mollier chart. To do this, the density also has to be determined. A Mollier chart for air that shows some constant density lines is therefore given in Fig. 15.13.

The use of the Mollier charts to calculate the changes across a normal shock wave is illustrated in the following example.

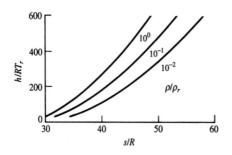

**FIGURE 15.13**
Constant density lines on Mollier chart for air ($T_r = 273$ K).

## EXAMPLE 15.5

A normal shock wave forms ahead of a body moving through the air at a velocity 7500 m/s. If the ambient temperature, density, and pressure are 230 K, 0.018 kg/m³, and 1.2 kPa respectively, find the pressure and temperature behind the shock wave.

### Solution

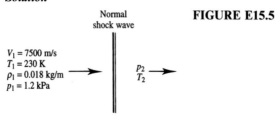

Normal shock wave

**FIGURE E15.5**

$V_1 = 7500$ m/s
$T_1 = 230$ K
$\rho_1 = 0.018$ kg/m
$p_1 = 1.2$ kPa

$p_2$
$T_2$

As shown earlier in this chapter—see eqs. (15.36) and (15.37)—the following equations apply across a normal shock wave:

$$p_2 = p_1 + \rho_1 V_1^2 \left[ 1 - \frac{\rho_1}{\rho_2} \right]$$

and:

$$\frac{V_1^2}{2} \left[ 1 - \left( \frac{\rho_1}{\rho_2} \right)^2 \right] = h_2 - h_1$$

Here $V_1 = 7500$ m/s, $p_1 = 1.2$ kPa, $\rho_1 = 0.018$ kg/m³, and $T_1 = 230$ K. Hence, the first of the above equations gives:

$$p_2 = 1200 + 0.018 \times 7500^2 \times \left[ 1 - \frac{\rho_1}{\rho_2} \right]$$

i.e.,

$$p_2 = 1200 + 1\,013\,000 \left[ 1 - \frac{\rho_1}{\rho_2} \right]$$

i.e.,

$$\frac{p_2}{p_r} = 0.012 + 10.0 \left[ 1 - \frac{\rho_1}{\rho_2} \right]$$

The second of the above equations gives:

$$\frac{(h_2 - h_1)}{RT_r} = 359 \left[ 1 - \left( \frac{\rho_1}{\rho_2} \right)^2 \right]$$

The simplest, although not very elegant, method of finding the solution is to use a trial-and-error approach. One possible such procedure involves the following steps:

1. Guess a value of $\rho_1/\rho_2$
2. Use the first of the above two equations to calculate the value of $p_2/p_r$
3. Use the second of the above two equations to calculate $h_2/RT$
4. These two values together define a point on the Mollier chart (Fig. 15.12). Establish the $x$- and $y$-coordinates of this point on the chart
5. Find the value of $\rho_2/\rho_r$ corresponding to this point on the Mollier chart by using the second of the two charts given above (Fig. 15.13)
6. Find the corresponding value of $\rho_1/\rho_2$ using:

$$\frac{\rho_1}{\rho_2} = \frac{\rho_1}{\rho_r}\frac{\rho_r}{\rho_2} = \frac{0.014}{\rho_2/\rho_r}$$

7. Compare the value of the density ratio so obtained with the initial guessed value
8. Repeat the procedure with different initial guessed values until the two values agree

Using this procedure gives $p_2 = 950$ kPa and $T_2 = 8000$ K. Due to the coarse scales used, it is not possible to get the result very accurately using the Mollier charts given above.

## 15.5
## NONEQUILIBRIUM EFFECTS

The discussion up to this point in this chapter has assumed that the gas is always in thermodynamic equilibrium, i.e., that the composition and properties depend only on the temperature and the pressure at the point in the flow being considered. In low temperature flows, this assumption is essentially always valid unless a gas–liquid or gas–solid phase change occurs. In high temperature flows, however, when the effects of vibrational excitation and dissociation and ionization are significant, this may not be the case. This is because the excitation of the vibrational modes of the gas molecules and dissociation and ionization proceed at a relatively slow rate, the rates being too slow to ensure that thermodynamic equilibrium always exists at all points in the flow. This situation is particularly likely to occur with flow through a shock wave. As discussed before, a shock wave is very thin. The time taken for the gas to pass through the shock wave may, therefore, be so short that there is no time for vibrational excitation or dissociation or ionization to occur to any significant extent, i.e., the nature of the gas remains essentially frozen in its upstream state during its passage through the shock. As a result of this, the flow immediately downstream of the shock can be calculated using the equations given in Chapter 5. In the gas flow downstream of the shock, the gas is, therefore, not in thermodynamic equilibrium and the vibrational excitation, dissociation, and ionization proceed at a finite rate and thermodynamic equi-

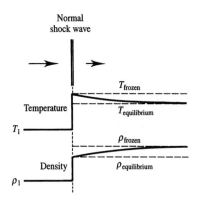

**FIGURE 15.14**
Nonequilibrium effects downstream of a shock wave.

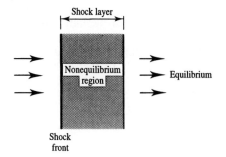

**FIGURE 15.15**
Shock layer in nonequilibrium flow.

librium is therefore only again attained at some distance downstream of the shock. The flow near the shock is, therefore, as illustrated in Fig. 15.14.

In such a case, it is more accurate to speak of a shock front and a shock layer rather than a shock wave. This is illustrated in Fig. 15.15. The overall change through the shock layer will be given by the methods of analyses discussed earlier in this chapter. If the thickness of the shock layer and the nature of the flow in this layer are required, rate equations for the non-equilibrium reactions have to be solved simultaneously with the flow equations.

## 15.6
## CONCLUDING REMARKS

In high temperature flows, which are usually associated with high Mach numbers, it has been shown that:

- Changes in the specific heats with temperature can occur as a result of the excitation of the vibrational modes of the molecules. This can have a significant effect on the flow and the analysis of flows incorporating this effect has been discussed.

- Deviations from the perfect gas law, as measured by the deviation of the factor $Z$ from 1, are unlikely to have a significant effect on the majority of flows encountered in practice.
- The effects of dissociation and ionization can become significant at high temperatures. Methods of calculating flows in which such effects are important have been discussed. A Mollier chart for the calculation of air flows was discussed.
- Nonequilibrium effects can be important if very rapid changes occur in the flow. The flow through a shock wave is an example of such a situation.

## PROBLEMS

**15.1.** During the entry of a space vehicle into the Earth's atmosphere, the Mach number at a given point on the trajectory is 38 and the atmospheric temperature is 0°C. Calculate the temperature at the stagnation point of the vehicle assuming that a normal shock wave occurs ahead of the vehicle and assuming that the air behaves as a calorically perfect gas with $\gamma = 1.4$. Do you think that the value so calculated is accurate? If not, why?

**15.2.** Oxygen, kept at a pressure of 10.1 kPa, is heated to a temperature of 4000 K. Determine the relative amounts of diatomic and monatomic oxygen that are present after the heating.

**15.3.** At a point in an air flow system at which the velocity is extremely low, the pressure is 10 MPa and the temperature is 8000 K. At some other point in the flow system, the pressure is 100 kPa. Assuming that the flow is isentropic, find the temperature and velocity at this second point.

**15.4.** Nitrogen at a static temperature of 800 K and a pressure of 70 kPa is flowing at Mach 3. Determine the pressure and temperature that would exist if the gas is brought to rest isentropically.

**15.5.** Air at a pressure of 101 kPa and a temperature of 20°C has its temperature raised to 4000 K in a constant-pressure process. Determine the composition of the air at this elevated temperature. Assume the air to initially consist of 3.76 mols of nitrogen per mol of oxygen.

**15.6.** As a result of an explosion, a normal shock wave moves at a velocity of 6000 m/s through still air at a pressure and temperature of 101 kPa and 25°C respectively. Find the pressure, temperature, and air velocity behind the wave.

**15.7.** A blunt-nosed body is moving through air at a velocity of 5000 m/s. The pressure and the temperature of the air are 22 kPa and 43°C respectively. The shock wave that exists ahead of the body can be assumed to be normal in the vicinity of the stagnation point. Find the pressure behind the shock wave.

# CHAPTER 16

# Low Density Flows

## 16.1
## INTRODUCTION

It has been assumed in all of the preceding chapters in this book that the gas behaves as a continuum, i.e., that the molecular nature of the gas does not have to be considered in analyzing the flow of the gas. However, it may not be possible to use this continuum assumption in the analysis of the flow when the density of the gas is very low. Flows in which the density is so low that noncontinuum effects become important are often termed rarefied gas flows.

The conditions under which noncontinuum effects become important and the nature of the changes in the flow produced by these effects is the subject of the present chapter. Noncontinuum effects can have an important influence on the flow over craft operating at high altitudes at high Mach numbers. They can also have an important influence on the flow in high vacuum systems. However, because this book is intended to give a broad introduction to compressible fluid flows, no more than a very brief introduction to the topic will be given here despite its significant practical importance.

## 16.2
## KNUDSEN NUMBER

A gas can be assumed to behave as a continuum if the mean free path, i.e., the average distance that a molecule moves before colliding with another molecule, $\lambda$, is small compared to the significant characteristic length, $L$, of the flow

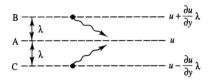

**FIGURE 16.1**
Layers considered in the analysis of viscosity.

system. The ratio of $\lambda/L$ is, of course, dimensionless and is called the Knudsen number, $Kn$, i.e.,

$$Kn = \frac{\lambda}{L} \tag{16.1}$$

In order to relate the Knudsen number to the dimensionless parameters used elsewhere in the study of compressible flows, it is convenient to be able to relate the coefficient of viscosity to the near free path. To do this, consider three layers distance $\lambda$ apart in the flow as shown in Fig. 16.1. Because molecules arriving at plane A from plane B have not collided with any other molecules over the distance $\lambda$, they arrive with an excess mean velocity of $\lambda \, \partial u/\partial y$. Similarly, molecules arriving at plane A from plane C arrive with a mean velocity deficit of $\lambda \, \partial u/\partial y$. When the molecules from planes B and C arrive at plane A, they collide with the molecules on this plane and attain the mean velocity on this plane. Because of the change in mean momentum that, therefore, occurs at plane A, there is effectively a shear force acting on plane A that is proportional to the excess or deficit of momentum with which the molecules arrive, i.e., proportional to $\lambda \, \partial u/\partial y$. The net force per unit area will then be proportional to the number of molecules arriving per unit area per unit time multiplied by $\lambda \, \partial u/\partial y$. Now the number of molecules arriving per unit area per unit time on plane A will depend on the number of molecules per unit volume, i.e., on the density, and on the mean speed of the molecules, $c_m$, i.e., the force per unit area, the shear stress, $\tau$, will be given by:

$$\tau \propto \rho c_m \lambda \frac{\partial u}{\partial y} \tag{16.2}$$

But, by definition, the coefficient of viscosity, $\mu$, is given by:

$$\tau = \mu \frac{\partial u}{\partial y} \tag{16.3}$$

Comparing these two equations indicates that:

$$\mu \propto \rho c_m \lambda \tag{16.4}$$

But the mean molecular speed is proportional to the speed of sound, $a$, since a sound wave is propagated as a result of molecular collisions. Hence, eq. (16.4) gives:

$$\mu \propto \rho a \lambda \tag{16.5}$$

from which it follows that:

$$\lambda \propto \frac{\mu}{\rho a} \tag{16.6}$$

A more complete analysis gives $\lambda = 1.26\sqrt{\bar{\gamma}}\mu/\rho a$.

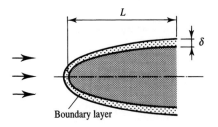

**FIGURE 16.2**
Boundary layer around body.

Using eq. (16.6), it will be seen from the definition of the Knudsen number given in eq. (16.1) that:

$$Kn \propto \frac{\mu}{\rho a L} = \frac{\mu}{\rho V L} \frac{V}{a}$$

i.e.,
$$Kn \propto \frac{M}{Re} \qquad (16.7)$$

From this equation, it follows that large Knudsen numbers will be associated with high Mach number, low Reynolds number flows. These are exactly the conditions that normally exist when a body, such as a re-entering orbital craft, is passing through the upper atmosphere.

For many purposes, the size of the body $L$ is suitable for use in defining the Knudsen number. However, if the Reynolds number is significantly above 1, a distinct boundary layer will exist adjacent to the surface of the body (see Fig. 16.2), and the thickness of this boundary layer, $\delta$ may be a more suitable length scale to compare with the near free path in defining the Knudsen number.

Now, because noncontinuum effects are associated with low density flows and because in such flows the Reynolds numbers will usually be relatively small, noncontinuum effects are usually likely to be important in flows in which the boundary layer is laminar, and in this case:

$$\delta \propto \frac{L}{Re^{0.5}} \qquad (16.8)$$

Therefore, in situations in which a distinct boundary layer exists, a more suitable Knudsen number to use is:

$$Kn = \frac{\lambda}{\delta} \propto \frac{\mu}{\rho a} \frac{Re^{0.5}}{L} = \frac{M}{Re^{0.5}} \qquad (16.9)$$

To conclude this section it will be noted that since for many gases $\mu$ is approximately proportional to $T^{0.5}$ and since $a$ is also proportional to $T^{0.5}$, eq. (16.6) shows that $\lambda$ is dominantly dependent on $1/\rho$. The interrelationship between $\rho$ and $\lambda$ will be seen by comparing the results given in Figs. 16.3 and 16.4. These two figures show the approximate variations of mean free path and air density ratio in the upper atmosphere, the air density at sea-level being approximately 1.2 kg/m$^3$. It will be seen from Fig. 16.3 that the mean free path is approximately 0.000 066 mm at sea-level and approximately 50 m at an

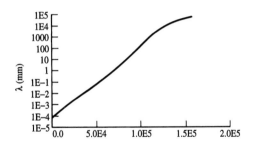

**FIGURE 16.3**
Variation of mean free path in upper atmosphere.

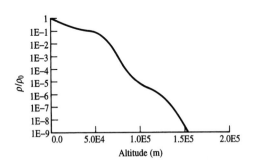

**FIGURE 16.4**
Variation of density ratio in upper atmosphere.

altitude of 150 km. Thus the ratio of the mean free path at sea-level to that at 150 km is approximately $1.3 \times 10^{-9}$ which, as will be seen from Fig. 16.4, is close to the ratio of the density at an altitude of 150 km to the sea-level density.

## 16.3
## LOW DENSITY FLOW REGIMES

As the Knudsen number in a flow over a body increases, the first observable noncontinuum flow effect is that there is an apparent jump in the velocity at the surface of the body, i.e., the gas velocity at the body surface can no longer be assumed to be equal to the surface velocity, i.e., there is a "slip" at the surface as indicated in Fig. 16.5. Flows in which this effect becomes important are termed "slip flows."

Slip flow occurs roughly in the following ranges:

$$\text{If } Re > 1: \qquad 0.01 < \frac{M}{Re^{0.5}} < 0.1$$

$$\text{If } Re < 1: \qquad 0.01 < \frac{M}{Re} < 0.1$$

(16.10)

Two criteria are necessary because, as explained in the previous section, if $Re$ is large (taken to imply $Re > 1$), the boundary layer thickness is the

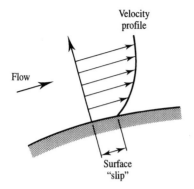

**FIGURE 16.5**
Velocity variation near a surface in slip flow.

important length scale while if *Re* is small (taken to imply $Re < 1$) the size of the body is the important length scale.

If the Knudsen number for the flow is very large, i.e., the mean free path is much greater than the size of the body, very infrequent intermolecular collisions occur. Molecules arriving at the surface of the body over which the gas is flowing will, essentially, therefore not have collided with any molecules reflected from the surface. The molecules, therefore, arrive at the surface with the full freestream velocity. In such flows there is then essentially no velocity gradient adjacent to the surface as shown in Fig. 16.6. Such flows in which there is essentially no interaction between the molecules leaving the surface and those approaching the surface are termed "free molecular flows." They can be assumed to exist if:

$$\frac{M}{Re} > 3 \tag{16.11}$$

A Knudsen number based on the boundary layer thickness has no meaning in such flows because the boundary layer concept is not applicable in such flows.

Between the slip flow region and the free molecular flow region, the flow is said to be in the transition region. When no low density effects are present, the flow is in the continuum region. Using eqs. (16.10) and (16.11), the Reynolds

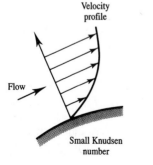

**FIGURE 16.6**
Effect of Knudsen number on velocity gradient in flow near a surface.

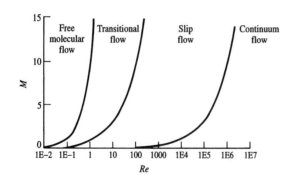

**FIGURE 16.7**
Effect of Mach and Reynolds numbers on low density flow regimes.

and Mach number regions for the various flow regions can be defined and are shown in Fig. 16.7.

A very brief introduction to the analysis of slip flows and of free molecular flows is presented in the next two sections.

**EXAMPLE 16.1**
The variation of velocity with altitude for an orbiting body during re-entry into the atmosphere is shown in Fig. E16.1. The body has a length of 4 m. Find the altitudes at which the flow over the body passes from the free molecular regime to the transition regime, from the transition to the slip flow regime, and from the slip flow to the continuum regime.

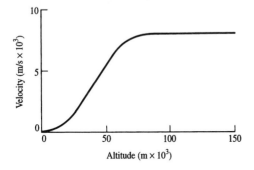

**FIGURE E16.1**

**Solution**
The velocity at various altitudes can be obtained from the above figure and some typical values are shown in the following table. Values of the density, the speed of sound and the viscosity at these altitudes are also shown in this table. These values have been derived from published information about the upper atmosphere. Using these values, the values of the Mach number, $M$, of the Reynolds number, $Re$, of $M/Re^{0.5}$, and of $M/Re$ have been derived and are also shown in Table E16.1.

From these results it can be seen that $M/Re^{0.5}$ becomes equal to 0.01 at an altitude of about 65 000 m and becomes equal to 0.1 at an altitude of about 110 000 m. This defines the slip flow regime. $M/Re^{0.5}$ can be used to define this regime because, as will be seen from the tabulated results, the Reynolds number is greater than 1 at all altitudes considered.

**TABLE E16.1**

| $H$ (m) | $V$ (m/s) | $\rho$ (kg/m³) | $a$ (m/s) | $\mu$ (kg/m·s) | $M$ | $Re$ | $M/Re^{0.5}$ | $M/Re$ |
|---|---|---|---|---|---|---|---|---|
| 30 000 | 2200 | $1.7\times10^{-1}$ | 290 | $1.6\times10^{-5}$ | 7.6 | $9.3\times10^{7}$ | 0.008 | — |
| 50 000 | 5500 | $2.6\times10^{-2}$ | 280 | $1.5\times10^{-5}$ | 19.6 | $3.8\times10^{7}$ | 0.003 | — |
| 70 000 | 7800 | $2.6\times10^{-3}$ | 260 | $1.3\times10^{-5}$ | 30.0 | $6.2\times10^{6}$ | 0.012 | — |
| 100 000 | 8000 | $3.9\times10^{-6}$ | 280 | $1.5\times10^{-5}$ | 28.2 | $8.3\times10^{3}$ | 0.310 | 0.003 |
| 120 000 | 8000 | $2.3\times10^{-7}$ | 400 | $2.5\times10^{-5}$ | 20.0 | $1.7\times10^{2}$ | 1.530 | 0.118 |
| 150 000 | 8000 | $2.0\times10^{-9}$ | 630 | $5.3\times10^{-5}$ | 12.6 | 1.2 | 11.50 | 10.50 |

It will also be seen from the tabulated results that $M/Re$ becomes equal to 3 at an altitude of about 140 000 m. Hence:

$H < 65\,000$ m: continuum flow
$65\,000$ m $< H < 110\,000$ m: slip flow
$110\,000$ m $< H < 140\,000$ m: transitional flow
$H > 140\,000$ m: free molecular flow

## 16.4
## SLIP FLOW

In this type of flow, the mean velocity of the molecules at the surface is significantly different from the velocity of the surface. Consider flow very close to a surface as shown in Fig. 16.8. Here, $u_s$ is the mean velocity of the molecules at the surface. Since the molecules move a mean distance $\lambda$ between collisions, the molecules arriving at the surface will do so with a mean gas velocity parallel to the surface of:

$$u_s + \lambda\frac{\partial u}{\partial y}\bigg|_{y=0} \tag{16.12}$$

When the molecules strike the surface they can be reflected from the surface either in a spectral manner or in a diffuse manner. In the former

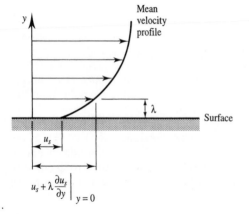

**FIGURE 16.8**
Slip flow near surface.

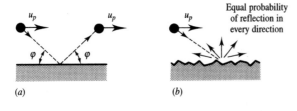

**FIGURE 16.9**
Spectral and diffuse reflection
of molecules at a surface.
(a) Spectral reflection
(velocity parallel to surface
unchanged by reflection);
(b) diffuse reflection (no mean
velocity parallel to surface
after reflection).

case, they are reflected with the same mean velocity parallel to the surface as
they had before reaching the surface, i.e., $\lambda \, \partial u/\partial y$, the derivatives being
evaluated at the surface. In the case of diffuse reflection, the molecules, on
the average, lose their mean velocity parallel to the surface. The two types of
reflection are illustrated in Fig. 16.9. With an actual surface, some of the
molecules are reflected in a diffuse manner and some are reflected in a spectral
manner, the relative fraction of molecules reflected in the two ways being
dependent on the nature of the surface. Now, all of the molecules arrive at
the surface with a mean velocity parallel to the surface that is given by eq.
(16.12). Let a fraction $d$ of these molecules leave the surface with, on the
average, no mean velocity parallel to the surface, i.e., $d$ is the fraction of
the molecules reflected diffusely. A fraction $(1 - d)$ of the incident molecules
will, therefore, leave the surface with a mean velocity parallel to the surface
given by eq. (16.12), i.e., with the same velocity parallel to the wall as they had
when they impinged on the wall. If all the molecules adjacent to the surface are
considered, on the average, half will be about to strike the surface and half will
be leaving the surface as a result of reflection. The mean velocity parallel to the
surface that the molecules adjacent to the surface have, $u_s$, is therefore given
by:

$$u_s = \frac{1}{2}\left[u_s + \lambda \frac{\partial u}{\partial y}\Big|_{y=0}\right] + \tfrac{1}{2}(1-d)\left[u_s + \lambda \frac{\partial u}{\partial y}\Big|_{y=0}\right]$$

i.e.,
$$u_s = \left[1 - \frac{d}{2}\right]\left[u_s + \lambda \frac{\partial u}{\partial y}\Big|_{y=0}\right] \tag{16.13}$$

This equation can then be rearranged to give the surface velocity $u_s$ as:

$$u_s = \left[\frac{2}{d} - 1\right]\lambda \frac{\partial u}{\partial y}\Big|_{y=0} \tag{16.14}$$

Thus, even in continuum flows, a slip velocity will exist. However, in such
flows, $\lambda$ is so small that $u_s$ is effectively zero.

As mentioned before, $d$ will depend on the nature of the surface. It will
also depend on the type of gas involved. Typical values for air flow are listed
in Table 16.1.

**TABLE 16.1**
**Diffuse reflection fractions**

| Gas and surface material | d |
| --- | --- |
| Air on machined brass | 1.0 |
| Air on glass | 0.89 |
| Air on oil | 0.90 |
| Hydrogen on oil | 0.93 |
| Helium on oil | 0.87 |

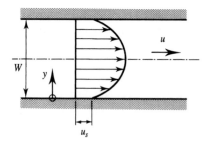

**FIGURE 16.10**
Slip flow situation considered.

   To illustrate how the effects of the slip velocity can be incorporated into the analysis of the viscous flow adjacent to a surface, consider fully developed two-dimensional slip flow between parallel plates as shown in Fig. 16.10. This flow situation is not of great practical importance but its analysis does illustrate, in a relatively simple manner, how surface slip can be incorporated into the analysis of a flow.
   Since the flow is fully developed, the velocity profile is not changing with $x$. This means that there is no velocity component in the $y$-direction and consequently no change in pressure in the $y$-direction at a given $x$.
   Consider the control volume shown in Fig. 16.11. Because the velocity is not changing with $x$, there is no change of momentum through the control volume so the forces acting on the control volume must balance, i.e.,

$$[p - (p + dp)]\, dy = [(\tau + d\tau) - \tau]\, dx$$

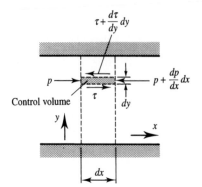

**FIGURE 16.11**
Control volume used in analyses.

This can be rearranged to give:

$$\frac{d\tau}{dy} = -\frac{dp}{dx} \tag{16.15}$$

But:

$$\tau = \mu\frac{du}{dy}$$

so eq. (16.15) gives:

$$\frac{d}{dy}\left[\mu\frac{du}{dy}\right] = -\frac{dp}{dx} \tag{16.16}$$

If the flow is assumed to be at a uniform temperature, $\mu$ will be a constant and this equation can then be written as:

$$\frac{d^2u}{dy^2} = -\frac{1}{\mu}\frac{dp}{dx} \tag{16.17}$$

Since $u$ is not a function of $x$, this indicates that, as is obvious, $dp/dx$ is constant.

Integrating equation (16.17) once gives:

$$\frac{du}{dy} = -\frac{1}{\mu}\frac{dp}{dx}y + C_1 \tag{16.18}$$

where $C_1$ is a constant of integration which will be determined by applying the boundary conditions.

Integrating again gives:

$$u = -\frac{1}{\mu}\frac{dp}{dx}\frac{y^2}{2} + C_1 y + C_2 \tag{16.19}$$

where $C_2$ is a second constant of integration which will also be determined by applying the boundary conditions.

Now the boundary conditions on the solution are:

$$\begin{aligned} y = 0, &\qquad u = u_s \\ y = W, &\qquad u = u_s \end{aligned} \tag{16.20}$$

where $W$ is the width of the channel and $u_s$ is the slip velocity, which will of course be the same on the top and bottom walls. Substituting these boundary conditions into eq. (16.19) then gives:

$$u_s = C_2$$

and:

$$u_s = -\frac{1}{\mu}\frac{dp}{dx}\frac{W^2}{2} + C_1 W + C_2$$

Solving between these two equations then gives:

$$C_1 = \frac{1}{\mu}\frac{dp}{dx}\frac{W}{2} \tag{16.21}$$

and:

$$C_2 = u_s \tag{16.22}$$

Substituting these two results into eqs. (16.18) and (16.19) then gives:

$$\frac{du}{dy} = \frac{1}{\mu}\frac{dp}{dx}\left[\frac{W}{2} - y\right] \tag{16.23}$$

and:

$$u = \frac{y}{2\mu}\frac{dp}{dx}(W - y) + u_s \tag{16.24}$$

Applying eq. (16.23) at the lower wall and substituting the result into eq. (16.14) then gives:

$$u_s = \left[\frac{2}{d} - 1\right]\lambda\frac{dp}{dx}\frac{W}{2\mu} \tag{16.25}$$

Substituting this into eq. (16.24) then gives:

$$u = \frac{1}{2\mu}\frac{dp}{dx}\left[Wy - y^2 + \left(\frac{2}{d} - 1\right)\lambda W\right] \tag{16.26}$$

The mean velocity through the channel is given by:

$$u_m = \frac{1}{W}\int_0^W u\,dy \tag{16.27}$$

Substituting eq. (16.26) into this equation and carrying out the integration gives:

$$u_m = \left(\frac{1}{2\mu}\frac{dp}{dx}\right)W^2\left[\frac{1}{6} + \left(\frac{2}{d} - 1\right)\left(\frac{\lambda}{W}\right)\right] \tag{16.28}$$

Dividing equation (16.26) by this equation then gives the velocity profile as:

$$\frac{u}{u_m} = \frac{(y/W) - (y/W)^2 + (2/d - 1)(\lambda/W)}{\frac{1}{6} + (2/d - 1)(\lambda/W)} \tag{16.29}$$

The wall shear stress is then given by:

$$\tau_w = \mu\frac{du}{dy}\bigg|_{y=0} = \left[\frac{\mu u_m/W}{\frac{1}{6} + (2/d - 1)(\lambda/W)}\right]$$

which can be rearranged to give:

$$\frac{\tau_w}{\rho u_m^2} = \frac{1}{Re}\left[\frac{1}{6} + \left(\frac{2}{d} - 1\right)\left(\frac{\lambda}{W}\right)\right] \tag{16.30}$$

The term:

$$\left(\frac{2}{d} - 1\right)\left(\frac{\lambda}{W}\right) = \left(\frac{2}{d} - 1\right)Kn$$

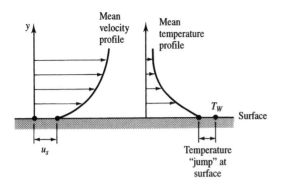

**FIGURE 16.12**
Temperature variation near a surface in slip flow.

represents the effect of wall slip on the solution. For slip effects to be negligible, therefore, it is necessary that:

$$\left(\frac{2}{d}-1\right)\left(\frac{\lambda}{W}\right) \ll \frac{1}{6}$$

i.e.,

$$\frac{\lambda}{W} \ll \frac{1}{6}\left(\frac{2}{d}-1\right) \tag{16.31}$$

Since $d$ is near 1 for most real surfaces, this means that wall slip effects will be negligible if:

$$\frac{\lambda}{W} \ll \frac{1}{6} \tag{16.32}$$

While the flow situation considered above is not of great practical significance, the analysis does indicate how the effects of wall slip can be incorporated into a viscous flow analysis.

It should also be noted that if there are temperature gradients in a slip flow there will be a temperature "jump" at the wall for the same reason that there is a velocity slip at the wall. This is illustrated in Fig. 16.12. In the analysis of heat transfer in slip flow it will then normally be necessary to account for the effects of this temperature jump at the wall.

### EXAMPLE 16.2
Consider air flow between two parallel plates as discussed above. The gap between the plates is 1 cm and the mean velocity is such that the Reynolds number is 2000 at an ambient pressure of 100 kPa and at a temperature of 20°C. If the pressure is reduced but the mean velocity and temperature are kept the same, estimate the pressure at which free molecular effects will become important.

### Solution
It will be assumed that slip effects are important when:

$$\frac{\lambda}{W} > \frac{1}{6}$$

But, as discussed above:

$$\lambda = 1.26\sqrt{\gamma}\frac{\mu}{\rho a}$$

Hence, it will be assumed that slip effects are important when:

$$1.26\sqrt{\gamma}\frac{\mu}{\rho a W} > \frac{1}{6}$$

Now, $\mu$ and $a$ can be assumed to depend only on temperature and so do not change as the pressure is reduced. Hence $a = \sqrt{\gamma R \times 293} = 343$ m/s and from tabulated properties of air $\mu = 180 \times 10^{-7}$ N·s/m². Also, because the temperature remains the same, the density will be proportional to the pressure, i.e., $\rho = p/RT = (100\,000/287 \times 293)(p/100) = 0.0119p$, where $p$ is in kPa. Hence, since $W = 0.01$ m, slip effects are important when:

$$1.26 \times \frac{\sqrt{1.4} \times 0.000\,018}{0.0119 \times p \times 343 \times 0.01} > \frac{1}{6}$$

i.e.,
$$\frac{0.000\,658}{p} > \frac{1}{6}, \quad \text{i.e.,} \quad p < 0.0040\,\text{kPa}$$

Therefore, the possibility that slip effects are important will have to be considered when the pressure is less than 4 Pa.

## 16.5
## FREE MOLECULAR FLOW

As discussed above, the free molecular flow regime is entered when the Knudsen number is large. In order to give a very simple introduction to the analysis of free molecular flows, the drag force on a flat plate placed at right angles to a flow will be considered, i.e., the flow situation shown in Fig. 16.13 will be considered. Since free molecular flow is being considered, the molecules reach the plate with the velocity they have well away from the plate. If there was no mean gas velocity, the molecules would arrive at the front and the rear sides of the plate with the same velocity and the same pressure would be exerted on the two sides of the plate and there would then, of course, be no drag force acting on the plate. However, when there is a mean gas velocity, the molecules reaching the front of the plate have a higher mean velocity than those reaching the back of the plate. There is, therefore, an increase in the pressure on the front of the plate and a decrease in the pressure on the back which leads the drag force.

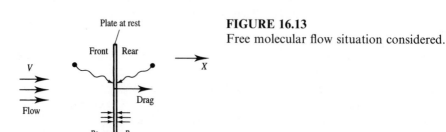

**FIGURE 16.13**
Free molecular flow situation considered.

Now the pressure force on the front of the plate will be equal to the loss of x-wise momentum that results from the fact that the plate is at rest, i.e.,

$$P_{\text{front}} = \dot{m}A(c_m + V) \qquad (16.33)$$

where $\dot{m}$ is the total mass of the molecules that strike the plate per unit time per unit area and $A$ is the frontal area of the plate. But $\dot{m}$ will be given by:

$$\dot{m} = Nm \qquad (16.34)$$

where $N$ is the number of molecules striking the plate per unit time per unit area and $m$ is the mass of one molecule. Hence, eq. (16.33) gives:

$$P_{\text{front}} = NmA(c_m + V) \qquad (16.35)$$

Similarly, the pressure force on the back surface will be given by:

$$P_{\text{back}} = NmA(c_m - V) \qquad (16.36)$$

The net force on the plate is therefore given by:

$$P = P_{\text{front}} - P_{\text{back}} = 2NmAV \qquad (16.37)$$

But $N$ will be proportional to the number of molecules by unit volume, $n$, and the mean molecular velocity $c_m$, i.e.,

$$N \propto nc_m \qquad (16.38)$$

Hence: $\qquad\qquad\qquad P \propto 2mnc_mAV \qquad (16.39)$

But: $\qquad\qquad\qquad\qquad \rho = mn \qquad (16.40)$

so eq. (16.39) gives:

$$P \propto 2\rho c_m AV \qquad (16.41)$$

Now, the drag coefficient, $C_D$, is defined by:

$$C_D = \frac{P}{\rho V^2 A/2} \qquad (16.42)$$

so eq. (16.41) gives:

$$C_D \propto 4\left(\frac{c_m}{V}\right) \qquad (16.43)$$

i.e., writing:

$$S = V/c_m \qquad (16.44)$$

eq. (16.43) gives:

$$C_D \propto \frac{4}{S} \qquad (16.45)$$

Since $c_m$ is proportional to the speed of sound, $S$ is dependent on the effective Mach number in the flow.

Equation (16.45) indicates that $C_DS$ will be a constant. If bodies other than a flat plate are considered, e.g., if flow over a cylinder is considered, the same form of result is obtained. The measured variation of $C_DS$ with $S$ for a cylinder is shown in Fig. 16.14. It will be seen from Fig. 16.14 that the

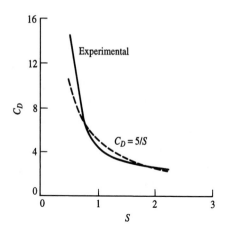

**FIGURE 16.14**
Drag coefficient variation for a cylinder in free molecular flow.

experimental results confirm that the product $C_D S$ is approximately constant in free molecular flow.

### EXAMPLE 16.3

If the mean molecular speed $c_m$ is assumed to be equal to $\sqrt{2RT}$, find the order of magnitude of the drag force per unit frontal area on a flat plate over which air is flowing at a velocity of 8000 m/s, the plate being normal to the flow. The density of the air is $10^{-10}$ times the standard sea-level density and the temperature of the air is 1000 K.

### Solution

At a temperature of 1000 K, the speed of sound is equal to $\sqrt{\gamma R 1000}$, i.e., 634 m/s, so the Mach number in the flow is $8000/634 = 12.6$. The viscosity of the air is approximately $420 \times 10^{-7}$ N·s/m² so the Reynolds number per meter is $\rho V/\mu$, i.e., since the air density at sea-level is 1.16 kg/m³ the Reynolds number per meter is $1.16 \times (10 \times 10^{-10}) \times 8000/420 \times 10^{-7} = 0.22$. Hence, the ratio of the Mach number to the Reynolds number per meter is equal to $12.6/0.22 = 57.2$. Therefore, unless the plate is extremely large, free molecular flow exists.

But in such flow:

$$C_D \propto 4\left(\frac{c_m}{V}\right)$$

and since:

$$c_m = \sqrt{2RT} = \sqrt{2 \times 287 \times 1000} = 758 \text{ m/s}$$

in the present case:

$$C_D \propto 4\left(\frac{758}{8000}\right) = 0.379$$

The drag force on the plate per unit plate area is:

$$C_D \rho V^2 A/2$$

Hence, again noting that the air density at sea-level is 1.16 kg/m³, the order of magnitude of the drag force per unit plate area is:

$$C_D \rho V^2/2 = 0.379 \times 1.16 \times 10^{-10} \times 8000^2/2 = 0.0014 \text{ N}$$

Therefore the order of magnitude of the drag force per unit area is 0.0014 N.

# 16.6
# CONCLUDING REMARKS

In low density flows it may not be possible to assume that the gas is a continuum. The Knudsen number is the parameter that is used to determine when noncontinuum effects start to become important. The relation between the Knudsen number and the Mach and Reynolds numbers has been discussed.

The first noncontinuum effect is that "slip" starts to be important at the surface, i.e., that the effective gas velocity at the surface of a body in a flow is nonzero. Flows in which this is the only noncontinuum effect are called slip flows. The range of flow conditions over which the slip flow region extends has been discussed.

When the gas density is very low, there are so few molecules present that the molecular motion is essentially unaffected by the presence of a body until the molecules reach the surface. This type of flow is called free molecular flow and the conditions under which it occurs have also been discussed.

Between the slip flow region and the free molecular flow region there is a region termed the transition region. The analysis of flow in this region is relatively complex and has not been discussed here.

# PROBLEMS

**16.1.** A small rocket probing the atmosphere has a length of 3 m. It is fired vertically upward through the atmosphere, its average velocity being 1000 m/s. Consider the flow over the rocket at this average velocity at altitudes of 30 000 m and 80 000 m. Can the air flow over the rocket be assumed to be continuous at these two altitudes? At an altitude of 30 000 m, the air has a temperature, pressure, and viscosity of $-55°C$, 120 Pa, and $1.5 \times 10^{-5}$ kg/m·s respectively while at an altitude of 80 000 m, the air has a temperature, pressure, and viscosity of $-34°C$, 0.013 Pa, and $1.7 \times 10^{-5}$ kg/m·s respectively.

**16.2.** A small research vehicle with a length of 4 ft travels at a Mach number of 15 at altitudes of 100 000 ft and 250 000 ft. Determine whether, at these two altitudes, the missile is in the continuum, slip, transition, or free molecular flow regimes. It can be assumed that at an altitude of 100 000 ft, the pressure, temperature, and viscosity are 22 psf, 340 R, and $96 \times 10^{-7}$ lbm/ft-sec respectively while at an altitude of 250 000 ft, they are 0.11 psf, 450 R, and $100 \times 10^{-7}$ lbm/ft-sec respectively.

**16.3.** Derive an expression for the velocity distribution in fully developed pipe flow in the slip flow regime.

**16.4.** Find the drag force per unit length on a 0.5 cm diameter cylinder placed in an air flow in which the temperature is 800 K, the density is $8 \times 10^{-9}$ kg/m³ and the velocity is 10 000 m/s.

# APPENDIX A

# Using the Software

**INTRODUCTION TO COMPROP (Version 1.0)**

COMPROP is an interactive DOS-based software program for the calculation of compressible flow properties. The program consists of five modules, **Isentropic Flow, Normal Shock Waves, Oblique Shock Waves, Fanno Flow** (adiabatic and isothermal flow with friction), and **Rayleigh Flow** (flow with heat transfer). This program was developed by Dr. A. J. Ghajar, Professor of Mechanical and Aerospace Engineering at Oklahoma State University, and his former student Dr. L. M. Tam, now of the University of Macau, to support the present book.

**Isentropic Flow**

This module allows the user to input one of the following five parameters:

1. Mach number ($M$)
2. Pressure ratio ($p_0/p$)
3. Temperature ratio ($T_0/T$)
4. Density ratio ($\rho_0/\rho$)
5. Area ratio ($A/A^*$)

and obtain the other four parameters and the Prandtl–Meyer angle (for $M \geq 1$). The program provides the results in a tabular form.

**Normal Shock Waves**

This module allows the user to input one of the following six parameters:

1. Mach number ahead of the shock wave ($M_1$)
2. Mach number behind the shock wave ($M_2$)
3. Stagnation pressure ratio across the shock wave ($p_{02}/p_{01}$)
4. Static pressure ratio across the shock wave ($p_2/p_1$)
5. Static temperature ratio across the shock wave ($T_2/T_1$)
6. Stagnation to static pressure ratio across the shock wave ($p_{02}/p_1$)

and obtain tabulated results for $M_1$, $p_2/p_1$, $\rho_2/\rho_1$, $T_2/T_1$, $p_{02}/p_{01}$, $p_{02}/p_1$, and $M_2$.

### Oblique Shock Waves

This module allows the user to input two of the following three combinations of parameters:

1. Mach number ahead of the shock wave ($M_1$) and the shock angle
2. Shock and turning angles
3. Turning angle and the Mach number ahead of the shock wave ($M_1$)

and obtain tabulated weak and/or strong shock solutions for shock angle, turning angle, $M_1$, $M_{1n}$, $M_{1t}$, $M_{2n}$, $M_{2t}$, $M_2$, $p_2/p_1$, $T_2/T_1$, and $p_{02}/p_{01}$. In addition, for the options 1 and 3 (above), the program gives for the specified conditions the maximum turning angle and the shock angle corresponding to the maximum turning angle.

### Fanno Flow

This module allows calculations for **ADIABATIC** and/or **ISOTHERMAL** friction in a constant area duct.

#### Adiabatic flow

This option allows the user to input one of the following four parameters:

1. Mach number ($M$)
2. Friction factor term ($4fL^*/D$)
3. Temperature ratio ($T/T^*$)
4. Pressure ratio ($p/p^*$)

and obtain tabulated results for $M$, $T/T^*$, $p/p^*$, $\rho/\rho^*$, $p_0/p_0^*$, and $4fL^*/D$.

#### Isothermal flow

This option allows the user to input one of the following four parameters:

1. Mach number ($M$)
2. Friction factor term ($4fL^*/D$)
3. Temperature ratio ($T_0/T_0^*$)
4. Pressure ratio ($p_0/p_0^*$)

and obtain tabulated results for $M$, $T_0/T_0^*$, $p_0/p_0^*$, $\rho/\rho^*$, and $4fL^*/D$.

## Rayleigh Flow

This module allows the user to input one of the following two parameters:

1. Mach number $(M)$
2. Temperature ratio $(T_0/T_0^*)$

and obtain tabulated results for $M$, $p/p^*$, $T/T^*$, $\rho/\rho^*$, $p_0/p_0^*$, and $T_0/T_0^*$.

### Additional Features of the Program

The program also gives the user the option of changing the value of the specific heat ratio $(\gamma)$, assigning a name to the output file to be generated, and printing the generated output file. This is accomplished in the **UTILITY** option of the program. If this option is not used, the default value for the specific heat ratio is 1.4 (air at standard conditions) and the default name for the output file is **COMPROP.DAT**.

### DIRECTIONS FOR USING COMPROP

**To run the program**, you will need an IBM-compatible 486 PC or higher, a color monitor, and a printer (optional).

**To start the program**, at the DOS prompt type **COMPROP** and strike the [ENTER] key. Now follow the directions on the screen. To access any of the seven options shown at the top of the screen (Isentropic, Normal shock, Oblique shock, Fanno, Rayleigh, Utility, and Exit), type the highlighted letter and a menu for each module will appear on the screen. To input the required information for the options given on the menu, type the highlighted number or letter and the input value followed by the [ENTER] key. To exit from the menu provided by any of the seven options use the [ESC] key.

**To generate a specific output file**, use the **UTILITY** option of the program. As soon as the program is accessed, the program sets up an output file with a default name of **COMPROP.DAT**. To change the output file name for a specific calculation, use the **UTILITY** option of the program. The name change should be made prior to the calculation of the flow properties.

**To print the generated output file**, use the **UTILITY** option of the program.

**To exit the program**, use the **EXIT** option of the program and follow the instructions on the screen.

**To edit the generated output file**, you must first terminate the program by using the **EXIT** option of the program and then use the **MS-DOS EDITOR**. The full-screen editor allows you to edit, save, and print the output file generated by COMPROP. For example, to edit the output file COMPROP.DAT, type the following at the DOS prompt and strike the [ENTER] key: **EDIT COMPROP.DAT**.

**Installation of the program files on the hard disk:** "COMPROP" program contains the main program as **COMPROP.EXE** and a DOS extender utility **DOSXMSF.EXE**. These two program files must be copied into the desired subdirectory by the usual copying procedure.

## GETTING A COPY OF THE SOFTWARE

The software is available free of charge to adopters of the book via the McGraw-Hill Web Page. The Web address to access the program is: http://www.mhcollege.com/engineering/mecheng.html.

# APPENDIX B

# Isentropic Flow Tables for $\gamma = 1.4$

| M | $T_0/T$ | $p_0/p$ | $\rho_0/\rho$ | $a_0/a$ | $A/A^*$ | $\theta$ |
|------|----------|----------|----------|----------|-----------|-----|
| 0.00 | 1.000 00 | 1.000 00 | 1.000 00 | 1.000 00 | — | — |
| 0.02 | 1.000 08 | 1.000 28 | 1.000 20 | 1.000 04 | 28.942 13 | — |
| 0.04 | 1.000 32 | 1.001 12 | 1.000 80 | 1.000 16 | 14.481 48 | — |
| 0.06 | 1.000 72 | 1.002 52 | 1.001 80 | 1.000 36 | 9.665 91 | — |
| 0.08 | 1.001 28 | 1.004 49 | 1.003 20 | 1.000 64 | 7.261 61 | — |
| 0.10 | 1.002 00 | 1.007 02 | 1.005 01 | 1.001 00 | 5.821 83 | — |
| 0.12 | 1.002 88 | 1.010 12 | 1.007 22 | 1.001 44 | 4.864 32 | — |
| 0.14 | 1.003 92 | 1.013 79 | 1.009 83 | 1.001 96 | 4.182 40 | — |
| 0.16 | 1.005 12 | 1.018 04 | 1.012 85 | 1.002 56 | 3.672 74 | — |
| 0.18 | 1.006 48 | 1.022 86 | 1.016 28 | 1.003 23 | 3.277 93 | — |
| 0.20 | 1.008 00 | 1.028 28 | 1.020 12 | 1.003 99 | 2.963 52 | — |
| 0.22 | 1.009 68 | 1.034 29 | 1.024 38 | 1.004 83 | 2.707 60 | — |
| 0.24 | 1.011 52 | 1.040 90 | 1.029 05 | 1.005 74 | 2.495 56 | — |
| 0.26 | 1.013 52 | 1.048 13 | 1.034 14 | 1.006 74 | 2.317 29 | — |
| 0.28 | 1.015 68 | 1.055 96 | 1.039 66 | 1.007 81 | 2.165 55 | — |
| 0.30 | 1.018 00 | 1.064 43 | 1.045 61 | 1.008 96 | 2.035 07 | — |
| 0.32 | 1.020 48 | 1.073 53 | 1.051 99 | 1.010 19 | 1.921 85 | — |
| 0.34 | 1.023 12 | 1.083 29 | 1.058 81 | 1.011 49 | 1.822 88 | — |
| 0.36 | 1.025 92 | 1.093 70 | 1.066 07 | 1.012 88 | 1.735 78 | — |
| 0.38 | 1.028 88 | 1.104 78 | 1.073 77 | 1.014 34 | 1.658 70 | — |
| 0.40 | 1.032 00 | 1.116 55 | 1.081 93 | 1.015 87 | 1.590 14 | — |
| 0.42 | 1.035 28 | 1.129 02 | 1.090 55 | 1.017 49 | 1.528 91 | — |
| 0.44 | 1.038 72 | 1.142 21 | 1.099 63 | 1.019 18 | 1.474 01 | — |
| 0.46 | 1.042 32 | 1.156 12 | 1.109 18 | 1.020 94 | 1.424 63 | — |
| 0.48 | 1.046 08 | 1.170 78 | 1.119 21 | 1.022 78 | 1.380 10 | — |
| 0.50 | 1.050 00 | 1.186 21 | 1.129 73 | 1.024 70 | 1.339 84 | — |
| 0.52 | 1.054 08 | 1.202 42 | 1.140 73 | 1.026 68 | 1.303 39 | — |
| 0.54 | 1.058 32 | 1.219 44 | 1.152 24 | 1.028 75 | 1.270 32 | — |
| 0.56 | 1.062 72 | 1.237 27 | 1.164 25 | 1.030 88 | 1.240 29 | — |
| 0.58 | 1.067 28 | 1.255 96 | 1.176 78 | 1.033 09 | 1.213 01 | — |
| 0.60 | 1.072 00 | 1.275 50 | 1.189 84 | 1.035 37 | 1.188 20 | — |
| 0.62 | 1.076 88 | 1.295 94 | 1.203 42 | 1.037 73 | 1.165 65 | — |
| 0.64 | 1.081 92 | 1.317 29 | 1.217 55 | 1.040 15 | 1.145 15 | — |

| M | $T_0/T$ | $p_0/p$ | $\rho_0/\rho$ | $a_0/a$ | $A/A^*$ | $\theta$ |
|------|-----------|-----------|-----------|-----------|-----------|-----------|
| 0.66 | 1.087 12 | 1.339 59 | 1.232 24 | 1.042 65 | 1.126 54 | — |
| 0.68 | 1.092 48 | 1.362 85 | 1.247 48 | 1.045 22 | 1.109 66 | — |
| 0.70 | 1.098 00 | 1.387 10 | 1.263 30 | 1.047 85 | 1.094 37 | — |
| 0.72 | 1.103 68 | 1.412 38 | 1.279 70 | 1.050 56 | 1.080 57 | — |
| 0.74 | 1.109 52 | 1.438 71 | 1.296 69 | 1.053 34 | 1.068 14 | — |
| 0.76 | 1.115 52 | 1.466 12 | 1.314 30 | 1.056 18 | 1.057 00 | — |
| 0.78 | 1.121 68 | 1.494 66 | 1.332 52 | 1.059 09 | 1.047 05 | — |
| 0.80 | 1.128 00 | 1.524 34 | 1.351 36 | 1.062 07 | 1.038 23 | — |
| 0.82 | 1.134 48 | 1.555 21 | 1.370 86 | 1.065 12 | 1.030 46 | — |
| 0.84 | 1.141 12 | 1.587 30 | 1.391 00 | 1.068 23 | 1.023 70 | — |
| 0.86 | 1.147 92 | 1.620 65 | 1.411 82 | 1.071 41 | 1.017 87 | — |
| 0.88 | 1.154 88 | 1.655 31 | 1.433 32 | 1.074 65 | 1.012 94 | — |
| 0.90 | 1.162 00 | 1.691 30 | 1.455 51 | 1.077 96 | 1.008 86 | — |
| 0.92 | 1.169 28 | 1.728 68 | 1.478 41 | 1.081 33 | 1.005 60 | — |
| 0.94 | 1.176 72 | 1.767 48 | 1.502 04 | 1.084 77 | 1.003 11 | — |
| 0.96 | 1.184 32 | 1.807 76 | 1.526 41 | 1.088 26 | 1.001 36 | — |
| 0.98 | 1.192 08 | 1.849 56 | 1.551 54 | 1.091 82 | 1.000 34 | — |
| 1.00 | 1.200 00 | 1.892 93 | 1.577 44 | 1.095 44 | 1.000 00 | — |
| 1.02 | 1.208 08 | 1.937 91 | 1.604 13 | 1.099 13 | 1.000 33 | 0.125 68 |
| 1.04 | 1.216 32 | 1.984 57 | 1.631 62 | 1.102 87 | 1.001 30 | 0.350 97 |
| 1.06 | 1.224 72 | 2.032 96 | 1.659 94 | 1.106 67 | 1.002 91 | 0.636 68 |
| 1.08 | 1.233 28 | 2.083 13 | 1.689 09 | 1.110 53 | 1.005 12 | 0.968 03 |
| 1.10 | 1.242 00 | 2.135 13 | 1.719 11 | 1.114 45 | 1.007 93 | 1.336 19 |
| 1.12 | 1.250 88 | 2.189 04 | 1.750 00 | 1.118 43 | 1.011 31 | 1.735 03 |
| 1.14 | 1.259 92 | 2.244 92 | 1.781 79 | 1.122 46 | 1.015 27 | 2.159 94 |
| 1.16 | 1.269 12 | 2.302 81 | 1.814 50 | 1.126 55 | 1.019 78 | 2.607 33 |
| 1.18 | 1.278 48 | 2.362 81 | 1.848 14 | 1.130 70 | 1.024 84 | 3.074 24 |
| 1.20 | 1.288 00 | 2.424 96 | 1.882 74 | 1.134 90 | 1.030 44 | 3.558 22 |
| 1.22 | 1.297 68 | 2.489 35 | 1.918 31 | 1.139 16 | 1.036 57 | 4.057 18 |
| 1.24 | 1.307 52 | 2.556 05 | 1.954 88 | 1.143 47 | 1.043 23 | 4.569 34 |
| 1.26 | 1.317 52 | 2.625 12 | 1.992 47 | 1.147 83 | 1.050 41 | 5.093 13 |
| 1.28 | 1.327 68 | 2.696 66 | 2.031 11 | 1.152 25 | 1.058 10 | 5.627 17 |
| 1.30 | 1.338 00 | 2.770 74 | 2.070 81 | 1.156 72 | 1.066 30 | 6.170 26 |
| 1.32 | 1.348 48 | 2.847 44 | 2.111 60 | 1.161 24 | 1.075 02 | 6.721 31 |
| 1.34 | 1.359 12 | 2.926 86 | 2.153 50 | 1.165 81 | 1.084 24 | 7.279 34 |
| 1.36 | 1.369 92 | 3.009 07 | 2.196 53 | 1.170 44 | 1.093 96 | 7.843 48 |
| 1.38 | 1.380 88 | 3.094 18 | 2.240 73 | 1.175 11 | 1.104 19 | 8.412 94 |
| 1.40 | 1.392 00 | 3.182 27 | 2.286 11 | 1.179 83 | 1.114 93 | 8.987 00 |
| 1.42 | 1.403 28 | 3.273 44 | 2.332 71 | 1.184 60 | 1.126 16 | 9.564 99 |
| 1.44 | 1.414 72 | 3.367 80 | 2.380 54 | 1.189 42 | 1.137 90 | 10.146 33 |
| 1.46 | 1.426 32 | 3.465 44 | 2.429 64 | 1.194 29 | 1.150 15 | 10.730 47 |
| 1.48 | 1.438 08 | 3.566 48 | 2.480 03 | 1.199 20 | 1.162 90 | 11.316 91 |
| 1.50 | 1.450 00 | 3.671 03 | 2.531 74 | 1.204 16 | 1.176 17 | 11.905 18 |
| 1.52 | 1.462 08 | 3.779 19 | 2.584 80 | 1.209 16 | 1.189 94 | 12.494 86 |
| 1.54 | 1.474 32 | 3.891 08 | 2.639 24 | 1.214 22 | 1.204 23 | 13.085 57 |
| 1.56 | 1.486 72 | 4.006 83 | 2.695 08 | 1.219 31 | 1.219 04 | 13.676 93 |
| 1.58 | 1.499 28 | 4.126 57 | 2.752 37 | 1.224 45 | 1.234 38 | 14.268 62 |
| 1.60 | 1.512 00 | 4.250 41 | 2.811 12 | 1.229 63 | 1.250 23 | 14.860 32 |
| 1.62 | 1.524 88 | 4.378 49 | 2.871 37 | 1.234 86 | 1.266 62 | 15.451 77 |
| 1.64 | 1.537 92 | 4.510 94 | 2.933 15 | 1.240 13 | 1.283 55 | 16.042 68 |
| 1.66 | 1.551 12 | 4.647 91 | 2.996 49 | 1.245 44 | 1.301 02 | 16.632 82 |
| 1.68 | 1.564 48 | 4.789 54 | 3.061 43 | 1.250 79 | 1.319 04 | 17.221 95 |
| 1.70 | 1.578 00 | 4.935 98 | 3.128 00 | 1.256 18 | 1.337 61 | 17.809 88 |

| M | $T_0/T$ | $p_0/p$ | $\rho_0/\rho$ | $a_0/a$ | $A/A^*$ | $\theta$ |
|------|----------|-----------|------------|----------|----------|-----------|
| 1.72 | 1.591 68 | 5.087 38 | 3.196 24 | 1.261 62 | 1.356 73 | 18.396 40 |
| 1.74 | 1.605 52 | 5.243 90 | 3.266 17 | 1.267 09 | 1.376 43 | 18.981 34 |
| 1.76 | 1.619 52 | 5.405 69 | 3.337 84 | 1.272 60 | 1.396 70 | 19.564 53 |
| 1.78 | 1.633 68 | 5.572 93 | 3.411 28 | 1.278 15 | 1.417 54 | 20.145 80 |
| 1.80 | 1.648 00 | 5.745 78 | 3.486 52 | 1.283 74 | 1.438 98 | 20.725 03 |
| 1.82 | 1.662 48 | 5.924 43 | 3.563 61 | 1.289 37 | 1.461 01 | 21.302 08 |
| 1.84 | 1.677 12 | 6.109 05 | 3.642 58 | 1.295 04 | 1.483 65 | 21.876 82 |
| 1.86 | 1.691 92 | 6.299 82 | 3.723 48 | 1.300 74 | 1.506 89 | 22.449 14 |
| 1.88 | 1.706 88 | 6.496 95 | 3.806 33 | 1.306 48 | 1.530 76 | 23.018 93 |
| 1.90 | 1.722 00 | 6.700 62 | 3.891 19 | 1.312 25 | 1.555 25 | 23.586 10 |
| 1.92 | 1.737 28 | 6.911 04 | 3.978 08 | 1.318 06 | 1.580 39 | 24.150 56 |
| 1.94 | 1.752 72 | 7.128 41 | 4.067 06 | 1.323 90 | 1.606 17 | 24.712 23 |
| 1.96 | 1.768 32 | 7.352 96 | 4.158 16 | 1.329 78 | 1.632 61 | 25.271 02 |
| 1.98 | 1.784 08 | 7.584 89 | 4.251 43 | 1.335 69 | 1.659 71 | 25.826 88 |
| 2.00 | 1.800 00 | 7.824 43 | 4.346 91 | 1.341 64 | 1.687 50 | 26.379 73 |
| 2.02 | 1.816 08 | 8.071 82 | 4.444 64 | 1.347 62 | 1.715 97 | 26.929 52 |
| 2.04 | 1.832 32 | 8.327 29 | 4.544 67 | 1.353 63 | 1.745 14 | 27.476 19 |
| 2.06 | 1.848 72 | 8.591 09 | 4.647 05 | 1.359 68 | 1.775 01 | 28.019 70 |
| 2.08 | 1.865 28 | 8.863 46 | 4.751 81 | 1.365 75 | 1.805 61 | 28.560 00 |
| 2.10 | 1.882 00 | 9.144 66 | 4.859 02 | 1.371 86 | 1.836 94 | 29.097 05 |
| 2.12 | 1.898 88 | 9.434 97 | 4.968 70 | 1.378 00 | 1.869 01 | 29.630 82 |
| 2.14 | 1.915 92 | 9.734 64 | 5.080 92 | 1.384 17 | 1.901 84 | 30.161 27 |
| 2.16 | 1.933 12 | 10.043 96 | 5.195 73 | 1.390 37 | 1.935 43 | 30.688 38 |
| 2.18 | 1.950 48 | 10.363 21 | 5.313 16 | 1.396 60 | 1.969 81 | 31.212 12 |
| 2.20 | 1.968 00 | 10.692 68 | 5.433 28 | 1.402 85 | 2.004 97 | 31.732 47 |
| 2.22 | 1.985 68 | 11.032 69 | 5.556 13 | 1.409 14 | 2.040 94 | 32.249 40 |
| 2.24 | 2.003 52 | 11.383 52 | 5.681 77 | 1.415 46 | 2.077 73 | 32.762 91 |
| 2.26 | 2.021 52 | 11.745 51 | 5.810 24 | 1.421 80 | 2.115 35 | 33.272 98 |
| 2.28 | 2.039 68 | 12.118 98 | 5.941 61 | 1.428 17 | 2.153 81 | 33.779 61 |
| 2.30 | 2.058 00 | 12.504 25 | 6.075 93 | 1.434 57 | 2.193 13 | 34.282 76 |
| 2.32 | 2.076 48 | 12.901 67 | 6.213 25 | 1.441 00 | 2.233 32 | 34.782 46 |
| 2.34 | 2.095 12 | 13.311 59 | 6.353 62 | 1.447 45 | 2.274 40 | 35.278 68 |
| 2.36 | 2.113 92 | 13.734 37 | 6.497 11 | 1.453 93 | 2.316 38 | 35.771 43 |
| 2.38 | 2.132 88 | 14.170 37 | 6.643 78 | 1.460 44 | 2.359 27 | 36.260 70 |
| 2.40 | 2.152 00 | 14.619 98 | 6.793 68 | 1.466 97 | 2.403 10 | 36.746 50 |
| 2.42 | 2.171 28 | 15.083 57 | 6.946 86 | 1.473 53 | 2.447 87 | 37.228 83 |
| 2.44 | 2.190 72 | 15.561 55 | 7.103 40 | 1.480 11 | 2.493 60 | 37.707 69 |
| 2.46 | 2.210 32 | 16.054 32 | 7.263 35 | 1.486 71 | 2.540 31 | 38.183 09 |
| 2.48 | 2.230 08 | 16.562 28 | 7.426 77 | 1.493 34 | 2.588 01 | 38.655 04 |
| 2.50 | 2.250 00 | 17.085 89 | 7.593 73 | 1.500 00 | 2.636 71 | 39.123 54 |
| 2.52 | 2.270 08 | 17.625 55 | 7.764 29 | 1.506 68 | 2.686 45 | 39.588 59 |
| 2.54 | 2.290 32 | 18.181 74 | 7.938 52 | 1.513 38 | 2.737 22 | 40.050 23 |
| 2.56 | 2.310 72 | 18.754 88 | 8.116 47 | 1.520 10 | 2.789 06 | 40.508 44 |
| 2.58 | 2.331 28 | 19.345 57 | 8.298 22 | 1.526 85 | 2.841 97 | 40.963 26 |
| 2.60 | 2.352 00 | 19.953 97 | 8.483 84 | 1.533 62 | 2.895 97 | 41.414 68 |
| 2.62 | 2.372 88 | 20.580 88 | 8.673 38 | 1.540 41 | 2.951 08 | 41.862 72 |
| 2.64 | 2.393 92 | 21.226 70 | 8.866 93 | 1.547 23 | 3.007 33 | 42.307 41 |
| 2.66 | 2.415 12 | 21.891 94 | 9.064 54 | 1.554 06 | 3.064 71 | 42.748 74 |
| 2.68 | 2.436 48 | 22.577 12 | 9.266 30 | 1.560 92 | 3.123 27 | 43.186 76 |
| 2.70 | 2.458 00 | 23.282 80 | 9.472 26 | 1.567 80 | 3.183 00 | 43.621 45 |
| 2.72 | 2.479 68 | 24.009 52 | 9.682 51 | 1.574 70 | 3.243 94 | 44.052 85 |
| 2.74 | 2.501 52 | 24.757 83 | 9.897 12 | 1.581 62 | 3.306 11 | 44.480 97 |
| 2.76 | 2.523 52 | 25.528 32 | 10.116 17 | 1.588 56 | 3.369 51 | 44.905 83 |

| M | $T_0/T$ | $p_0/p$ | $\rho_0/\rho$ | $a_0/a$ | $A/A^*$ | $\theta$ |
|---|---------|---------|---------------|---------|---------|----------|
| 2.78 | 2.545 68 | 26.321 58 | 10.339 71 | 1.595 52 | 3.434 17 | 45.327 46 |
| 2.80 | 2.568 00 | 27.138 21 | 10.567 85 | 1.602 50 | 3.500 12 | 45.745 86 |
| 2.82 | 2.590 48 | 27.978 82 | 10.800 64 | 1.609 50 | 3.567 36 | 46.161 06 |
| 2.84 | 2.613 12 | 28.844 06 | 11.038 18 | 1.616 51 | 3.635 93 | 46.573 09 |
| 2.86 | 2.635 92 | 29.734 55 | 11.280 53 | 1.623 55 | 3.705 83 | 46.981 95 |
| 2.88 | 2.658 88 | 30.650 97 | 11.527 78 | 1.630 61 | 3.777 11 | 47.387 68 |
| 2.90 | 2.682 00 | 31.593 98 | 11.780 02 | 1.637 68 | 3.849 76 | 47.790 28 |
| 2.92 | 2.705 28 | 32.564 27 | 12.037 31 | 1.644 77 | 3.923 82 | 48.189 80 |
| 2.94 | 2.728 72 | 33.562 55 | 12.299 75 | 1.651 88 | 3.999 31 | 48.586 24 |
| 2.96 | 2.752 32 | 34.589 54 | 12.567 42 | 1.659 01 | 4.076 25 | 48.979 62 |
| 2.98 | 2.776 08 | 35.645 97 | 12.840 41 | 1.666 16 | 4.154 65 | 49.369 97 |
| 3.00 | 2.800 00 | 36.732 60 | 13.118 80 | 1.673 32 | 4.234 56 | 49.757 32 |
| 3.05 | 2.860 50 | 39.586 34 | 13.838 97 | 1.691 30 | 4.441 01 | 50.712 67 |
| 3.10 | 2.922 00 | 42.646 09 | 14.594 84 | 1.709 38 | 4.657 30 | 51.649 72 |
| 3.15 | 2.984 50 | 45.924 97 | 15.387 84 | 1.727 57 | 4.883 82 | 52.568 81 |
| 3.20 | 3.048 00 | 49.436 84 | 16.219 45 | 1.745 85 | 5.120 95 | 53.470 31 |
| 3.25 | 3.112 50 | 53.196 26 | 17.091 19 | 1.764 23 | 5.369 08 | 54.354 57 |
| 3.30 | 3.178 00 | 57.218 57 | 18.004 60 | 1.782 69 | 5.628 63 | 55.221 95 |
| 3.35 | 3.244 50 | 61.519 89 | 18.961 31 | 1.801 25 | 5.900 02 | 56.072 81 |
| 3.40 | 3.312 00 | 66.117 20 | 19.962 94 | 1.819 89 | 6.183 68 | 56.907 49 |
| 3.45 | 3.380 50 | 71.028 34 | 21.011 22 | 1.838 61 | 6.480 06 | 57.726 36 |
| 3.50 | 3.450 00 | 76.272 00 | 22.107 85 | 1.857 42 | 6.789 60 | 58.529 74 |
| 3.55 | 3.520 50 | 81.867 87 | 23.254 64 | 1.876 30 | 7.112 79 | 59.317 99 |
| 3.60 | 3.592 00 | 87.836 57 | 24.453 42 | 1.895 26 | 7.450 09 | 60.091 43 |
| 3.65 | 3.664 50 | 94.199 76 | 25.706 06 | 1.914 29 | 7.802 01 | 60.850 41 |
| 3.70 | 3.738 00 | 100.980 10 | 27.014 50 | 1.933 39 | 8.169 05 | 61.595 26 |
| 3.75 | 3.812 50 | 108.201 36 | 28.380 72 | 1.952 56 | 8.551 72 | 62.326 27 |
| 3.80 | 3.888 00 | 115.888 43 | 29.806 73 | 1.971 80 | 8.950 56 | 63.043 78 |
| 3.85 | 3.964 49 | 124.067 40 | 31.294 63 | 1.991 10 | 9.366 12 | 63.748 09 |
| 3.90 | 4.041 99 | 132.765 49 | 32.846 52 | 2.010 47 | 9.798 95 | 64.439 50 |
| 3.95 | 4.120 49 | 142.011 32 | 34.464 62 | 2.029 90 | 10.249 62 | 65.118 31 |
| 4.00 | 4.199 99 | 151.834 58 | 36.151 13 | 2.049 39 | 10.718 72 | 65.784 80 |
| 4.05 | 4.280 50 | 162.266 74 | 37.908 39 | 2.068 94 | 11.206 86 | 66.439 29 |
| 4.10 | 4.362 00 | 173.340 19 | 39.738 71 | 2.088 54 | 11.714 64 | 67.082 01 |
| 4.15 | 4.444 50 | 185.088 88 | 41.644 47 | 2.108 20 | 12.242 69 | 67.713 26 |
| 4.20 | 4.528 00 | 197.548 26 | 43.628 13 | 2.127 91 | 12.791 66 | 68.333 28 |
| 4.25 | 4.612 50 | 210.755 45 | 45.692 19 | 2.147 67 | 13.362 19 | 68.942 34 |
| 4.30 | 4.698 01 | 224.748 89 | 47.839 20 | 2.167 49 | 13.954 95 | 69.540 70 |
| 4.35 | 4.784 51 | 239.568 98 | 50.071 81 | 2.187 35 | 14.570 63 | 70.128 59 |
| 4.40 | 4.872 01 | 255.257 69 | 52.392 68 | 2.207 26 | 15.209 95 | 70.706 26 |
| 4.45 | 4.960 51 | 271.858 55 | 54.804 52 | 2.227 22 | 15.873 59 | 71.273 94 |
| 4.50 | 5.050 01 | 289.417 21 | 57.310 17 | 2.247 22 | 16.562 31 | 71.831 86 |
| 4.60 | 5.232 02 | 327.599 15 | 62.614 28 | 2.287 36 | 18.017 96 | 72.919 30 |
| 4.70 | 5.418 02 | 370.205 87 | 68.328 57 | 2.327 66 | 19.583 04 | 73.970 28 |
| 4.80 | 5.608 03 | 417.672 52 | 74.477 59 | 2.368 13 | 21.263 98 | 74.986 45 |
| 4.90 | 5.802 03 | 470.468 96 | 81.086 90 | 2.408 74 | 23.067 45 | 75.969 35 |
| 5.00 | 6.000 04 | 529.101 93 | 88.183 07 | 2.449 50 | 25.000 39 | 76.920 43 |
| 5.10 | 6.202 04 | 594.117 25 | 95.793 76 | 2.490 39 | 27.070 04 | 77.841 09 |
| 5.20 | 6.408 05 | 666.102 11 | 103.947 69 | 2.531 41 | 29.283 89 | 78.732 67 |
| 5.30 | 6.618 06 | 745.687 19 | 112.674 64 | 2.572 56 | 31.649 71 | 79.596 42 |
| 5.40 | 6.832 06 | 833.549 62 | 122.005 55 | 2.613 82 | 34.175 57 | 80.433 50 |
| 5.50 | 7.050 07 | 930.415 10 | 131.972 47 | 2.655 20 | 36.869 84 | 81.245 07 |

# APPENDIX C

# Normal Shock Tables for $\gamma = 1.4$

| $M_1$ | $M_2$ | $p_2/p_1$ | $T_2/T_1$ | $\rho_2/\rho_1$ | $p_{02}/p_{01}$ | $p_{02}/p_1$ |
|-------|-------|-----------|-----------|-----------------|-----------------|--------------|
| 1.00 | 1.000 00 | 1.000 00 | 1.000 00 | 1.000 00 | 1.000 00 | 1.892 93 |
| 1.02 | 0.980 52 | 1.047 13 | 1.013 25 | 1.033 44 | 0.999 99 | 1.937 90 |
| 1.04 | 0.962 03 | 1.095 20 | 1.026 34 | 1.067 09 | 0.999 92 | 1.984 42 |
| 1.06 | 0.944 45 | 1.144 20 | 1.039 31 | 1.100 92 | 0.999 75 | 2.032 45 |
| 1.08 | 0.927 71 | 1.194 13 | 1.052 17 | 1.134 92 | 0.999 43 | 2.081 94 |
| 1.10 | 0.911 77 | 1.245 00 | 1.064 94 | 1.169 08 | 0.998 93 | 2.132 85 |
| 1.12 | 0.896 56 | 1.296 80 | 1.077 63 | 1.203 38 | 0.998 21 | 2.185 13 |
| 1.14 | 0.882 04 | 1.349 53 | 1.090 27 | 1.237 79 | 0.997 26 | 2.238 77 |
| 1.16 | 0.868 16 | 1.403 20 | 1.102 87 | 1.272 31 | 0.996 05 | 2.293 72 |
| 1.18 | 0.854 88 | 1.457 80 | 1.115 44 | 1.306 93 | 0.994 57 | 2.349 98 |
| 1.20 | 0.842 17 | 1.513 33 | 1.127 99 | 1.341 61 | 0.992 80 | 2.407 50 |
| 1.22 | 0.829 99 | 1.569 80 | 1.140 54 | 1.376 36 | 0.990 73 | 2.466 28 |
| 1.24 | 0.818 30 | 1.627 20 | 1.153 09 | 1.411 16 | 0.988 36 | 2.526 29 |
| 1.26 | 0.807 09 | 1.685 53 | 1.165 66 | 1.445 99 | 0.985 68 | 2.587 53 |
| 1.28 | 0.796 31 | 1.744 80 | 1.178 25 | 1.480 84 | 0.982 68 | 2.649 96 |
| 1.30 | 0.785 96 | 1.805 00 | 1.190 87 | 1.515 69 | 0.979 37 | 2.713 59 |
| 1.32 | 0.776 00 | 1.866 13 | 1.203 53 | 1.550 55 | 0.975 75 | 2.778 40 |
| 1.34 | 0.766 41 | 1.928 20 | 1.216 24 | 1.585 38 | 0.971 82 | 2.844 38 |
| 1.36 | 0.757 18 | 1.991 20 | 1.229 00 | 1.620 18 | 0.967 58 | 2.911 52 |
| 1.38 | 0.748 29 | 2.055 13 | 1.241 81 | 1.654 94 | 0.963 04 | 2.979 80 |
| 1.40 | 0.739 71 | 2.120 00 | 1.254 69 | 1.689 65 | 0.958 19 | 3.049 23 |
| 1.42 | 0.731 44 | 2.185 80 | 1.267 64 | 1.724 30 | 0.953 06 | 3.119 80 |
| 1.44 | 0.723 45 | 2.252 53 | 1.280 66 | 1.758 88 | 0.947 65 | 3.191 49 |
| 1.46 | 0.715 74 | 2.320 20 | 1.293 76 | 1.793 37 | 0.941 96 | 3.264 30 |
| 1.48 | 0.708 29 | 2.388 80 | 1.306 95 | 1.827 77 | 0.936 00 | 3.338 23 |
| 1.50 | 0.701 09 | 2.458 33 | 1.320 22 | 1.862 07 | 0.929 79 | 3.413 27 |
| 1.52 | 0.694 13 | 2.528 80 | 1.333 57 | 1.896 26 | 0.923 32 | 3.489 42 |
| 1.54 | 0.687 39 | 2.600 20 | 1.347 03 | 1.930 33 | 0.916 62 | 3.566 66 |
| 1.56 | 0.680 87 | 2.672 53 | 1.360 57 | 1.964 27 | 0.909 70 | 3.645 01 |
| 1.58 | 0.674 55 | 2.745 80 | 1.374 22 | 1.998 08 | 0.902 55 | 3.724 44 |
| 1.60 | 0.668 44 | 2.820 00 | 1.387 97 | 2.031 75 | 0.895 20 | 3.804 97 |

| $M_1$ | $M_2$ | $p_2/p_1$ | $T_2/T_1$ | $\rho_2/\rho_1$ | $p_{02}/p_{01}$ | $p_{02}/p_1$ |
|-------|-------|-----------|-----------|-----------------|-----------------|--------------|
| 1.62 | 0.662 51 | 2.895 13 | 1.401 82 | 2.065 26 | 0.887 65 | 3.886 58 |
| 1.64 | 0.656 77 | 2.971 20 | 1.415 78 | 2.098 63 | 0.879 92 | 3.969 28 |
| 1.66 | 0.651 19 | 3.048 20 | 1.429 85 | 2.131 83 | 0.872 01 | 4.053 05 |
| 1.68 | 0.645 79 | 3.126 13 | 1.444 03 | 2.164 86 | 0.863 94 | 4.137 90 |
| 1.70 | 0.640 54 | 3.205 00 | 1.458 33 | 2.197 72 | 0.855 72 | 4.223 83 |
| 1.72 | 0.635 45 | 3.284 80 | 1.472 74 | 2.230 40 | 0.847 36 | 4.310 83 |
| 1.74 | 0.630 51 | 3.365 53 | 1.487 27 | 2.262 89 | 0.838 86 | 4.398 90 |
| 1.76 | 0.625 70 | 3.447 20 | 1.501 92 | 2.295 20 | 0.830 24 | 4.488 04 |
| 1.78 | 0.621 04 | 3.529 80 | 1.516 69 | 2.327 31 | 0.821 51 | 4.578 24 |
| 1.80 | 0.616 50 | 3.613 33 | 1.531 58 | 2.359 22 | 0.812 68 | 4.669 51 |
| 1.82 | 0.612 09 | 3.697 80 | 1.546 59 | 2.390 93 | 0.803 76 | 4.761 84 |
| 1.84 | 0.607 80 | 3.783 20 | 1.561 73 | 2.422 44 | 0.794 76 | 4.855 24 |
| 1.86 | 0.603 63 | 3.869 53 | 1.577 00 | 2.453 73 | 0.785 69 | 4.949 69 |
| 1.88 | 0.599 57 | 3.956 80 | 1.592 39 | 2.484 81 | 0.776 55 | 5.045 20 |
| 1.90 | 0.595 62 | 4.045 00 | 1.607 91 | 2.515 68 | 0.767 36 | 5.141 77 |
| 1.92 | 0.591 77 | 4.134 13 | 1.623 57 | 2.546 32 | 0.758 12 | 5.239 40 |
| 1.94 | 0.588 02 | 4.224 20 | 1.639 35 | 2.576 75 | 0.748 84 | 5.338 08 |
| 1.96 | 0.584 37 | 4.315 20 | 1.655 27 | 2.606 95 | 0.739 54 | 5.437 81 |
| 1.98 | 0.580 82 | 4.407 13 | 1.671 32 | 2.636 92 | 0.730 21 | 5.538 60 |
| 2.00 | 0.577 35 | 4.500 00 | 1.687 50 | 2.666 67 | 0.720 87 | 5.640 44 |
| 2.02 | 0.573 97 | 4.593 80 | 1.703 82 | 2.696 18 | 0.711 53 | 5.743 32 |
| 2.04 | 0.570 68 | 4.688 53 | 1.720 27 | 2.725 46 | 0.702 18 | 5.847 26 |
| 2.06 | 0.567 47 | 4.784 19 | 1.736 86 | 2.754 51 | 0.692 84 | 5.952 25 |
| 2.08 | 0.564 33 | 4.880 80 | 1.753 59 | 2.783 32 | 0.683 51 | 6.058 29 |
| 2.10 | 0.561 28 | 4.978 33 | 1.770 45 | 2.811 90 | 0.674 20 | 6.165 37 |
| 2.12 | 0.558 29 | 5.076 79 | 1.787 45 | 2.840 24 | 0.664 92 | 6.273 50 |
| 2.14 | 0.555 38 | 5.176 19 | 1.804 59 | 2.868 34 | 0.655 67 | 6.382 68 |
| 2.16 | 0.552 54 | 5.276 53 | 1.821 87 | 2.896 21 | 0.646 45 | 6.492 90 |
| 2.18 | 0.549 77 | 5.377 79 | 1.839 30 | 2.923 83 | 0.637 27 | 6.604 16 |
| 2.20 | 0.547 06 | 5.479 99 | 1.856 86 | 2.951 22 | 0.628 14 | 6.716 47 |
| 2.22 | 0.544 41 | 5.583 13 | 1.874 56 | 2.978 36 | 0.619 05 | 6.829 83 |
| 2.24 | 0.541 82 | 5.687 19 | 1.892 41 | 3.005 27 | 0.610 02 | 6.944 23 |
| 2.26 | 0.539 30 | 5.792 19 | 1.910 40 | 3.031 93 | 0.601 05 | 7.059 67 |
| 2.28 | 0.536 83 | 5.898 13 | 1.928 53 | 3.058 36 | 0.592 14 | 7.176 15 |
| 2.30 | 0.534 41 | 6.004 99 | 1.946 80 | 3.084 55 | 0.583 30 | 7.293 67 |
| 2.32 | 0.532 05 | 6.112 79 | 1.965 22 | 3.110 49 | 0.574 52 | 7.412 24 |
| 2.34 | 0.529 74 | 6.221 53 | 1.983 78 | 3.136 20 | 0.565 81 | 7.531 84 |
| 2.36 | 0.527 49 | 6.331 19 | 2.002 48 | 3.161 67 | 0.557 18 | 7.652 49 |
| 2.38 | 0.525 28 | 6.441 79 | 2.021 33 | 3.186 90 | 0.548 62 | 7.774 18 |
| 2.40 | 0.523 12 | 6.553 33 | 2.040 33 | 3.211 89 | 0.540 14 | 7.896 91 |
| 2.42 | 0.521 00 | 6.665 79 | 2.059 47 | 3.236 65 | 0.531 75 | 8.020 67 |
| 2.44 | 0.518 94 | 6.779 19 | 2.078 76 | 3.261 17 | 0.523 44 | 8.145 48 |
| 2.46 | 0.516 91 | 6.893 53 | 2.098 19 | 3.285 46 | 0.515 21 | 8.271 32 |
| 2.48 | 0.514 93 | 7.008 79 | 2.117 77 | 3.309 51 | 0.507 07 | 8.398 21 |
| 2.50 | 0.512 99 | 7.124 99 | 2.137 50 | 3.333 33 | 0.499 02 | 8.526 13 |
| 2.52 | 0.511 09 | 7.242 12 | 2.157 37 | 3.356 92 | 0.491 05 | 8.655 09 |
| 2.54 | 0.509 23 | 7.360 19 | 2.177 39 | 3.380 28 | 0.483 18 | 8.785 08 |
| 2.56 | 0.507 41 | 7.479 19 | 2.197 56 | 3.403 41 | 0.475 40 | 8.916 12 |
| 2.58 | 0.505 62 | 7.599 12 | 2.217 88 | 3.426 31 | 0.467 72 | 9.048 19 |
| 2.60 | 0.503 87 | 7.719 99 | 2.238 34 | 3.448 98 | 0.460 12 | 9.181 30 |
| 2.62 | 0.502 16 | 7.841 79 | 2.258 95 | 3.471 43 | 0.452 63 | 9.315 44 |
| 2.64 | 0.500 48 | 7.964 52 | 2.279 71 | 3.493 65 | 0.445 22 | 9.450 63 |
| 2.66 | 0.498 83 | 8.088 19 | 2.300 62 | 3.515 65 | 0.437 92 | 9.586 84 |

| $M_1$ | $M_2$ | $p_2/p_1$ | $T_2/T_1$ | $\rho_2/\rho_1$ | $p_{02}/p_{01}$ | $p_{02}/p_1$ |
|------|----------|-----------|-----------|-----------|-----------|-----------|
| 2.68 | 0.497 22 | 8.212 79 | 2.321 68 | 3.537 43 | 0.430 71 | 9.724 10 |
| 2.70 | 0.495 63 | 8.338 32 | 2.342 89 | 3.558 99 | 0.423 59 | 9.862 39 |
| 2.72 | 0.494 08 | 8.464 79 | 2.364 25 | 3.580 33 | 0.416 57 | 10.001 71 |
| 2.74 | 0.492 56 | 8.592 19 | 2.385 75 | 3.601 46 | 0.409 65 | 10.142 08 |
| 2.76 | 0.491 07 | 8.720 52 | 2.407 41 | 3.622 37 | 0.402 83 | 10.283 47 |
| 2.78 | 0.489 60 | 8.849 79 | 2.429 22 | 3.643 06 | 0.396 10 | 10.425 91 |
| 2.80 | 0.488 17 | 8.979 99 | 2.451 17 | 3.663 55 | 0.389 46 | 10.569 37 |
| 2.82 | 0.486 76 | 9.111 12 | 2.473 28 | 3.683 83 | 0.382 93 | 10.713 88 |
| 2.84 | 0.485 38 | 9.243 19 | 2.495 53 | 3.703 89 | 0.376 49 | 10.859 41 |
| 2.86 | 0.484 02 | 9.376 19 | 2.517 94 | 3.723 75 | 0.370 14 | 11.005 99 |
| 2.88 | 0.482 69 | 9.510 12 | 2.540 50 | 3.743 41 | 0.363 89 | 11.153 59 |
| 2.90 | 0.481 38 | 9.644 99 | 2.563 21 | 3.762 86 | 0.357 73 | 11.302 23 |
| 2.92 | 0.480 10 | 9.780 79 | 2.586 06 | 3.782 11 | 0.351 67 | 11.451 91 |
| 2.94 | 0.478 84 | 9.917 52 | 2.609 07 | 3.801 17 | 0.345 70 | 11.602 62 |
| 2.96 | 0.477 60 | 10.055 19 | 2.632 23 | 3.820 02 | 0.339 82 | 11.754 36 |
| 2.98 | 0.476 38 | 10.193 79 | 2.655 55 | 3.838 68 | 0.334 04 | 11.907 14 |
| 3.00 | 0.475 19 | 10.333 32 | 2.679 01 | 3.857 14 | 0.328 34 | 12.060 95 |
| 3.05 | 0.472 30 | 10.686 24 | 2.738 33 | 3.902 46 | 0.314 50 | 12.450 00 |
| 3.10 | 0.469 53 | 11.044 99 | 2.798 60 | 3.946 61 | 0.301 21 | 12.845 51 |
| 3.15 | 0.466 89 | 11.409 57 | 2.859 82 | 3.989 61 | 0.288 46 | 13.247 48 |
| 3.20 | 0.464 35 | 11.779 98 | 2.921 99 | 4.031 49 | 0.276 23 | 13.655 90 |
| 3.25 | 0.461 92 | 12.156 23 | 2.985 11 | 4.072 29 | 0.264 51 | 14.070 78 |
| 3.30 | 0.459 59 | 12.538 32 | 3.049 19 | 4.112 02 | 0.253 28 | 14.492 12 |
| 3.35 | 0.457 35 | 12.926 23 | 3.114 22 | 4.150 71 | 0.242 52 | 14.919 91 |
| 3.40 | 0.455 20 | 13.319 98 | 3.180 20 | 4.188 40 | 0.232 23 | 15.354 15 |
| 3.45 | 0.453 14 | 13.719 56 | 3.247 15 | 4.225 11 | 0.222 37 | 15.794 84 |
| 3.50 | 0.451 15 | 14.124 98 | 3.315 05 | 4.260 87 | 0.212 95 | 16.241 98 |
| 3.55 | 0.449 25 | 14.536 23 | 3.383 91 | 4.295 70 | 0.203 93 | 16.695 57 |
| 3.60 | 0.447 41 | 14.953 31 | 3.453 72 | 4.329 62 | 0.195 31 | 17.155 61 |
| 3.65 | 0.445 65 | 15.376 23 | 3.524 50 | 4.362 67 | 0.187 07 | 17.622 10 |
| 3.70 | 0.443 95 | 15.804 98 | 3.596 24 | 4.394 86 | 0.179 19 | 18.095 04 |
| 3.75 | 0.442 31 | 16.239 56 | 3.668 94 | 4.426 23 | 0.171 67 | 18.574 43 |
| 3.80 | 0.440 73 | 16.679 98 | 3.742 60 | 4.456 79 | 0.164 47 | 19.060 26 |
| 3.85 | 0.439 21 | 17.126 22 | 3.817 22 | 4.486 57 | 0.157 60 | 19.552 54 |
| 3.90 | 0.437 74 | 17.578 31 | 3.892 81 | 4.515 58 | 0.151 03 | 20.051 26 |
| 3.95 | 0.436 33 | 18.036 22 | 3.969 36 | 4.543 86 | 0.144 75 | 29.556 44 |
| 4.00 | 0.434 96 | 18.499 97 | 4.046 87 | 4.571 43 | 0.138 76 | 21.068 05 |
| 4.05 | 0.433 64 | 18.969 57 | 4.125 35 | 4.598 29 | 0.133 03 | 21.586 12 |
| 4.10 | 0.432 36 | 19.444 99 | 4.204 79 | 4.624 48 | 0.127 56 | 22.110 64 |
| 4.15 | 0.431 13 | 19.926 26 | 4.285 20 | 4.650 02 | 0.122 33 | 22.641 61 |
| 4.20 | 0.429 94 | 20.413 35 | 4.366 57 | 4.674 91 | 0.117 33 | 23.179 01 |
| 4.25 | 0.428 78 | 20.906 28 | 4.448 91 | 4.699 19 | 0.112 56 | 23.722 86 |
| 4.30 | 0.427 67 | 21.405 04 | 4.532 22 | 4.722 86 | 0.108 00 | 24.273 16 |
| 4.35 | 0.426 59 | 21.909 64 | 4.616 49 | 4.745 95 | 0.103 64 | 24.829 90 |
| 4.40 | 0.425 54 | 22.420 06 | 4.701 73 | 4.768 48 | 0.099 48 | 25.393 08 |
| 4.45 | 0.424 53 | 22.936 33 | 4.787 93 | 4.790 45 | 0.095 50 | 25.962 70 |
| 4.50 | 0.423 55 | 23.458 42 | 4.875 10 | 4.811 88 | 0.091 70 | 26.538 76 |
| 4.60 | 0.421 68 | 24.520 12 | 5.052 34 | 4.853 22 | 0.084 59 | 27.710 22 |
| 4.70 | 0.419 92 | 25.605 14 | 5.233 46 | 4.892 59 | 0.078 08 | 28.907 45 |
| 4.80 | 0.418 26 | 26.713 51 | 5.418 45 | 4.930 11 | 0.072 14 | 30.130 45 |
| 4.90 | 0.416 70 | 27.845 20 | 5.607 30 | 4.965 88 | 0.066 70 | 31.379 21 |
| 5.00 | 0.415 23 | 29.000 23 | 5.800 04 | 5.000 01 | 0.061 72 | 32.653 73 |

# APPENDIX D

# Tables for One-Dimensional Adiabatic Flow with Friction for $\gamma = 1.4$

| $M$ | $p/p^*$ | $T/T^*$ | $\rho/\rho^*$ | $V/V^*$ | $p_0/p_0^*$ | $4fl^*/D_H$ |
|------|-----------|-----------|-----------|-----------|-----------|-----------|
| 0.10 | 10.943 51 | 1.197 60 | 9.137 83 | 0.109 44 | 5.821 83 | 66.921 55 |
| 0.12 | 9.115 59 | 1.196 55 | 7.618 20 | 0.131 26 | 4.864 32 | 45.407 96 |
| 0.14 | 7.809 32 | 1.195 31 | 6.533 27 | 0.153 06 | 4.182 40 | 32.511 31 |
| 0.16 | 6.829 07 | 1.193 89 | 5.720 03 | 0.174 82 | 3.672 74 | 24.197 83 |
| 0.18 | 6.066 18 | 1.192 27 | 5.087 91 | 0.196 54 | 3.277 93 | 18.542 65 |
| 0.20 | 5.455 45 | 1.190 48 | 4.582 57 | 0.218 22 | 2.963 52 | 14.533 26 |
| 0.22 | 4.955 37 | 1.188 50 | 4.169 45 | 0.239 84 | 2.707 60 | 11.596 05 |
| 0.24 | 4.538 29 | 1.186 33 | 3.825 47 | 0.261 41 | 2.495 56 | 9.386 48 |
| 0.26 | 4.185 05 | 1.183 99 | 3.534 70 | 0.282 91 | 2.317 29 | 7.687 56 |
| 0.28 | 3.881 99 | 1.181 47 | 3.285 71 | 0.304 35 | 2.165 55 | 6.357 21 |
| 0.30 | 3.619 06 | 1.178 78 | 3.070 17 | 0.325 72 | 2.035 06 | 5.299 25 |
| 0.32 | 3.388 74 | 1.175 92 | 2.881 79 | 0.347 01 | 1.921 85 | 4.446 74 |
| 0.34 | 3.185 29 | 1.172 88 | 2.715 78 | 0.368 22 | 1.822 88 | 3.751 95 |
| 0.36 | 3.004 22 | 1.169 68 | 2.568 41 | 0.389 35 | 1.735 78 | 3.180 12 |
| 0.38 | 2.842 00 | 1.166 32 | 2.436 73 | 0.410 39 | 1.658 70 | 2.705 45 |
| 0.40 | 2.695 82 | 1.162 79 | 2.318 41 | 0.431 33 | 1.590 14 | 2.308 49 |
| 0.42 | 2.563 38 | 1.159 11 | 2.211 51 | 0.452 18 | 1.528 90 | 1.974 37 |
| 0.44 | 2.442 81 | 1.155 27 | 2.114 49 | 0.472 93 | 1.474 01 | 1.691 53 |
| 0.46 | 2.332 56 | 1.151 28 | 2.026 06 | 0.493 57 | 1.424 63 | 1.450 91 |
| 0.48 | 2.231 35 | 1.147 14 | 1.945 14 | 0.514 10 | 1.380 10 | 1.245 34 |
| 0.50 | 2.138 09 | 1.142 86 | 1.870 83 | 0.534 52 | 1.339 84 | 1.069 06 |
| 0.52 | 2.051 87 | 1.138 43 | 1.802 37 | 0.554 83 | 1.303 39 | 0.917 42 |
| 0.54 | 1.971 92 | 1.133 87 | 1.739 10 | 0.575 01 | 1.270 32 | 0.786 63 |
| 0.56 | 1.897 55 | 1.129 18 | 1.680 47 | 0.595 07 | 1.240 29 | 0.673 57 |
| 0.58 | 1.828 20 | 1.124 35 | 1.626 00 | 0.615 01 | 1.213 01 | 0.575 68 |
| 0.60 | 1.763 36 | 1.119 40 | 1.575 27 | 0.634 81 | 1.188 20 | 0.490 82 |
| 0.62 | 1.702 61 | 1.114 33 | 1.527 92 | 0.654 48 | 1.165 65 | 0.417 20 |

| $M$ | $p/p^*$ | $T/T^*$ | $\rho/\rho^*$ | $V/V^*$ | $p_0/p_0^*$ | $4fl^*/D_H$ |
|------|---------|---------|---------|---------|---------|---------|
| 0.64 | 1.645 56 | 1.109 14 | 1.483 64 | 0.674 02 | 1.145 15 | 0.353 30 |
| 0.66 | 1.591 87 | 1.103 83 | 1.442 13 | 0.693 42 | 1.126 54 | 0.297 85 |
| 0.68 | 1.541 26 | 1.098 42 | 1.403 16 | 0.712 68 | 1.109 65 | 0.249 78 |
| 0.70 | 1.493 45 | 1.092 90 | 1.366 51 | 0.731 79 | 1.094 37 | 0.208 14 |
| 0.72 | 1.448 23 | 1.087 27 | 1.331 98 | 0.750 76 | 1.080 57 | 0.172 15 |
| 0.74 | 1.405 37 | 1.081 55 | 1.299 41 | 0.769 58 | 1.068 14 | 0.141 12 |
| 0.76 | 1.364 70 | 1.075 73 | 1.268 63 | 0.788 25 | 1.057 00 | 0.114 47 |
| 0.78 | 1.326 06 | 1.069 82 | 1.239 51 | 0.806 77 | 1.047 05 | 0.091 67 |
| 0.80 | 1.289 28 | 1.063 83 | 1.211 92 | 0.825 14 | 1.038 23 | 0.072 29 |
| 0.82 | 1.254 23 | 1.057 75 | 1.185 75 | 0.843 35 | 1.030 46 | 0.055 93 |
| 0.84 | 1.220 80 | 1.051 60 | 1.160 90 | 0.861 40 | 1.023 70 | 0.042 26 |
| 0.86 | 1.188 88 | 1.045 37 | 1.137 28 | 0.879 29 | 1.017 87 | 0.030 97 |
| 0.88 | 1.158 35 | 1.039 07 | 1.114 80 | 0.897 03 | 1.012 94 | 0.021 79 |
| 0.90 | 1.129 13 | 1.032 70 | 1.093 38 | 0.914 60 | 1.008 86 | 0.014 51 |
| 0.92 | 1.101 14 | 1.026 27 | 1.072 95 | 0.932 01 | 1.005 60 | 0.008 91 |
| 0.94 | 1.074 30 | 1.019 78 | 1.053 46 | 0.949 25 | 1.003 11 | 0.004 82 |
| 0.96 | 1.048 54 | 1.013 24 | 1.034 84 | 0.966 33 | 1.001 36 | 0.002 06 |
| 0.98 | 1.023 79 | 1.006 64 | 1.017 04 | 0.983 25 | 1.000 34 | 0.000 49 |
| 1.00 | 1.000 00 | 1.000 00 | 1.000 00 | 1.000 00 | 1.000 00 | 0.000 00 |
| 1.02 | 0.977 11 | 0.993 31 | 0.983 69 | 1.016 58 | 1.000 33 | 0.000 46 |
| 1.04 | 0.955 07 | 0.986 58 | 0.968 06 | 1.033 00 | 1.001 30 | 0.001 77 |
| 1.06 | 0.933 83 | 0.979 82 | 0.953 06 | 1.049 25 | 1.002 91 | 0.003 84 |
| 1.08 | 0.913 35 | 0.973 02 | 0.938 68 | 1.065 33 | 1.005 12 | 0.006 58 |
| 1.10 | 0.893 59 | 0.966 18 | 0.924 86 | 1.081 24 | 1.007 93 | 0.009 93 |
| 1.12 | 0.874 51 | 0.959 32 | 0.911 59 | 1.096 98 | 1.011 31 | 0.013 82 |
| 1.14 | 0.856 08 | 0.952 44 | 0.898 83 | 1.112 56 | 1.015 27 | 0.018 19 |
| 1.16 | 0.838 27 | 0.945 54 | 0.886 55 | 1.127 97 | 1.019 78 | 0.022 98 |
| 1.18 | 0.821 04 | 0.938 61 | 0.874 73 | 1.143 21 | 1.024 84 | 0.028 14 |
| 1.20 | 0.804 36 | 0.931 68 | 0.863 35 | 1.158 28 | 1.030 44 | 0.033 64 |
| 1.22 | 0.788 22 | 0.924 73 | 0.852 38 | 1.173 18 | 1.036 57 | 0.039 43 |
| 1.24 | 0.772 58 | 0.917 77 | 0.841 81 | 1.187 92 | 1.043 23 | 0.045 47 |
| 1.26 | 0.757 43 | 0.910 80 | 0.831 61 | 1.202 49 | 1.050 41 | 0.051 74 |
| 1.28 | 0.742 74 | 0.903 83 | 0.821 76 | 1.216 90 | 1.058 10 | 0.058 20 |
| 1.30 | 0.728 48 | 0.896 86 | 0.812 26 | 1.231 14 | 1.066 30 | 0.064 83 |
| 1.32 | 0.714 65 | 0.889 89 | 0.803 08 | 1.245 21 | 1.075 02 | 0.071 61 |
| 1.34 | 0.701 22 | 0.882 92 | 0.794 21 | 1.259 12 | 1.084 24 | 0.078 50 |
| 1.36 | 0.688 18 | 0.875 96 | 0.785 63 | 1.272 86 | 1.093 96 | 0.085 50 |
| 1.38 | 0.675 51 | 0.869 01 | 0.777 34 | 1.286 45 | 1.104 19 | 0.092 59 |
| 1.40 | 0.663 20 | 0.862 07 | 0.769 31 | 1.299 87 | 1.114 93 | 0.099 74 |
| 1.42 | 0.651 22 | 0.855 14 | 0.761 54 | 1.313 13 | 1.126 16 | 0.106 94 |
| 1.44 | 0.639 58 | 0.848 22 | 0.754 02 | 1.326 23 | 1.137 90 | 0.114 19 |
| 1.46 | 0.628 25 | 0.841 33 | 0.746 73 | 1.339 17 | 1.150 15 | 0.121 46 |
| 1.48 | 0.617 22 | 0.834 45 | 0.739 67 | 1.351 95 | 1.162 90 | 0.128 75 |
| 1.50 | 0.606 48 | 0.827 59 | 0.732 83 | 1.364 58 | 1.176 17 | 0.136 05 |
| 1.52 | 0.596 02 | 0.820 75 | 0.726 19 | 1.377 05 | 1.189 94 | 0.143 35 |
| 1.54 | 0.585 83 | 0.813 93 | 0.719 76 | 1.389 36 | 1.204 23 | 0.150 63 |
| 1.56 | 0.575 91 | 0.807 15 | 0.713 51 | 1.401 52 | 1.219 04 | 0.157 90 |
| 1.58 | 0.566 23 | 0.800 38 | 0.707 45 | 1.413 53 | 1.234 37 | 0.165 14 |
| 1.60 | 0.556 79 | 0.793 65 | 0.701 56 | 1.425 39 | 1.250 23 | 0.172 36 |
| 1.62 | 0.547 59 | 0.786 95 | 0.695 84 | 1.437 10 | 1.266 62 | 0.179 53 |
| 1.64 | 0.538 62 | 0.780 28 | 0.690 29 | 1.448 66 | 1.283 55 | 0.186 67 |
| 1.66 | 0.529 86 | 0.773 63 | 0.684 90 | 1.460 08 | 1.301 02 | 0.193 77 |
| 1.68 | 0.521 31 | 0.767 03 | 0.679 65 | 1.471 35 | 1.319 04 | 0.200 81 |

| M | $p/p^*$ | $T/T^*$ | $\rho/\rho^*$ | $V/V^*$ | $p_0/p_0^*$ | $4fl^*/D_H$ |
|------|---------|---------|---------|---------|---------|---------|
| 1.70 | 0.512 97 | 0.760 46 | 0.674 55 | 1.482 47 | 1.337 61 | 0.207 80 |
| 1.72 | 0.504 82 | 0.753 92 | 0.669 59 | 1.493 45 | 1.356 73 | 0.214 74 |
| 1.74 | 0.496 86 | 0.747 42 | 0.664 77 | 1.504 29 | 1.376 43 | 0.221 62 |
| 1.76 | 0.489 09 | 0.740 96 | 0.660 07 | 1.514 99 | 1.396 70 | 0.228 44 |
| 1.78 | 0.481 49 | 0.734 54 | 0.655 50 | 1.525 55 | 1.417 54 | 0.235 19 |
| 1.80 | 0.474 07 | 0.728 16 | 0.651 05 | 1.535 98 | 1.438 98 | 0.241 89 |
| 1.82 | 0.466 81 | 0.721 81 | 0.646 72 | 1.546 26 | 1.461 01 | 0.248 51 |
| 1.84 | 0.459 72 | 0.715 51 | 0.642 50 | 1.556 42 | 1.483 65 | 0.255 07 |
| 1.86 | 0.452 78 | 0.709 25 | 0.638 39 | 1.566 44 | 1.506 89 | 0.261 56 |
| 1.88 | 0.446 00 | 0.703 04 | 0.634 39 | 1.576 33 | 1.530 76 | 0.267 98 |
| 1.90 | 0.439 36 | 0.696 86 | 0.630 48 | 1.586 09 | 1.555 25 | 0.274 33 |
| 1.92 | 0.432 87 | 0.690 74 | 0.626 68 | 1.595 72 | 1.580 39 | 0.280 61 |
| 1.94 | 0.426 51 | 0.684 65 | 0.622 97 | 1.605 22 | 1.606 17 | 0.286 81 |
| 1.96 | 0.420 30 | 0.678 61 | 0.619 35 | 1.614 60 | 1.632 61 | 0.292 95 |
| 1.98 | 0.414 21 | 0.672 62 | 0.615 82 | 1.623 86 | 1.659 71 | 0.299 01 |
| 2.00 | 0.408 25 | 0.666 67 | 0.612 37 | 1.632 99 | 1.687 50 | 0.305 00 |
| 2.02 | 0.402 41 | 0.660 76 | 0.609 01 | 1.642 00 | 1.715 97 | 0.310 91 |
| 2.04 | 0.396 70 | 0.654 91 | 0.605 73 | 1.650 90 | 1.745 14 | 0.316 76 |
| 2.06 | 0.391 10 | 0.649 10 | 0.602 53 | 1.659 67 | 1.775 01 | 0.322 53 |
| 2.08 | 0.385 62 | 0.643 34 | 0.599 40 | 1.668 33 | 1.805 61 | 0.328 22 |
| 2.10 | 0.380 24 | 0.637 62 | 0.596 35 | 1.676 87 | 1.836 94 | 0.333 85 |
| 2.12 | 0.374 98 | 0.631 95 | 0.593 37 | 1.685 30 | 1.869 01 | 0.339 40 |
| 2.14 | 0.369 82 | 0.626 33 | 0.590 45 | 1.693 62 | 1.901 84 | 0.344 89 |
| 2.16 | 0.364 76 | 0.620 76 | 0.587 60 | 1.701 82 | 1.935 43 | 0.350 30 |
| 2.18 | 0.359 80 | 0.615 23 | 0.584 82 | 1.709 92 | 1.969 81 | 0.355 64 |
| 2.20 | 0.354 94 | 0.609 76 | 0.582 10 | 1.717 91 | 2.004 97 | 0.360 91 |
| 2.22 | 0.350 17 | 0.604 33 | 0.579 44 | 1.725 79 | 2.040 94 | 0.366 11 |
| 2.24 | 0.345 50 | 0.598 95 | 0.576 84 | 1.733 57 | 2.077 73 | 0.371 24 |
| 2.26 | 0.340 91 | 0.593 61 | 0.574 30 | 1.741 25 | 2.115 35 | 0.376 31 |
| 2.28 | 0.336 41 | 0.588 33 | 0.571 82 | 1.748 82 | 2.153 81 | 0.381 30 |
| 2.30 | 0.332 00 | 0.583 09 | 0.569 38 | 1.756 29 | 2.193 13 | 0.386 23 |
| 2.32 | 0.327 67 | 0.577 90 | 0.567 00 | 1.763 66 | 2.233 32 | 0.391 09 |
| 2.34 | 0.323 42 | 0.572 76 | 0.564 67 | 1.770 93 | 2.274 40 | 0.395 89 |
| 2.36 | 0.319 25 | 0.567 67 | 0.562 40 | 1.778 11 | 2.316 38 | 0.400 62 |
| 2.38 | 0.315 16 | 0.562 62 | 0.560 16 | 1.785 19 | 2.359 27 | 0.405 29 |
| 2.40 | 0.311 14 | 0.557 62 | 0.557 98 | 1.792 18 | 2.403 10 | 0.409 89 |
| 2.42 | 0.307 20 | 0.552 67 | 0.555 84 | 1.799 07 | 2.447 87 | 0.414 43 |
| 2.44 | 0.303 32 | 0.547 77 | 0.553 75 | 1.805 87 | 2.493 60 | 0.418 91 |
| 2.46 | 0.299 52 | 0.542 91 | 0.551 70 | 1.812 58 | 2.540 30 | 0.423 32 |
| 2.48 | 0.295 79 | 0.538 10 | 0.549 69 | 1.819 21 | 2.588 01 | 0.427 68 |
| 2.50 | 0.292 12 | 0.533 33 | 0.547 72 | 1.825 74 | 2.636 71 | 0.431 98 |
| 2.52 | 0.288 52 | 0.528 62 | 0.545 79 | 1.832 19 | 2.686 45 | 0.436 21 |
| 2.54 | 0.284 98 | 0.523 94 | 0.543 91 | 1.838 55 | 2.737 22 | 0.440 39 |
| 2.56 | 0.281 50 | 0.519 32 | 0.542 05 | 1.844 83 | 2.789 06 | 0.444 51 |
| 2.58 | 0.278 08 | 0.514 74 | 0.540 24 | 1.851 03 | 2.841 97 | 0.448 58 |
| 2.60 | 0.274 73 | 0.510 20 | 0.538 46 | 1.857 14 | 2.895 97 | 0.452 59 |
| 2.62 | 0.271 43 | 0.505 72 | 0.536 72 | 1.863 18 | 2.951 08 | 0.456 54 |
| 2.64 | 0.268 18 | 0.501 27 | 0.535 01 | 1.869 13 | 3.007 33 | 0.460 44 |
| 2.66 | 0.265 00 | 0.496 87 | 0.533 33 | 1.875 01 | 3.064 71 | 0.464 29 |
| 2.68 | 0.261 86 | 0.492 51 | 0.531 69 | 1.880 81 | 3.123 27 | 0.468 08 |
| 2.70 | 0.258 78 | 0.488 20 | 0.530 07 | 1.886 53 | 3.183 00 | 0.471 82 |
| 2.72 | 0.255 76 | 0.483 93 | 0.528 49 | 1.892 18 | 3.243 94 | 0.475 51 |
| 2.74 | 0.252 78 | 0.479 71 | 0.526 94 | 1.897 75 | 3.306 11 | 0.479 15 |

| M | $p/p^*$ | $T/T^*$ | $\rho/\rho^*$ | $V/V^*$ | $p_0/p_0^*$ | $4fl^*/D_H$ |
|---|---------|---------|---------------|---------|-------------|-------------|
| 2.76 | 0.249 85 | 0.475 53 | 0.525 42 | 1.903 25 | 3.369 51 | 0.482 73 |
| 2.78 | 0.246 97 | 0.471 39 | 0.523 92 | 1.908 68 | 3.434 17 | 0.486 27 |
| 2.80 | 0.244 14 | 0.467 29 | 0.522 46 | 1.914 04 | 3.500 12 | 0.489 76 |
| 2.82 | 0.241 35 | 0.463 24 | 0.521 02 | 1.919 33 | 3.567 36 | 0.493 21 |
| 2.84 | 0.238 61 | 0.459 22 | 0.519 60 | 1.924 55 | 3.635 93 | 0.496 60 |
| 2.86 | 0.235 92 | 0.455 25 | 0.518 21 | 1.929 70 | 3.705 83 | 0.499 95 |
| 2.88 | 0.233 26 | 0.451 32 | 0.516 85 | 1.934 79 | 3.777 10 | 0.503 26 |
| 2.90 | 0.230 66 | 0.447 43 | 0.515 51 | 1.939 81 | 3.849 76 | 0.506 52 |
| 2.92 | 0.228 09 | 0.443 58 | 0.514 20 | 1.944 77 | 3.923 82 | 0.509 73 |
| 2.94 | 0.225 56 | 0.439 77 | 0.512 91 | 1.949 66 | 3.999 31 | 0.512 90 |
| 2.96 | 0.223 07 | 0.436 00 | 0.511 64 | 1.954 49 | 4.076 25 | 0.516 03 |
| 2.98 | 0.220 63 | 0.432 26 | 0.510 40 | 1.959 25 | 4.154 65 | 0.519 12 |
| 3.00 | 0.218 22 | 0.428 57 | 0.509 18 | 1.963 96 | 4.234 56 | 0.522 16 |
| 3.05 | 0.212 36 | 0.419 51 | 0.506 21 | 1.975 47 | 4.441 01 | 0.529 59 |
| 3.10 | 0.206 72 | 0.410 68 | 0.503 37 | 1.986 61 | 4.657 30 | 0.536 78 |
| 3.15 | 0.201 30 | 0.402 08 | 0.500 65 | 1.997 40 | 4.883 82 | 0.543 72 |
| 3.20 | 0.196 08 | 0.393 70 | 0.498 04 | 2.007 86 | 5.120 94 | 0.550 44 |
| 3.25 | 0.191 05 | 0.385 54 | 0.495 54 | 2.017 99 | 5.369 08 | 0.556 94 |
| 3.30 | 0.186 21 | 0.377 60 | 0.493 14 | 2.027 81 | 5.628 63 | 0.563 23 |
| 3.35 | 0.181 54 | 0.369 86 | 0.490 84 | 2.037 33 | 5.900 02 | 0.569 32 |
| 3.40 | 0.177 04 | 0.362 32 | 0.488 63 | 2.046 56 | 6.183 68 | 0.575 21 |
| 3.45 | 0.172 20 | 0.354 98 | 0.486 50 | 2.055 51 | 6.480 06 | 0.580 91 |
| 3.50 | 0.168 51 | 0.347 83 | 0.484 45 | 2.064 19 | 6.789 60 | 0.586 43 |
| 3.55 | 0.164 46 | 0.340 86 | 0.482 48 | 2.072 61 | 7.112 79 | 0.591 78 |
| 3.60 | 0.160 55 | 0.334 08 | 0.480 59 | 2.080 77 | 7.450 09 | 0.596 95 |
| 3.65 | 0.156 78 | 0.327 47 | 0.478 77 | 2.088 70 | 7.802 01 | 0.601 97 |
| 3.70 | 0.153 13 | 0.321 03 | 0.477 01 | 2.096 39 | 8.169 04 | 0.606 84 |
| 3.75 | 0.149 61 | 0.314 75 | 0.475 32 | 2.103 86 | 8.551 72 | 0.611 55 |
| 3.80 | 0.146 20 | 0.308 64 | 0.473 68 | 2.111 11 | 8.950 56 | 0.616 12 |
| 3.85 | 0.142 90 | 0.302 69 | 0.472 11 | 2.118 15 | 9.366 12 | 0.620 55 |
| 3.90 | 0.139 71 | 0.296 88 | 0.470 59 | 2.124 99 | 9.798 94 | 0.624 85 |
| 3.95 | 0.136 62 | 0.291 23 | 0.469 12 | 2.131 63 | 10.249 62 | 0.629 02 |
| 4.00 | 0.133 63 | 0.285 71 | 0.467 71 | 2.138 09 | 10.718 72 | 0.633 06 |
| 4.05 | 0.130 73 | 0.280 34 | 0.466 34 | 2.144 36 | 11.206 86 | 0.636 99 |
| 4.10 | 0.127 93 | 0.275 10 | 0.465 02 | 2.150 46 | 11.714 64 | 0.640 80 |
| 4.15 | 0.125 21 | 0.270 00 | 0.463 74 | 2.156 39 | 12.242 69 | 0.644 51 |
| 4.20 | 0.122 57 | 0.265 02 | 0.462 50 | 2.162 15 | 12.791 66 | 0.648 10 |
| 4.25 | 0.120 01 | 0.260 16 | 0.461 31 | 2.167 76 | 13.362 18 | 0.651 59 |
| 4.30 | 0.117 53 | 0.255 43 | 0.460 15 | 2.173 21 | 13.954 95 | 0.654 99 |
| 4.35 | 0.115 13 | 0.250 81 | 0.459 03 | 2.178 52 | 14.570 63 | 0.658 28 |
| 4.40 | 0.112 79 | 0.246 30 | 0.457 94 | 2.183 68 | 15.209 95 | 0.661 49 |
| 4.45 | 0.110 53 | 0.241 91 | 0.456 89 | 2.188 71 | 15.873 59 | 0.664 61 |
| 4.50 | 0.108 33 | 0.237 62 | 0.455 87 | 2.193 60 | 16.562 31 | 0.667 64 |
| 4.60 | 0.104 11 | 0.229 36 | 0.453 93 | 2.203 00 | 18.017 95 | 0.673 45 |
| 4.70 | 0.100 13 | 0.221 48 | 0.452 10 | 2.211 92 | 19.583 03 | 0.678 95 |
| 4.80 | 0.096 37 | 0.213 98 | 0.450 37 | 2.220 38 | 21.263 97 | 0.684 17 |
| 4.90 | 0.092 81 | 0.206 82 | 0.448 75 | 2.228 43 | 23.067 45 | 0.689 11 |
| 5.00 | 0.089 44 | 0.200 00 | 0.447 21 | 2.236 07 | 25.000 40 | 0.693 80 |
| 5.10 | 0.086 25 | 0.193 48 | 0.445 76 | 2.243 34 | 27.070 04 | 0.698 26 |
| 5.20 | 0.083 22 | 0.187 26 | 0.444 39 | 2.250 26 | 29.283 88 | 0.702 50 |
| 5.30 | 0.080 34 | 0.181 32 | 0.443 09 | 2.256 85 | 31.649 70 | 0.706 52 |
| 5.40 | 0.077 61 | 0.175 64 | 0.441 86 | 2.263 14 | 34.175 56 | 0.710 36 |
| 5.50 | 0.075 01 | 0.170 21 | 0.440 70 | 2.269 13 | 36.869 83 | 0.714 01 |

# APPENDIX E

# Tables for One-Dimensional Flow with Heat Exchange for $\gamma = 1.4$

| $M$ | $T_0/T_0^*$ | $T/T^*$ | $p/p^*$ | $p_0/p_0^*$ | $V/V^*$ | $\rho/\rho^*$ |
|------|------------|---------|---------|-------------|---------|---------------|
| 0.06 | 0.017 12 | 0.020 53 | 2.387 96 | 1.264 70 | 0.008 60 | 116.324 07 |
| 0.08 | 0.030 22 | 0.036 21 | 2.378 69 | 1.262 26 | 0.015 22 | 65.687 50 |
| 0.10 | 0.046 78 | 0.056 02 | 2.366 86 | 1.259 15 | 0.023 67 | 42.250 00 |
| 0.12 | 0.066 61 | 0.079 70 | 2.352 57 | 1.255 39 | 0.033 88 | 29.518 52 |
| 0.14 | 0.089 47 | 0.106 95 | 2.335 90 | 1.251 03 | 0.045 78 | 21.841 84 |
| 0.16 | 0.115 11 | 0.137 43 | 2.316 96 | 1.246 08 | 0.059 31 | 16.859 37 |
| 0.18 | 0.143 24 | 0.170 78 | 2.295 86 | 1.240 59 | 0.074 39 | 13.443 41 |
| 0.20 | 0.173 55 | 0.206 61 | 2.272 73 | 1.234 60 | 0.090 91 | 11.000 00 |
| 0.22 | 0.205 74 | 0.244 52 | 2.247 70 | 1.228 14 | 0.108 79 | 9.192 15 |
| 0.24 | 0.239 48 | 0.284 11 | 2.220 91 | 1.221 26 | 0.127 92 | 7.817 13 |
| 0.26 | 0.274 46 | 0.324 96 | 2.192 50 | 1.214 00 | 0.148 21 | 6.747 04 |
| 0.28 | 0.310 35 | 0.366 67 | 2.162 63 | 1.206 42 | 0.169 55 | 5.897 96 |
| 0.30 | 0.346 86 | 0.408 87 | 2.131 44 | 1.198 55 | 0.191 83 | 5.212 96 |
| 0.32 | 0.383 69 | 0.451 19 | 2.099 08 | 1.190 45 | 0.214 95 | 4.652 34 |
| 0.34 | 0.420 56 | 0.493 27 | 2.065 69 | 1.182 15 | 0.238 79 | 4.187 72 |
| 0.36 | 0.457 23 | 0.534 82 | 2.031 42 | 1.173 71 | 0.263 27 | 3.798 35 |
| 0.38 | 0.493 46 | 0.575 53 | 1.996 41 | 1.165 17 | 0.288 28 | 3.468 84 |
| 0.40 | 0.529 03 | 0.615 15 | 1.960 78 | 1.156 58 | 0.313 73 | 3.187 50 |
| 0.42 | 0.563 76 | 0.653 46 | 1.924 68 | 1.147 96 | 0.339 51 | 2.945 39 |
| 0.44 | 0.597 48 | 0.690 25 | 1.888 22 | 1.139 36 | 0.365 56 | 2.735 54 |
| 0.46 | 0.630 07 | 0.725 38 | 1.851 51 | 1.130 82 | 0.391 78 | 2.552 46 |
| 0.48 | 0.661 39 | 0.758 71 | 1.814 66 | 1.122 38 | 0.418 10 | 2.391 78 |
| 0.50 | 0.691 36 | 0.790 12 | 1.777 78 | 1.114 05 | 0.444 44 | 2.250 00 |
| 0.52 | 0.719 90 | 0.819 55 | 1.740 95 | 1.105 88 | 0.470 75 | 2.124 26 |
| 0.54 | 0.746 95 | 0.846 95 | 1.704 26 | 1.097 89 | 0.496 96 | 2.012 23 |
| 0.56 | 0.772 49 | 0.872 27 | 1.667 78 | 1.090 11 | 0.523 02 | 1.911 99 |
| 0.58 | 0.796 48 | 0.895 52 | 1.631 59 | 1.082 56 | 0.548 87 | 1.821 94 |
| 0.60 | 0.818 92 | 0.916 70 | 1.595 75 | 1.075 25 | 0.574 47 | 1.740 74 |
| 0.62 | 0.839 82 | 0.935 84 | 1.560 31 | 1.068 22 | 0.599 78 | 1.667 27 |
| 0.64 | 0.859 20 | 0.952 98 | 1.525 32 | 1.061 47 | 0.624 77 | 1.600 59 |

| M | $T_0/T_0^*$ | $T/T^*$ | $p/p^*$ | $p_0/p_0^*$ | $V/V^*$ | $\rho/\rho^*$ |
|---|---|---|---|---|---|---|
| 0.66 | 0.877 08 | 0.968 15 | 1.490 83 | 1.055 03 | 0.649 41 | 1.539 87 |
| 0.68 | 0.893 50 | 0.981 44 | 1.456 88 | 1.048 90 | 0.673 66 | 1.484 43 |
| 0.70 | 0.908 50 | 0.992 90 | 1.423 49 | 1.043 10 | 0.697 51 | 1.433 67 |
| 0.72 | 0.922 12 | 1.002 60 | 1.390 69 | 1.037 64 | 0.720 93 | 1.387 09 |
| 0.74 | 0.934 42 | 1.010 62 | 1.358 51 | 1.032 53 | 0.743 92 | 1.344 23 |
| 0.76 | 0.945 46 | 1.017 06 | 1.326 96 | 1.027 77 | 0.766 45 | 1.304 71 |
| 0.78 | 0.955 28 | 1.021 98 | 1.296 06 | 1.023 37 | 0.788 53 | 1.268 19 |
| 0.80 | 0.963 95 | 1.025 48 | 1.265 82 | 1.019 34 | 0.810 13 | 1.234 38 |
| 0.82 | 0.971 52 | 1.027 63 | 1.236 25 | 1.015 69 | 0.831 25 | 1.203 00 |
| 0.84 | 9.978 07 | 1.028 53 | 1.207 34 | 1.012 41 | 0.851 90 | 1.173 85 |
| 0.86 | 0.983 63 | 1.028 26 | 1.179 11 | 1.009 51 | 0.872 07 | 1.146 70 |
| 0.88 | 0.988 28 | 1.026 89 | 1.151 54 | 1.006 99 | 0.891 75 | 1.121 38 |
| 0.90 | 0.992 07 | 1.024 52 | 1.124 65 | 1.004 86 | 0.910 96 | 1.097 74 |
| 0.92 | 0.995 06 | 1.021 20 | 1.098 42 | 1.003 11 | 0.929 70 | 1.075 61 |
| 0.94 | 0.997 29 | 1.017 02 | 1.072 85 | 1.001 75 | 0.947 97 | 1.054 89 |
| 0.96 | 0.998 83 | 1.012 05 | 1.047 93 | 1.000 78 | 0.965 77 | 1.035 45 |
| 0.98 | 0.999 71 | 1.006 36 | 1.023 65 | 1.000 19 | 0.983 11 | 1.017 18 |
| 1.00 | 1.000 00 | 1.000 00 | 1.000 00 | 1.000 00 | 1.000 00 | 1.000 00 |
| 1.02 | 0.999 73 | 0.993 04 | 0.976 98 | 1.000 19 | 1.016 45 | 0.983 82 |
| 1.04 | 0.998 95 | 0.985 54 | 0.954 56 | 1.000 78 | 1.032 45 | 0.968 57 |
| 1.06 | 0.997 69 | 0.977 56 | 0.932 75 | 1.001 75 | 1.048 04 | 0.954 17 |
| 1.08 | 0.996 01 | 0.969 13 | 0.911 52 | 1.003 11 | 1.063 20 | 0.940 56 |
| 1.10 | 0.993 92 | 0.960 31 | 0.890 87 | 1.004 86 | 1.077 95 | 0.927 69 |
| 1.12 | 0.991 48 | 0.951 15 | 0.870 78 | 1.006 99 | 1.092 30 | 0.915 50 |
| 1.14 | 0.988 71 | 0.941 69 | 0.851 23 | 1.009 52 | 1.106 26 | 0.903 95 |
| 1.16 | 0.985 64 | 0.931 96 | 0.832 22 | 1.012 43 | 1.119 84 | 0.892 98 |
| 1.18 | 0.982 30 | 0.922 00 | 0.813 74 | 1.015 73 | 1.133 05 | 0.882 58 |
| 1.20 | 0.978 72 | 0.911 85 | 0.795 76 | 1.019 41 | 1.145 89 | 0.872 69 |
| 1.22 | 0.974 92 | 0.901 53 | 0.778 27 | 1.023 49 | 1.158 38 | 0.863 28 |
| 1.24 | 0.970 92 | 0.891 08 | 0.761 27 | 1.027 95 | 1.170 52 | 0.854 32 |
| 1.26 | 0.966 75 | 0.880 52 | 0.744 73 | 1.032 80 | 1.182 33 | 0.845 78 |
| 1.28 | 0.962 43 | 0.869 88 | 0.728 65 | 1.038 03 | 1.193 82 | 0.837 65 |
| 1.30 | 0.957 98 | 0.859 17 | 0.713 01 | 1.043 66 | 1.204 99 | 0.829 88 |
| 1.32 | 0.953 41 | 0.848 43 | 0.697 80 | 1.049 67 | 1.215 85 | 0.822 47 |
| 1.34 | 0.948 73 | 0.837 66 | 0.683 01 | 1.056 08 | 1.226 42 | 0.815 38 |
| 1.36 | 0.943 98 | 0.826 89 | 0.668 63 | 1.062 88 | 1.236 69 | 0.808 61 |
| 1.38 | 0.939 14 | 0.816 13 | 0.654 64 | 1.070 07 | 1.246 69 | 0.802 13 |
| 1.40 | 0.934 25 | 0.805 39 | 0.641 03 | 1.077 65 | 1.256 41 | 0.795 92 |
| 1.42 | 0.929 31 | 0.794 69 | 0.627 79 | 1.085 63 | 1.265 87 | 0.789 97 |
| 1.44 | 0.924 34 | 0.784 05 | 0.614 91 | 1.094 01 | 1.275 07 | 0.784 27 |
| 1.46 | 0.919 33 | 0.773 46 | 0.602 37 | 1.102 78 | 1.284 02 | 0.778 80 |
| 1.48 | 0.914 31 | 0.762 94 | 0.590 18 | 1.111 96 | 1.292 73 | 0.773 56 |
| 1.50 | 0.909 28 | 0.752 50 | 0.578 31 | 1.121 54 | 1.301 20 | 0.768 52 |
| 1.52 | 0.904 24 | 0.742 15 | 0.566 77 | 1.131 53 | 1.309 45 | 0.763 68 |
| 1.54 | 0.899 21 | 0.731 89 | 0.555 53 | 1.141 93 | 1.317 48 | 0.759 02 |
| 1.56 | 0.894 18 | 0.721 74 | 0.544 58 | 1.152 74 | 1.325 30 | 0.754 55 |
| 1.58 | 0.889 17 | 0.711 68 | 0.533 93 | 1.163 97 | 1.332 91 | 0.750 24 |
| 1.60 | 0.884 19 | 0.701 74 | 0.523 56 | 1.175 61 | 1.340 31 | 0.746 09 |
| 1.62 | 0.879 22 | 0.691 90 | 0.513 46 | 1.187 68 | 1.347 53 | 0.742 10 |
| 1.64 | 0.874 29 | 0.682 19 | 0.503 63 | 1.200 17 | 1.354 55 | 0.738 25 |
| 1.66 | 0.869 39 | 0.672 59 | 0.494 05 | 1.213 09 | 1.361 39 | 0.734 54 |
| 1.68 | 0.864 53 | 0.663 12 | 0.484 72 | 1.226 44 | 1.368 06 | 0.730 96 |
| 1.70 | 0.859 71 | 0.653 77 | 0.475 62 | 1.240 23 | 1.374 55 | 0.727 51 |

| M | $T_0/T_0^*$ | $T/T^*$ | $p/p^*$ | $p_0/p_0^*$ | $V/V^*$ | $\rho/\rho^*$ |
|---|---|---|---|---|---|---|
| 1.72 | 0.854 93 | 0.644 55 | 0.466 77 | 1.254 47 | 1.380 88 | 0.724 18 |
| 1.74 | 0.850 19 | 0.635 45 | 0.458 13 | 1.269 15 | 1.387 05 | 0.720 96 |
| 1.76 | 0.845 51 | 0.626 49 | 0.449 72 | 1.284 28 | 1.393 06 | 0.717 85 |
| 1.78 | 0.840 87 | 0.617 65 | 0.441 52 | 1.299 87 | 1.398 91 | 0.714 84 |
| 1.80 | 0.836 28 | 0.608 94 | 0.433 53 | 1.315 92 | 1.404 62 | 0.711 93 |
| 1.82 | 0.831 74 | 0.600 36 | 0.425 73 | 1.332 44 | 1.410 19 | 0.709 12 |
| 1.84 | 0.827 26 | 0.591 91 | 0.418 13 | 1.349 43 | 1.415 62 | 0.706 40 |
| 1.86 | 0.822 83 | 0.583 60 | 0.410 72 | 1.366 90 | 1.420 92 | 0.703 77 |
| 1.88 | 0.818 46 | 0.575 40 | 0.403 49 | 1.384 85 | 1.426 08 | 0.701 22 |
| 1.90 | 0.814 14 | 0.567 34 | 0.396 43 | 1.403 30 | 1.431 12 | 0.698 75 |
| 1.92 | 0.809 87 | 0.559 41 | 0.389 55 | 1.422 24 | 1.436 04 | 0.696 36 |
| 1.94 | 0.805 67 | 0.551 60 | 0.382 83 | 1.441 68 | 1.440 83 | 0.694 04 |
| 1.96 | 0.801 52 | 0.543 92 | 0.376 28 | 1.461 63 | 1.445 51 | 0.691 80 |
| 1.98 | 0.797 42 | 0.536 36 | 0.369 88 | 1.482 10 | 1.450 08 | 0.689 62 |
| 2.00 | 0.793 39 | 0.528 93 | 0.363 64 | 1.503 09 | 1.454 55 | 0.687 50 |
| 2.02 | 0.789 41 | 0.521 61 | 0.357 54 | 1.524 62 | 1.458 90 | 0.685 45 |
| 2.04 | 0.785 49 | 0.514 42 | 0.351 58 | 1.546 68 | 1.463 15 | 0.683 46 |
| 2.06 | 0.781 62 | 0.507 35 | 0.345 77 | 1.569 28 | 1.467 31 | 0.681 52 |
| 2.08 | 0.777 82 | 0.500 40 | 0.340 09 | 1.592 44 | 1.471 36 | 0.679 64 |
| 2.10 | 0.774 06 | 0.493 56 | 0.334 54 | 1.616 16 | 1.475 33 | 0.677 82 |
| 2.12 | 0.770 37 | 0.486 84 | 0.329 12 | 1.640 44 | 1.479 20 | 0.676 04 |
| 2.14 | 0.766 73 | 0.480 23 | 0.323 82 | 1.665 31 | 1.482 98 | 0.674 32 |
| 2.16 | 0.763 14 | 0.473 73 | 0.318 65 | 1.690 76 | 1.486 68 | 0.672 64 |
| 2.18 | 0.759 61 | 0.467 34 | 0.313 59 | 1.716 80 | 1.490 29 | 0.671 01 |
| 2.20 | 0.756 14 | 0.461 06 | 0.308 64 | 1.743 44 | 1.493 83 | 0.669 42 |
| 2.22 | 0.752 71 | 0.454 88 | 0.303 81 | 1.770 70 | 1.497 28 | 0.667 88 |
| 2.24 | 0.749 34 | 0.448 82 | 0.299 08 | 1.798 58 | 1.500 66 | 0.666 37 |
| 2.26 | 0.746 02 | 0.442 85 | 0.294 46 | 1.827 08 | 1.503 96 | 0.664 91 |
| 2.28 | 0.742 76 | 0.436 99 | 0.289 93 | 1.856 22 | 1.507 19 | 0.663 49 |
| 2.30 | 0.739 54 | 0.431 22 | 0.285 51 | 1.886 02 | 1.510 35 | 0.662 10 |
| 2.32 | 0.736 38 | 0.425 55 | 0.281 18 | 1.916 47 | 1.513 44 | 0.660 75 |
| 2.34 | 0.733 26 | 0.419 98 | 0.276 95 | 1.947 59 | 1.516 46 | 0.659 43 |
| 2.36 | 0.730 20 | 0.414 51 | 0.272 81 | 1.979 38 | 1.519 42 | 0.658 14 |
| 2.38 | 0.727 18 | 0.409 13 | 0.268 75 | 2.011 87 | 1.522 32 | 0.656 89 |
| 2.40 | 0.724 21 | 0.403 84 | 0.264 78 | 2.045 05 | 1.525 15 | 0.655 67 |
| 2.42 | 0.721 29 | 0.398 64 | 0.260 90 | 2.078 94 | 1.527 93 | 0.654 48 |
| 2.44 | 0.718 42 | 0.393 52 | 0.257 10 | 2.113 56 | 1.530 65 | 0.653 32 |
| 2.46 | 0.715 59 | 0.388 50 | 0.253 37 | 2.148 90 | 1.533 31 | 0.652 19 |
| 2.48 | 0.712 80 | 0.383 56 | 0.249 73 | 2.184 99 | 1.535 91 | 0.651 08 |
| 2.50 | 0.710 06 | 0.378 70 | 0.246 15 | 2.221 83 | 1.538 46 | 0.650 00 |
| 2.52 | 0.707 36 | 0.373 92 | 0.242 66 | 2.259 43 | 1.540 96 | 0.648 95 |
| 2.54 | 0.704 71 | 0.369 23 | 0.239 23 | 2.297 81 | 1.543 41 | 0.647 92 |
| 2.56 | 0.702 10 | 0.364 61 | 0.235 87 | 2.336 98 | 1.545 81 | 0.646 91 |
| 2.58 | 0.699 53 | 0.360 07 | 0.232 58 | 2.376 95 | 1.548 16 | 0.645 93 |
| 2.60 | 0.697 00 | 0.355 61 | 0.229 36 | 2.417 74 | 1.550 46 | 0.644 97 |
| 2.62 | 0.694 51 | 0.351 22 | 0.226 20 | 2.459 34 | 1.552 72 | 0.644 03 |
| 2.64 | 0.692 06 | 0.346 91 | 0.223 10 | 2.501 79 | 1.554 93 | 0.643 12 |
| 2.66 | 0.689 64 | 0.342 66 | 0.220 07 | 2.545 09 | 1.557 10 | 0.642 22 |
| 2.68 | 0.687 27 | 0.338 49 | 0.217 09 | 2.589 25 | 1.559 22 | 0.641 35 |
| 2.70 | 0.684 94 | 0.334 39 | 0.214 17 | 2.634 28 | 1.561 31 | 0.640 49 |
| 2.72 | 0.682 64 | 0.330 35 | 0.211 31 | 2.680 21 | 1.563 35 | 0.639 65 |
| 2.74 | 0.680 37 | 0.326 38 | 0.208 50 | 2.727 03 | 1.565 36 | 0.638 83 |
| 2.76 | 0.678 15 | 0.322 48 | 0.205 75 | 2.774 78 | 1.567 32 | 0.638 03 |

| M | $T_0/T_0^*$ | $T/T^*$ | $p/p^*$ | $p_0/p_0^*$ | $V/V^*$ | $\rho/\rho^*$ |
|---|---|---|---|---|---|---|
| 2.78 | 0.675 95 | 0.318 64 | 0.203 05 | 2.823 45 | 1.569 25 | 0.637 25 |
| 2.80 | 0.673 80 | 0.314 86 | 0.200 40 | 2.873 07 | 1.571 14 | 0.636 48 |
| 2.82 | 0.671 67 | 0.311 14 | 0.197 80 | 2.923 65 | 1.573 00 | 0.635 73 |
| 2.84 | 0.669 58 | 0.307 49 | 0.195 25 | 2.975 20 | 1.574 82 | 0.634 99 |
| 2.86 | 0.667 52 | 0.303 89 | 0.192 75 | 3.027 74 | 1.576 61 | 0.634 27 |
| 2.88 | 0.665 50 | 0.300 35 | 0.190 29 | 3.081 29 | 1.578 36 | 0.633 57 |
| 2.90 | 0.663 50 | 0.296 87 | 0.187 88 | 3.135 85 | 1.580 08 | 0.632 88 |
| 2.92 | 0.661 54 | 0.293 44 | 0.185 52 | 3.191 44 | 1.581 77 | 0.632 20 |
| 2.94 | 0.659 60 | 0.290 07 | 0.183 19 | 3.248 08 | 1.583 43 | 0.631 54 |
| 2.96 | 0.657 70 | 0.286 75 | 0.180 91 | 3.305 78 | 1.585 06 | 0.630 89 |
| 2.98 | 0.655 83 | 0.283 49 | 0.178 67 | 3.364 57 | 1.586 66 | 0.630 25 |
| 3.00 | 0.653 98 | 0.280 28 | 0.176 47 | 3.424 44 | 1.588 24 | 0.629 63 |
| 3.05 | 0.649 49 | 0.272 46 | 0.171 14 | 3.579 04 | 1.592 04 | 0.628 12 |
| 3.10 | 0.645 16 | 0.264 95 | 0.166 04 | 3.740 84 | 1.595 68 | 0.626 69 |
| 3.15 | 0.641 00 | 0.257 73 | 0.161 17 | 3.910 10 | 1.599 17 | 0.625 33 |
| 3.20 | 0.636 99 | 0.250 78 | 0.156 49 | 4.087 11 | 1.602 50 | 0.624 02 |
| 3.25 | 0.633 13 | 0.244 10 | 0.152 02 | 4.272 14 | 1.605 70 | 0.622 78 |
| 3.30 | 0.629 41 | 0.237 66 | 0.147 73 | 4.465 48 | 1.608 77 | 0.621 59 |
| 3.35 | 0.625 82 | 0.231 46 | 0.143 61 | 4.667 43 | 1.611 70 | 0.620 46 |
| 3.40 | 0.622 36 | 0.225 49 | 0.139 67 | 4.878 29 | 1.614 53 | 0.619 38 |
| 3.45 | 0.619 02 | 0.219 74 | 0.135 87 | 5.098 38 | 1.617 23 | 0.618 34 |
| 3.50 | 0.615 81 | 0.214 19 | 0.132 23 | 5.328 02 | 1.619 83 | 0.617 35 |
| 3.55 | 0.612 70 | 0.208 85 | 0.128 73 | 5.567 54 | 1.622 33 | 0.616 40 |
| 3.60 | 0.609 70 | 0.203 69 | 0.125 37 | 5.817 28 | 1.624 74 | 0.615 48 |
| 3.65 | 0.606 81 | 0.198 71 | 0.122 13 | 6.077 59 | 1.627 05 | 0.614 61 |
| 3.70 | 0.604 01 | 0.193 90 | 0.119 01 | 6.348 83 | 1.629 28 | 0.613 77 |
| 3.75 | 0.601 31 | 0.189 26 | 0.116 01 | 6.631 35 | 1.631 42 | 0.612 96 |
| 3.80 | 0.598 70 | 0.184 78 | 0.113 12 | 6.925 55 | 1.633 48 | 0.612 19 |
| 3.85 | 0.596 17 | 0.180 45 | 0.110 34 | 7.231 79 | 1.635 47 | 0.611 44 |
| 3.90 | 0.593 73 | 0.176 27 | 0.107 65 | 7.550 48 | 1.637 39 | 0.610 73 |
| 3.95 | 0.591 37 | 0.172 22 | 0.105 06 | 7.882 02 | 1.639 24 | 0.610 04 |
| 4.00 | 0.589 09 | 0.168 31 | 0.102 56 | 8.226 83 | 1.641 03 | 0.609 38 |
| 4.05 | 0.586 87 | 0.164 53 | 0.100 15 | 8.585 32 | 1.642 75 | 0.608 74 |
| 4.10 | 0.584 73 | 0.160 86 | 0.097 82 | 8.957 94 | 1.644 41 | 0.608 12 |
| 4.15 | 0.582 66 | 0.157 32 | 0.095 57 | 9.345 11 | 1.646 02 | 0.607 53 |
| 4.20 | 0.580 65 | 0.153 88 | 0.093 40 | 9.747 30 | 1.647 57 | 0.606 95 |
| 4.25 | 0.578 70 | 0.150 56 | 0.091 30 | 10.164 96 | 1.649 07 | 0.606 40 |
| 4.30 | 0.576 82 | 0.147 34 | 0.089 27 | 10.598 58 | 1.650 52 | 0.605 87 |
| 4.35 | 0.574 99 | 0.144 21 | 0.087 30 | 11.048 62 | 1.651 93 | 0.605 35 |
| 4.40 | 0.573 21 | 0.141 19 | 0.085 40 | 11.515 60 | 1.653 29 | 0.604 86 |
| 4.45 | 0.571 49 | 0.138 25 | 0.083 56 | 12.000 00 | 1.654 60 | 0.604 37 |
| 4.50 | 0.569 82 | 0.135 40 | 0.081 77 | 12.502 35 | 1.655 88 | 0.603 91 |
| 4.60 | 0.566 63 | 0.129 96 | 0.078 37 | 13.563 00 | 1.658 31 | 0.603 02 |
| 4.70 | 0.563 62 | 0.124 83 | 0.075 17 | 14.701 89 | 1.660 59 | 0.602 20 |
| 4.80 | 0.560 78 | 0.119 99 | 0.072 17 | 15.923 56 | 1.662 74 | 0.601 42 |
| 4.90 | 0.558 09 | 0.115 43 | 0.069 34 | 17.232 69 | 1.664 76 | 0.600 69 |
| 5.00 | 0.555 56 | 0.111 11 | 0.066 67 | 18.634 19 | 1.666 67 | 0.600 00 |
| 5.10 | 0.553 15 | 0.107 03 | 0.064 15 | 20.133 13 | 1.668 47 | 0.599 35 |
| 5.20 | 0.550 88 | 0.103 16 | 0.061 77 | 21.734 80 | 1.670 17 | 0.598 74 |
| 5.30 | 0.548 72 | 0.099 49 | 0.059 51 | 23.444 67 | 1.671 78 | 0.598 17 |
| 5.40 | 0.546 67 | 0.096 02 | 0.057 38 | 25.268 42 | 1.673 30 | 0.597 62 |
| 5.50 | 0.544 72 | 0.092 72 | 0.055 36 | 27.211 95 | 1.674 74 | 0.597 11 |

# APPENDIX F

# Tables for One-Dimensional Isothermal Flow with Friction for $\gamma = 1.4$

| M | $p/p^*$ | $T_0/T_0^*$ | $\rho/\rho^*$ | $V/V^*$ | $p_0/p_0^*$ | $4fl^*/D_H$ |
|---|---------|-------------|---------------|---------|-------------|-------------|
| 0.10 | 8.451 54 | 0.876 75 | 8.451 54 | 0.118 32 | 5.333 36 | 66.159 87 |
| 0.12 | 7.042 95 | 0.877 52 | 7.042 95 | 0.141 99 | 4.458 15 | 44.699 12 |
| 0.14 | 6.036 82 | 0.878 43 | 6.036 82 | 0.165 65 | 3.835 16 | 31.847 39 |
| 0.16 | 5.282 21 | 0.879 48 | 5.282 21 | 0.189 31 | 3.369 82 | 23.573 09 |
| 0.18 | 4.695 30 | 0.880 67 | 4.695 30 | 0.212 98 | 3.009 61 | 17.952 73 |
| 0.20 | 4.225 77 | 0.882 00 | 4.225 77 | 0.236 64 | 2.722 99 | 13.974 73 |
| 0.22 | 3.841 61 | 0.883 47 | 3.841 61 | 0.260 31 | 2.489 92 | 11.066 18 |
| 0.24 | 3.521 48 | 0.885 08 | 3.521 48 | 0.283 97 | 2.297 01 | 8.883 03 |
| 0.26 | 3.250 59 | 0.886 83 | 3.250 59 | 0.307 64 | 2.135 03 | 7.208 68 |
| 0.28 | 3.018 41 | 0.888 72 | 3.018 41 | 0.331 30 | 1.997 35 | 5.901 33 |
| 0.30 | 2.817 18 | 0.890 75 | 2.817 18 | 0.354 96 | 1.879 14 | 4.865 03 |
| 0.32 | 2.641 11 | 0.892 92 | 2.641 11 | 0.378 63 | 1.776 76 | 4.033 05 |
| 0.34 | 2.485 75 | 0.895 23 | 2.485 75 | 0.402 29 | 1.687 44 | 3.357 80 |
| 0.36 | 2.347 65 | 0.897 68 | 2.347 65 | 0.425 96 | 1.609 01 | 2.804 63 |
| 0.38 | 2.224 09 | 0.900 27 | 2.224 09 | 0.449 62 | 1.539 77 | 2.347 88 |
| 0.40 | 2.112 89 | 0.903 00 | 2.112 89 | 0.473 29 | 1.478 37 | 1.968 18 |
| 0.42 | 2.012 27 | 0.905 87 | 2.012 27 | 0.496 95 | 1.423 70 | 1.650 71 |
| 0.44 | 1.920 81 | 0.908 88 | 1.920 81 | 0.520 61 | 1.374 85 | 1.384 00 |
| 0.46 | 1.837 29 | 0.912 03 | 1.837 29 | 0.544 28 | 1.331 10 | 1.159 06 |
| 0.48 | 1.760 74 | 0.915 32 | 1.760 74 | 0.567 94 | 1.291 81 | 0.968 73 |
| 0.50 | 1.690 31 | 0.918 75 | 1.690 31 | 0.591 61 | 1.256 48 | 0.807 32 |
| 0.52 | 1.625 30 | 0.922 32 | 1.625 30 | 0.615 27 | 1.224 67 | 0.670 21 |
| 0.54 | 1.565 10 | 0.926 03 | 1.565 10 | 0.638 94 | 1.196 00 | 0.553 64 |
| 0.56 | 1.509 20 | 0.929 88 | 1.509 20 | 0.662 60 | 1.170 15 | 0.454 53 |
| 0.58 | 1.457 16 | 0.933 87 | 1.457 16 | 0.686 26 | 1.146 86 | 0.370 34 |
| 0.60 | 1.408 59 | 0.938 00 | 1.408 59 | 0.709 93 | 1.125 89 | 0.298 95 |
| 0.62 | 1.363 15 | 0.942 27 | 1.363 15 | 0.733 59 | 1.107 03 | 0.238 58 |
| 0.64 | 1.320 55 | 0.946 68 | 1.320 55 | 0.757 26 | 1.090 10 | 0.187 76 |
| 0.66 | 1.280 54 | 0.951 23 | 1.280 54 | 0.780 92 | 1.074 96 | 0.145 22 |
| 0.68 | 1.242 87 | 0.955 92 | 1.242 87 | 0.804 59 | 1.061 46 | 0.109 88 |

| $M$ | $p/p^*$ | $T_0/T_0^*$ | $\rho/\rho^*$ | $V/V^*$ | $p_0/p_0^*$ | $4fl^*/D_H$ |
|------|---------|-------------|---------------|---------|-------------|-------------|
| 0.70 | 1.207 36 | 0.960 75 | 1.207 36 | 0.828 25 | 1.049 48 | 0.080 85 |
| 0.72 | 1.173 83 | 0.965 72 | 1.173 83 | 0.851 92 | 1.038 92 | 0.057 33 |
| 0.74 | 1.142 10 | 0.970 83 | 1.142 10 | 0.875 58 | 1.029 69 | 0.038 66 |
| 0.76 | 1.112 05 | 0.976 08 | 1.112 05 | 0.899 24 | 1.021 70 | 0.024 24 |
| 0.78 | 1.083 53 | 0.981 47 | 1.083 53 | 0.922 91 | 1.014 87 | 0.013 59 |
| 0.80 | 1.056 44 | 0.987 00 | 1.056 44 | 0.946 57 | 1.009 15 | 0.006 26 |
| 0.82 | 1.030 68 | 0.992 67 | 1.030 68 | 0.970 24 | 1.004 48 | 0.001 86 |
| 0.84 | 1.006 14 | 0.998 48 | 1.006 14 | 0.993 90 | 1.000 79 | 0.000 08 |
| 0.86 | 0.982 74 | 1.004 43 | 0.982 74 | 1.017 57 | 0.998 06 | 0.000 60 |
| 0.88 | 0.960 40 | 1.010 52 | 0.960 40 | 1.041 23 | 0.996 23 | 0.003 18 |
| 0.90 | 0.939 06 | 1.016 75 | 0.939 06 | 1.064 89 | 0.995 28 | 0.007 59 |
| 0.92 | 0.918 65 | 1.023 12 | 0.918 65 | 1.088 56 | 0.995 16 | 0.013 62 |
| 0.94 | 0.899 10 | 1.029 63 | 0.899 10 | 1.112 22 | 0.995 85 | 0.021 10 |
| 0.96 | 0.880 37 | 1.036 28 | 0.880 37 | 1.135 89 | 0.997 32 | 0.029 88 |
| 0.98 | 0.862 40 | 1.043 07 | 0.862 40 | 1.159 55 | 0.999 56 | 0.039 80 |
| 1.00 | 0.845 15 | 1.050 00 | 0.845 15 | 1.183 22 | 1.002 53 | 0.050 76 |
| 1.02 | 0.828 58 | 1.057 07 | 0.828 58 | 1.206 88 | 1.006 23 | 0.062 63 |
| 1.04 | 0.812 65 | 1.064 28 | 0.812 65 | 1.230 54 | 1.010 64 | 0.075 31 |
| 1.06 | 0.797 32 | 1.071 63 | 0.797 32 | 1.254 21 | 1.015 75 | 0.088 72 |
| 1.08 | 0.782 55 | 1.079 12 | 0.782 55 | 1.277 87 | 1.021 54 | 0.102 78 |
| 1.10 | 0.768 32 | 1.086 75 | 0.768 32 | 1.301 54 | 1.028 01 | 0.117 41 |
| 1.12 | 0.754 60 | 1.094 52 | 0.754 60 | 1.325 20 | 1.035 14 | 0.132 55 |
| 1.14 | 0.741 36 | 1.102 43 | 0.741 36 | 1.348 87 | 1.042 94 | 0.148 15 |
| 1.16 | 0.728 58 | 1.110 48 | 0.728 58 | 1.372 53 | 1.051 39 | 0.164 14 |
| 1.18 | 0.716 23 | 1.118 67 | 0.716 23 | 1.396 19 | 1.060 50 | 0.180 49 |
| 1.20 | 0.704 30 | 1.127 00 | 0.704 30 | 1.419 86 | 1.070 26 | 0.197 15 |
| 1.22 | 0.692 75 | 1.135 47 | 0.692 75 | 1.443 52 | 1.080 66 | 0.214 07 |
| 1.24 | 0.681 58 | 1.144 08 | 0.681 58 | 1.467 19 | 1.091 72 | 0.231 24 |
| 1.26 | 0.670 76 | 1.152 83 | 0.670 76 | 1.490 85 | 1.103 43 | 0.248 61 |
| 1.28 | 0.660 28 | 1.161 72 | 0.660 28 | 1.514 52 | 1.115 79 | 0.266 16 |
| 1.30 | 0.650 12 | 1.170 75 | 0.650 12 | 1.538 18 | 1.128 80 | 0.283 85 |
| 1.32 | 0.640 27 | 1.179 92 | 0.640 27 | 1.561 84 | 1.142 47 | 0.301 68 |
| 1.34 | 0.630 71 | 1.189 23 | 0.630 71 | 1.585 51 | 1.156 81 | 0.319 61 |
| 1.36 | 0.621 44 | 1.198 68 | 0.621 44 | 1.609 17 | 1.171 81 | 0.337 62 |
| 1.38 | 0.612 43 | 1.208 27 | 0.612 43 | 1.632 84 | 1.187 49 | 0.355 71 |
| 1.40 | 0.603 68 | 1.218 00 | 0.603 68 | 1.656 50 | 1.203 85 | 0.373 85 |
| 1.42 | 0.595 18 | 1.227 87 | 0.595 18 | 1.680 17 | 1.220 90 | 0.392 02 |
| 1.44 | 0.586 91 | 1.237 88 | 0.586 91 | 1.703 83 | 1.238 65 | 0.410 22 |
| 1.46 | 0.578 87 | 1.248 03 | 0.578 87 | 1.727 49 | 1.257 10 | 0.428 44 |
| 1.48 | 0.571 05 | 1.258 32 | 0.571 05 | 1.751 16 | 1.276 27 | 0.446 65 |
| 1.50 | 0.563 44 | 1.268 75 | 0.563 44 | 1.774 82 | 1.296 17 | 0.464 86 |
| 1.52 | 0.556 02 | 1.279 32 | 0.556 02 | 1.798 49 | 1.316 80 | 0.483 05 |
| 1.54 | 0.548 80 | 1.290 03 | 0.548 80 | 1.822 15 | 1.338 18 | 0.501 22 |
| 1.56 | 0.541 77 | 1.300 88 | 0.541 77 | 1.845 82 | 1.360 32 | 0.519 35 |
| 1.58 | 0.534 91 | 1.311 87 | 0.534 91 | 1.869 48 | 1.383 24 | 0.537 45 |
| 1.60 | 0.528 22 | 1.323 00 | 0.528 22 | 1.893 14 | 1.406 94 | 0.555 50 |
| 1.62 | 0.521 70 | 1.334 27 | 0.521 70 | 1.916 81 | 1.431 44 | 0.573 49 |
| 1.64 | 0.515 34 | 1.345 68 | 0.515 34 | 1.940 47 | 1.456 76 | 0.591 44 |
| 1.66 | 0.509 13 | 1.357 23 | 0.509 13 | 1.964 14 | 1.482 91 | 0.609 32 |
| 1.68 | 0.503 07 | 1.368 92 | 0.503 07 | 1.987 80 | 1.509 90 | 0.627 14 |
| 1.70 | 0.497 15 | 1.380 75 | 0.497 15 | 2.011 47 | 1.537 76 | 0.644 89 |
| 1.72 | 0.491 37 | 1.392 72 | 0.491 37 | 2.035 13 | 1.566 50 | 0.662 56 |
| 1.74 | 0.485 72 | 1.404 83 | 0.485 72 | 2.058 79 | 1.596 13 | 0.680 17 |

| M | $p/p^*$ | $T_0/T_0^*$ | $\rho/\rho^*$ | $V/V^*$ | $p_0/p_0^*$ | $4fl^*/D_H$ |
|---|---------|-------------|---------------|---------|-------------|-------------|
| 1.76 | 0.480 20 | 1.417 08 | 0.480 20 | 2.082 46 | 1.626 68 | 0.697 69 |
| 1.78 | 0.474 81 | 1.429 47 | 0.474 81 | 2.106 12 | 1.658 17 | 0.715 14 |
| 1.80 | 0.469 53 | 1.442 00 | 0.469 53 | 2.129 79 | 1.690 60 | 0.732 50 |
| 1.82 | 0.464 37 | 1.454 67 | 0.464 37 | 2.153 45 | 1.724 01 | 0.749 78 |
| 1.84 | 0.459 32 | 1.467 48 | 0.459 32 | 2.177 12 | 1.758 41 | 0.766 98 |
| 1.86 | 0.454 38 | 1.480 43 | 0.454 38 | 2.200 78 | 1.793 82 | 0.784 09 |
| 1.88 | 0.449 55 | 1.493 52 | 0.449 55 | 2.224 44 | 1.830 27 | 0.801 11 |
| 1.90 | 0.444 82 | 1.506 75 | 0.444 82 | 2.248 11 | 1.867 78 | 0.818 04 |
| 1.92 | 0.440 18 | 1.520 12 | 0.440 18 | 2.271 77 | 1.906 37 | 0.834 88 |
| 1.94 | 0.435 65 | 1.533 63 | 0.435 65 | 2.295 44 | 1.946 06 | 0.851 63 |
| 1.96 | 0.431 20 | 1.547 28 | 0.431 20 | 2.319 10 | 1.986 87 | 0.868 29 |
| 1.98 | 0.426 85 | 1.561 07 | 0.426 85 | 2.342 77 | 2.028 84 | 0.884 86 |
| 2.00 | 0.422 58 | 1.575 00 | 0.422 58 | 2.366 43 | 2.071 99 | 0.901 34 |
| 2.02 | 0.418 39 | 1.589 07 | 0.418 39 | 2.390 09 | 2.116 34 | 0.917 72 |
| 2.04 | 0.414 29 | 1.603 28 | 0.414 29 | 2.413 76 | 2.161 91 | 0.934 01 |
| 2.06 | 0.410 27 | 1.617 63 | 0.410 27 | 2.437 42 | 2.208 74 | 0.950 20 |
| 2.08 | 0.406 32 | 1.632 12 | 0.406 32 | 2.461 09 | 2.256 86 | 0.966 31 |
| 2.10 | 0.402 45 | 1.646 75 | 0.402 45 | 2.484 75 | 2.306 28 | 0.982 32 |
| 2.12 | 0.398 66 | 1.661 52 | 0.398 66 | 2.508 42 | 2.357 05 | 0.998 23 |
| 2.14 | 0.394 93 | 1.676 43 | 0.394 93 | 2.532 08 | 2.409 19 | 1.014 05 |
| 2.16 | 0.391 28 | 1.691 48 | 0.391 28 | 2.555 74 | 2.462 72 | 1.029 78 |
| 2.18 | 0.387 69 | 1.706 67 | 0.387 69 | 2.579 41 | 2.517 69 | 1.045 42 |
| 2.20 | 0.384 16 | 1.722 00 | 0.384 16 | 2.603 07 | 2.574 12 | 1.060 97 |
| 2.22 | 0.380 70 | 1.737 47 | 0.380 70 | 2.626 74 | 2.632 04 | 1.076 42 |
| 2.24 | 0.377 30 | 1.753 08 | 0.377 30 | 2.650 40 | 2.691 49 | 1.091 78 |
| 2.26 | 0.373 96 | 1.768 83 | 0.373 96 | 2.674 07 | 2.752 50 | 1.107 05 |
| 2.28 | 0.370 68 | 1.784 72 | 0.370 68 | 2.697 73 | 2.815 11 | 1.122 23 |
| 2.30 | 0.367 46 | 1.800 75 | 0.367 46 | 2.721 39 | 2.879 35 | 1.137 31 |
| 2.32 | 0.364 29 | 1.816 92 | 0.364 29 | 2.745 06 | 2.945 25 | 1.152 31 |
| 2.34 | 0.361 18 | 1.833 23 | 0.361 18 | 2.768 72 | 3.012 86 | 1.167 22 |
| 2.36 | 0.358 12 | 1.849 68 | 0.358 12 | 2.792 39 | 3.082 20 | 1.182 04 |
| 2.38 | 0.355 11 | 1.866 27 | 0.355 11 | 2.816 05 | 3.153 33 | 1.196 77 |
| 2.40 | 0.352 15 | 1.883 00 | 0.352 15 | 2.839 72 | 3.226 26 | 1.211 42 |
| 2.42 | 0.349 24 | 1.899 87 | 0.349 24 | 2.863 38 | 3.301 06 | 1.225 97 |
| 2.44 | 0.346 37 | 1.916 88 | 0.346 37 | 2.887 04 | 3.777 75 | 1.240 44 |
| 2.46 | 0.343 56 | 1.934 03 | 0.343 56 | 2.910 71 | 3.456 38 | 1.254 83 |
| 2.48 | 0.340 79 | 1.951 32 | 0.340 79 | 2.934 37 | 3.536 98 | 1.269 12 |
| 2.50 | 0.338 06 | 1.968 75 | 0.338 06 | 2.958 04 | 3.619 61 | 1.283 34 |
| 2.52 | 0.335 38 | 1.986 32 | 0.335 38 | 2.981 70 | 3.704 31 | 1.297 47 |
| 2.54 | 0.332 74 | 2.004 03 | 0.332 74 | 3.005 37 | 3.791 11 | 1.311 51 |
| 2.56 | 0.330 14 | 2.021 88 | 0.330 14 | 3.029 03 | 3.880 06 | 1.325 48 |
| 2.58 | 0.327 58 | 2.039 87 | 0.327 58 | 3.052 70 | 3.971 22 | 1.339 36 |
| 2.60 | 0.325 06 | 2.058 00 | 0.325 06 | 3.076 36 | 4.064 62 | 1.353 16 |
| 2.62 | 0.322 58 | 2.076 27 | 0.322 58 | 3.100 02 | 4.160 32 | 1.366 88 |
| 2.64 | 0.320 13 | 2.094 68 | 0.320 13 | 3.123 69 | 4.258 37 | 1.380 51 |
| 2.66 | 0.317 73 | 2.113 23 | 0.317 73 | 3.147 35 | 4.358 80 | 1.394 07 |
| 2.68 | 0.315 36 | 2.131 92 | 0.315 36 | 3.171 02 | 4.461 68 | 1.407 55 |
| 2.70 | 0.313 02 | 2.150 75 | 0.313 02 | 3.194 68 | 4.567 05 | 1.420 96 |
| 2.72 | 0.310 72 | 2.169 72 | 0.310 72 | 3.218 34 | 4.674 97 | 1.434 28 |
| 2.74 | 0.308 45 | 2.188 83 | 0.308 45 | 3.242 01 | 4.785 49 | 1.447 53 |
| 2.76 | 0.306 22 | 2.208 08 | 0.306 22 | 3.265 67 | 4.898 66 | 1.460 70 |
| 2.78 | 0.304 01 | 2.227 47 | 0.304 01 | 3.289 34 | 5.014 55 | 1.473 80 |
| 2.80 | 0.301 84 | 2.247 00 | 0.301 84 | 3.313 00 | 5.133 19 | 1.486 82 |

| M | $p/p^*$ | $T_0/T_0^*$ | $\rho/\rho^*$ | $V/V^*$ | $p_0/p_0^*$ | $4fl^*/D_H$ |
|------|----------|------------|------------|----------|------------|------------|
| 2.82 | 0.299 70 | 2.266 67 | 0.299 70 | 3.336 67 | 5.254 66 | 1.499 76 |
| 2.84 | 0.297 59 | 2.286 48 | 0.297 59 | 3.360 33 | 5.379 01 | 1.512 64 |
| 2.86 | 0.295 51 | 2.306 43 | 0.295 51 | 3.384 00 | 5.506 30 | 1.525 44 |
| 2.88 | 0.293 46 | 2.326 52 | 0.293 46 | 3.407 66 | 5.636 59 | 1.538 17 |
| 2.90 | 0.291 43 | 2.346 75 | 0.291 43 | 3.431 32 | 5.769 93 | 1.550 83 |
| 2.92 | 0.289 44 | 2.367 12 | 0.289 44 | 3.454 99 | 5.906 40 | 1.563 41 |
| 2.94 | 0.287 47 | 2.387 63 | 0.287 47 | 3.478 65 | 6.046 05 | 1.575 93 |
| 2.96 | 0.285 53 | 2.408 28 | 0.285 53 | 3.502 32 | 6.188 96 | 1.588 37 |
| 2.98 | 0.283 61 | 2.429 07 | 0.283 61 | 3.525 98 | 6.335 18 | 1.600 75 |
| 3.00 | 0.281 72 | 2.450 00 | 0.281 72 | 3.549 64 | 6.484 77 | 1.613 06 |
| 3.05 | 0.277 10 | 2.502 93 | 0.277 10 | 3.608 81 | 6.874 01 | 1.643 54 |
| 3.10 | 0.272 63 | 2.556 75 | 0.272 63 | 3.667 97 | 7.285 88 | 1.673 60 |
| 3.15 | 0.268 30 | 2.611 43 | 0.268 30 | 3.727 13 | 7.721 52 | 1.703 26 |
| 3.20 | 0.264 11 | 2.667 00 | 0.264 11 | 3.786 29 | 8.182 11 | 1.732 53 |
| 3.25 | 0.260 05 | 2.723 43 | 0.260 05 | 3.845 45 | 8.668 87 | 1.761 41 |
| 3.30 | 0.256 11 | 2.780 75 | 0.256 11 | 3.904 61 | 9.183 06 | 1.789 91 |
| 3.35 | 0.252 29 | 2.838 93 | 0.252 29 | 3.963 77 | 9.726 02 | 1.818 04 |
| 3.40 | 0.248 57 | 2.898 00 | 0.248 57 | 4.022 93 | 10.299 12 | 1.845 81 |
| 3.45 | 0.244 97 | 2.957 93 | 0.244 97 | 4.082 09 | 10.903 78 | 1.873 23 |
| 3.50 | 0.241 47 | 3.018 75 | 0.241 47 | 4.141 25 | 11.541 48 | 1.900 31 |
| 3.55 | 0.238 07 | 3.080 43 | 0.238 07 | 4.200 41 | 12.213 77 | 1.927 04 |
| 3.60 | 0.234 77 | 3.143 00 | 0.234 77 | 4.259 57 | 12.922 23 | 1.953 45 |
| 3.65 | 0.231 55 | 3.206 43 | 0.231 55 | 4.318 73 | 13.668 52 | 1.979 54 |
| 3.70 | 0.228 42 | 3.270 75 | 0.228 42 | 4.377 90 | 14.454 35 | 2.005 31 |
| 3.75 | 0.225 37 | 3.335 93 | 0.225 37 | 4.437 06 | 15.281 50 | 2.030 78 |
| 3.80 | 0.222 41 | 3.402 00 | 0.222 41 | 4.496 22 | 16.151 80 | 2.055 94 |
| 3.85 | 0.219 52 | 3.468 93 | 0.219 52 | 4.555 38 | 17.067 17 | 2.080 81 |
| 3.90 | 0.216 71 | 3.536 75 | 0.216 71 | 4.614 54 | 18.029 56 | 2.105 39 |
| 3.95 | 0.213 96 | 3.605 43 | 0.213 96 | 4.673 70 | 19.041 02 | 2.129 68 |
| 4.00 | 0.211 29 | 3.675 00 | 0.211 29 | 4.732 86 | 20.103 66 | 2.153 70 |
| 4.05 | 0.208 68 | 3.745 43 | 0.208 68 | 4.792 02 | 21.219 69 | 2.177 45 |
| 4.10 | 0.206 14 | 3.816 75 | 0.206 14 | 4.851 18 | 22.391 32 | 2.200 94 |
| 4.15 | 0.203 65 | 3.888 94 | 0.203 65 | 4.910 35 | 23.620 89 | 2.224 16 |
| 4.20 | 0.201 23 | 3.962 00 | 0.201 23 | 4.969 51 | 24.910 82 | 2.247 13 |
| 4.25 | 0.198 86 | 4.035 94 | 0.198 86 | 5.028 67 | 26.263 57 | 2.269 86 |
| 4.30 | 0.196 55 | 4.110 75 | 0.196 55 | 5.087 83 | 27.681 71 | 2.292 33 |
| 4.35 | 0.194 29 | 4.186 44 | 0.194 29 | 5.146 99 | 29.167 89 | 2.314 57 |
| 4.40 | 0.192 08 | 4.263 01 | 0.192 08 | 5.206 16 | 30.724 84 | 2.336 58 |
| 4.45 | 0.189 92 | 4.340 45 | 0.189 92 | 5.265 32 | 32.355 37 | 2.358 35 |
| 4.50 | 0.187 81 | 4.418 76 | 0.187 81 | 5.324 48 | 34.062 39 | 2.379 90 |
| 4.60 | 0.183 73 | 4.578 02 | 0.183 73 | 5.442 81 | 37.717 94 | 2.422 35 |
| 4.70 | 0.179 82 | 4.740 77 | 0.179 82 | 5.561 13 | 41.716 55 | 2.463 94 |
| 4.80 | 0.176 07 | 4.907 02 | 0.176 07 | 5.679 45 | 46.084 77 | 2.504 71 |
| 4.90 | 0.172 48 | 5.076 78 | 0.172 48 | 5.797 78 | 50.850 77 | 2.544 70 |
| 5.00 | 0.169 03 | 5.250 03 | 0.169 03 | 5.916 10 | 56.044 35 | 2.583 93 |
| 5.10 | 0.165 72 | 5.426 79 | 0.165 72 | 6.034 43 | 61.697 04 | 2.622 42 |
| 5.20 | 0.162 53 | 5.607 04 | 0.162 53 | 6.152 75 | 67.842 14 | 2.660 21 |
| 5.30 | 0.159 46 | 5.790 80 | 0.159 46 | 6.271 08 | 74.514 83 | 2.697 32 |
| 5.40 | 0.156 51 | 5.978 05 | 0.156 51 | 6.389 40 | 81.752 20 | 2.733 78 |
| 5.50 | 0.153 66 | 6.168 81 | 0.153 66 | 6.507 72 | 89.593 34 | 2.769 59 |

# Oblique Shock Wave Charts for $\gamma = 1.4$

The graphs in this Appendix are from NACA Report 1135, *Equations, Tables and Charts for Compressible Flow*, Ames Research Staff, 1953.

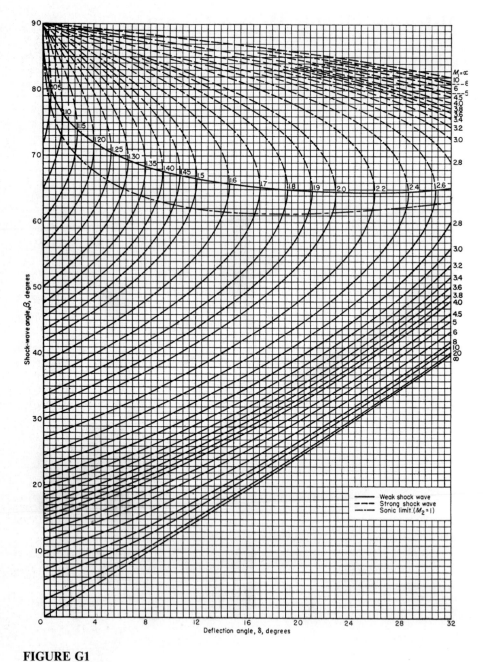

**FIGURE G1**
Variation of oblique shock wave angle with flow deflection angle for various upstream Mach numbers.

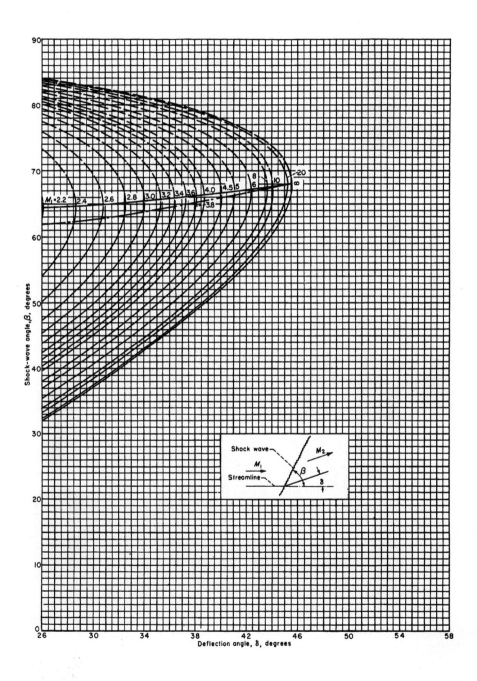

**FIGURE G1** (*continued*)

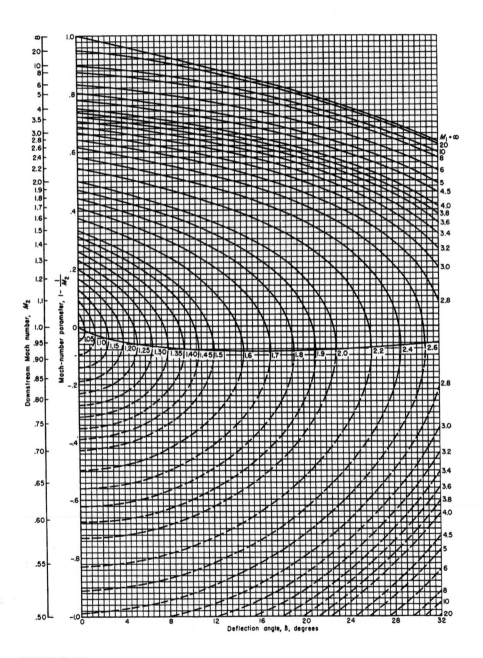

**FIGURE G2**
Variation of Mach number downstream of oblique shock with flow deflection angle for various upstream Mach numbers.

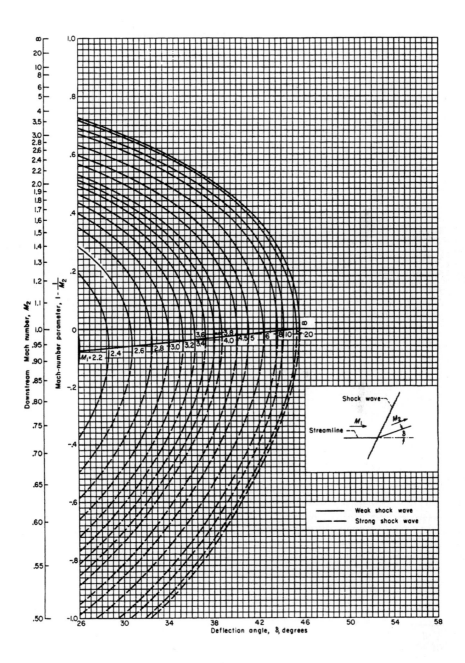

**FIGURE G2** (*continued*)

# APPENDIX H

# Approximate Properties of the Standard Atmosphere

| $H$ (m) | $T$ (K) | $p$ (Pa × 10$^5$) | $\rho$ (kg/m$^3$) | $a$ (m/s) | $\nu$ (m$^2$/s × 10$^{-5}$) |
|---|---|---|---|---|---|
| 0 | 288.16 | 1.0133 | 1.2250 | 340.29 | 1.4610 |
| 200 | 286.86 | 0.9895 | 1.2017 | 339.52 | 1.4841 |
| 400 | 285.56 | 0.9662 | 1.1787 | 338.76 | 1.5077 |
| 600 | 284.26 | 0.9433 | 1.1560 | 337.98 | 1.5318 |
| 800 | 282.97 | 0.9209 | 1.1337 | 337.21 | 1.5564 |
| 1000 | 281.67 | 0.8989 | 1.1118 | 336.44 | 1.5814 |
| 1200 | 280.37 | 0.8773 | 1.0901 | 335.66 | 1.6070 |
| 1400 | 279.07 | 0.8561 | 1.0688 | 334.88 | 1.6331 |
| 1600 | 277.77 | 0.8354 | 1.0478 | 334.10 | 1.6598 |
| 1800 | 276.47 | 0.8151 | 1.0271 | 333.32 | 1.6870 |
| 2000 | 275.17 | 0.7952 | 1.0067 | 332.54 | 1.7148 |
| 2200 | 273.87 | 0.7756 | 0.9866 | 331.75 | 1.7431 |
| 2400 | 272.58 | 0.7565 | 0.9669 | 330.96 | 1.7721 |
| 2600 | 271.28 | 0.7377 | 0.9474 | 330.17 | 1.8017 |
| 2800 | 269.98 | 0.7194 | 0.9283 | 329.38 | 1.8319 |
| 3000 | 268.68 | 0.7014 | 0.9094 | 328.59 | 1.8627 |
| 3200 | 267.38 | 0.6837 | 0.8909 | 327.79 | 1.8943 |
| 3400 | 266.08 | 0.6665 | 0.8726 | 327.00 | 1.9265 |
| 3600 | 264.78 | 0.6495 | 0.8546 | 326.20 | 1.9594 |
| 3800 | 263.49 | 0.6330 | 0.8369 | 325.40 | 1.9930 |
| 4000 | 262.19 | 0.6167 | 0.8195 | 324.59 | 2.0273 |
| 4200 | 260.89 | 0.6008 | 0.8023 | 322.98 | 2.0983 |
| 4600 | 258.29 | 0.5701 | 0.7689 | 322.17 | 2.1350 |
| 4800 | 256.99 | 0.5552 | 0.7526 | 321.36 | 2.1725 |
| 5000 | 255.69 | 0.5406 | 0.7365 | 320.55 | 2.2109 |
| 5200 | 254.39 | 0.5263 | 0.7207 | 319.73 | 2.2501 |
| 5400 | 253.10 | 0.5123 | 0.7052 | 318.92 | 2.2902 |
| 5600 | 251.80 | 0.4987 | 0.6899 | 317.28 | 2.3732 |
| 6000 | 249.20 | 0.4722 | 0.6602 | 316.45 | 2.4161 |

| H (m) | T (K) | p (Pa × 10⁵) | ρ (kg/m³) | a (m/s) | v (m²/s × 10⁻⁵) |
|---|---|---|---|---|---|
| 6200 | 247.90 | 0.4594 | 0.6456 | 315.63 | 2.4600 |
| 6400 | 246.60 | 0.4469 | 0.6314 | 314.80 | 2.5049 |
| 6600 | 245.30 | 0.4347 | 0.6173 | 313.97 | 2.5509 |
| 6800 | 244.01 | 0.4227 | 0.6035 | 313.14 | 2.5980 |
| 7000 | 242.71 | 0.4110 | 0.5900 | 312.30 | 2.6462 |
| 7200 | 241.41 | 0.3996 | 0.5767 | 311.47 | 2.6955 |
| 7400 | 240.11 | 0.3884 | 0.5636 | 310.63 | 2.7460 |
| 7600 | 238.81 | 0.3775 | 0.5507 | 309.79 | 2.7977 |
| 7800 | 237.51 | 0.3668 | 0.5381 | 308.94 | 2.8506 |
| 8000 | 236.21 | 0.3564 | 0.5257 | 308.10 | 2.9048 |
| 8200 | 234.91 | 0.3462 | 0.5135 | 307.25 | 2.9604 |
| 8400 | 233.62 | 0.3363 | 0.5015 | 306.40 | 3.0173 |
| 8600 | 232.32 | 0.3266 | 0.4898 | 305.55 | 3.0756 |
| 8800 | 231.02 | 0.3171 | 0.4782 | 304.69 | 3.1353 |
| 9000 | 229.72 | 0.3079 | 0.4669 | 303.83 | 3.1966 |
| 9200 | 228.42 | 0.2988 | 0.4448 | 302.11 | 3.3237 |
| 9600 | 225.82 | 0.2814 | 0.4341 | 301.25 | 3.3896 |
| 9800 | 224.53 | 0.2730 | 0.4236 | 300.38 | 3.4573 |
| 10 000 | 223.23 | 0.2648 | 0.4132 | 299.51 | 3.5266 |
| 10 200 | 221.93 | 0.2568 | 0.4031 | 298.64 | 3.5978 |
| 10 400 | 220.63 | 0.2490 | 0.3932 | 297.76 | 3.6708 |
| 10 600 | 219.33 | 0.2414 | 0.3834 | 296.88 | 3.7456 |
| 10 800 | 218.03 | 0.2340 | 0.3738 | 296.00 | 3.8225 |
| 11 000 | 216.73 | 0.2267 | 0.3644 | 295.12 | 3.9013 |
| 11 200 | 216.66 | 0.2200 | 0.3537 | 295.07 | 4.0188 |
| 11 400 | 216.66 | 0.2131 | 0.3427 | 295.07 | 4.1475 |
| 11 600 | 216.66 | 0.2065 | 0.3321 | 295.07 | 4.2802 |
| 11 800 | 216.66 | 0.2001 | 0.3218 | 295.07 | 4.4172 |
| 12 000 | 216.66 | 0.1939 | 0.3118 | 295.07 | 4.5586 |
| 12 200 | 216.66 | 0.1879 | 0.3021 | 295.07 | 4.7045 |
| 12 400 | 216.66 | 0.1821 | 0.2928 | 295.07 | 4.8551 |
| 12 600 | 216.66 | 0.1764 | 0.2837 | 295.07 | 5.0105 |
| 12 800 | 216.66 | 0.1710 | 0.2749 | 295.07 | 5.1708 |
| 13 000 | 216.66 | 0.1657 | 0.2664 | 295.07 | 5.3363 |
| 13 200 | 216.66 | 0.1605 | 0.2581 | 295.07 | 5.5071 |
| 13 400 | 216.66 | 0.1555 | 0.2501 | 295.07 | 5.6834 |
| 13 600 | 216.66 | 0.1507 | 0.2423 | 295.07 | 5.8653 |
| 13 800 | 216.66 | 0.1460 | 0.2348 | 295.07 | 6.0530 |
| 14 000 | 216.66 | 0.1415 | 0.2275 | 295.07 | 6.2467 |
| 14 200 | 216.66 | 0.1371 | 0.2205 | 295.07 | 6.4467 |
| 14 400 | 216.66 | 0.1329 | 0.2137 | 295.07 | 6.6530 |
| 14 600 | 216.66 | 0.1288 | 0.2070 | 295.07 | 6.8659 |
| 14 800 | 216.66 | 0.1248 | 0.2006 | 295.07 | 7.0857 |
| 15 000 | 216.66 | 0.1209 | 0.1944 | 295.07 | 7.3125 |
| 15 400 | 216.66 | 0.1135 | 0.1825 | 295.07 | 7.7881 |
| 15 800 | 216.66 | 0.1066 | 0.1714 | 295.07 | 8.2946 |
| 16 200 | 216.66 | 0.1001 | 0.1609 | 295.07 | 8.8340 |
| 16 600 | 216.66 | 0.0940 | 0.1511 | 295.07 | 9.4086 |
| 17 000 | 216.66 | 0.0882 | 0.1419 | 295.07 | 10.0205 |
| 17 400 | 216.66 | 0.0828 | 0.1332 | 295.07 | 10.6722 |
| 17 800 | 216.66 | 0.0778 | 0.1251 | 295.07 | 11.3662 |
| 18 200 | 216.66 | 0.0730 | 0.1174 | 295.07 | 12.1055 |

| $H$ (m) | $T$ (K) | $p$ (Pa $\times 10^5$) | $\rho$ (kg/m$^3$) | $a$ (m/s) | $\nu$ (m$^2$/s $\times 10^{-5}$) |
|---|---|---|---|---|---|
| 18 600 | 216.66 | 0.0686 | 0.1102 | 295.07 | 12.8928 |
| 19 000 | 216.66 | 0.0644 | 0.1035 | 295.07 | 13.7313 |
| 19 400 | 216.66 | 0.0604 | 0.0972 | 295.07 | 14.6243 |
| 19 800 | 216.66 | 0.0568 | 0.0913 | 295.07 | 15.5754 |
| 20 000 | 216.66 | 0.0550 | 0.0884 | 295.07 | 16.0739 |

# APPENDIX I

# Properties of Dry Air at Atmospheric Pressure

| $T$ (°C) | $\rho$ (kg/m$^3$) | $c_p$ (kJ/kg · K) | $\mu$ (kg/m · s × 10$^{-5}$) | $\nu$ (m$^2$/s × 10$^{-6}$) | $k$ (W/m · K) | $Pr$ |
|---|---|---|---|---|---|---|
| −100 | 2.039 | 1.010 | 1.16 | 5.69 | 0.0163 | 0.75 |
| −50 | 1.582 | 1.006 | 1.46 | 9.25 | 0.0200 | 0.73 |
| 0 | 1.292 | 1.006 | 1.72 | 13.31 | 0.0249 | 0.72 |
| 50 | 1.092 | 1.007 | 1.96 | 17.92 | 0.0278 | 0.71 |
| 100 | 0.946 | 1.011 | 2.18 | 23.02 | 0.0313 | 0.70 |
| 150 | 0.834 | 1.017 | 2.38 | 28.58 | 0.0346 | 0.70 |
| 200 | 0.746 | 1.025 | 2.58 | 34.57 | 0.0378 | 0.70 |
| 300 | 0.616 | 1.045 | 2.94 | 47.72 | 0.0440 | 0.70 |
| 500 | 0.457 | 1.093 | 3.57 | 78.22 | 0.0554 | 0.70 |
| 1000 | 0.277 | 1.185 | 4.82 | 173 | 0.0755 | 0.71 |

$Pr = \nu/\alpha$.
$c_p$, $\mu$, and $k$ are approximately independent of pressure.
At pressure $p$, density = density at atmospheric pressure $\times (p/p_{\text{atm}})$.
At pressure $p$, $\nu = \nu$ at atmospheric pressure $\times (p_{\text{atm}}/p)$.

# APPENDIX J

# Constants, Conversion Factors, and Units

## CONSTANTS

| | | |
|---|---|---|
| Universal gas constant | $R$ | $= 8.314 \text{ kg/kmol·K}$ |
| | | $= 1545.33 \text{ ft lbf/lbmol°R}$ |
| Atmospheric pressure | $p_{atm}$ | $= 0.101\,325 \text{ MPa}$ |
| | | $= 101.325 \text{ kPa}$ |
| | | $= 1.013\,25 \text{ bar}$ |
| Speed of light in a vacuum | $c_0$ | $= 2.998 \times 10^8 \text{ m/s}$ |
| Gravitational acceleration at sea-level | $g$ | $= 9.807 \text{ m/s}^2$ |
| | | $= 32.17 \text{ ft/sec}^2$ |

## CONVERSION FACTORS

| | | |
|---|---|---|
| Area ($A$) | $1 \text{ m}^2$ | $= 10.764 \text{ ft}^2$ |
| | $1 \text{ ft}^2$ | $= 0.0929 \text{ m}^2$ |
| Density ($\rho$) | $1 \text{ kg/m}^3$ | $= 0.062\,428 \text{ lbm/ft}^3$ |
| | $1 \text{ lbm/ft}^3$ | $= 16.019 \text{ kg/m}^3$ |
| Energy, work | $1 \text{ kJ}$ | $= 737.56 \text{ ft·lbf}$ |
| | $1 \text{ ft·lbf}$ | $= 1.3558 \text{ J}$ |
| | $1 \text{ btu}$ | $= 778.17 \text{ ft·lbf}$ |
| | $1 \text{ btu}$ | $= 1.0551 \text{ kJ}$ |
| Force | $1 \text{ N}$ | $= 0.224\,81 \text{ lbf}$ |
| | $1 \text{ lbf}$ | $= 4.4482 \text{ N}$ |
| Heat flux ($q$) | $1 \text{ W/m}^2$ | $= 0.317\,00 \text{ btu/hr·ft}^2$ |
| | $1 \text{ btu/hr·ft}^2$ | $= 3.154\,69 \text{ W/m}^2$ |

| | | |
|---|---|---|
| Heat transfer coefficient ($h$) | $1\ \text{W/m}^2\cdot\text{K}$ | $= 5.6783\ \text{btu/hr}\cdot\text{ft}^2\cdot{}^\circ\text{F}$ |
| | $1\ \text{btu/hr}\cdot\text{ft}^2\cdot{}^\circ\text{F}$ | $= 0.1761\ \text{W/m}^2\cdot\text{K}$ |
| Heat transfer rate ($Q$) | $1\ \text{W}$ | $= 3.4118\ \text{btu/hr}$ |
| | $1\ \text{btu/hr}$ | $= 0.293\,07\ \text{W}$ |
| Kinematic viscosity ($\nu$) | $1\ \text{m}^2/\text{s}$ | $= 38\,750\ \text{ft}^2/\text{hr}$ |
| | $1\ \text{ft}^2/\text{hr}$ | $= 2.581 \times 10^{-5}\ \text{m}^2/\text{s}$ |
| Length | $1\ \text{m}$ | $= 3.2808\ \text{ft}$ |
| | $1\ \text{ft}$ | $= 0.304\,80\ \text{m}$ |
| Mass | $1\ \text{kg}$ | $= 2.2046\ \text{lbm}$ |
| | $1\ \text{lbm}$ | $= 0.453\,59\ \text{kg}$ |
| Pressure, stress ($p, \tau$) | $1\ \text{kPa}$ | $= 0.145\,04\ \text{lbf/in}^2$ |
| | $1\ \text{lbf/in}^2$ | $= 6.894\,75\ \text{kPa}$ |
| Power | $1\ \text{W}$ | $= 0.073\,756\ \text{ft}\cdot\text{lbf/sec}$ |
| | $1\ \text{btu/hr}$ | $= 0.029\,307\ \text{W}$ |
| | $1\ \text{hp}$ | $= 0.074\,570\ \text{kW}$ |
| Specific heat ($c_p, c_v$) | $1\ \text{kJ/kg}\cdot{}^\circ\text{C}$ | $= 0.238\,85\ \text{btu/lbm}\cdot{}^\circ\text{F}$ |
| | $1\ \text{btu/lbm}\cdot{}^\circ\text{F}$ | $= 4.1868\ \text{kJ/kg}\cdot{}^\circ\text{C}$ |
| Temperature ($T$) | $\text{T (K)}$ | $= \text{T }({}^\circ\text{C}) + 273.15$ |
| | $\text{T }({}^\circ\text{C})$ | $= 5/9(\text{T }({}^\circ\text{F}) - 32)$ |
| | $\text{T }({}^\circ\text{R})$ | $= \text{T }({}^\circ\text{F}) + 459.67$ |
| | $\text{T }({}^\circ\text{F})$ | $= 1.8\,\text{T }({}^\circ\text{C}) + 32$ |
| Temperature difference ($\Delta T$) | $1{}^\circ\text{C}$ | $= 1\ \text{K}$ |
| | $1{}^\circ\text{C}$ | $= 1.8{}^\circ\text{F}$ |
| | $1{}^\circ\text{F}$ | $= 1{}^\circ\text{R}$ |
| | $1{}^\circ\text{F}$ | $= 0.555\,56{}^\circ\text{C}$ |
| Thermal conductivity ($k$) | $1\ \text{W/m}\cdot\text{K}$ | $= 0.5782\ \text{btu/hr}\cdot\text{ft}\cdot{}^\circ\text{F}$ |
| | $1\ \text{btu/hr}\cdot\text{ft}\cdot{}^\circ\text{F}$ | $= 1.7295\ \text{W/m}\cdot\text{K}$ |
| Thermal diffusivity ($\alpha$) | $1\ \text{m}^2/\text{s}$ | $= 38\,750\ \text{ft}^2/\text{hr}$ |
| | $1\ \text{ft}^2/\text{hr}$ | $= 2.5807 \times 10^{-5}\ \text{m}^2/\text{s}$ |
| Velocity | $1\ \text{m/s}$ | $= 3.2808\ \text{ft/sec}$ |
| | $1\ \text{ft/sec}$ | $= 0.304\,80\ \text{m/s}$ |
| Viscosity ($\mu$) | $1\ \text{N}\cdot\text{s/m}^2$ | $= 2419.1\ \text{lbm/ft}\cdot\text{hr}$ |
| | | $= 5.8016 \times 10^{-6}\ \text{lbf}\cdot\text{hr/ft}^2$ |
| Volume ($V$) | $1\ \text{m}^3$ | $= 35.315\ \text{ft}^3$ |
| | $1\ \text{ft}^3$ | $= 0.028\,317\ \text{m}^3$ |

# Optical Methods in Compressible Flows

## INTRODUCTION

A number of photographs showing various features of compressible flows, such as shock waves, are given in the main body of this book. An example of such a photograph is shown in Fig. K1. A very brief discussion of the methods used to obtain such photographs will be presented in this Appendix.

There are basically three such methods:

- Shadowgraph
- Schlieren
- Interferometer

All of these methods utilize the fact that the speed of light through a gas varies with the density of the gas, i.e., the fact that the refractive index, $n$, which is the ratio of the speed of light in a vacuum to the speed of light in the gas, is a function of density, i.e.,

$$n = \text{function } (\rho) \tag{K1}$$

where:

$$n = \frac{c_0}{c} \tag{K2}$$

$c_0$ being the speed of light in a vacuum and $c$ being the speed of light at some point in the gas.

The relation between $n$ and $\rho$ is approximately linear and is usually written as:

$$n = 1 + \beta \frac{\rho}{\rho_s} \tag{K3}$$

**FIGURE K1**
Typical Schlieren
photograph of supersonic
flow over a body.

TABLE K1

| Gas | $\beta$ |
|---|---|
| Air | 0.000 292 |
| Nitrogen | 0.000 297 |
| Oxygen | 0.000 271 |
| Water vapor | 0.000 254 |
| Carbon dioxide | 0.000 451 |

where $\rho_s$ is the density of the gas at 0°C and standard atmospheric pressure and $\beta$ is a constant that depends on the type of gas. Typical values of $\beta$ are given in Table K1. These values strictly only apply at a particular wavelength of light.

Equation (K3) is sometimes written in terms of the Gladstone–Dale constant, $K$, such that:

$$n = 1 + K\rho \qquad \text{(K4)}$$

so:

$$K = \beta/\rho_s \qquad \text{(K5)}$$

Because the speed of light depends on the density of the gas through which it is passing, it follows that if the density changes in the gas, the speed of light will be different in different parts of the gas. However, there is another related effect produced by the change in refractive index. If a beam of light passes through a gas in which there is a density gradient normal to the direction of the beam, the light will be turned in the direction of increasing density. This is shown schematically in Fig. K2.

The angle through which the light ray is turned is dependent on the gradient of density normal to the direction of the light, i.e., for the situation shown in Fig. K2, on $d\rho/dy$ which by virtue of eq. (K3) will be proportional to the gradient of the refractive index, i.e., on $dn/dy$.

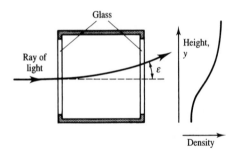

**FIGURE K2**
Bending of light beam in the presence of a density gradient.

## SHADOWGRAPH SYSTEM

Consider a series of light rays passing through a gas in which there is a vertical gradient of density as indicated in Fig. K3.

Because of the deflection of the light rays resulting from the density variations, if a screen is placed in such a way that it intercepts the light rays that have passed through the gas, the rays will be crowded together in some places and spread apart in other places as indicated in Fig. K3. When the rays are crowded together, the screen appears lighter than average, whereas when the rays are spread apart, the screen appears darker than average. Hence, because of the density gradients in the gas, regions of light and dark will appear on the screen as indicated in Fig. K4.

If the deflection of the light rays shown in Fig. K3 is considered, it will be seen that if there is a uniform vertical gradient of the density, all of the rays of light will be deflected by the same amount and there will still be a uniform illumination of the screen. The shadowgraph is, therefore, sensitive to the second derivative of density, i.e., to $d^2\rho/dy^2$ and hence to the second derivative of $n$, i.e., to $d^2n/dy^2$. Because of this, it is the least sensitive of the three

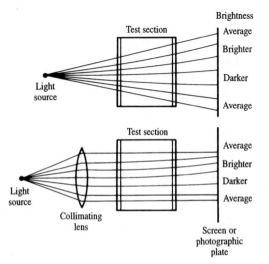

**FIGURE K3**
Shadowgraph system.

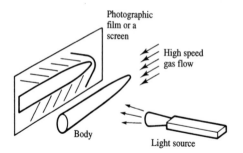

**FIGURE K4**
Formation of a shadowgraph.

systems considered here. It does, however, give a good indication of regions of high rates of density change, e.g., it gives a clear indication of the presence of shock waves. Of course, density changes in any direction normal to the rays will produce an image. The vertical $y$-direction was used only for illustrative purposes.

The shimmer about a hot roof in summer and the visible "smokeless" flow of "heat" out of a chimney in winter are basically the result of the shadow-graph effect, the sun in these cases being the light source.

## SCHLIEREN SYSTEM

The basic layout of a Schlieren system is shown in Fig. K5. It will be seen that the light from the source is passed through the gas and then focused onto a knife-edge before it is projected onto the screen. The knife-edge, whose orientation can usually be selected, is usually adjusted so that when there are no density changes in the gas, half the light is intercepted by the knife-edge. As a result, the knife-edge produces a uniform darkening of the image. If, as a result of the flow of the gas, density changes occur, the light rays will strike either less or more of the knife-edge, as indicated in Fig. K6. There will, therefore, be either a lightening or a darkening of the image depending on the angle through which the light is turned. The intensity of the image will, therefore, depend on the angle of turning, i.e., on the gradient of density, $d\rho/dy$, i.e., on $dn/dy$ if the knife-edge is horizontal or on $d\rho/dx$, i.e., on $dn/dx$, if the knife-edge is vertical.

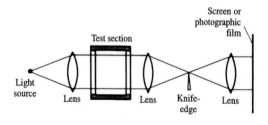

**FIGURE K5**
Basic Schlieren system
arrangement.

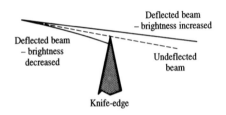

**FIGURE K6**
Light rays near knife-edge.

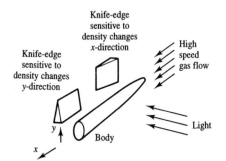

**FIGURE K7**
Effect of changing the direction of the knife-edge.

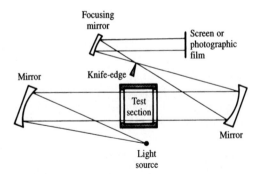

**FIGURE K8**
Schlieren system with mirrors.

As mentioned above, the direction of the knife-edge can usually be adjusted so that the density gradients in different directions can be examined. This is illustrated in Fig. K7. In actual Schlieren systems, mirrors rather than lenses are usually used for practical reasons, the system then being as shown in Fig. K8.

## INTERFEROMETER SYSTEM

To understand the principle on which the interferometer works, consider a ray of light which is split into two by a beam splitter as shown in Fig. K9. The splitter plate can be a partially mirrored surface which reflects some of the light striking it and transmits the remainder of the light. The two light rays formed in this way follow different paths and are then superimposed on the

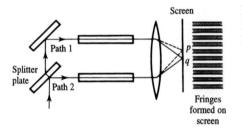

**FIGURE K9**
Basic interferometer system using a beam splitter.

screen. Now there will be a phase shift between the beams because they follow paths of different length and as a result the two rays can reinforce each other giving a bright area on the screen, or they can tend to cancel each other out giving a dark area on the screen. Whether the two rays reinforce or annul each other depends only on the difference between the lengths of the paths taken by the rays if, on these paths, the rays pass through the same media. A series of fringes is, therefore, formed on the screen because the relative lengths of the paths followed by the two rays will depend on the position on the screen being considered as indicated by points $q$ and $p$ shown in Fig. K9.

Now consider what happens if there is a density change in the gas through which one of the rays from the beam splitter passes. The times taken by the two rays to traverse their respective paths now differ from the times taken when the media were the same along the two paths. As a result there is a change in the fringe pattern due to the density differences along the two paths. The changes in the fringe pattern will depend on the density differences and, in fact, by measuring the fringe shift a measure of the density differences in the flow can be obtained.

Actually interferometers usually use mirrors rather than lenses, the set-up then resembling that shown schematically in Fig. K10. Shadowgraph, Schlieren, and interferometer photographs of the same flow are shown in Fig. K11.

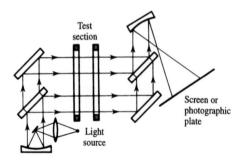

**FIGURE K10**
Mach–Zehnder interferometer.

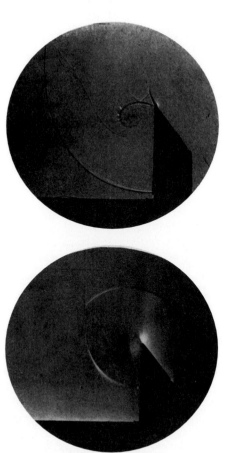

**FIGURE K11**
(*a*) Shadowgraph, (*b*) Schlieren, and (*c*) interferometer photographs. (From H. F. Waldron, An experimental study of a spiral vortex formed by shock-wave diffraction, Univ. Toronto, Inst. Aerophysics Tech. Note 2, September 1954.)

# APPENDIX L

# Simple BASIC Programs for Compressible Flow

As discussed in Chapters 4, 5, 9, and 10, it is now often convenient to use a programmable calculator to find the flow property ratios that in the past have mainly been obtained using tables. To illustrate how this can be done, a series of simple BASIC programs are presented in this Appendix. These programs apply to flows considered in Chapters 4, 5, 9, and 10.

## ONE-DIMENSIONAL ISENTROPIC FLOW

A very simple BASIC program for one-dimensional isentropic flow (see Chapter 4) that allows any one of $M$, $p_0/p$, or $T_0/T$ to be entered and then finds the value of the other two quantities is listed below. This program is easily extended to apply to all the variables in standard isentropic tables.

```
 10  CLS
 20  INPUT "SPECIFIC HEAT RATIO"; GA
 30  GF = (GA-1)/2
 40  GP = GA/(GA-1)
 50  CLS: PRINT" ": PRINT "ENTER ONE OF THE FOLLOWING;"
 60  PRINT" "
 70  PRINT "(1) MACH NUMBER (m OR M)"
 80  PRINT "(2) PRESSURE RATIO, p0/p (p OR P)"
 90  PRINT "(3) TEMPERATURE RATIO, T0/T (t OR T)"
100  PRINT" "
110  INPUT "WHICH VARIABLE (s OR S TO END)" AS$
120  IF AS$ = "M" OR AS$ = "m" THEN GOTO 180
130  IF AS$ = "P" OR AS$ = "p" THEN GOTO 300
140  IF AS$ = "T" OR AS$ = "t" THEN GOTO 420
150  IF AS$ = "S" OR AS$ = "s" THEN GOTO 540
160  PRINT "********** SELECTION INVALID **********"
170  GOTO 110
180  CLS
```

```
190   INPUT "MACH NUMBER ";M
200   FT = 1 + GF*M*M
210   FP = FT ^ GP
220   PRINT "T0/T = ";FT
230   PRINT "p0/p = ";FP
240   A$ = " "
250   PRINT " ": PRINT " "
260   PRINT "ANY KEY TO CONTINUE"
270   A$ = INKEY$
280   IF A$ = " " THEN GOTO 270
290   GOTO 50
300   CLS
310   INPUT "PRESSURE RATIO (p0/p)";PR
320   TR = PR ^ (1/GP)
330   M = SQR((RT-1)/GF)
340   PRINT "M = ";M
350   PRINT "T0/T = ";TR
360   A$ = ""
370   PRINT" ": PRINT" "
380   PRINT" ANY KEY TO CONTINUE"
390   A$ = INKEY$
400   IF A$ = "" THEN GOTO 390
410   GOTO 50
420   CLS
430   INPUT "TEMPERATURE RATIO (T0/T)";TR
440   M = SQR((TR-1)/GF)
450   PR = TR < UP ARROW > GP
460   PRINT " M = ";M
470   PRINT "p0/p = ";PR
480   A$ = ""
490   PRINT" ": PRINT" "
500   PRINT" ANY KEY TO CONTINUE"
510   A$ = INKEY$
520   IF A$ = "" THEN GOTO 510
530   GOTO 50
540   STOP
550   CLS
560   END
```

## NORMAL SHOCK WAVES

A very simple BASIC program for normal shock waves (see Chapter 5) that allows $M_1$ or $p_2/p_1$ to be entered and which finds the values of some other changes across a normal shock wave is listed below.

```
10    CLS
20    PRINT " ": PRINT " "
30    PRINT "NORMAL SHOCK WAVE CHARACTERISTICS": PRINT " ": PRINT " "
40    INPUT" SPECIFIC HEAT RATIO";GAM
50    IF GAM < = 1.01 THEN GOTO 40
60    CLS
70    INPUT" DO YOU WISH TO ENTER THE UPSTREAM MACH NUMBER
      (M) OR THE PRESSURE RATIO ACROSS THE SHOCK (P);AS$
80    IF AS$ = "M" OR AS$ = "m" THEN GOTO 120
90    IF AS$ = "P" OR AS$ = "p" THEN GOTO 170
100   PRINT " ": PRINT " "
110   PRINT "************** INVALID SELECTION **************": GOTO 70
120   CLS
130   INPUT"UPSTREAM MACH NUMBER M1";M
140   IF M < 1 THEN GOTO 130
150   P = (2*GAM*M*M-(GAM-1))/(GAM + 1)
160   GOTO 210
170   CLS
```

```
180   INPUT" PRESSURE RATIO ACROSS NORMAL SHOCK P2/P1";P
190   IF P < 1 THEN GOTO 180
200   M = SQR(((GAM + 1)*P + (GAM-1))/(2*GAM))
210   M2 = SQR(((GAM-1)*M*M + 2)/(2*GAM*M*M-(GAM-1)))
220   TR = ((2*GAM*M*M-(GAM-1))*(2 + (GAM-1)*M*M))/((GAM + 1)*M) ^ 2
230   P01 = (1 + ((GAM-1)/2)*M*M) ^ (GAM/(GAM-1))
240   P02 = (1 + ((GAM-1)/2)*M2*M2) ^ (GAM/(GAM-1))
250   P021 = (P02/P01)*P
260   P02P = P02*P
270   CLS
280   PRINT " SPECIFIC HEAT RATIO =";GAM
290   PRINT " UPSTREAM MACH NUMBER =";M
300   PRINT " PRESSURE RATIO (P2/P1) =";P
310   PRINT " DOWNSTREAM MACH NUMBER =";M2
320   PRINT " TEMPERATURE RATIO =";TR
330   PRINT " STAGNATION PRESSURE RATIO (P20/P10) =";P021
340   PRINT " PITOT TUBE PRESSURE RATIO (P20/P1) =";P02P
350   PRINT                                                          "
      _____"
360   PRINT " ": PRINT " "
370   INPUT" C TO CONTINUE OR E TO END";CE$
380   IF CE$ = "E" OR CE$ = "e" THEN GOTO 410
390   IF CE$ = "C" OR CE$ = "c" THEN GOTO 60
400   PRINT " ************** INVALID SELECTION **************": GOTO 370
410   CLS
420   STOP
430   END
```

## ADIABATIC FLOW IN A DUCT WITH FRICTION

A very simple BASIC program for one-dimensional adiabatic flow with friction (see Chapter 9) is listed below. This program allows the value of $M$ to be entered and then finds the value of the other quantities.

```
10    CLS
20    INPUT "SPECIFIC HEAT RATIO"; GA
30    GF = (GA-1)/2
40    GP = (GA + 1)/(2*GA)
50    GJ = (GA + 1)/(2*(GA-1))
60    GK = (GA + 1)/2
70    INPUT "MACH NUMBER (NEG. NUMBER TO END)";M
80    IF M < 0 THEN GOTO 240
90    M2 = M*M
100   FL = ((1-M2)/(GA*M2)) + GP*(LOG(((GA + 1)*M2)/(2*(1 + GF*M2))))
110   PR = ((GK/(1 + GF*M2)) ^ 0.5)/M
120   TR = GK/(1 + GF*M2)
130   SR = (((1 + GF*M2)/GK) ^ GJ)/M
140   PRINT "4fl*/D = ";FL
150   PRINT "p/p* = ";PR
160   PRINT "T/T* = ";TR
170   PRINT "p0/p0* = ";SR
180   A$ = " "
190   PRINT " ": PRINT " "
200   PRINT "ANY KEY TO CONTINUE"
210   A$ = INKEY$
220   IF A$ = " " THEN GOTO 200
230   GOTO 70
240   CLS
250   STOP
260   END
```

## FLOW IN A CONSTANT AREA DUCT WITH HEAT TRANSFER

The flow variables for one-dimensional flow in a constant area duct with heat transfer (see Chapter 10) can also easily be found using a programmable calculator, the following simple BASIC program illustrating the procedure used.

```
 10  CLS
 20  PRINT" ": PRINT" "
 30  PRINT "RAYLEIGH FLOW"
 40  PRINT" ": PRINT" "
 50  INPUT "SPECIFIC HEAT RATIO"; GA
 60  PRINT" ": PRINT " "
 70  GF = (GA-1)/2
 80  GP = (GA + 1)
 90  GJ = GA/(GA-1)
100  GK = (GA + 1)/2
110  PRINT" ": PRINT" "
120  INPUT "MACH NUMBER (NEG. NUMBER TO END)";M
130  PRINT" "
140  IF M < 0 THEN GOTO 350
150  M2 = M*M
160  F1 = (1 + (GA*M2))
170  TOR = 2*GP*M2*(1 + GF*M2)/(F1*F1)
180  PR = GP/F1
190  TR = GP*GP*M2/(F1*F1)
200  SR = (GP/F1)*(((1 + GF*M2)/GK)^GJ)
210  PRINT"gamma = ";:PRINT USING "##.####";GA
220  PRINT" "
230  PRINT"M = ";:PRINT USING "##.#####"; M
240  PRINT"T0/T0* = ";:PRINT USING "###.#####" ;TOR
250  PRINT"p/p* = ";:PRINT USING "###.#####" ;PR
260  PRINT"T/T* = ";:PRINT USING "###.#####" ;TR
270  PRINT"p0/p0* = ";:PRINT USING "###.#####" ;SR
280  A$ = " "
290  PRINT" ": PRINT" "
300  PRINT"ANY KEY TO CONTINUE"
310  A$ = INKEY$
320  IF A$ = " " THEN GOTO 310
330  GOTO 110
340  CLS
350  STOP
360  END
```

## ISOTHERMAL FLOW WITH FRICTION IN A CONSTANT AREA DUCT

The properties of this type of flow (see Chapter 10) can also be easily obtained using a programmable calculator, the procedure used in such a case being illustrated by the following simple BASIC program:

```
 10  CLS
 20  INPUT "SPECIFIC HEAT RATIO"; GA
 30  PRINT " ": PRINT " "
 40  PRINT " ": PRINT " "
 50  INPUT "MACH NUMBER (NEG. NUMBER TO END)";M
 60  PRINT " "
 70  IF M < 0 THEN GOTO 240
 80  M2 = M*M
```

```
 90   FL = ((1-GA*M2)/(GA*M2)) + LOG(GA*M2)
100   PR = 1/(M*SQR(GA))
110   PRINT "gamma = ";:PRINT USING "##.###";GA
120   PRINT " "
130   PRINT "M = ";:PRINT USING "###.####";M
140   PRINT "4f1*/D = ";:PRINT USING "###.####";FL
150   PRINT "p/p* = ";:PRINT USING "###.####";PR
160   A$ = " "
170   PRINT " ": PRINT " "
180   PRINT "ANY KEY TO CONTINUE"
190   A$ = INKEY$
200   IF A$ = " " THEN GOTO 190
210   GOTO 40
220   CLS
230   STOP
240   END
```

# Bibliography

The items in the following list of books, papers, and reports have not been directly referred to in the text. Their titles in almost all cases, however, make it clear what subject they deal with and the reader who requires more information on any particular topic should have little difficulty in identifying suitable references from this list.

Ames Research Staff (1953) *Equations, Tables and Charts for Compressible Flow.* NACA Report 1135.

Anderson, D. A., Tannehill, J. C., and Pletcher, R. H. (1984) *Computational Fluid Mechanics and Heat Transfer.* McGraw-Hill, New York.

Anderson, J. D., Jr. (1976) *Gas Dynamic Lasers: An Introduction.* Academic Press, New York.

Anderson, J. D., Jr. (1985) *Introduction to Flight*, 2nd edn. McGraw-Hill, New York.

Anderson, J. D., Jr. (1989) *Hypersonic and High Temperature Gas Dynamics.* McGraw-Hill, New York.

Anderson, J. D., Jr. (1990) *Modern Compressible Flow with Historical Perspective*, 2nd edn. McGraw-Hill, New York.

Anderson, J. D., Jr. and Flugge-Lotz, I. (1964) Second-order boundary layer effects in hypersonic flow past axisymmetric blunt bodies. *Journal of Fluid Mechanics* **20** (4), 593–623.

Anderson, J. D., Jr. (1964) Nonequilibrium laminar boundary-layer flow of ionized air. *AIAA Journal* **2** (11), 1921–7.

Anderson, J. D., Jr., Savin, R. C., and Syvertson, C. A. (1955) The generalized shock-expansion method and its application to bodies travelling at high supersonic airspeeds. *Journal of the Aeronautical Sciences* **22**, 231–8.

Anderson, J. D., Jr. and Cleary, J. W. (1970) Theoretical and experimental study of supersonic steady flow around inclined bodies of revolution. *AIAA Journal* **8** (3), 511–18.

Anderson, J. D., Jr. (1970) A time-dependent analysis for vibrational and chemical nonequilibrium nozzle flows. *AIAA Journal* **8** (3), 545–50.

Anon. (1957) *Mollier Chart for Air in Dissociated Equilibrium at Temperatures of 2000 K to 15000 K.* NAVORD Report 4446, U. S. Naval Ordnance Lab, White Oak, MD.

Back, L. H., Massier, P. F., and Grier, H. L. (1965) Comparison of measured and predicted flows through conical supersonic nozzles, with emphasis on the transonic region. *AIAA Journal* **3** (9), 1606–14.

Baradell, D. L. and Bertram, M. H. (1960) *The Blunt Plate in Hypersonic Flow.* NASA TN-D-408.

Beans, E. W. (1970) Computer solution to generalized one-dimensional flow. *Journal of Spacecraft and Rockets* **7** (12), 1460–4.

Bertram, M. H. and Baradell, D. L. (1957) A note on the sonic-wedge leading edge approximation in hypersonic flow. *Journal of Aeronautical Sciences* **24** (8), 627–9.

Benedict, R. P. (1980) *Fundamentals of Pipe Flow.* John Wiley, New York.

Benedict, R. P. (1983) *Fundamentals of Gas Dynamics.* John Wiley, New York.

Berman, A. I. (1961) *The Physical Principles of Astronautics.* John Wiley, New York.

Bers, L. (1958) *Mathematical Aspects of Subsonic and Transonic Gas Dynamics.* John Wiley, New York.

Bertin, J. J. and Smith, M. L. (1989) *Aerodynamics for Engineers*, 2nd edn. Prentice Hall, Englewood Cliffs, NJ.

Billig, F. S. (1967) Shock-wave shapes around spherical and cylindrical-nosed bodies. *Journal of Spacecraft and Rockets* **4** (6), 822–3.

Black, J. (1947) *An Introduction to Aerodynamic Compressibility.* Bunhill Publications, London.

Blottner, F. G. (1970) Finite difference methods of solution of the boundary-layer equations. *AIAA Journal* **8** (2), 193–205.

Blottner, F. G. (1964) Chemical nonequilibrium boundary layer. *AIAA Journal* **2** (2), 232–40.

Bohachevsky, I. O. and Rubin, E. L. (1966) A direct method for computation of nonequilibrium flows with detached shock waves. *AIAA Journal* **4** (4), 600–7.

Brower, W. B. (1990) *Theory, Tables, and Data for Compressible Flow.* Hemisphere, New York.

Bryson, A. E., Jr. (1951) *An Experimental Investigation of Transonic Flow Past Two-Dimensional Wedge and Circular-Arc Sections Using a Mach–Zehnder Interferometer.* NASA TN 2560.

Cambel, A. B. and Jennings, B. H. (1958) *Gas Dynamics.* McGraw-Hill, New York.

Cebeci, T. and Smith, A. M. O. (1972) *Analysis of Turbulent Boundary Layers.* Academic Press, New York.

Chapman, A. J. and Walker, W. F. (1971) *Introductory Gas Dynamics.* Holt, Rinehart & Winston, New York.

Chapman, S. and Cowling, T. G. (1958) *The Mathematical Theory of Nonuniform Gases*, 2nd edn. Cambridge University Press, New York.

Cheers, F. (1963) *Elements of Compressible Flow.* John Wiley, New York.

Chernyi, G. G. (1961) *Introduction to Hypersonic Flow.* Academic Press, New York.

Churchill, S. W. (1977) Friction factor equation spans all fluid flow regimes. *Chemical Engineering*, 7 November, 91–2.

Chushkin, P. I. (1968) *Numerical Method of Characteristics for Three-Dimensional Supersonic Flows. Progress in Aeronautical Sciences*, Vol. 9 (ed. D. Kuchemann), Pergamon, Elmsford, New York.

Courant, R. and Friedrichs, K. O. (1948) *Supersonic Flow and Shock Waves.* Interscience, New York (Reprinted in 1977 by Springer-Verlag, New York).

Cox, R. N. and Crabtree, L. F. (1965) *Elements of Hypersonic Aerodynamics.* Academic Press, New York.

Cuffel, R. F., Back, L. H., and Massier, P. F. (1969) The transonic flowfield in a supersonic nozzle with small throat radius of curvature. *AIAA Journal* **7** (7), 1364–6.

Daneshyar, H. (1976) *One-Dimensional Compressible Flow.* Pergamon Press, New York.

Davis, R. T. (1970) Numerical solution of the hypersonic viscous shock-layer equations. *AIAA Journal* **8** (5), 843–51.

Dorrance, W. H. (1962) *Viscous Hypersonic Flow.* McGraw-Hill, New York.

Doty, R. T. and Rasmussen, M. L. (1973) Approximation for hypersonic flow past an inclined cone. *AIAA Journal* **11** (9), 1310–15.

Eggers, A. J. and Cyvertson, C. A. (1952) *Inviscid Flow About Airfoils at High Supersonic Speeds.* NACA TN 2646.

Emanuel, G. (1986) *Gasdynamics: Theory and Applications.* AIAA, Washington.

Erickson, W. D. (1963) *Vibrational Nonequilibrium Flow of Nitrogen in Hypersonic Nozzles.* NASA TN D-1810.

Evans, J. S., Schexnayder, C. J., and Huber, P. W. (1970) Computation of ionization in re-entry flow fields. *AIAA Journal* **8** (6), 1082–9.

Fay, J. A. and Riddell, F. R. (1958) Theory of stagnation point heat transfer in dissociated air. *Journal of the Aeronautical Sciences* **25** (2), 73–85.

Ferri, A. (1949) *Elements of Aerodynamics of Supersonic Flows.* Macmillan, New York.

Fowler, R. H. and Guggenheim, E. A. (1952) *Statistical Thermodynamics.* Cambridge University Press, New York.

Fox, R. W. and McDonald, A. T. (1978) *Introduction to Fluid Mechanics*, 2nd edn. John Wiley, New York.

Fox, R. W. and McDonald, A. T. (1992) *Introduction to Fluid Mechanics*, 4th edn. John Wiley, New York.

Gallagher, R. J. (1970) *Investigation of a Digital Simulation of the XB-70 Inlet and Its Application to Flight-Experienced Free-Stream Disturbances at Mach Numbers of 2.4 to 2.6.* NASA TN D-5827.

Gerhart, P. M., Gross, R. J., and Hochstein, J. I. (1992) *Fundamentals of Fluid Mechanics*, 2nd edn. Addison-Wesley, Reading, MA.

Glass, I. I. (1958) *Theory and Performances of Simple Shock Tubes.* Institute of Aerophysics, University of Toronto, UTIA Review No. 12.

Glass, I. I. (1974) *Shock Waves and Man.* University of Toronto Press, Toronto.

Goodrich, W. D., Li, C. P., Houston, D. K., Chiu, P. B., and Olmedo, L. (1977) Numerical computations of orbiter flow fields and laminar heating rates. *Journal of Spacecraft and Rockets* **14** (5), 257–64.

Haaland, S. E. (1983) Simple and explicit formulas for the friction factor in turbulent flow. *Transactions of the ASME, Journal of Fluid Engineering* **105** (3), 89–90.

Haberman, W. L. and John, J. E. A. (1980) *Engineering Thermodynamics.* Allyn and Bacon, Boston.

Hall, U. M. (1962) Transonic flow in two-dimensional and axially-symmetric nozzles. *Quarterly Journal of Mechanics and Applied Mathematics* **XV** (4), 487–508.

Hall, J. G., Eschenroeder, A. A., and Marrone, P. V. (1962) Blunt-nosed inviscid airflows with coupled nonequilibrium processes. *Journal of the Aerospace Sciences* **29** (9), 1038–51.

Hall, J. G. and Russo, A. L. (1959) *Studies of Chemical Nonequilibrium in Hypersonic Nozzle Flows.* Cornell Aeronautical Laboratory (now CALSPAN) Report no. AF-1118-A-6.

Hall, J. G. and Treanor, C. E. (1968) *Nonequilibrium Effects in Supersonic Nozzle Flows.* AGARDograph No. 124.

Hansen, C. F. (1959) *Approximation for the Thermodynamic and Transport Properties of High Temperature Air.* NASA TR-R-50.

Hayes, W. D. and Probstein, R. F. (1966) *Hypersonic Flow Theory.* Academic Press, New York.

Hayes, W. D. (1947) On hypersonic similitude. *Quarterly of Applied Mathematics* **5** (1), 105–6.

Hill, P. G. and Peterson, C. R. (1992) *Mechanics and Thermodynamics of Propulsion,* 2nd edn. Addison-Wesley, Reading, MA.

Hilsenrath, J. *et al.* (1955) *Tables of Thermodynamic and Transport Properties.* NBS Circular 564; reprinted (1960) by Pergamon Press, New York.

Hilsenrath, J. and Klein, M. (1965) *Tables of Thermodynamic Properties of Air in Chemical Equilibrium Including Second Virial Corrections from 1500 to 15,000 K.* AEDC-TR-65-68.

Hilton, W. F. (1951) *High-Speed Aerodynamics.* Longman, London.

Hodge, B. K. (1990) Generalized one-dimensional compressible flow matrix inverse. *Journal of Spacecraft and Rockets* **27** (4), 446–7.

Hodge, B. K. and Koenig, K. (1995) *Compressible Fluid Dynamics with Personal Computer Applications.* Prentice Hall, Englewood Cliffs, NJ.

Hodgson, J. W. and Lamas, J. A. (1990) A simple basic algorithm for determining the equilibrium composition of gaseous C/H/O/N systems involving ten species. *Proceedings of the 1990 ASEE Southeastern Section Meeting,* ed. J. M. Biedenbach.

Hoerner, S. F. (1965) *Fluid-Dynamic Drag.* Hoerner Fluid Dynamics, Brick Town, NJ.

Holman, J. P. (1980) *Thermodynamics,* 3rd edn. McGraw-Hill, New York.

Howarth, L. (Editor) (1953) *Modern Developments in Fluid Dynamics High Speed Flows,* Vols. 1 and 2. Oxford University Press, London.

Howell, J. R. and Buckius, R. O. (1992) *Fundamentals of Engineering Thermodynamics,* 2nd edn. McGraw-Hill, New York.

Huber, P. W. (1963) *Hypersonic Shock-Heated Flow Parameters for Velocities to 46,000 Feet per Second and Altitudes to 323,000 Feet.* NASA TR R-163.

Hudgins, H. E. (1964) *Supersonic Flow about Right-Circular Cones at Zero Yaw in Air at Thermodynamic Equilibrium. Parts II and III. Tables of Data.* TM 1493, Picatinny Arsenal.

Hudgins, H. E. (1965) *Supersonic Flow about Right-Circular Cones at Zero Yaw in Air at Chemical Equilibrium. Part I. Correlation of Flow Properties.* TM 1493, Picatinny Arsenal.

Imric, B. W. (1974) *Compressible Fluid Flow.* Halstead, New York.

Inouye, M. and Lomax, H. (1962) *Comparison of Experimental and Numerical Results for the Flow of a Perfect Gas About Blunt-Nosed Bodies.* NASA TN D-1426.

Ivey, R. H., Klunker, E. B., and Bowen, E. N. (1948) *A Method for Determining the Aerodynamic Characteristics of Two and Three-Dimensional Shapes at Hypersonic Speeds.* NACA TN 1613.

Jeans, J. H. (1940) *The Dynamical Theory of Gases*, 4th edn. Cambridge University Press, New York; reprinted (1960) by Dover Publications, New York.

Johannsen, N. H. (1952) Experiments on two-dimensional supersonic flow in corners and over concave surface. *Philosophical Magazine* **43**, 568–80.

John, J. E. A. and Haberman, W. L. (1980) *Introduction to Fluid Mechanics*, 2nd edn. Prentice-Hall, Englewood Cliffs, NJ.

John, J. E. A. (1984) *Gas Dynamics*, 2nd edn. Allyn and Bacon, Boston.

Keenan, J. H. and Kaye, J. (1980) *Gas Tables*, 2nd edn. John Wiley, New York.

Keenan, J. H. and Neumann, E. P. (1946) Measurements of friction in a pipe for subsonic and supersonic flow of air. *Journal of Applied Mechanics* **13** (2), A-91.

Kemp, N. H. R., Rose, H., and Detra, R. W. (1959) Laminar heat transfer around blunt bodies in dissociated air. *Journal of the Aerospace Sciences* **26** (7), 421–30.

Kennard, E. H. (1938) *Kinetic Theory of Gases*. McGraw-Hill, New York.

Kim, M. D., Swaminathan, S., and Lewis, C. H. (1984) Three-dimensional non-equilibrium viscous shock-layer flow over the space shuttle orbiter. *Journal of Spacecraft and Rockets* **21**, 29–35.

Kliegel, J. R. and Levine, J. N. (1969) Transonic flow in small radius of curvature nozzles. *AIAA Journal* **7** (7), 1375–8.

Kopal, Z. (1947) *Tables of Supersonic Flow Around Cones*. M.I.T. Center of Analysis Tech., Report no. 1, U.S. Government Printing Office, Washington, DC.

Kutler, P., Warming, R. F., and Lomax, H. (1973) Computation of space shuttle flowfields using noncentered finite-difference schemes. *AIAA Journal* **11** (2), 196–204.

Lee, J. F., Sears, F. W., and Turcotte, D. L. (1973) *Statistical Thermodynamics*, 2nd edn. Addison-Wesley, Reading, MA.

Lees, L. and Kubota, T. (1957) Inviscid hypersonic flow over blunt-nosed slender bodies. *Journal of the Aeronautical Sciences* **24**, 195–202.

Lees, L. (1956) Laminar heat transfer over blunt-nosed bodies at hypersonic flight speeds. *Jet Propulsion* **26**, 259–69.

Li, C. P. (1972) Time-dependent solutions of nonequilibrium dissociating gases past a blunt body. *Journal of Spacecraft and Rockets* **9** (8), 571–2.

Liepmann, H. W. and Roshko, A. (1957) *Elements of Gas Dynamics*. John Wiley, New York.

Liu, Y. and Vinokur, M. (1989) *Equilibrium Gas Flow Computations. I. Accurate and Efficient Calculation of Equilibrium Gas Properties*. AIAA Paper 89-1736.

Lomax, H. and Inouye, M. (1964) *Numerical Analysis of Flow Properties About Blunt Bodies Moving at Supersonic Speeds in an Equilibrium Gas*. NASA TR-R-204.

Maccoll, J. W. (1937) The conical shock wave formed by a cone moving at high speed. *Proceedings of the Royal Society* A **159**, 459–72.

Marrone, P. V. (1962) *Normal Shock Waves in Air: Equilibrium Composition and Flow Parameters for Velocities from 26,000 to 50,000 ft/s*. Cornell Aeronautical Laboratory Report AG-1729-A-2.

Martin, R. A. (1970) *Dynamic Analysis of XB-70 Inlet Pressure Fluctuations During Takeoff and Prior to a Compressor Stall at Mach 2.5*. NASA TN D-5826.

Mascitti, V. R. (1969) A closed-form solution to oblique shock-wave properties. *Journal of Spacecraft and Rockets* **6** (1), 66.

Maslen, S. H. (1964) Inviscid hypersonic flow past smooth symmetric bodies. *AIAA Journal* **2** (6), 1055–61.

Maus, J. R., Griffith, B. J., Szema, K. Y., and Best, J. T. (1984) Hypersonic real Mach number and real gas effects on space shuttle orbiter aerodynamics. *Journal of Spacecraft and Rockets* **21** (2), 136–41.

McBride, B. J., Heimel, S., Ehlers, J. G., and Gordon, S. (1963) *Thermodynamic Properties to 6000 K for 210 Substances Involving the First 18 Elements*. NASA SP-3001.

Miles, E. R. (1950) *Supersonic Aerodynamics*. Dover, New York.

Milne-Thomson, L. M. (1952) *Theoretical Aerodynamics*. Macmillan, New York.

Milne-Thomson, L. M. (1968) *Theoretical Hydrodynamics*, 5th edn. Macmillan, New York.

Mironer, A. (1979) *Engineering Fluid Mechanics*. McGraw-Hill, New York.

Moeckel, W. E. (1957) *Oblique-Shock Relations at Hypersonic Speeds for Air in Chemical Equilibrium*. NACA TN-3985.

Moretti, G. and Abbet, R. (1966) A time-dependent computational method for blunt-body flows. *AIAA Journal* **4** (12), 2136–41.

Moss, J. N. and Bird, G. A. (1984) *Direct Simulation of Transitional Flow for Hypersonic Reentry Conditions*. AIAA Paper no. 84-0223.

Nickerson, G. R., Coats, D. E., Dang, A. L., Dunn, S. S., and Kehtarnavaz, H. (1989) *Two-Dimensional Kinetic (TDK) Nozzle Performance Computer Program*. Software and Engineering Associates, Carson City.

Owatitsch, K. (1956) *Gas Dynamics*. Academic Press, New York.

Owczarek, K. (1964) *Fundamentals of Gas Dynamics*. International Textbook Company, Scranton, PA.

Pai, S. (1959) *Introduction to the Theory of Compressible Flow*. Van Nostrand, Princeton, NJ.

Pai, S. and Luo, S. (1991) *Theoretical and Computational Dynamics of a Compressible Flow*. Van Nostrand Reinhold, New York.

Palmer, G. (1987) *An Implicit Flux-Split Algorithm to Calculate Hypersonic Flowfields in Chemical Equilibrium*. AIAA Paper 87-1580.

Park, C. (1990) *Nonequilibrium Hypersonic Aerothermodynamics*. John Wiley, New York.

Patterson, G. N. (1956) *Molecular Flow of Gases*. John Wiley, New York.

Pope, A. (1958) *Aerodynamics of Supersonic Flight*, 2nd edn. Pitman, New York.

Pope, A. Y. and Goin, K. L. (1965) *High Speed Wind Tunnel Testing*. John Wiley, New York.

Prabhu, D. K. and Tannehill, J. C. (1986) Numerical solution of space shuttle orbiter flowfield including real-gas effects. *Journal of Spacecraft and Rockets* **23** (3), 264–72.

Present, R. D. (1958) *Kinetic Theory of Gases*. McGraw-Hill, New York.

Probstein, R. F. and Bray, K. N. C. (1955) Hypersonic similarity and the tangent cone approximation for unyawed bodies of revolution. *Journal of the Aeronautical Sciences* **22** (1), 66–8.

Rakich, J. V. (1969) *A Method of Characteristics for Steady Three-Dimensional Supersonic Flow with Application to Inclined Bodies of Revolution*. NASA TN D-5341.

Rakich, J. V. and Cleary, J. W. (1970) Theoretical and experimental study of supersonic steady flow around inclined bodies of revolution. *AIAA Journal* **8** (3), 511–18.

Rakich, J. V. and Lanfranco, M. J. (1977) Numerical computation of space shuttle laminar heating and surface streamlines. *Journal of Spacecraft and Rockets* **14** (5), 265–72.

Rakich, J. V., Bailey, H. E., and Park, C. (1983) Computation of nonequilibrium, supersonic three-dimensional inviscid flow over blunt-nosed bodies. *AIAA Journal* **21** (6), 834–41.

Rasmussen, M. L. (1967) On hypersonic flow past an unyawed cone. *AIAA Journal* **5** (8), 1495–7.

Reynolds, W. C. and Perkins, H. C. (1977) *Engineering Thermodynamics*, 2nd edn. McGraw-Hill, New York.

Robertson, J. M. (1965) *Hydrodynamics in Theory and Applications*. Prentice-Hall, Englewood Cliffs, NJ.

Roe, P. L. (1972) Thin shock-layer theory in aerodynamic problems of hypersonic vehicles. *AGARD Lecture Series* **42**, Vol. I, 4-1–4-26.

Romere, P. O. and Whitnah, A. M. (1983) Space shuttle entry longitudinal aerodynamic comparisons with flights 1–4 with preflight predictions. In *Shuttle Performance: Lessons Learned*, NASA CP-2283, ed. J. P. Arrington and J. J. Jones, pp. 283–307.

Romig, M. F. (1960) *Conical Flow Parameters for Air in Dissociating Equilibrium*. Convair Scientific Research Laboratory Research Report No. 7.

Rosenhead, L. (1954) *A Selection of Graphs for Use in Calculations of Compressible Flow*. Clarendon Press, Oxford.

Saad, M. A. (1992) *Compressible Fluid Flow*, 2nd edn. Prentice-Hall, Englewood Cliffs, NJ.

Sarli, V. J., Burwell, W. G. and Zupnik, T. F. (1964) *Investigation of Nonequilibrium Flow Effects in High Expansion Ratio Nozzles*. NASA CR-54221.

Sauer, R. (1947a) *General Characteristics of the Flow through Nozzles at Near Critical Speeds*. NACA TM 1147.

Sauer, R. (1947b) *Introduction to Theoretical Gas Dynamics*. Edwards, Ann Arbor, MI.

Sauerwein, H. (1967) Numerical calculation of multidimensional and unsteady flows by the method of characteristics. *Journal of Computational Physics* **1** (1), 406–32.

Schaaf, S. A. and Chambre, P. L. (1958) Flow of rarefied gases. Section H in *Fundamentals of Gas Dynamics*, ed. H. W. Emmons, p. 689. Princeton University Press, Princeton, NJ.

Schreier, S. (1982) *Compressible Flow*. Wiley Interscience, New York.

Sears, F. W. (1953) *An Introduction to Thermodynamics, the Kinetic Theory of Gases, and Statistical Mechanics*. Addison-Wesley, Reading, MA.

Seddon, J. and Goldsmith, E. L. (1985) *Intake Aerodynamics*. AIAA, Washington.

Shames, I. H. (1992) *Mechanics of Fluids*, 3rd edn. McGraw-Hill, New York.

Shang, J. S. and Scherr, S. J. (1986) Navier–Stokes solution for a complete re-entry configuration. *Journal of Aircraft* **23** (12), 881–8.

Shapiro, A. H. (1953) *The Dynamics and Thermodynamics of Compressible Fluid Flow*, Vols. 1 and 2. Ronald Press, New York.

Sherman, F. S. (1990) *Viscous Flow*. McGraw-Hill, New York.

Shevell, R. S. (1989) *Fundamentals of Flight*, 2nd edn. Prentice Hall, Englewood Cliffs, NJ.

Sims, J. (1964) *Tables for Supersonic Flow around Right Circular Cones at Zero Angle of Attack*. NASA SP-3004.

Spurk, J. H., Gerber, N., and Sedney, R. (1966) Characteristic calculation of flowfields with chemical reactions. *AIAA Journal* **4** (1), 30–7.

Streeter, V. L. and Wylie, E. B. (1975) *Fluid Mechanics*, 6th edn. McGraw-Hill, New York.

Sutton, G. P. (1992) *Rocket Propulsion Elements*, 6th edn. John Wiley, New York.

Tannehill, J. C. and Mugge, P. H. (1974) *Improved Curve Fits for the Thermodynamic Properties of Equilibrium Air Suitable for Numerical Computation Using Time-Dependent or Shock-Capturing Methods*. NASA CR-2470.

Taylor, G. I. and Maccoll, J. W. (1933) The air pressure on a cone moving at high speeds. *Proceedings of the Royal Society* A **139**, 278–311.

Thompson, P. A. (1972) *Compressible-Fluid Dynamics*. McGraw-Hill, New York.

Tien, C. L. and Lienhard, J. H. (1971) *Statistical Thermodynamics*, p. 51. Holt, Rinehart and Winston, New York.

Truitt, R. W. (1959) *Hypersonic Aerodynamics*. Ronald Press, New York.

Tsien, H. S. (1946) Similarity laws of hypersonic flows. *Journal of Mathematics and Physics* **25**, 247–51.

Vamos, J. S. and Anderson, J. D., Jr. (1973) Time-dependent analysis of nonequilibrium nozzle flows with complex chemistry. *Journal of Spacecraft and Rockets* **10** (4), 225–6.

Viegas, J. R. and Howe, J. T. (1962) *Thermodynamic and Transport Property Correlation Formulas for Equilibrium Air from 1000 K to 15,000 K*. NASA TN D-1429.

Von Karman, T. (1947) Supersonic aerodynamics—principles and applications. *Journal of Aeronautical Sciences* **14** (7), 373.

Von Mises, R. (1958) *Mathematical Theory of Compressible Flow*. Academic Press, New York.

Vincenti, W. G. and Kruger, C. H. (1965) *Introduction to Physical Gas Dynamics*. John Wiley, New York.

Wark, K. (1983) *Thermodynamics*, 4th edn. McGraw-Hill, New York.

Warsi, Z. U. A. (1993) *Fluid Dynamics—Theoretical and Computational Approaches*. CRC Press, Boca Raton, FL.

White, F. M. (1986) *Fluid Mechanics*, 2nd edn. McGraw-Hill, New York.

White, F. M. (1991) *Viscous Fluid Flow*, 2nd edn. McGraw-Hill, New York.

Wilson, D. G. (1984) *The Design of High-Efficiency Turbomachinery and Gas Turbines*. MIT Press, Cambridge, MA.

Wilson, J. K., Schofield, D., and Lapworth, K. C. (1967) *A Computer Program for Nonequilibrium Convergent–Divergent Nozzle Flow*. National Physical Laboratory Report no. 1250.

Wittliff, C. E. and Curtiss, J. T. (1961) *Normal Shock Wave Parameters in Equilibrium Air*. Cornell Aeronautical Laboratory Report CAL-111.

Woods, W. C., Arrington, J. P., and Hamilton, H. H. (1983) A review of real gas effects on space shuttle aerodynamics characteristics. In *Shuttle Performance: Lessons Learned*, NASA CP-2283, ed. J. P. Arrington and J. J. Jones, pp. 309–46.

Wray, K. L. (1962) Chemical kinetics of high-temperature air. In *Hypersonic Flow Research*, ed. F. Riddell, pp. 181–204. Academic Press, New York.

Yahya, S. M. (1982) *Fundamentals of Compressible Flow*. Wiley Eastern, New Delhi.

Young, F. M. (1993) Generalized one-dimensional, steady, compressible flow. *AIAA Journal* **31** (1), 204–8.

Zel'dovich, Y. B. and Raizer, Y. P. (1968) *Elements of Gasdynamics and the Classical Theory of Shock Waves*. Academic Press, New York.

Zemansky, M. W. (1957) *Heat and Thermodynamics*, 4th edn. McGraw-Hill, New York.

Zoby, E. V., Moss, J. N., and Sutton, K. (1981) Approximate convective-heating equations for hypersonic flows. *Journal of Spacecraft and Rockets* **18** (1), 64–70.

Zucker, R. D. (1977) *Fundamentals of Gas Dynamics*. Matrix Publishers, Portland, OR.

Zucrow, M. J. and Hoffman, J. D. (1976) *Gas Dynamics*. John Wiley, New York.

# Index